International Federation of Automatic Control

ANALYSIS, DESIGN AND EVALUATION OF MAN-MACHINE SYSTEMS

NOTICE TO READERS

IFAC Related Titles

ANALYSIS, DESIGN AND EVALUATION OF MAN-MACHINE SYSTEMS

Proceedings of the IFAC/IFIP/IFORS/IEA Conference
Baden-Baden, Federal Republic of Germany, 27-29 September 1982

Edited by

G. JOHANNSEN

Universität-Gesamthochschule Kassel, Federal Republic of Germany

and

J. E. RIJNSDORP

Twente University of Technology, The Netherlands

Published for the

INTERNATIONAL FEDERATION OF AUTOMATIC CONTROL

by

PERGAMON PRESS

OXFORD · NEW YORK · TORONTO · SYDNEY · PARIS · FRANKFURT

U.K.	Pergamon Press Ltd., Headington Hill Hall, Oxford OX3 0BW, England
U.S.A.	Pergamon Press Inc., Maxwell House, Fairview Park, Elmsford, New York 10523, U.S.A.
CANADA	Pergamon Press Canada Ltd., Suite 104, 150 Consumers Road, Willowdale, Ontario M2J 1P9, Canada
AUSTRALIA	Pergamon Press (Aust.) Pty. Ltd., P.O. Box 544, Potts Point, N.S.W. 2011, Australia
FRANCE	Pergamon Press SARL, 24 rue des Ecoles, 75240 Paris, Cedex 05, France
FEDERAL REPUBLIC OF GERMANY	Pergamon Press GmbH, Hammerweg 6, D-6242 Kronberg-Taunus, Federal Republic of Germany

First edition 1983

Library of Congress Cataloging in Publication Data

Main entry under title:
Analysis, design & evaluation of man-machine systems.
(IFAC proceedings series)
Papers presented at the IFAC/IFIP/IFORS/IEA
Conference on Analysis, Design, and Evaluation of Man-
Machine Systems.
1. Man-machine systems—Congresses.   I. Johannsen, G.
II. Rijnsdorp, John E.   III. IFAC/IFIP/IFORS/IEA
Conference on Analysis, Design, and Evaluation of Man-
Machine Systems (1982: Baden-Baden, Germany)
IV. International Federation of Automatic Control.
V. Title: Analysis, design, and evaluation of man-
machine systems.   VI. Series.
TA167.A53   1983   620.8'2   83-8029

British Library Cataloguing in Publication Data

Analysis, design & evaluation of man-machine
systems. — (IFAC proceedings)
1. Man-machine systems—Congresses
1. Johannsen, G. II. Rijnsdorp. J
III. International Federation of Automatic
Control IV. Series
620.8'2   TA167

ISBN 0-08-029348-4

Printed in Great Britain by A. Wheaton & Co. Ltd., Exeter

IFAC/IFIP/IFORS/IEA CONFERENCE ON ANALYSIS, DESIGN AND EVALUATION OF MAN-MACHINE SYSTEMS

Organized by
VDI/VDE-Gesellschaft Mess- und Regelungstechnik (GMR)
P.O.B. 1139, D-4000 Duesseldorf 1

Sponsored by
International Federation of Automatic Control (IFAC)
Technical Committee on Systems Engineering
Technical Committee on Social Effects of Automation
Technical Committee on Economic and Management Systems

Co-sponsored by
International Federation for Information Processing (IFIP)
International Federation of Operational Research Societies (IFORS)
International Ergonomics Association (IEA)

International Program Committee (IPC)
G. Johannsen, F.R.G. (Joint Chairman)
J. E. Rijnsdorp, Netherlands (Joint Chairman)
K. Bindewald, F.R.G.
P. B. Checkland, U.K.
W. J. Edwards, Australia
F. Filippazzi, Italy
R. Genser, Austria
R. Haller, F.R.G.
J. Hatvany, Hungary
Mrs. V. de Keyser, Belgium
N. Malvache, France
F. Margulies, Austria
Mrs. L. Mårtensson, Sweden
J. Rasmussen, Denmark
W. Rohmert, F.R.G.
W. B. Rouse, U.S.A.
M. L. Schneider, USA
R. Seifert, F.R.G.
B. Shackel, U.K.
T. B. Sheridan, U.S.A.
T. Terano, Japan
D. Waye, Canada

National Organizing Committee (NOC)
R. Haller (Chairman)
G. Johannsen
H. Wiefels
L. Zühlke

SESSION CHAIRMEN (AND SECRETARIES — S)

H. Akashi (J)	(Papers 1.1S - 1.5T)
T. Terano (J) - S: B. Döring (D)	(Papers 1.6I - 1.7I)
W.E. Miller (USA) - S: R. Grimm (D)	(Papers 2.1T - 2.5I)
J.E. Rijnsdorp (NL)	(Papers 2.8I - 2.9I)
L.P. Goodstein (DK)	(Papers 2.10I - 2.12I)
W. Rohmert (D)	(Papers 3.1S - 3.3T)
L. Mårtensson (S) - S: K.R. Kimmel (D)	(Papers 3.4I - 3.6I)
S. Baron (USA)	(Papers 4.1S - 4.5T)
T.B. Sheridan (USA)	(Papers 4.6I - 4.8I)
W.B. Rouse (USA)	(Papers 5.1S - 5.5T)
J. Charwat (D)	(Papers 5.6I)
J. Hatvany (H) - S: D. Arnold (D)	(Papers 5.7I - 5.8I)
G.H. Geiser (D)	(Papers 5.10I - 5.11I)
R.C. Williges (USA)	(Papers 5.12I - 5.15I)
G. Johannsen (D)	(Papers 6.1S - 6.2T)
R. Seifert (D) - S: K. Brauser (D)	(Papers 7.1I - 7.5I)
R. Genser (A)	(Papers 7.6I - 7.8I)

PREFACE

The field of man-machine systems has grown rapidly during the last decade. It is now recognized and still expanding as the major interdisciplinary contribution of research and development to the improvement of the interaction of humans with all kinds of technological systems that more and more frequently include computers as an integral system component. Various aspects of human-machine interaction in such systems have emerged. These are reflected in the different sessions of the Conference. Task issues of interest are, among others, controlling, monitoring, decision making, fault management, problem solving, and planning.

The aim of the Conference is to present, discuss, and summarize recent advances in theory, experimental and analytical research, and applications, related to man-machine systems. The International Program Committee has made a careful selection of papers for the Conference. Three types of papers have been distinguished: Survey (S), Technical (T), and Interactive (I). All papers are included in the Proceedings in sequence of the 7 sessions of the Conference.

We hope that the result will be beneficial to all engineers and scientists in automatic control, information processing, operational research, and ergonomics who are actively working or strongly interested in the young field of man-machine systems.

G. Johannsen

J.E. Rijnsdorp

Editors

CONTENTS

ROUND-TABLE DISCUSSIONS

MAN–MACHINE SYSTEMS — INTRODUCTION AND BACKGROUND

G. Johannsen

*Research Institute for Human Engineering (FGAN/FAT), Königstrasse,
Wachtberg-Werthhoven, Federal Republic of Germany*

Abstract. This paper is an introduction to the IFAC/IFIP/IFORS/IEA Conference on Analysis, Design, and Evaluation of Man-Machine Systems. It serves as
an umbrella for the survey papers and topic areas of the conference. Therefore, it is very broad in its scope and condensed in its exposition. The man-
machine system is defined, its general purpose explained, and the multitude
of application areas stated. The historical and scientific background of the
field is briefly outlined. Human task categories in man-machine systems are
described.

Keywords. Man-machine systems; manual control; optimal control model; supervisory control; operations research methodologies; problem solving; human-
computer interaction; software ergonomics; human reliability; social effects
of automation.

DEFINITION AND PURPOSE OF MAN-MACHINE SYSTEMS

A man-machine system is defined as a functional synthesis between a biological/psycho-
logical/social system (the man or a group of
people) and a technological system (the machine) characterized predominantly by the
interaction and functional interdependence
between these two. All kinds of technological
systems regardless of degree of complexity
may be part of a man-machine system, e.g.,
industrial plants, vehicles, manipulators,
prostheses, computers or management information systems. For the interaction with such
systems, mostly psychological but also social
aspects are of concern. Task categories like
controlling and problem solving describe typical human activities in man-machine systems.
Later on, these task categories will be explained in more detail.

The overall purpose of any man-machine system
is to provide a certain function, product or
service as an output with reasonable costs,
even under conditions of disturbances influencing man, machine or both (see Fig. 1). The
main goals or inputs of a man-machine system
are expected values of performance, costs,
reliability, and safety. At least since some
spectacular accidents have occurred with aircraft and nuclear power plants, reliability
and safety have become vitally important op-·
erational as well as design goals in addition
to performance and costs; see Sheridan (1982).
Also, an acceptable level of workload and job
satisfaction of the man should be maintained;
see, e.g., Moray (1979).

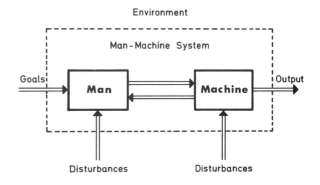

Fig. 1. Man-Machine System (MMS)

Some of these goals are in conflict with each
other. Such conflicts have to be resolved in
the most favorable manner by the designers
of a particular man-machine system. Any deficiencies left to the human user of such systems may cause poorer performance, job satisfaction, and safety.

The interaction between man and machine is
the essential aspect of a man-machine system.
Classical ergonomic aspects like knobs and
dials design, anthropometry, lighting, or
adverse environmental factors have intensively
been investigated. Many results are available,
although not always applied appropriately. In
contrast, the focus of attention has centered
on informational aspects in the last years.
Questions of concern to a successful interaction between man and machine are:

What kind of information is needed?

How should the information be organized?

Which information should be preprocessed?

How should the information be transmitted?

All of these or similar questions can arise in different application areas. The questions relate to the control of technological systems, namely, to the degree of automation as well as to the design of computer-generated displays with preprocessing capabilities. Further, they relate to all kinds of human-computer interaction as well as to management tasks on different organizational levels. The importance of the shifting from hardware and environmental aspects to software considerations is nowadays expressed by the new term software ergonomics.

From the preceding discussion, one can see that a wide range of technical areas is involved and contributing to the field of man-machine systems. Therefore, a conference dealing with the subject will necessarily be interdisciplinary in nature. Consequently, this conference is sponsored by four international federations which represent the most important disciplines concerned with the field of man-machine systems, namely IFAC (automatic control), IFIP (information processing), IFORS (operational research), and IEA (ergonomics).

HISTORICAL AND SCIENTIFIC BACKGROUND

A brief outline of the historical and scientific background of man-machine systems may further illustrate the growing importance of the field. The first existence of man-machine systems can be traced back into the early days of simple machines powered by men. With

respect to human use, these were designed intuitively by experience. This is even today a very common method. With more complex and faster responding technological systems however, it turns out to be more and more mandatory in many application areas to use analytical and consciously applied methodologies and systematic techniques for the design of the man-machine system as a whole.

For about 40 years, methodological knowledge has been gathered and systematic techniques have been elaborated. Most of the first investigations of man-machine systems were concerned with manual control tasks, often applied to aircraft piloting, later also to ship steering, car driving, and industrial process control. This work was done either by experimental psychologists or by control, systems, and application-oriented engineers. Overviews and literature surveys have been given in several books: Kelley (1968), Oppelt and Vossius (1970), Edwards and Lees (1974), Sheridan and Ferrell (1974), and Johannsen et al. (1977).

Many control theoretic models were developed to describe the behavior of the human operator in manual control tasks. They have successfully been applied as design tools for automatic control systems which are better adapted to the human operator, for unburdening displays, etc. The most sophisticated and well validated model is the optimal control model shown in Fig. 2 in its basic form; see also, e.g., Johannsen and Govindaraj (1980) and Pew and Baron (1982). It is structured into (1) a perception and attention allocation part, (2) a central information processing part with an internal representation of the system to be controlled, and

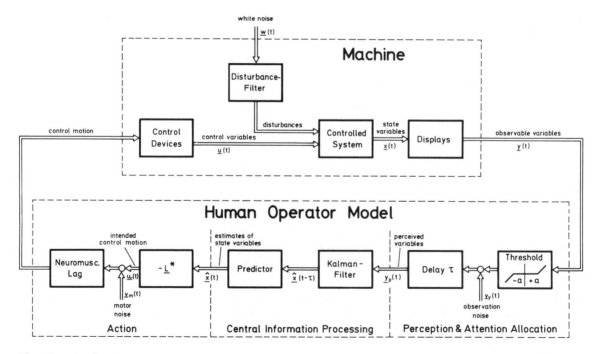

Fig. 2. Optimal Control Model

(3) an action part generating optimal control signals with respect to a cost criterion and based on the estimates of all systems states.

With slower responding systems like ships and industrial process plants, it became obvious that it is more difficult to explain the human operator behavior by well established control theoretic methods. The human control behavior is highly nonlinear and intermittent in these cases. Intrinsic monitoring, decision-making, and supervisory control behavior became evident and attracted the attention of several investigators; see Sheridan and Johannsen (1976). Some methodologies from operations research like network analysis and queueing theory have been adopted and comprehensive extensions of the optimal control model have been developed to describe broader human operator tasks as well as the whole design process for complex man-machine systems; see, e.g., Siegel and Wolf (1969), Pritsker and Pegden (1979), Rouse (1980), Moraal and Kraiss (1981), and Pew and Baron (1982).

Another root for the field of man-machine systems came out of what is nowadays called cognitive science, a combination of cognitive psychology and computer science. These sciences developed without any strong relationship to man-machine systems. Models of the brain, theories for memory and thought as well as human and artificial intelligence and problem solving have been investigated; see, e.g., Newell and Simon (1972), Klix (1979).

Task analyses show that problem solving tasks are more important than control tasks in many man-machine systems. Therefore, methodologies from cognitive sciences have been adopted for the analysis and design of these systems, but only since a few years ago; see Rouse (1982).

Technological advances such as computers and electronic displays have changed and will continue to change man-machine systems in almost all application areas. This is true for industrial plants used for production or power generation as well as for vehicles and transportation systems. In addition, office systems and information systems for observation, management, and command tasks in business, defense, and medicine are today similarly influenced. Not only the operation of technological systems by highly skilled personnel, but also its use by inexperienced people like in mass transport, as well as the maintenance and design of systems are aided by computers.

A common problem to all these applications is the design of the human-computer interaction; see, e.g., Rouse (1981), Hatvany and Guedj (1982), Williges and Williges (1982). The possibly adaptive task allocation between man and computer, the dialog design including the use of natural language, and other software ergonomic aspects are nowadays especially important topics of research and development.

Also, the design of knowledge-based systems will lead to helpful tools in such areas as computer-aided decision-making, information retrieval, and fault diagnosis.

With more computerization and higher degrees of automation, new social effects and perspectives become evident. The advanced technologies allow a more flexible work organization with higher user acceptance and job satisfaction. However, this advantage can only be achieved if the social implications are considered early enough by the designers of future computerized man-machine systems; see Margulies and Zemanek (1982).

HUMAN TASK CATEGORIES

All tasks of human personnel in man-machine systems can be condensed into only two categories:

(1) controlling and

(2) problem solving.

These human task categories are fairly general. Fig. 3 shows an attempt to integrate them into a schematic block diagram.

Controlling shall here be understood in a broader sense than, e.g., in control theory; see also Johannsen and Rouse (1979). It comprises controlling in the narrower sense (including open-loop vs. closed-loop and linear vs. intermittent controlling) but also all other action-oriented tasks such as reaching and discrete-event acting (e.g., switching, typing). Only through controlling, outputs of the man-machine system to the environment can be produced (see Fig. 3).

In contrast to controlling, problem solving is an internal process on a higher cognitive level. It comprises different tasks, mainly, fault managing (especially fault diagnosing) and planning. Fault managing is concerned with solving problems in actual failure situations, thereby using and updating certain rules with the objective of returning to a good state of the overall system; see also Rasmussen and Rouse (1981) and Johannsen (1981). Planning is concerned with solving possible future problems in the sense of mentally generating a sequence of appropriate alternatives or rules for reaching future states under different foreseeable and unforeseeable conditions; see also Johannsen and Rouse (1982). In all these problem solving tasks, the rules are stored in the knowledge base after their generation or modification. From there, they can be utilized in the lower-level process of controlling (see Fig. 3).

All other tasks in man-machine systems such as, e.g., monitoring and communicating, can be classified as subtasks or supporting tasks of controlling and problem solving. Communicating comprises two very different types of communication between the members of the group jointly responsible within a

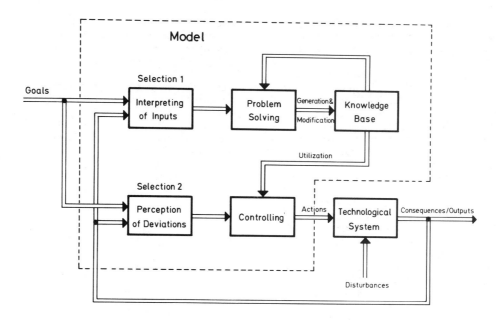

Fig. 3. Controlling and Problem Solving

man-machine system. The second type is the communication of the man with the technological system and the environment, sometimes identical with the human-computer dialog. It includes sensing, perceiving, and interpreting of input information and has been considered by Selection 1 and Selection 2 in Fig. 3. The man selects only those inputs from the generally available information source which he wants to use. This selection is handled differently for the purpose of problem solving where inputs are interpreted, and the purpose of controlling where, at a much higher pace, deviations are perceived.

Many of the topics mentioned in this introduction will be further elaborated in the survey papers and throughout the seven topic areas of this conference.

REFERENCES

Edwards, E., and F.P. Lees (Eds.) (1974). The Human Operator in Process Control. Taylor and Francis, London.

Hatvany, J., and R.A. Guedj (1982). Man-machine interaction in computer-aided design systems. Proc. IFAC/IFIP/IFORS/IEA Conf. Analysis, Design, and Evaluation of Man-Machine Systems, Baden-Baden, Sept. Pergamon Press, Oxford.

Johannsen, G., H.E. Boller, E. Donges, and W. Stein (1977). Der Mensch im Regelkreis - Lineare Modelle. Oldenbourg Verlag, München.

Johannsen, G., and W.B. Rouse (1979). Mathematical concepts for modeling human behavior in complex man-machine systems. Human Factors, 21, 733-747.

Johannsen, G., and T. Govindaraj (1980). Optimal control model predictions of systems performance and attention allocation and their experimental validation in a display design study. IEEE Trans, Systems,

Man, Cybernetics, SMC-10, 249-261.

Johannsen, G. (1981). Human-computer interaction in decentralized control and fault management of dynamic systems. Proc. IFAC 8th Triennial World Congress, Session 74, Kyoto, August. Pergamon Press, Oxford.

Johannsen, G., and W.B. Rouse (1982). Studies of planning behavior of aircraft pilots in normal, abnormal, and emergency situations. IEEE Trans. Systems, Man, Cybernetics (submitted).

Kelley, C.R. (1968). Manual and Automatic Control. Wiley, New York.

Klix, F. (Ed.) (1979). Human and Artificial Intelligence. North Holland, Amsterdam.

Margulies, F., and H. Zemanek (1982). Man's role in man-machine systems. Proc. IFAC/IFIP/IFORS/IEA Conf. Analysis, Design, and Evaluation of Man-Machine Systems, Baden-Baden, Sept. Pergamon Press, Oxford.

Moraal, J., and K.-F. Kraiss (Eds.) (1981). Manned Systems Design - Methods, Equipment, and Applications. Plenum Press, New York.

Moray, N. (Ed.) (1979). Mental Workload: Its Theory and Measurement. Plenum Press, New York.

Newell, A., and H.A. Simon (1972). Human Problem Solving. Prentice-Hall, Englewood-Cliffs, New Jersey.

Oppelt, W., and G. Vossius (Eds.) (1970). Der Mensch als Regler. VEB Verlag Technik, Berlin.

Pew, R.W., and S. Baron (1982). Perspectives on human performance modelling. Proc. IFAC/IFIP/IFORS/IEA Conf. Analysis, Design, and Evaluation of Man-Machine Systems, Baden-Baden, Sept. Pergamon Press, Oxford.

Pritsker, A.A.B., and C.D. Pegden (1979). Introduction to Simulation and SLAM. Wiley, New York.

Rasmussen, J., and W.B. Rouse (Eds.) (1981). Human Detection and Diagnosis of System

Failures. Plenum Press, New York.

Rouse, W.B. (1980). Systems Engineering Models of Human-Machine Interaction. North Holland, New York.

Rouse, W.B. (1981). Human-computer interaction in the control of dynamic systems. ACM Computing Surveys, 13, 71-99.

Rouse, W.B. (1982). Models of human problem solving: detection, diagnosis, and compensation for system failures. Proc. IFAC/IFIP/IFORS/IEA Conf. Analysis, Design, and Evaluation of Man-Machine Systems, Baden-Baden, Sept. Pergamon Press, Oxford.

Sheridan, T.B., and W.R. Ferrell (1974). Man-Machine Systems: Information, Control, and Decision Models of Human Performance. MIT Press, Cambridge, Mass.

Sheridan, T.B., and G. Johannsen (Eds.) (1976). Monitoring Behavior and Supervisory Control. Plenum Press, New York.

Sheridan, T.B. (1982). Measuring, modeling, and augmenting reliability of man-machine systems. Proc. IFAC/IFIP/IFORS/IEA Conf. Analysis, Design, and Evaluation of Man-Machine Systems, Baden-Baden, Sept. Pergamon Press, Oxford.

Siegel, A.I., and J.J. Wolf (1969). Man-Machine Simulation Models. Wiley, New York.

Williges, R.C., and B.H. Williges (1982). Human-computer dialogue design considerations. Proc. IFAC/IFIP/IFORS/IEA Conf. Analysis, Design, and Evaluation of Man-Machine Systems, Baden-Baden, Sept. Pergamon Press, Oxford.

PERSPECTIVES ON HUMAN PERFORMANCE MODELLING

R. W. Pew and S. Baron

Bolt Beranek and Newman, Inc., Cambridge, MA, USA

INTRODUCTION

The importance of including consideration of the contribution of human performance to overall success of a complex man-machine system during the design stage is receiving wider recognition in recent years. There are differing views about how this best can be accomplished, but we consider it as an iterative process encompassing four kinds of activities; (1) Task Analysis, (2) Modelling and Prediction, (3) Simulation and Test, and (4) Functional Specification.

Task Analysis produces an in-depth statement of how the activities are accomplished in the current system in a way that focuses on the human operator's tasks. It involves the examination of written procedures, collection of interviews, direct observation and protocol data from individuals experienced in the tasks. The product is a set of flow charts and supporting material that documents the behavioral features of the task.

Modelling builds from the task-analytic output to produce a formal, often quantitative, description of the behavior of one or more people in interaction with equipment. A model of human performance requires first a model or representation of the system and environment with which the people are to function.

Simulation involves building up preprototype versions of promising system concepts, formulating specific scenarios in which they would be applicable, and testing human operators directly with the opportunity to measure their performance and to interact with the test subjects to understand the conceptual basis for their performance.

Functional Specification is the process of distilling out of the other three steps those system requirements that are essential to meet the predefined system goals and, at the same time, are responsive to the human-performance capacities and limitations that are uncovered.

The iterative and synergistic nature of these activities must be emphasized. While modelling assists in the definition of what concepts to test and what parameter boundaries are critical, testing provides empirical validation for the models as well as the basis for improving their scope and validity. The statement of functional specification delimits the range of conditions of interest and requires modelling and empirical validation as well. Modelling and testing also provide a much expanded understanding of the task that is being performed and can contribute to improved task analysis and specification for purposes of personnel selection, job design and training development.

Without intending to minimize the importance of the other activities, this paper will focus on the subject of human performance modelling:

Why do it?

What are the current methodologies?

A tempting synthesis

Behavioral and engineering scientists have differing definitions of the term model. To an engineer, a model is an abstraction that involves an explicit mathematical or computer-based formalism. The psychologist or social scientist often speaks of verbal-analytical models and uses the term in a way that is virtually synonymous with the term, theory. While subtle distinctions may be derived, we adopt the position that there is no useful distinction between models and theories. We assert that there is a continuum along which models vary that has loose verbal analogy and metaphor at one end and closed-form mathematical equations at the other, and that most models lie somewhere in-between.

We also believe that to ask whether a model is right or wrong is not the proper question. The purpose for testing the validity of a model is to decide how and under what conditions it is useful and whether its usefulness can be improved, not whether it is correct. Models are generally not disproved. They are only shown to be of restricted utility and then ultimately extended, abandoned or replaced.

One final general point: A model of human performance implies the existence of a model of the environment or system in which that performance takes place. The more specific and quantitative the description of the behavior of that environment, the better. Success in quantitative modelling of human performance has been related to how constraining that environment is. Modelling human signal-detection performance and tracking skill are relatively successful because these tasks impose severe constraints on acceptable behavior. The operator is left with little discretionary time as features of the environment continually demand attention. In highly unstructured procedural tasks in which much freedom of choice is available, modelling is much more difficult, and has tended to be less quantitative. However, in this regime, it also may prove to be more useful. It is in this domain that much current effort in model development is being addressed.

Why Develop Models?

The ultimate reason for building models in general and human performance models in particular is that they can serve as an aid to the designer's, scientist's or user's thinking about the problem being addressed. This extension of and aid to intuition can take many forms:

1. A systematic framework around which to organize facts that reduces the memory load of the investigator.

2. A framework which prompts the investigator not to overlook a feature of the problem that will later be important.

3. A basis for extrapolating from the information given to draw new insights and new testable or observable inferences about system or component behavior.

4. System design tools that permit the generation of design solutions directly.

5. An embodiment of concepts or derived parameters that are useful as measures of performance in the simulated or real environment.

6. A system component to be used in the operational setting to generate behavior, for comparison with the actual operator behavior to anticipate a display of needed data, to introduce alternative strategies or to monitor operator performance.

7. The very act of generating or implementing a model is an intellectual exercise that forces consideration of otherwise neglected or obscure aspects of the problem.

Design is a creative process. No amount of formalism is likely to replace the serendipity that is necessary to bring the proper elements into consideration at the right time and the right place. But models can exercise the intellect in ways that facilitate this intrinsically creative process.

What are the Current Methodologies?. There are, of course, as many different types of models and modelling methodologies as there are inventive minds. At the risk of omitting many approaches that are both useful and valid we will discuss four classes of model development derived from a more psychological perspective and several models that arise from a control theoretic background. An earlier review encompassing a broader set of models is given in Pew, Baron, Feehrer and Miller (1977).

PSYCHOLOGICALLY-BASED MODELS

The psychologically-based models do not come from a single frame of reference as do those derived from control theoretic perspectives. However, they do all start from a task analysis which tends to focus attention on the behavioral activities to be performed. As will be noted, this is in contrast with an intrinsically dynamical system's view that dominates the control theoretic models.

The psychologically based models to be discussed are:

> Reliability Models
>
> Network Models
>
> Information Processing Models
>
> Problem Solving Models

Reliability Models

While the idea of predicting system reliability by aggregating component reliability estimates in appropriate networks has been of interest for many years, it is only in the last 20 years that there has been interest shown in incorporating reliability estimates for the human component. If two system components are connected in series, their reliabilities multiply while the reliability of components in parallel, all of which are required, is equal to the reliability of the weakest link. It is not difficult, in principle, to assign reliability = (1-probability of error) to the human component in systems. Crude data bases of human reliability have been generated from time to time (Payne and Altman, 1962).

The state-of-the-art is reflected in a recent handbook prepared by Swain and Guttman (1980) for the U. S. Nuclear Regulatory Commission. It is specifically adapted for estimation of human reliability in nuclear power plant operations, but it reports most of the available data and methodology developed to date. The approach that is taken is to provide a table of data for estimating the probability of error for the operation of specific controls, the reading of specific displays and the carrying out of preplanned procedures. The conditions under which the operation is performed are taken into account by assigning a multiplier to each case based on the performance shaping factors that are judged by the analyst to be relevant to that case. These factors include the full range of conditions including listening in noisy environments, discriminating among overly similar controls and taking into account task-induced stress or excessive workload. It is as if the influence of human factors in design has been incorporated into multipliers on human reliability coefficients.

A severe limitation of the method is the lack of objectively-derived data on the frequency of human error over a wide enough range of conditions. Some relative frequency data exist, but most of the data are extrapolated by experts like Swain himself from other related conditions. Proposals for massive data collection efforts have been put forward but to date only one has been funded. On behalf of Sandia Labs, General Physics is collecting together and codifying existing data bases.

In spite of the data limitations, it is probably true that using such estimates is better than neglecting the human component altogether. Where it is clear that proper control actuation and equipment operation are both intimately linked to successful operation, the assignment of a reliability to the control actuation will undoubtedly improve the overall estimate of system reliability. In fact, since the reliability of switch actuation is likely to range from one error per hundred to one error per thousand operations, it can easily "swamp" the typically higher equipment reliabilities. Nevertheless, it is very difficult to validate the estimates for such low probability events.

In our opinion the more serious problem is that failure to operate a switch, close a valve or execute a procedure in the proper sequence, while important, does not begin to capture the full role that the human operator plays in system operations. It is in the detection, diagnosis and action selection activities that human capacities and limitations are especially important. In systems such as nuclear power plants where most errors in specific control actions are reversible without catastrophic failure, the role of the operator in recovering from such mistakes is even more important than the occurrence of the mistakes. No human reliability modelling approach contributes to a thoughtful analysis of how to design to assist system operators in their broader and more important decision-making function. This is, after all, the central role of a control center in contributing to the ultimate criteria of effectiveness; safety, productivity, and mission success.

SAINT and Other Network Models

Network models refer to a modelling procedure rather than to models themselves. When one builds a network model of a task, the first step is to construct a flow chart of subtask components indicating their logical interconnections. Pert charts provide examples of the kind of flow charts one might construct. When networks are used to model human performance, subtask elements performed by people are included in the overall task flow charts. For each element the investigator specifies either the completion time or a statistical distribution of completion times together with probability of successful completion or reliability. Sometimes the criticality of a subtask is also specified. When all of the network elements and their interconnections have been described, then initial conditions are selected and a Monte Carlo simulation is run to estimate the parameters relating to overall system performance. Reliability models may be thought of as a subset of network models in which the only modelled parameter is the probability of error.

The first and best known network models of human performance were those of Siegel and Wolf (1969). In addition to the procedures indicated above, these authors introduced the concept of task-induced stress into their model. All the subtask elements incorporated a multiplicative adjustment. When the time remaining was less than the estimated time to complete the overall task, <u>all</u> subtask parameters were adjusted to reduce the probability of successful completion of each subtask thus degrading overall performance.

The success of this approach stimulated the development of a special purpose simulation language called SAINT (Systems Analysis of Integrated Networks of Tasks) (Pritzker, Wortman Seum, Chubb and Seifert, 1974). This language facilitated the development of Network models specifically for human performance evaluation. Since that time a variety of applications of the method have been described ranging from Evaluation of the Digital Avionics Information system (Kuperman, Hann and Berisford, 1977) to simulation of the operations of a hot strip mill. (Buck and Maltas, 1979).

The advantages of network models are their intrinsic generality and the ability to formulate them at any desired level of detail. The weaknesses are twofold: First, to apply the approach, it is necessary to analyze tasks into isolated modules having well-defined inputs and outputs. While, in principle, if interactions between two or more modules are known, they can be modelled, in practice highly interacting modules lead to a level of complexity that makes the check out and validation almost prohibitive. On the other hand assuming the independence of modules can lead to inaccurate results.

Second, the quality of the network models, as discussed in relation to reliability models, is only as good as the quality of the data for providing estimates of completion time and probability of successful completion of the individual subtask modules. In many circumstances, these data are either not available or are derived from severe extrapolations from other contexts. As in any other application the models are only as good as the data on which they are based.

Information Processing Models

There are many descriptions of models in the psychology literature that claim to predict human performance in one or, at most, a modest class of experimental paradigms. Some are little more than block diagrams. (See Pew, 1974 for an example) Others include quantitative specifications, but do not encompass the full range of perceptual, central and motor processes. (Atkinson and Shiffern, 1968; Norman and Rumelhart, 1970; Rumelhart and Norman, 1982). In this section we will discuss in more detail the Human Operator Simulator (HOS), not because it is particularly representative of the application of information processing psychology, but because it is the most comprehensive, and perhaps the only, attempt to take this approach and to apply it to develop a model that will predict the behavior of an individual operator in a system context. It shares two features with its more theoretically-based heritage. (1) It is intrinsically a bottom-up model in the sense that it begins with behavioral components and principles such as movement, information absorption and memory and systematically builds to a model that can mimic task-oriented behavior. (2) It assumes the human operator can only do one thing at a time; the limited- capacity single-channel processing assumption.

The HOS is actually a collection of computer programs which, together, can simulate a system including its human operator. It includes a language for specifying task-oriented behavior, which, in turn, issues subroutine calls to the software representing the micro modelling elements. It also provides for the programming specification of the task environment, including equipment, standard task procedures, target patterns, etc. A comprehensive but readable summary of the work on HOS is reported by Lane, Strieb, Glenn and Wherry (1980).

The HOS includes a micromodel for Information Absorption which is characterized as a change in the operator's knowledge of the state of a display or control. The Information Recall micromodel provides for complete recall of the spatial layout of controls and displays. It assumes probabilistic recall of information from short term memory with two parameters; time-in-store and confidence

level at the time of last input. The mental computation micromodel assumes that the operator assembles the component elements in the calculation and then executes it. It utilizes the aggregate of the information absorption activities required to assemble the components and then requires the analyst to estimate the additional time cost to execute the computation. The decision making micromodel is restricted to decisions about how to accomplish a procedural step and the assignment of priorities for what procedure to undertake. Finally, an elaborate Anatomy Movement micromodel together with the Control Manipulator micromodel account for the time required to complete movements and actuate controls.

The only stochastic element in the model is in the memory process, and in that sense the model does not predict errors that are due only to human randomness. On the other hand, errors that result from the inability to complete activities in time or with sufficient accuracy will lead to the full range of variability in the quality of performance. Errors can also be intentionally inserted and their consequences examined explicitly, a far cleaner procedure, than ad hoc randomization.

The HOS also has a sophisticated data display package that permits summarization of both intrinsic performance parameters as well as specific operator-centered parameters such as average memory load or workload. These measures can be plotted as a function of time with respect to the scenario if desired.

The HOS has been applied to a wide range of task complexity from simple instrument reading to the simulation of a sensor operator on a U. S. Navy patrol aircraft reconnaissance mission (Lane Strieb and Leyland, 1979).

HOS appears to work quite well on small task segments in analysis of individual workplace layout, where the specifics of control design and placement are important. However, it may represent overkill for complex tasks. As Lane et.al. (1980) suggest, in complex tasks, the predictions it makes are not all that different than could be obtained with a much more global representation. These authors also cite the areas of control and monitoring behavior where micromodels are in need of further development and refinement.

On balance, of all the current psychologically-oriented models, HOS comes the closest to fulfilling the needs of systems modelling.

Problem Solving Models

Two distinct lines of development of models appear to be coalescing into a new and important modelling direction; the development of formal, computer-based representations of qualitative information.

One thread of this development stems from behavioral decision theory. In the 1960's Edwards and his colleagues popularized a Bayesian approach to the description of human decision making that was normative, and quantitative. (Edwards, Hays, Phillips and Goodman, 1968). While Edwards recognized the typical stages in formal decision making beginning with hypothesis formulation and ending with action selection and execution, his own work focused on a Bayesian view of the problem of diagnosis and employed subjectively expected utility as a prescription of what individuals should do when making decisions.

Subsequently, many investigators have argued against the generality of Bayesian models and have proposed many descriptive, mostly qualitative alternatives. A particularly recent and relevant one is that of Rasmussen (1980). It is illustrated in Figure 1. Rasmussen includes the stages in the decision making process that are typical of most analyses of human problem solving behavior. However, as indicated, he uses a vocabulary that is compatible with state-space or control theoretic analyses.

A unique feature of Rasmussen's description is that he admits short cuts. He argues that an observer may detect a problem, collect limited data, and conclude immediately that a specific control action must be executed. Such a short cut he calls skill-based behavior because the specific features have been experienced together frequently before and the response is more or less automatic.

Alternatively, the control room operator may detect an alarm, collect data, identify the system state and then immediately select and execute a procedure that results in an action sequence. This is called rule-based behavior by Rasmussen and reflects a higher level short-cut.

Finally, when the circumstances are new and the specific combination of conditions and procedures experienced previously do not match current ones, then the full range of problem solving behavior illustrated in the diagram is called forth, called knowledge-based behavior by Rasmussen. In this case the operator must draw on more fundamental knowledge of system operation and synthesize new combinations of procedures to solve the problem.

While these types of behavior have fuzzy boundaries and may be difficult to isolate in specific cases, the distinctions are useful in pointing out the range of behavior that must be accounted for in more formal models that attempt to predict or augment human behavior generally.

The second thread can be traced to the early work attempting to produce computer simulations of human performance. A milestone in this early work was the General Problem Solver of Newell and Simon (1963).

Here, for the first time, was a formalism that attempted to structure a problem so that it was amenable to solution in a way that mirrored the interview protocols taken from human subjects working on the same problem. This work led in two directions: (1) more sophisticated computer simulations of behavior such as Feigenbaum (1961); or Rumelhart and Norman (1982) and (2) development of knowledge-based representations such as SCHOLAR (Carbonnel, 1970) and MYCIN (Shortliffe, 1976). These are computer-based systems in which a data-base of information is represented in an associative structure analogous to (but with no commitment to being exactly like) the human structures. SCHOLAR was first demonstrated with the Geography of South America. There were associative links between countries and descriptions of their climate, terrain, government, economics etc. Similarly, there were built-in links among the economics of the various countries. A question generating subroutine selected items from throughout this data base to quiz students. Students, in turn, could ask questions of the program. It could parse the question, enter the associative network and give back the highest frequency association. MYCIN performs similar question answering about medical diagnosis.

While these programs are not models of human performance, they may be thought of as formal qualitative models of the environment in which human performance takes place. This framework would be entirely compatible with the kind of qualitative model of performance described by Rasmussen. Indeed, in circumstances where it is desired to study knowledge-based behavior it is difficult to see how this could be handled without such a knowledge base.

Perhaps the closest we have come to a knowledge-based model of human performance is Wesson's (1981) production system description of air traffic control planning. Wesson has build a knowledge-based system that can plan traffic flow, anticipate critical incidents, and issue orders to pilots. In a series of scenarios, in most cases it performed better than real controllers.

CONTROL THEORETIC APPROACH TO MODELLING HUMAN PERFORMANCE

The control theoretic approach to human performance modelling is based, fundamentally, on three interrelated perspectives. First, the major technical concern is in predicting total person-machine system performance as a means for analyzing and designing systems[1] in which humans play a central control role. In effect, the

---

[1] In general, the systems of interest are dynamic in nature, but they may be continuous, discrete, or hybrid.

approach begins with a consideration of the system, not the human, and this has a direct impact on the characteristics and goals of the models that emerge. It tends to direct attention away from the traditional "knobs and dials" concerns of human engineering and towards issues relating to the information requirements, the response of the system and, especially, the cognitive and response capabilities and limitations of the operator. It imposes the need for characterizing these limitations in terms relevant to the system context, so that the model can be used to compute system performance. In this view, it also becomes highly desirable that these limitations be specified at the processing level rather than directly at the performance level as is done in most network models. Such a specification allows development of a model that unwraps human performance from the particulars of the task and, thereby, is general enough to predict performance in other situations.

The second idea permeating the control theoretic approach is the view of the human operator as an information processing and control/decision element in a closed-loop (sometimes referred to as the cybernetic view of the human). Feedback is a central idea and it implies the comparison of actual responses with predicted and/or desired responses. It also leads to a focus on the role of information processing to "filter" noise and to predict future responses. An integral aspect of this view is that trained human operators are expected to exhibit many of the characteristics of a 'good", or even optimal, inanimate system performing the same functions in the "loop". It must be emphasized that this view of the operator is far more general than the simple conception of an error-correcting device or servomechanism.

The third perspective underlying this approach is the so-called top-down view of the problem. Thus, this type of modelling starts from a description of the system that includes its goals and sub-goals and attempts to model the human component in modular fashion at the task or function level. This approach contrasts with bottom-up methods which attempt to synthesize human performance from a sequence of fundamental activities such as anatomical movements, memory recalls, etc. Top-down models may, or may not, provide structural analogs of human activities, as they do not start from such a basis. On the other hand, the goals of the task are imbedded in thes models, either explicitly or implicitly. Moreover, the general thrust is to develop normative models that prescribe operator performance with respect to these goals (and constrained by operator limitations). If normative solutions and models cannot be derived because of the complexity of the problem, then sub-optimal models or ones employing goal-oriented heuristics are employed.

Within the framework of these general perspectives there is much room for a variety of approaches to human performance modelling. The variations tend to reflect the tastes and training of the modellers and, more importantly, the mathematical developments in control and systems theory which, themselves, mirror changes in technology and in the problems of interest. Thus, we have seen an evolution in interest from problems in which skilled psychomotor performance in single-axis continuous control tasks was of prime concern to the supervisory control problems of major interest today. And, we have also witnessed the application of different mathematical and analytical methods from control theory (and/or systems theory) to modelling human performance. Rather than review a variety of such techniques here, we will focus on the so-called modern control approach to modelling and will illustrate how it provides a sufficiently general framework to support an orderly evolution of models from those used for relatively simple, continuous control problems to ones suitable for analysis of the more complex information processing and decision-making tasks associated with supervisory control.

By modern control, we mean the concepts and techniques that have emerged in control theory in and since the late fifties. This period in control theoretical development has been characterized by a number of significant trends and advancements, but from the standpoint of importance for this discussion the major ones were the following:

i)   A shift toward state-space description of multi-variable systems and the corresponding emphasis on internal, rather than input-output, models of system behavior.
ii) An increased concern with, and ability to analyze, stochastic processes in systems.
iii) The consideration of state estimation as integral, and often essential, to the control process.
iv) An increased emphasis on optimization of performance as a control goal.
v)   A shift toward time-domain analysis.

It is these ideas that infused the development of the optimal control model (OCM) and its relatives and descendants.

A General Closed-Loop View of Human Functions in Control Tasks

An analysis of fundamental human operator functions in a variety of tasks related to system operation suggests a basic commonality that is summarized in the following list:

1. Observation of the display variables, according to an attention allocation strategy dictated by the needs of the task.
2. Assessment of the present situation, based on the monitored information, the

operator's inherent knowledge of system operation, and a knowledge of alternative situations.

3. Deciding to take some action (or not), and making this decision on the basis of the assessed situation, the basic task objectives, the procedural and other means available to effect this decision, and the expected consequences of the various actions.

4. Action to implement the decision, either in terms of communicating the intent of the decision, or by directly observing or controlling the system in an appropriate manner.

This verbal description can be given somewhat more substance by casting it in a block diagram, or flow chart form, as shown in figure 2. As drawn, the "system" comprises the upper portion, while the "operator" comprises the lower. The system is affected by both the operator's own direct actions, and by the actions of others (e.g., the other crew members). The system displays not only provide an indication of the system states, but also provide the communications channel for obtaining relevant information from other operators, or from the system itself (e.g., an auditory alarm). The portion of the diagram representing the operator is deliberately simplified, to show the basic structure of functional processing. Thus, it should be recognized that various pathways can interconnect the several functions shown, and allow the result of one function (e.g., decision-making) to affect how another is carried out (e.g., monitoring). In addition, the diagram does not show how either human limitations or perceived goals may affect the various functions or their parameters.

In what follows we show how this generic structure and view is applicable to several models of human performance including the OCM, some related decision-making models and a recent supervisory control model.

The Model for Continuous Control (OCM)

The OCM has been described at length in many references (see Baron and Levison (1980) for an extensive bibliography), so we provide only a brief overview of the model emphasizing its relation to the generic approach and structure described above.

A block diagram for the OCM is given in Figure 3. This diagram is different than that usually shown in the literature to emphasize the connection to the generic structure of Figure 2 and to supervisory control models to be discussed subsequently. Here, the perceptual processing limitations embodied in the OCM (observation noise, delay and attention-sharing) are incorporated in the "monitoring block". The logic for selecting information for operator attention is also included in this block. Estimation of the system state at the current time is seen as a special (and somewhat limited) form of situation assessment germane to the

control problem. However, it is a cognitive process involving the use of internal models of the plant to operate on incomplete and noisy information and to predict ahead to compensate for delay. Finally, the "decision-making" process is one of selecting the appropriate control procedure (i.e., the gains or weights on the estimated states) to generate commanded controls aimed at accomplishing the desired performance objective (optimizing an appropriate criterion). A monitoring decision as to which display to fixate (or how to share attention fractionally) so as to optimize this criterion is also part of the process. The "action" outputs are the monitoring requests (i.e., the fractional attention assignments) and the control inputs to the system. These controls are the "commanded" inputs, and are modified by the human's motor response limitations.

Thus, the OCM is an explicit realization (perhaps, a simple one) of the general structure described above. More importantly, its success in predicting human performance in control tasks demonstrates the inherent value of the general approach taken (albeit in a limited domain). In particular, one should note that in the OCM human limitations are modelled in control theoretic terms (i.e., in system terms) and, essentially, at the process level (i.e., by processing limitations such as observation noise rather than by task-specific, performance limits). This has allowed performance to be predicted across a variety of changes in task parameters, such as changes in vehicle dynamics, displays, etc.

The OCM has proven to be capable of predicting or matching, with precision, human performance in a variety of continuous control tasks. It predicts (or simulates) both the goal-oriented control activity and the random response of the operator. It contains within it the mechanisms for adaptation to changes in vehicle dynamics, displays, etc. and has been used extensively in analyzing pilot/vehicle system problems. It has also been applied successfully in the design of experiments (Junker and Levison, 1978) and in the analysis of experimental data (see, e.g., Baron, (1976)). In the design of experiments it has provided a mechanism for pretesting so as to focus simulator experiments and reduce the experimental space that must be investigated. In data analysis, it has provided compact, generalizable and insightful measures of human performance. Thus, in the domain of continuous control, the OCM managed to fulfill many, if not all, of the goals for models of human performance.

Some Control Models for Detection and Decision Making

Despite its success, the OCM will not be of direct utility in modelling a majority of supervisory control tasks because of the relative unimportance of continuous control in such problems. However, the modern control-theoretic approach provides the means for developing models of monitoring, detections and decision-making that are of direct relevance to supervisory control. Moreover, these models generally share with the OCM the same fundamental information processing structure and, indeed, the same models for human display processing or monitoring limitations. This is due, in large measure, to a focus on the detection of state-related events as a prelude to any discrete decision. The events may correspond to situations of interest or components of such situations. For example, until recently, the application of prime interest for this modelling was the detection of system failures. The addition of event detection logic to the information processing portion of the control-theoretic models may be viewed as an extension of the situation assessment function.

A diversity in the various decision-making models derived from the modern control approach exists as a result of using different features of the optimal state estimator. The optimal estimator processes the available observations to generate a minimum variance estimate of the system state and the covariance matrix of the error in that estimate.[2] The manner in which this estimate is generated is important, too. In particular, an internal model of the system is used to predict an estimate of the state and of the observation. The predicted state estimate is updated or corrected using the observation "residual", which is defined as the difference between the actual observation and the expected observation. When the estimation process is optimal, which requires that the internal model be identical to the true system model, the residuals are a white noise process; in other words, all information about the state of the system has been extracted from the measurements.

Thus, the estimator generates outputs that can be used to detect events that are specified directly in terms of system state variables. Alternatively, the residuals generated by the estimator can be processed to detect the occurrence of an unexpected response; i.e., one that deviates significantly from that of the internal model.

The estimator outputs may be used to derive either descriptive or prescriptive models of decision-making. For example, a descriptive model based on the state-estimate itself could involve decision regions in the state space. Detection of when the state is in a given region could be modelled as a deterministic decision based on the estimate alone, or it could be a probabilistic

assessment based on the probability density function. A descriptive detection model using the residuals could be based on a test of the "whiteness" of the residual process with significant deviations from the ideal indicating an "event". For prescriptive models of decision-making, optimal decisions must be generated. This, of course, involves specification of decision criteria.

The first use of the OCM information processing structure in modelling human decision-making was by Levison and Tanner (1971). They studied the problem of how well subjects could determine whether a signal embedded in added noise was within specified tolerances (a continuous, visual analog of classical signal detection experiments). They modelled this situation by assuming that the operator is an optimal decision maker in the sense of maximizing expected utility. For equal penalties on missed detections and false alarms, this rule reduces to one of minimizing expected decision error. The decision rule is, simply, a likelihood ratio test that, effectively, uses the densities generated by optimal estimator. Levison and Tanner were able to validate the model for a single decision tasks and for two such identical, but interfering, tasks.

In a theoretical paper, Kleinman and Phatak (1972) first suggested the use of the estimator residuals for failure detection. They also indicated how "multiple" internal models could be used for simultaneous detection and identification. Gai and Curry (1976) and Wewerinke (1981), in a somewhat more extensive study, have used residual monitoring schemes to model detection of instrument failures in monitoring of complex, multivariable systems. In each of these models, the perceptual and information-processing portions of the OCM, including the attention-sharing model, were used. In both cases, "optimal" sequential decision algorithms were used for decision-making. The models were shown to be capable of predicting experimental results collected in validation efforts.

The above studies provide further, independent validation of the display processing and state estimation models developed for the OCM. They also extend the notion of situation assessment for control-theoretic models, beyond that of pure state estimation, to include detection of state-dependent events. Thus, "regions" in the state-space can correspond to, or be "maps" of, different events or situations. Or, "whiteness" or other tests on the residuals may be seen as tests of the hypothesis that the situation is "normal"; and, clearly, multiple models may be used to define different situations.

Two, more recent models, go still further in extending this approach to modelling human decision-making. These are the Dynamic Decision Model (Pattipati, Ephrath and Kleinman, 1979, 1980) and the DEMON model (Muralidharan and Baron, 1978-1980).

---

[2]Given the assumptions generally made for these problems, this is usually sufficient to specify the conditional probability density for the state (conditioned on the observations).

Pattipati, Ephrath and Kleinman (1979, 1980) have developed a dynamic decision making model (DDM) for predicting human task sequencing that is similar in several respects to the other models discussed here. It, too, contains the same information processing structure as in the OCM. As in the original Siegel and Wolf models, situation assessment in the DDM involves estimation of the time available and time required for task completion. These variables are obtained from a memoryless transformation of the estimated system state variables.

The DDM does differ from the other decision-making models discussed here in one significant way, namely in the introduction of human randomness in the decision-making algorithm itself. This is accomplished by assuming a distribution for the payoff values (attractiveness measures) and then incorporating Luce's stochastic choice algorithm. Though the practical value of including this randomness in performing system design and analysis may be argued, it is clear that such randomness in human decision-making does exist. Moreover, in the context of the relatively simple paradigm to which the DDM was applied, the introduction of the stochastic choice axiom allows the DDM to be used to compute performance statistics analytically, rather than by Monte Carlo simulation. Of course, significant deviations from the simplifying assumptions used in the basic DDM validation task could compromise the possibilities for analytic solutions.

A major contribution of the DDM work is the experimental validation of the model. By constraining the experimental paradigm to a situation that can be treated carefully in an experimental environment, it becomes possible to test model hypotheses with reasonable cost and control. These tests show that the ideas underlying the DDM are essentially sound. They also provide further validation for the other control-theoretic models discussed herein.

DEMON is a decision-monitoring and control model for analyzing the en route control of multiple remotely piloted vehicles. Very briefly, the en route operator's task is to monitor the trajectories and estimated times of arrival (ETAs) of N vehicles, to decide if the lateral deviation from the desired preprogrammed flight path or the ETA error of any of them exceeds some tolerance threshold and to correct the paths of those that deviate excessively by issuing appropriate control commands ("patches"). Path deviations arise from navigation errors and disturbances. In addition, the operator must decide when to "pop-up" or "hand-off" the RPV's under his control. Display information is updated at discrete times, with ETA and lateral deviation errors presented separately and only a single RPV available for observation at a given time.

Prior to a frame update, the RPV en route operator with N RPVs under control must decide which among 2N+1 displays to monitor.[3] The operator's monitoring choice is assumed to be a rationale decision governed by his knowledge of the situation, his goals and priorities and his instructions. These factors are incorporated in an expected net gain (ENG) criterion for each display and, it is assumed the operator selects the display with the highest net gain.

Situation assessment in DEMON is a process which involves detecting whether any of the RPVs have deviated from their desired trajectories by more than the tolerable amount and/or whether any are sufficiently close to an important waypoint to warrant an action. The information processor utilized in this assessment is the same as that used in the OCM and the previously mentioned detection models.

Decision-making involves selection among control and monitoring options. Control of an RPV involves a pop-up, hand-off or patch (correction) command. As with monitoring, control decisions are assumed to be made on the basis of expected net gain criteria.

DEMON extended previous control-based approaches to a multi-task environment that involved essentially discrete monitoring and control decisions (Note that the task interference model of Levison, Elkind and Ward (1971) treated multiple continuous-control tasks). Control of each RPV represented a separate task with a payoff for maintaining errors within tolerance and for timely pop-ups and hand-offs. Inasmuch as only one RPV (and only one RPV-state) could be observed at any time, the DEMON operator had to rely on memory and prediction (based on the state-estimates) to assess the relative priorities for "servicing" of the RPVs under his control.

A TEMPTING SYNTHESIS

The DEMON model was an important integrative step in the development of supervisory control models based on a control-theoretic approach. Other such models developed by Kok and Stassen (1980) and Govindaraj and Rousse (1979) but not discussed herein have also advanced this development. However, these models, as well as the more "single decision" models discussed above, have not incorporated a number of very significant features common to many supervisory control problems. For example, these models do not consider the detection of events not explicitly related to the system state variables. Nor do they model multi-operator situations and the effects of communication among such operators. The models discussed this far

---

[3] An additional display option is included to account for possibilities not included in the basic state-space model.

also fail to account for the time to complete discrete tasks. Perhaps the chief shortcoming is that the models do not include the procedural activities of the operators. Interestingly, these neglected features are often the prime concern of psychologically-oriented models. In this section, we describe a more recent supervisory control model (PROCRU) that begins to integrate and synthesize features of the two general modelling approaches we have been discussing.

PROCRU (Procedure-Oriented Crew Model), a model that has been developed recently for analyzing flight crew procedures in a commercial ILS approach-to-landing, demonstrates how some of the additional features required to model supervisory control may be added to control-theoretic models in a reasonably integrated fashion (Baron, Zacharias, Muralidharan and Lancraft, 1981). PROCRU includes a system model and a model for each crew member, where the crew is assumed to be composed of a pilot flying (PF), a pilot not flying (PNF) and a second officer (SO). In the present implementation of PROCRU, the SO model does not include any information processing or decision-making components. Rather, the SO is modelled as a purely deterministic program that responds to events and generates requests. PF and PNF, on the other hand, are each represented by complex supervisory control models which have the same general form but differ in detail.

Briefly, PF and PNF are each assumed to have a set of "procedures" or tasks to perform. The procedures include both routines established "by the book" (such as checklists) and tasks to be performed in some "optimizing" fashion (such as flying the airplane). The particular task chosen at a given instant in time is the one perceived to have the highest expected gain for execution at that time. The gain is a function of mission priorities and of the perceived estimate of the state-of-the-world at that instant. This estimate is based on monitoring of the displays, the external visual scene and auditory inputs from other crew members.

The basic structure of the PROCRU model for either PF or PNF is shown in Figure 4. Note the correspondence between this flight task-related model and the generic model of Figure 2. The system and system displays of the generic model are broken out laterally in the PROCRU model, to illustrate the system states relevant to the problem, and the display cues available to the crew. The monitoring function of the generic model is partitioned to separate processing of auditory and non-auditory cues. The situation assessment function includes both discrete (event detection) and continuous (state estimation) information processing; the "list" of possible events here includes some that are not explicitly dependent on the vehicle state variables. Finally, the generic decision making function is, in this

procedure oriented model, called procedure selection.

The monitoring and information processing portions of PROCRU are not unlike those of the OCM or other models discussed above, though they have some novel features and extensions (the reader is referred to Baron, Zacharias, Muralidharan and Lancraft (1981) for a discussion). However, the procedure selection portion of the model is sufficiently different and important to warrant discussion here.

As noted previously, the operator is assumed to have a number of procedures or tasks that may be performed at each instant. The definition of these procedures is an essential step in the formulation of PROCRU. All crew actions, except for the decision as to which procedure to execute, are determined by the procedures. We emphasize that we use the term procedure here to apply to tasks in general; a procedure in these terms could have considerably more cognitive content than might normally be considered to be the case.

It is assumed that the PROCRU operator knows what is to be done and, essentially, how to accomplish the objective. However, only one procedure may be executed at a given time (single channel assumption) and he must decide which one to do next. This is a decision among alternatives and the procedure selected is assumed to be the one with the highest expected gain for execution at that time. The expected gain for executing a procedure is a function that is selected to reflect the urgency or priority of that procedure as well as its intrinsic "value".

For procedures that are triggered by the operator's internal assessment of a condition related to the vehicle state-vector, the expected gain functions are appropriate subjective probabilities, as determined by the state estimation portion of the model. Procedures that are triggered by events external to the operator, such as ATC commands, communications from the crew, etc., are characterized by expected gains that are explicit functions of time. For either type of function, the gain for performing a procedure will increase, subsequent to the perception of the triggering event, until the procedure is performed or until a time such that the procedure is assumed to be "missed" or no longer appropriate for execution.

The model for procedure selection captures many important aspects of human performance in a multi-task environment, and is directly relevant to investigating the efficacy of flight crew procedures. It allows for procedures to be missed and/or interrupted: even flying the airplane may be neglected, as can happen in real flight. Although sub-procedural steps will not be performed out of order with this modelling approach, it is possible to preprogram such errors if desired.

The selection and execution of a procedure will result in an action or a sequence of actions. Three types of actions are considered: control actions, monitoring requests and communications. The control actions include continuous manual flight control inputs to the aircraft and discrete control settings (switches, flap settings, etc.). Monitoring requests result from procedural requirements for specific information and, therefore, raise the attention allocated to the particular information source. (Note that verifying that a variable is within limits may not require an actual instrument check, if the operator already has a "confident" internal estimate of that variable.) Communications are verbal requests or responses as demanded by a procedure. They include callouts, requests or commands, and communications to the ATC. Verbal communication is modelled directly as the transfer of either state, command or event information.

Associated with each procedural ation is a time to complete the required action. (Just as with network models, it is possible to modify PROCRU to allow for a probabilistic distribution of action times). When the operator decides to execute a specific procedure, it is assumed that he is "locked in" to the appropriate mode for a specified time. For example, if the procedure requires "checking" a particular instrument and it is assumed that it takes T seconds to accomplish the check, then the "monitor" will not attend to other information for that period, nor will another procedure be executed.

In PROCRU procedural implementation is modelled as essentially error free. However, errors in execution of procedures can occur because of improper decisions that result from a lack of information (quantity or quality) due to perceptual, procedural and workload limitations. If the effects of action errors are also to be analyzed, this is accomplished by deliberately inserting such errors directly into the scenario.

PROCRU generates a number of outputs that are useful for analyzing crew performance and workload. First, one can obtain full trajectory information. In addition to this information, one can obtain each crew member's estimate of the state and the standard deviation of the estimation error, the attentional allocation at that time and PF's control inputs. These data, along with a time-record of significant events, are tabulated in a file. In addition to the trajectory output, PROCRU produces activity time lines. It should be emphasized that the time lines generated by PROCRU are dynamically-generated time lines, unlike those normally used in human factors analyses. That is, actions are not completely preprogrammed but depend on previous responses, disturbances, etc. Thus, one can change a system or human parameter in PROCRU and automatically generate a new, different time line.

DISCUSSION

PROCRU, while only a beginning, capitalizes on many of the best features of the full range of models reviewed in earlier sections. As we think ahead to further developments in supervisory control modelling, it appears that still further integration of those approaches will develop.

Supervisory control activities tend to reduce the emphasis on continuous control and increase the emphasis on discrete procedures. Although in PROCRU, monitoring and control procedures, involving continuous actions or processing, maintain their own time-scale, discrete procedures are formulated in a way that is analogous to the treatment of subtasks in the network approach. Thus, a discrete procedure captures attention for a prespecified duration (which must be estimated by the analyst using any available data base). On the other hand one can formulate the initiating conditions for these procedures in the form IF (situation) THEN (action), that is, as production rules, but in PROCRU the Boolean IF condition is evaluated probabilistically. In addition, selection among productions or procedures whose conditions are satisfied simultaneously is based on maximization of expected net gain. These factors allow human processing limitations and the goals of system generation to influence the output of the production system.

Another point of convergence between PROCRU and the psychological models, is the use of a preliminary task analysis for initial definition of the procedures. We expect the kind of task and protocol analysis utilized by cognitive psychologists and AI specialists to become an important aspect of premodelling analysis for system-oriented supervisory control models. Indeed, the push toward more sophisticated procedures has engendered an interesting debate among us concerning exactly where the boundaries are between dynamic system formalisms and AI formalisms. We believe supervisory control modelling is a fertile ground for integration across these domains.

The major developments arising from the control-theoretic models that are important in PROCRU, and will continue to be in subsequent developments, are the adaptive, goal-oriented display sampling, monitoring, and decision-making routines and the formalized priority structure possible with these models. PROCRU implements the selection of activities to which it will assign processing resources on the basis of maximizing the expected net gain. This makes it possible to choose among monitoring, controlling and executing discrete procedures within a common, dynamically changing frame of reference, yet within a unitary goal structure. As we examine more complex supervisory systems, we anticipate the need to develop goal structures that are more hierarchical in nature.

PROCRU did not deal in detail with the high level diagnostic processes that are an important component of many supervisory tasks. However, some of the control-theoretic models do involve interesting and sophisticated diagnostic algorithms. These algorithms may be embedded in a PROCRU-like supervisory control model because of the top-down modular approach used. It is possible to envision a sequential and hierarchical approach to modelling diagnosis (or, more generally, situation assessment) that includes both simple, formal diagnostic procedures and more elaborate algorithms. The higher level algorithms would use structural knowledge of the plant (the internal model) and may be viewed as a system-theoretic form of knowledge-based behavior. Or, the algorithms may employ heuristic, more qualitative diagnostic procedures. In any case, when dealing with these kinds of problem-solving problems, we would expect a convergence of several of the approaches discussed above.

In order for human performance models to be utilized more widely than just in research, the user interface for them must be simplified so that they are accessible to a wider community. The HOS has made a start in this respect in that it clearly partitioned the subsystems and created a user-specification interface that minimizes concern with the inner workings of the model. As model complexity increases, this area of development will receive significantly more attention.

Human performance modelling has made major advances in the last ten years as the cost and complexity of live simulation is increasing. We are entering a period when there is a stronger demand for models and their results. The potential for bringing together the work in psychology, AI and control is an exciting prospect.

REFERENCES

Atkinson, R. C., and R.M. Shiffrin (1968). Human memory: A Proposed System and its Control Processes. In K. W. Spence and J. T. Spence (Eds.), The Psychology of Learning and Motivation, Vol. 2, Academic Press, New York. pp. 89-195.

Baron, S. (1976). A Model for Human Control and Monitoring Based on Modern Control Theory, Journal of Cybernetics and Information Sciences, Spring.

Baron, S. and W. H. Levison (1980). The Optimal Control Model: Status and Future Directions. Proc. of IEEE Conf. on Cybernetics and Society, Cambridge, MA, October.

Baron, S., G. Zacharias, R. Muralidharan, and R. Lancraft (1980). PROCRU: A Model for Analyzing Flight Crew Procedures in

Approach to Landing, Proc. of Eight IFAC World Congress, Tokyo, Japan, August (Also, BBN Report No. 4374, April).

Blanchard, R. E. (1973). Requirements, Concept, and Specification for a Navy Human Performance Data Store. Behavior metrics Report No. 102-2, April.

Buck, J.R. and K.L. Maltas, (1979). Simulation of Industrial Man-Machine Systems, Ergonomics, 22, No. 7, pp. 785-797.

Carbonell, J. R. (1970). AI in CAI: An Artificial Intelligence Approach to Computer Assisted Instruction. IEEE Transaction on Man-Machine Systems. MMS-11 No. 4, Dec.

Edwards, W., L.D. Phillips, W.L. Hays, and B.C. Goodman (1968). Probabilistic Information Processing Systems: Design and Evaluation. IEEE Transactions on Systems Science and Cybernetics, SSC-4, 248-265.

Feigenbaum, E. (1961). The Simulation of Verbal Learning Behavior. Proceedings of the Western Joint Computer Conference, 9, 121-132.

Gai, E. G. and R. E. Curry (1976). A Model of the Human Observer in Failure Detection Tasks, IEEE Trans. on Systems, Man and Cybernetics, SMC-6, pp. 85-94.

Govindaraj, T. and W. B. Rouse (1981). Modeling the Human Controller in Environments that Include Continuous and Discrete Tasks, IEEE Trans. on Systems Man and Cybernetics, SMC 11, No. 6, pp. 410-417.

Junker, A. M. and W.H. Levison (1978). Recent Advances in Modelling the Effects of Roll Motion on the Human Operator, Aviation Space and Environmental Medicine Vol. 49, No. 1, Section 11.

Kok, J. J. and H. G. Stassen (1980). Human Operator Control of Slowly Responding Systems: Supervisory Control, Journal of Cybernetics and Information Science, Special Issue on Man-Machine Systems, Vol. 3, Nos. 1-4.

Kuperman, G. G., R.L. Hann, K.M. Berisford (1977). Refinement of a Computer Simulation Model for Evaluating DAIS Display Concepts. Proceedings of the Human Factors Society 21st Annual Meeting, pp. 305-310.

Lane, N., M.I. Strieb, W.E. Leyland, (1979). Modelling the Human Operator: Applications to Cost Effectiveness. In Modelling and Simulation of Avionics Systems and Command Control and Communications Systems NATO/AGARD Conf. Proceedings, CP-268.

Levison, W. H., J.I. Elkind, and J.L. Ward, (1971). Studies of Multi-Variable Control Systems: A Model for Task Interference, NASA CR-1746, May.

Levison, W. H. and R. B. Tanner (1971). A Control Theory Model for Human Decision Making, NASA CR-1953, Dec.

Muralidharan, R. and S. Baron, (1979). DEMON: A Human Operator Model for Decision Making, Monitoring and Control. Proc. of Fifteenth Annual Conf. on Manual Control, AFFDL-TR-79-3134, WPAFB, Ohio, Nov. (also Journal of Cybernetics and Information, Vol. 3, 1980).

Newell, A., and H.A. Simon, (1963). GPS, a Program that Simulates Human Thoughts. In E. Feigenbaum and J. Feldman (Eds.), Computers and Thought. New York: McGraw Hill.

Norman, D. A. and D.E. Rumelhart (1970). A System for Perception and Memory. In D. A. Norman (Ed.), Models of Human Memory. New York: Academic Press.

Pattipati, K. R., A.R. Ephrath, and D.L. Kleinman, (1979). Analysis of Human Decision Making in Multi-Task Environments, Tech. Rept. EECS TR-79-15, Univ. of Conn. Nov. (Also, Proc. of Int. Conf. on Cybernetics and Society, 1980).

Pattipati, K. R. and D.L. Keinman, (1979). A Survey of the Theories of Individual Choice Behavior, Tech. Rept. EECS TR-79-12, Univ. of Conn., August.

Payne, D., and J.W. Altman, (1962). An Index of Electronic Equipment Operability: Report of Development. American Institutes for Research, January.

Pew, Richard W. (1974). Human Perceptual-Motor Performance, in B. H. Kantowitz (Ed.), Human Information Processing: Tutorials in Performance and Cognition. Lawrence Erlbaum Associates, Hillsdale, New Jersey.

Pew, Richard W., S. Baron, C.E. Feehrer and D.C. Miller (1977). Critical Review and Analysis of Performance Models Applicable to Man-Machine Systems Evaluation. Report No. 3446, Bolt Beranek and Newman, Inc. Cambridge, MA, March.

Phatak, A. and D.L. Kleinman (1972). Current Statues of Models for the Human Operator as a Controller and Decision Maker in Manned Aerospace Systems, Automation in Manned Aerospace Systems, AGARD Proceedings No. 114, Dayton, Ohio, Oct.

Pritsker, A.B., D.B. Wortman (1974). Seum, C. S., Chubb, G. pp., and Seifert, D. J. SAINT: Volume I. Systems Analysis of an Integrated Network of Tasks. AMRL-TR-73-126, April.

Rasmussen, J. (1980). The Human as a Systems Component, in H. T. Smith and T. R. G. Green (Eds.), Human Interaction with Computers. London, Academic Press.

Rumelhart, D. E., D.A. Norman, (1982). Simulating a Skilled TypiSt: a study of Skilled Cognitive-Motor Performance. Cognitive Science, pp. 1-24.

Shortliffe, E. H. (1976). Computer-Based Medical Consultations: MYCIN. New York, Elsevier/North Holland Publishing Co.

Siegel, A. I., and J.J. Wolf, (1969). Man-Machine Simulation Models. New York, John Wiley and Sons, Inc.

Swain, A. D. and H.E. Guttmann, (Eds.), Handbook of Human Reliability Analysis with Emphasis on Nuclear Power Plant Applications. U. S. Nuclear Regulatory Commission. Nureg/DR-1278.

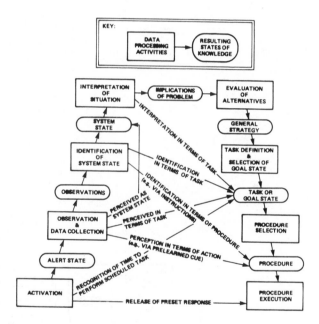

Figure 1.   Decision-Making Model (Adapted from Rasmussen)

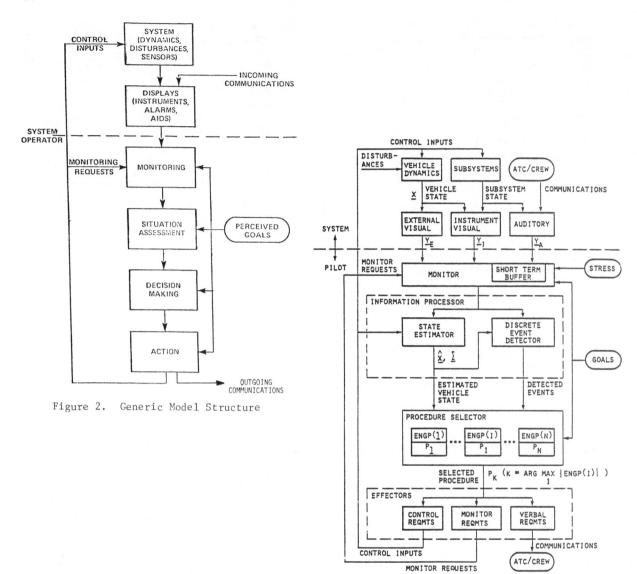

Figure 2.   Generic Model Structure

Figure 4.   PROCRU Model Structure for
Individual Crew Member

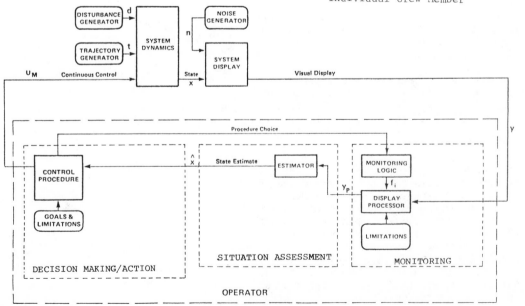

Figure 3.   Optimal Control Model

VISUAL SCAN AND PERFORMANCE — INDICATORS OF MAN'S WORKLOAD?

K. R. Kimmel

*Forschungsinstitut für Anthropotechnik, 5307 Wachtberg-Werthhoven,
Federal Republic of Germany*

Abstract. The feasibility of the concept of the human as an information
processing channel and its performance in terms of transinformation was
studied in a three-axis compensatory manual control task for the evaluation
of four different display configurations. Two experiments are described in
which the experimental parameters were 1) tracking bandwidth in input
random tracking and controlled element dynamics and 2) additional load of
subjects by a visual or auditory side task. The study describes an investigation of the relationship between the subjects' eye-scan behaviour and
their performance level. The findings are confirmed by the measures on
subjective effort expressed in terms of the standard deviations of the
control stick signals.

Keywords. Man-machine systems; human operator models; supervisory control;
manual control; mental workload; human instrument monitoring; human factors
design methodology.

INTRODUCTION

In several areas of man-machine interaction
there is an increased interest in workload
effects of display layout with control or
monitoring tasks. Predominantly, this interest stems from human ability to process information, knowledge of which is essential
for man-machine interface design. Errors
caused by improper systems design can impose
high penalties. In many complex systems
visual scan behaviour plays an important
role, since the operator usually has to monitor several instruments in order to obtain
necessary control information. In controlling
systems the human functions as planner, manual controller, monitor or event detector.

Procedures for the layout of the man-machine
systems known at present do hardly permit the
a priori design of such systems with full
details by means of model concepts and task
requirements, but one is rather still depending on systems evaluation by experimental
testing of different alternatives. Existing
models for the description of the attention
distribution in complex control- and monitoring tasks (cf. Levison, Elkind, and Ward
(1969), Kleinman and Curry (1976), or
Kleinman (1976)) have a more normative character, since they derive their strategies
from some global (performance-) criterion.
Trials to model the human visual scan behaviour in monitoring tasks for instance by the
use of queuing theory (cf. Carbonell (1966a))
or an inner model (cf. Palmer (1977),
Smallwood (1966)) are also to be seen in this
context.

One may ascribe to the operator a certain
limited ability to perform required activities which has the character of a total
capacity. This is a hypothetic bound regarding the parallel execution of several tasks
by the human operator. It is immediately
plausible that this hypothetical bound depends upon the type and extent of the tasks.
There are also individual differences. In
this view the concept of capacity is a crucial criterion for the design of man-machine
systems. An individual who works at 100 %
of his capacity has no spare capability for
performing additional tasks.

Clearly, one should avoid designs requiring
workload as high as 100 %, because of the
likelihood of unexpected demands, as experience reveals. In the case of very low mental
loads, errors can result from the low attention. Ideally, the operator's tasks should
be designed so that an appropriate fraction
of his capacity is required. This design
goal cannot easily be met, since there is no
agreement on the size of this capacity but
only that it exists. Clearly, this capacity
could reasonably be accepted to have a certain numerical value at any time.

A main problem is that of finding a measure
of "capacity" which is applicable to all
kinds of tasks. The effect of all tasks on
the operator is to elicit a task-performing
set such that he behaves as an information processing system. The nature of information
used by an operator can be deduced from its
effect on his tasks behaviour (Kampé de
Fériet, 1973). Consequently, man's capacity

can be seen in terms of an information processing channel with a so-called "channel
capacity" or maximum transinformation rate
for all possible codings of input information. From this theoretical standpoint different displays become different codings of
the input information. Display performance
can then be measured in terms of transinformation rate. The displays can be evaluated
and selected on the basis of such scores and
appropriate criteria including lowest workload, highest performance, lowest timing
requirements, etc.

Any investigation of the task workload must
include several dimensions such as sensory,
mental and motor functions. There is little
agreement among scientists in how to concep-
tualize workload. To arrive at a functional,
accurate definition of workload several
questions must be addressed. Does workload
refer to the task demands imposed on the
operator, or is workload the operator's effort required to satisfy these task demands?
Jahns (1973) defines workload as the extent
to which an operator is occupied by a task.
Focussing on task performance measurements,
Levison, Elkind and Ward (1973) define workload as the fraction of the controller's
capacity that is required to perform a given
task.

The information theoretic approach in combination with a side-task offers the possibility to deal with both definitions, provided
that the human operator has to work near his
capacity bound. This may be implied if the
insertion of a side-task results in a performance decrease.

Aside from the information capacity approach
there are others including, for example
electrocardiogram, sinusarrhythmia, heart
rate irregularities (Strasser, 1974), electro-
encephalogram (Wickens and others, 1976),
electrodermal activity (Zeier, 1979); subjective workload measurements such as rating
scales, questionnaires and so on (Pfendler
and Johannsen, 1974). More details are to be
found in the survey of Wierwille and Williges
(1978).

In order to evaluate the utility of transinformation rate as a performance measure, another measure was required for comparison.
Particularly useful with displays, especially
in multiple display configurations with
multi-task situation, are the measures obtained with eye-movement recordings, such as
the percentage of time spent looking at a
display feature or the average time duration
for each look. It was expected, that conclusions concerning subjects' strategies in performing the tasks could be drawn from eye-
movement recordings. This is supported by
Ellis and Stark (1981) who suggested that the
extent to which the scan patterns of pilots
in instrument scanning deviates from that
expected by the assumption of statistical
independence (Senders and others, 1965) may
reflect different information load on the
pilot.

METHOD

Experimental Setup

Two experiments were conducted. In the first
experiment a compensatory three-axis control
task was used. The parameters of the control-
loop were varied by employing two types of
controlled elements (K and K/s) and by variation of the bandwidth of the forcing functions. In the second experiment there was
only one parameter configuration, but the
subjects were loaded additionally by a discrete side-task. The information for the side-
task was presented either visually or auditorily. The experimental program is given by
the following TABLE 1.

TABLE 1   Experimental Design

Dynamics		Forcing Function	Case Design.	Side Task
Exp. 1	K (prop. or P)	.42Hz(III)	P-III	-
		.33Hz (II)	P-II	
		.22Hz (I)	P-I	
	K/s (Integ. or J)	.42Hz(III)	J-III	-
		.33Hz (II)	J-II	
		.22Hz (I)	J-I	
Exp. 2	K	.33Hz (II)	P-II	visual
				auditory

The forcing functions designated with III,
II, and I, had an equivalent bandwidth of
0.42 Hz (high), 0.33 Hz (medium), and 0.22 Hz
(low), respectively. For all three control
axes the forcing functions had the same statistical parameters, in order to avoid in-
homogeneities resulting from this respect.
The three forcing functions were mutually
uncorrelated.

The experiments were carried out with a fixed
seat experimental console. Both one-axis and
two-axis hand control sticks of about 10 cm
height were installed in the experimental
console. The one-axis control was mounted
vertically on the horizontal plane on the
left and used exclusively for the left hand.
It was assigned to the z-axis and moved towards or away from the screen. The right-
hand control element (two-axis) could be
moved freely in all directions on the horizontal plane. The displacement range of both
spring-centered controls was approximately
±30 degrees with a spring constant of about
5 p/cm. The electrical feed was chosen such
that the voltage level at full deflection of
the controls was about 150 % of the maximal
values of the forcing functions.

Displays were generated electronically on a
CRT-screen with a special computer-controlled
picture generator. Data were sampled with a
rate of 20 (respectively 50) sec^{-1} and stored
digitally on magnetic tape. Time duration

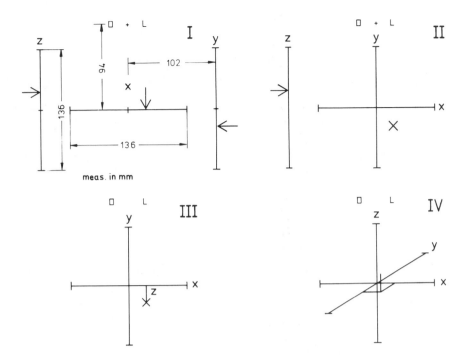

Fig. 1. The four used display configurations

was 210 sec, being a reasonable compromise between a desirable length of about 6 minutes in favor of good statistical confidence in the results with transinformation scores and a duration as short as possible in order to minimize fatigue.

Figure 1 shows the four display configurations used. To ascertain the proportions of display I its dimensions are given. Dimensions of the other display configurations are basically the same except for necessary changes. The deviations of the three pointers from the neutral points in the middle of the scales indicate the control errors from the nominal zero value. Orientations and positions of the scales on the screen and the operating directions of the control elements described above are compatible inasmuch as the left control corresponds to the z-axis and the right control to the x- and y-axes. The display configurations I-III represent different levels of integration, insofar as at the transition from I to II the scales x and y are joined to a cross scale. Accordingly, the indication of control-error is now mediated by a little cross. At the transition from II to III the z-deviation is represented by a stroke going up or down out of the centre of the cross according to the sign of the z-errors. In configuration IV the three errors are represented by a vector in a three-dimensional coordinate system, with the special feature that this vector is not represented itself but rather by its three cartesian components (the three deviations), in order to achieve a unique discriminability. The visual angle in the horizontal direction amounted to about ±10 degrees, vertically about 9 degrees, taken from the middle of the display. In the upper middle of the displays the symbols 0 and L are to be found.

This is used only in the second experiment with the visual side task. The two symbols designate two binary signal sources with an entropy of 1 bit/sec, respectively, since the states 0 and L were equally distributed. Both signal sources were uncorrelated, so that their entropies could simply be added. As a reaction to the binary events the subjects had to operate two pedals mounted on an inclined foot-rest.

In the case of the auditory side task the information was mediated by two earphones. The event corresponding to the symbol 0 of the visual presentation was marked by the lack of a tone (silence) on both earphones, whereas the symbol L was auditorily coded on the left earphone by a sinus tone of 200 Hz and on the right one by a 500-Hertz tone. In the second experiment eye-movement recordings with the use of a Honeywell oculometer were made. This device furnishes the angular coordinates for the viewing direction in both dimensions of the picture plane.

Experimental Procedure and Evaluation

For the first experiment subjects were trained on only the two most difficult tasks (namely the parameter configurations P-III and J-III, see TABLE 1) to save time. Twelve subjects were used for the first experiment and four subjects for the second one. Each subject had to perform the task with each of the four displays two times. The subjects used were institute employees and aged 21-52 with a technical background and with some tracking experience.

Subjects were trained at the beginning of the experimental series until there was practically no more improvement. One experimental session consisted in the running of the task, with all four display configurations, short rest pauses of about 1 minute between trials included. The sequence of presentation of the four displays was reversed in the two runs for each subject in order to compensate for fatigue influences. A random order was not used because in an earlier pilot experiment no differences were found between a random order group and a group using the experimental sequence. The results of both runs for each display were averaged.

The transinformation rates on the three control axes were computed by using the well-known formula

$$H_T = \int_0^{\infty} 1d \; \frac{S_{uu}(f)}{S_{uu}(f) - \frac{\left|S_{nu}(f)\right|^2}{S_{nn}(f)}} \; df, \qquad (1)$$

(Sheridan and Ferrel, 1974), wherein $S_{uu}(f)$ means the auto-power-density spectrum of the output signal of the final control element (the control stick), $S_{nn}(f)$ that of the noise input and $S_{nu}(f)$ the cross-power-density spectrum of those signals.

Additionally, the values from the computation of transinformation were supported by the rms errors in the three control axes. Transinformation rates for the discrete side tasks were found by means of

$$H_T = H(X) + H(Y) - H(X;Y) \qquad (2)$$

(Fano (1966) or Gallager (1968)). In Equ. (2) X and Y stand for the input and output ensemble of events at the communication channel, respectively.

RESULTS

We consider first the results of the pure 3-axis control task in the first experiment. Figure 2 shows the overall performances for the three bandwidths of the forcing functions and the two dynamics of the controlled elements used. The shaded areas in Fig. 2 designate the dispersion of the respective measures. It can be seen that there is an identical order of ranking for the four display configurations under all six parameter constellations. Furthermore, we see that displays II/III are significantly better (<.001) than configurations I/IV.

Significance tests between configurations II and III show that differences cannot be assumed between them. Another aspect, to be noticed concerns the rank order of displays I and IV which varies with the bandwidth of the forcing functions (P- and J-dynamics): Display IV is nearly significantly "worse" than I for the disturbance III (High), whereas it is significantly better (<.05) for the

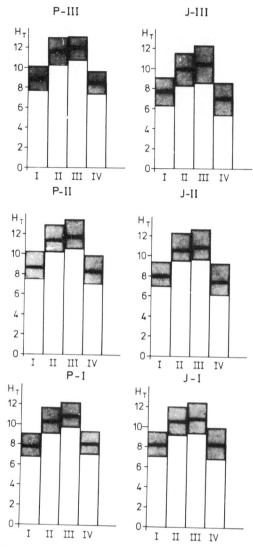

Fig. 2. Total mean values of transinformation H_T in bit/sec for the six parameter configurations (see TABLE 1)

low band forcing function. The situation is principally similar, if we consider integral dynamic controlled elements. In the J-II case I is almost significantly better than IV, whereas in case J-I there is no significant difference between I and IV.[1]

Finally, we consider the fact that transinformation rates are increased in the transition from high to low frequency forcing functions, despite the decrease in information quantitiy with lowered bandwidth. In order to explain this, one has to consider two facts. At first, the amount of information offered to the information processing

---

[1] Apparently, the rank order of alternative display designs may be depending on frequency. This is supported by the results of a pilot study (see Kimmel, 1976), where a significant superiority (1 % level) of configuration IV over I was found with an equivalent bandwidth of the forcing functions of 0.09 Hz.

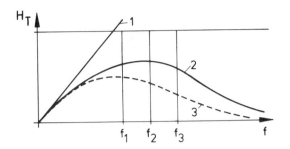

Fig. 3. Information transfer as a function
        of bandwidth (schematically)

human channel is altered by bandwidth (see
curve 1 in Fig. 3), and on the other hand the
transfer behaviour of the human (curve 2 in
Fig. 3) is altered as a function of the information offered which affects the ratio
between the linear correlated part of the
human control signal and the noise (the remnant), dependent again upon bandwidth. A
reasonable assumption for the description
of the noise level in the control signal is
that the increase of noise is proportional
to the noise level itself. In other words,
this approach says that if one takes the
amount of noise in the control signal as a
measure of task difficulty, this difficulty
is increasing proportionally to the respective level of task difficulty.

Therewith the observed higher transfer rates
for a lower bandwidth can be explained: The
signal attenuation by the integral dynamics
controlled element results in a lowered information offer. This corresponds to a flat-
ter curve with the peak shifted to the left
(see curve 3 in Fig. 3). This results in the
three forcing functions being to the right
of the peak with a lower transinformation
rate decreasing with increased frequency.
Experimental verifications of these reflections are to be found for instance for one-
axis manual control processes in Elkind and
Sprague (1961), see Fig. 4. In the context

similar curves for the information transfer
in discrete information processing tasks.

The results of experiment 2 with the visual
and auditory side task are contained in
TABLE 2. Therewith the first two lines refer
to the visual side task, and the next two
lines to the control task with auditory side
task. For comparison, the results of the 3-
axis control without side task with the same
(P-II) controlled element is given in the
last line of TABLE 2.

TABLE 2 Comparison of Display Results with
         the Continuous 3-axis Control Task
         with and without Side Task

Display No.	I	II	III	IV
$\overline{H}_{T\,dv}$ bit/s	0.534	0.554	0.444	0.401
$\overline{H}_{T\,3v}$ bit/s	5.50	7.21	7.51	5.23
$\overline{H}_{T\,da}$ bit/s	0.487	0.429	0.466	0.424
$\overline{H}_{T\,3a}$ bit/s	6.78	8.77	9.24	6.20
$\overline{H}_{T\,3}$ bit/s	8.66	11.52	11.98	8.25

Results show significant decreases in performance when the side tasks are added. This
supports the observation made above that the
three-axis control task imposes a full workload on the subjects. The fact that transinformation rates with the visual side task
are diminished was expected, but we have also
to note that transinformation rates with the
auditory side task are not reduced as much as
with the visual side task, but are still
lower than with no side task. This result is
in accordance with the findings of Freides
(1974), Jennings and Chiles (1977), and
Oatman (1975). Obviously, no enhancement of
information transfer rate can be achieved by
the use of multiple sensory channels, when
the information processing capacity is already fully loaded in one sensory modality.

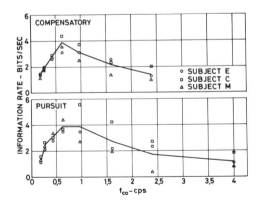

Fig. 4. Information transfer-rate as a
        function of bandwidth in one-axis
        manual control (from Elkind and
        Sprague (1961))

of the investigation of the capacity of the
human short-time memory, Slak (1976) obtained

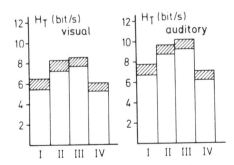

Fig. 5. Total mean values of transinforma-
        tion rates with visual and auditory
        side task. Shaded areas show trans-
        information rates for the side tasks.

Figure 5 shows the relative ranking of display configurations with both side tasks

(cf. Fig. 2). Similar rankings are also seen when the percentages of main task performance are calculated: With the auditory side task the performance is lowered to 77 %, in the visual case 63 % are reached. With both side tasks the ratio performance between side and main tasks vary with display configuration (see TABLE 2). There is a large difference in performance between the first two display configurations (I and II) and the last two configurations (III and IV) with the visual side task (first line in TABLE 2). Since this effect did not occur with the auditory side task, it can be suggested that it is connected with the visual character of the main or the side task, respectively. The results can be explained by the assumption that tasks III and IV impose a higher visual workload on the subjects than the other two. In order to pursue this question it is necessary to use the distribution of the eye-movements on the picture plane.

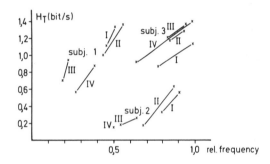

Fig. 6. Information transfer in the visual
         side task as a function of relative
         frequency of fixations

Figure 6 depicts the information and performance scores as a function of the relative frequencies of looking at the side task for three subjects and all four display configurations. The straight lines connect the two trials of each display configuration. We find very different results depending upon the respective displays and subjects. Furthermore, subjects show strong individually different behaviour. Subject no. 3, e.g., achieves high transmission rates by a relatively high fixation rate of the side task. Obviously, S tried to perform the main task with peripheral vision. The result of this "strategy" was the worst performance of all subjects. (In one trial there was an rms-error in one control axis of 140 % of doing nothing which is in accordance with the meaning of Sanders (1970) that peripheral perception is affected mainly by cognitive factors rather than by visual factors. According to Mackworth (1965) there is nothing which indicates the use of peripheral vision in search tasks.)

It can be seen in Fig. 6 (for subjects no 1 and 2) that control with displays I and II permits a higher information transfer in the side task than with III and IV. This implies that the latter displays impose a higher workload by causing subjects to fixate on

the main task longer, despite the fact that the "scanning effort" for the eyes is eliminated. The percentages formed with the informational measures of TABLE 2 show this clearly (see TABLE 3), as the percentages of the side task decrease to 95 % with both integrated displays. This decrease is in contradiction with expectations since performances in the side tasks with the integrated displays should increase without the requirement of eye scan movements in the main task.

TABLE 3   Ratios of the Performance Measures
          for the Side and Main Task

Display No.	I	II	III	IV
$\overline{H}_{Tdv}/\overline{H}_{Tda}$	1.10	1.29	0.95	0.95
$\overline{H}_{T3v}/\overline{H}_{T3a}$	0.81	0.82	0.82	0.84

As can be seen from TABLE 3, just the opposite was true. A first hint on why this is so, is given by the overall relative frequencies of looking at the side task, where it can be seen that subjects 1 and 2 have given less attention to the side task with configurations III and IV.

In order to better judge the behaviour of subjects on the side tasks we consider the frequency distributions of the length n of the eye-movement sequences which both subjects direct to the two discrete events of the side task. Figure 7 shows two typical

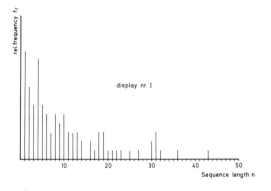

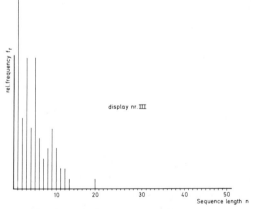

Fig. 7. Distribution of sequence lengths
         for two single trials

examples for the distribution of such eye-movement sequences. In the upper figure is a distribution for display I; the lower figure for display III. It can be seen that the distribution for the integrated display no. III is considerably more clustered on the shorter lengths thus supporting our suggestion that displays III and IV impose higher mental workloads on the subjects than the other two.

TABLE 4 Mean Standard Deviations of the Eye-Movement Sequence Lengths (σ_{vis}), Relative Control Effort (σ_{st}) and Transinformation Rates

Display No.	I	II	III	IV
σ_{vis}	14.13	13.33	9.71	11.30
σ_{st}	1.34	1.43	1.45	1.42
H_{T3} bit/s	8.66	11.52	11.98	8.25

TABLE 4 shows the standard deviations of eye-movement sequence lengths for all displays. We observe a decrease of mean sequence length from display I to III and with display IV an increase. In order to check those results we divided mean standard deviations of stick signals on the main control task (as a measure of subjective effort, see Pfendler (1981)), by the standard deviations of the forcing functions. We find just the opposite trend, i.e., the decrease of fixation sequence length from display I to III in the side task is accompanied by increasing effort in the main task.

The subjective ratings of the subjects which were taken for the 3-axis control task with no side task were nearly unanimous with the performance ranking of III, II, IV and I, where display no. III was felt to be the most convenient and best mastered. If we compare this with the values in TABLE 4 we can see that it corresponds to the subjective effort, but also (nearly) with the average length of eye-movement sequences. If we keep in mind that the performance index of transinformation is a means to designate the subjects effectivity, we can form a combined index of workload W in the following way:

$$W = \frac{\sigma_{st} H_T}{\sigma_{vis}} \qquad (3)$$

The above results are in accordance with Bird (1981) who found that as task difficulty increases subject effort, stick output standard deviation, and scale ratings also increased.

CONCLUSIONS

The available reports on the physiological assessment of operator workload in tracking tasks show that even though some physiological indicators are suitable to detect large differences such as between rest and tracking, they are unable to discriminate between different levels of tracking difficulty, e.g., heart rate variability (Hyndman and Gregory, 1975) and event related potentials (Isreal and others, 1980). But a useful measure of operator workload must be able to detect or measure smaller variations in effort, such as between qualitatively different mental operations as they were imposed in this study by the four display designs used. The ability to discriminate between different solutions of one and the same task is, by far, the more important one in human factors system design. It was shown that the measure of transinformation is a suitable tool for the evaluation of display configurations, being rather independent from control loop parameters as e.g., bandwidth and controlled element dynamics. This superiority over other possible evaluation measures might be founded in its general definition being a metric free measure (in the real sense), enabling us to compare results of very different tasks.

The information theoretic approach with its notion of capacity led us to the suggestion that the informational scores might also be used for workload measurements. Since we found a unique relation between stick activity, transinformation rate, and mean sequence length of fixation of the side task, we were induced to combine these three measures to an index of workload, comprising the aspects of effort, effectivity and attention. These results being based on a relatively small number of subjects, are confirmed by the findings of Ephrath and others (1980) who also investigated the scanning behaviour of four pilots in a simulated landing approach, controlling mental workload by a side task, finding a monotone trend between fixation sequences and workload.

If one leaves the concept of side tasks and interprets all parts of a given task as equivalent fractions of a whole, the method of eye-movement sequence analysis might also be a promising tool in the case that single parts of a display are to be mutually evaluated. In order to develop the method it might be helpful to test it against physiological measures of workload, as e.g., pupillometry which provides an indication of momentary fluctuations in central nervous excitability that occur as mental operations (Beatty, 1981) or other indicators like heart rate and event related potentials.

REFERENCES

Beatty, J. (1981). Pupillometric measurement of operator workload. Proc. Ann. Conf. Manual Control, JPL 81-95, Pasadena, USA, 1-6.

Bird, K.L. (1981). Subjective rating scales as a workload assessment technique. Proc. Ann. Conf. Manual Control, JPL 81-95, Pasadena, USA, 33-39.

Carbonell, J.R. (1966a). A queuing model for visual sampling. IEEE Trans. Human Factors Electr., HFE-7, 82-86.

Carbonell, J.R. (1966b). Queuing model of many instrument visual sampling. IEEE Trans. Human Factors Electr., HFE-7, 157-164.

Elkind, J.L., and L.T. Sprague (1961). Transmission of information in simple manual control systems. IRE Trans. Human Factors Electr., HFE-2, 58-60.

Ellis, S.R., and L. Stark (1981). Pilot scanning patterns while viewing cockpit displays of traffic information. Proc. Ann. Conf. Manual Control, JPL 81-95, Pasadena, USA, 517-524.

Ephrath, A.R., J.R. Tole, A.T. Stephens, and L.R. Young (1980). Instrument scan - is it an indicator of pilot's workload? Proc. Human Factors Soc. - 24th Ann. Meeting, 257-258.

Fano, R.M. (1966). Informationsübertragung - eine statistische Theorie der Nachrichtenübertragung. R. Oldenbourg, München-Wien.

Freides, D. (1964). Human Information processing and sensory modality: Cross-modal functions, information complexity, memory, and deficit. Psychol. Bull, 81, 284-310.

Gallager, R. (1968). Information Theory and Reliable Communication. Wiley, New York.

Hyndman, B.W., and T.R. Gregory (1975). Spectral analysis of sinusarrythmia during mental loading. Ergonomics, 18, 255-270.

Isreal, J., G.L. Chesney, C.D. Wickens, and E. Donchin (1980). P300 and tracking difficulty: Evidence for multiple resources in dual-task performance. Psychophysiol., 17, 255-270.

Jahns, D.W. (1973). A Concept of Operator Workload in Manual Vehicle Operations. Report No. 14, FAT, Werthhoven, Germany.

Jennings, A.E., and W.D. Chiles (1977). An investigation of time-sharing ability as a factor in complex performance. Human Factors, 19, 535-547.

Kampé de Fériet, J. (1973). La théorie généralisée de l'information et la mesure subjective de l'information. In A. Dold and B. Eckmann (Eds.), Theories de l'Information, Lecture Notes in Mathematics, Vol. 398, Springer Verlag, New York, pp. 1-35.

Kimmel, K.R. (1976). Evaluation of information displays in control and monitoring tasks. In T.B. Sheridan and G. Johannsen (Eds.), Monitoring Behavior and Supervisory Control, Plenum Press, New York, 51-58.

Kleinman, D.L., and R.E. Curry (1976). An equivalence between two representations for human attention sharing. IEEE Trans. Syst., Man. & Cybern., SMC-6, 650-652.

Kleinman, D.L. (1976). Solving the optimal attention allocation problem in manual control. IEEE Trans. Autom. Control, AC-21, 813-821.

Levison, W.H., J.I. Elkind, and J.L. Ward (1969). Studies of Multivariable Manual Control Systems: A Model for Task Interference, Report No. 1892, Bolt, Beranek & Newman, Inc., Cambridge, USA

Mackworth, N.H. (1965). Visual noise causes tunnel vision. Psychonomic Science, 3, 67-68.

Oatman, L.C. (1975). Simultaneous Presentation of Bisensory Information. Human Engineering Laboratories, Aberdeen Proving Grounds, Md., USA.

Palmer, E. (1977). Interrupted monitoring of a stochastic process. Proc. 13th Ann. Conf. Manual Control, MIT Cambridge, USA, 237-245.

Pfendler, C., and G. Johannsen (1977). Beiträge zur Beanspruchungsmessung und zum Lernverhalten in simulierten STOL-Anflügen. Report No. 30, FAT, Werthhoven, Germany.

Pfendler, C. (1981). Vergleichende Bewertung von Methoden zur Messung der mentalen Beanspruchung bei einer vereinfachten simulierten Kfz-Führungsaufgabe. Report No. 51, FAT, Werthhoven, Germany.

Sanders, A.F. (1970). Some aspects of the selection process in the functional visual field. Ergonomics, 13, 101-117.

Senders, J.W., J.E. Elkind, M.C. Grignetti, and R.P. Smallwood (1965). An Investigation of the Visual Sampling Behavior of Human Observers. NASA CR 434.

Sheridan, T.B., and W.R. Ferrell (1974). Man-Machine Systems. The MIT Press, Cambridge, USA.

Slak, S. (1976). Information-transmission method for short-term memory span as channel capacity. Perceptual and Motor Skills, 42, 491-496.

Smallwood, R.E. (1966). Internal models and the human instrument monitor. IEEE Trans. Human Factors Electr., HFE-8, 181-187.

Strasser, H. (1974). Beurteilung ergonomischer Fragestellungen mit Herzfrequenz und Sinusarrhythmie. Int. Arch. Arbeitsmed., 32, 261-287.

Wickens, C., J. Isreal, G. McCarthy, D. Gopher, and E. Donchin (1976). The use of event-related potentials in the enhancement of system performance. Proc. 12th Ann. NASA-University Conf. Manual Control, 124-134.

Wierwille, W.W., and R.C. Williges (1978). Survey and Analysis of Operator Workload Assessment Techniques. Report No. 5-78-101, Systemetrics, Inc., Blacksburg, Va., USA.

Zeier, H. (1979). Cuncurrent physiological activity of driver and passenger when driving with and without automatic transmission in heavy city traffic. Ergonomics, 22, 799-810.

CONSTRUCTION AND APPLICATION OF A COMBINED NETWORK AND PRODUCTION SYSTEMS MODEL OF PILOT BEHAVIOUR ON AN ILS-APPROACH

B. Döring and A. Knäuper

*Forschungsinstitut für Anthropotechnik, 5307 Wachtberg-Werthhoven,
Federal Republic of Germany*

Abstract. A simulation study was conducted for determining information flow
requirements of a pilot-cockpit interface during an ILS-approach. The steps
of the study for modeling, simulating, and analyzing system processes are
discussed. Starting with the definition of the problem which includes the
relevant flight processes, the network of pilot tasks, and the performance
measures, the conceptual model of processes is mathematically described by
means of algebraic and difference equations as well as production systems.
To transform the model into a simulation program, the simulation language
SLAM is used. SLAM elements utilized for modeling flight processes and pilot
activities as well as the procedure of model validation are discussed. The
analysis of simulation output data results in the determination of important
information requirements useful in a design project such as range of variables
to be indicated or affected, frequency and sequence of variables used by the
pilot, pilot tasks, duration of task variables, and the average utilization
rate of cockpit display and control components.

Keywords. Human factors; man-machine systems; modeling; simulation;
systems analysis.

INTRODUCTION

In modern industrial and military man-machine
systems, such as nuclear power plants, ships,
and aircrafts, the man shares key functions
with the computer. This in turn changes the
nature of some human operator tasks. Operators
now are increasingly involved with monitoring
and supervisory control tasks in addition to
or instead of continuous control tasks.
Sheridan and Johannsen (1976) define monitoring as the systematic observation by a human
operator of multiple sources of information to
determine whether operations are normal and
proceeding as desired, and to diagnose difficulties in the case of abnormality or undesirable outcomes. The authors characterize
supervisory control as the control of a computer by a human operator which, at a lower
level, is controlling a dynamic systems. The
computer-control normally operates continuously
or at high data rates. By contrast the human
operator normally signals or reprograms the
computer intermittently or at a much slower
pace.

For modeling human monitoring and supervisory
control tasks, different approaches have been
applied. Kok and Van Wijk (1977) and Muralidharan and Baron (1979) developed models of
human supervisory control behavior based on
control theory. For analyzing human decision
making with flight management tasks, queuing
theory models were applied by Rouse (1977),
Walden and Rouse (1978), and Chu and Rouse
(1979). Human problem solving performance in
fault diagnosis tasks has been modeled with a

fuzzy set theoretical approach by Rouse
(1978,1979). Kraiss (1981) used a combination
of a task network model und a control theory
model for analyzing human control and monitoring tasks required for submarine control.
Wesson (1977) and Goldstein and Grimson (1977)
modeled human supervisory control tasks in air
traffic control and attitude instrument flying
by means of a production system model, an
approach developed in the field of artificial
intelligence.

In the following, a study will be described in
which the pilot's monitoring and supervisory
control tasks during an automatic ILS-approach
are modeled by combining a task network with
production systems. The model was used in
connection with a digital computer simulation
to determine information requirements for a
pilot-cockpit interface.

PROBLEM DEFINITION

The problem definition of a modeling process
includes a statement of the phenomena of interest as well as a choice of performance measures (Rouse, 1980). The phenomena of interest
are processes of an aircraft system during an
automatic ILS-approach. The aircraft considered
was a twin engined HFB 320 Hansa Executive Jet
manufactured by Messerschmitt-Bölckow-Blohm,
Hamburger Flugzeugbau, and equipped with the
Collings AP-104 autopilot and the FD-109 flight
director. Runway 25 of the Cologne-Bonn airport, and its approach area were selected.

ADE-B*

During an ILS-approach, the course of events depends on aircraft motion relative to ground stations and the runway of the airport. Variables which characterize the aircraft motion are, e.g., heading, indicated air speed, vertical speed, and flight attitude. These variables determine the position of the aircraft in x, y, and z dimension over time. Relevant ground stations of the approach area are VOR- and NDB-stations, outer and middle marker, localizer, glide path transmitter, and tower. By relating the aircraft position to ground stations and runway, actual bearings and distances between them and the aircraft can be determined. Furthermore, weather conditions that affect the flight processes have to be considered. For the ILS-approach, we assumed visibility to be restricted because of clouds down to 600 feet with the runway visible below that altitude.

To identify the flight processes, a top down analysis was accomplished. Consequently the ILS-approach was partitioned into an intermediate and a final phase which then were further divided to identify system functions and pilot tasks belonging to each function (Döring, 1976b). Supervisory tasks required to control the flight processes during the ILS-approach were categorized into five classes. These are:

- Adjusting tasks, e.g., setting up the autopilot with new desired course or heading values.
- Activating tasks, e.g., changing autopilot mode, flaps, or gear state.
- Monitoring tasks, e.g., the systematic pilot observation of the indicated altitude during descent or heading during approach to an interception heading.
- Checking tasks, e.g., comparing current and desired values of indicated air speed, heading, altitude, etc.
- Special tasks, such as pilot activities which cannot be categorized as above, e.g., verbal communication when performing outer marker check or receiving landing clearance.

By using these categories with the ILS-approach, 37 pilot tasks were identified. Pilot tasks performance was considered to be normative, i.e., what the pilot should do during the landing approach (Sheridan, 1976). This normative approach is especially useful during early stages of system development when, e.g., required behavioral data is not yet available. Each task was described by that behavioral verb, which was used in the task classification and which indicated the nature of the activity being performed during the task, and by that information or state variable which was being acted on by the pilot (Döring, 1976a). Some examples for the task description are: adjust heading marker position (AD HMP), activate autopilot lateral mode (AC LM), monitor heading (MO HD), check indicated air speed (CH IAS), perform outer marker check (PF OMC). By considering the course of events during the ILS-approach, predecessor/successor-relations among the identified tasks were determined. Some of the relations, which define the approach dependent structure of some tasks are shown in Fig. 1.

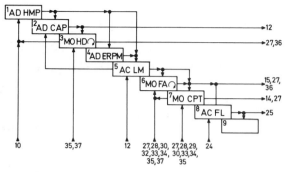

Fig. 1. Section from the task network

For each task identified, the time required by the pilot for performing that task was estimated. Basic data for time estimation were taken from Miller (1976). Additionally, operating times, e.g., for adjusting heading marker, course arrow, and vertical speed were measured in the HFB 320 flight simulator facility available in our institute and described in detail by Johannsen, Rouse and Hillmann (1981). This flight simulator was evaluated by experienced HFB 320 pilots and judged to be rather realistic.

On account of the structure of pilot tasks identified (Fig. 1), a network technique was used for modeling pilot behavior during the ILS-approach. Performance measures typical with network techniques are, e.g., the time required to complete certain tasks aggregates and the probability that the aggregates will be completed successfully (Pew and others, 1977). Therefore, to choose model performance measures, 17 approach related events were defined and the times of their occurance were regarded as performance measures. Some events are, e.g., "Activating autopilot lateral mode VOR" (LM.VOR), "Capturing the VOR radial 150 degrees" (CPT.150), "Reaching at indicated air speed 160 kts" (IAS.160 kts), "Reaching flight attitude level" (FA.LEV), "Activating autopilot vertical mode VERTICAL SPEED HOLD" (VM.VSH). These events were used for model validation described later. Because of the normative point of view taken for establishing pilot tasks, only ILS-approaches that were been completed successfully were considered.

PROBLEM REPRESENTATION

The term problem representation refers to the process of developing a formal mathematical-logical description of the problem (Rouse, 1980). This description is called a conceptual model (Schlesinger and others, 1979). For developing that model, flight processes were described in very simplified form. Because the goal of the study was to determine information flow requirements for the pilot-cockpit interface, only those state variables displayed in the cockpit and relevant to the pilot were modeled. For describing those variables and their value changes during the ILS-approach, a set of difference equations was used. For example, the values of heading (hd), indicated air speed (ias), and vertical speed (vs) at time t_n were described by the equations:

$$hd(t_n)= hd(t_{n-1})+\Delta t* \; hdv(t_{n-1}) \qquad (1)$$

$$ias(t_n)=ias(t_{n-1})+\Delta t*iasv(t_{n-1}) \qquad (2)$$

$$vs(t_n)= vs(t_{n-1})+\Delta t* \; vsv(t_{n-1}) \qquad (3)$$

where $hd(t_{n-1})$, $ias(t_{n-1})$, and $vs(t_{n-1})$ are representing the values of hd, ias, and vs at the last computation time t_{n-1}, Δt is the time interval (t_n-t_{n-1}), and $hdv(t_{n-1})$, $iasv(t_{n-1})$, and $vsv(t_{n-1})$ are the variation rates of hd, ias, and vs at time t_{n-1}. These variation rates were determined by measuring the changes of the related variables in the flight simulator, e.g., heading, engines revolutions per minute, or altitude change. Only the roll attitude was considered for flight attitude and was deduced from heading rate which represents the rate of turn. Since a roll attitude of 25 degrees results approximately in a heading rate of 3 deg/s, roll attitude (ra) at time t_n was approximated by

$$ra(t_n)= \pm \; 25*hdv(t_n)/3. \qquad (4)$$

The aircraft position at time t_n was described by the equations:

$$x.ac(t_n)=x.ac(t_{n-1})+\Delta t*ias(t_{n-1})*\sin(hd(t_{n-1})) \qquad (5)$$

$$y.ac(t_n)=y.ac(t_{n-1})+\Delta t*ias(t_{n-1})*\cos(hd(t_{n-1})) \qquad (6)$$

$$z.ac(t_n)=z.ac(t_{n-1})+\Delta t* \; vs(t_{n-1}) \qquad (7)$$

For computing bearings and distances from the aircraft to the ground stations mentioned previously, the coordinates x.vor, y.vor of the VOR-station, x.ndb, y.ndb of the NDB-station, x.om, y.om of the outer marker, x.mm, y.mm of the middle marker, x.gpt, y.gpt of the glide path transmitter were taken from the instrument approach chart of the runway. For instance, the bearing b.ac.vor from the aircraft to the VOR-station at time t_n was described by the equation:

$$b.ac.vor(t_n)= \arctan((x.ac(t_n)-x.vor(t_n))/(y.ac(t_n)-y.vor(t_n))) \qquad (8)$$

The distance d.ac.gpt between aircraft and glide path transmitter at time t_n is given by:

$$d.ac.gpt(t_n)=\sqrt{((x.ac(t_n)-x.gpt(t_n))^2+\ldots}$$
$$\overline{\ldots +(y.ac(t_n)-y.gpt(t_n))^2)} \qquad (9)$$

The course deviation cd is given by the difference between the bearings to the localizer or VOR-station and the corresponding courses divided by 1.25 (Collins Radio Company, 1974):

$$cd(t_n)=(b.ac.loc(t_n)-249.)/1.25 \qquad (10)$$

The glide path angle gpa at time t_n is given approximately by

$$gpa(t_n)=\arctan(z.ac(t_n)/d.ac.gpt(t_n)). \qquad (11)$$

The glide slope deviation gsd represents the difference between the glide path angle of the runway (3.1 deg) and the actual glide path angle gpa divided by 0.25 (Collins Radio Company, (1974).

$$gsd(t_n)=(3.1-gpa(t_n))/0.25 \qquad (12)$$

The capture of the VOR radial 150 by the autopilot is assumed to happen approximately 9 s before the radial is reached. Considering that the bearing to the VOR-station (b.ac.vor) is changing according to the aircraft motion, the capture event occurs approximately at that time when the following condition is satisfied:

$$b.ac.vor(t_n)=150+9*(b.ac.vor(t_{n-1})-$$
$$b.ac.vor(t_n))/\Delta t \qquad (13)$$

The capture of the localizer occurs when aircraft heading is approximately 1.9 dots from final course on the course deviation indicator. The capture of the glide slope occurs when the slope is intersected, i.e., when the glide slope deviation indicator reads very close to 0 dots.

For modeling pilot tasks, a method derived from the field of artificial intelligence called production systems was applied. The structure and operation of production systems are discussed extensively, e.g., by Davis and King (1977) and Nilsson (1980). Rouse (1980) defines a production as a situation-action pair where the situation side is a list of things to watch for and the action side is a list of things to do. A production system is a rank-ordered set of productions where the actions resulting from one production can result in situations that cause other productions to execute.

Applying the production scheme to a pilot task, the situation side represents the task specific values of pilot inputs and the action side represents the corresponding values of pilot outputs. Thereby, it is assumed that inputs and outputs are separated in time by the task duration, i.e., by that time which the pilot needs for performing the task. Because activating tasks are similar to adjusting tasks and monitoring tasks are similar to checking tasks, only an adjusting task and a monitoring task will be described in detail.

For example, the task "adjust heading marker position" (ad HMP; Fig. 1) appears during the approach at distinct points of time t at which distinct actions, i.e. the values 90, 170, 190 deg of the heading marker position hmp, have to be applied to the autopilot. At those points, the situations can be characterized by specific values hd(t) of the heading hd at time t. Depending on the situation, the successor task (ST) is "adjust course arrow position" (AD CAP), "adjust engines revolutions per minute" (AD ERPM), or "activate autopilot lateral mode" (AC LM) (Fig. 1). Therefore, a situation dependent selection of ST has to be performed additionally by the pilot with each task. Denoting the duration of the task AD HMP by D(AD HMP), the following productions can be established:

If hd(t)=263 deg
    then "adjust hmp(t+D(ad HMP))=170 deg"
        and perform ST "AD CAP"

If 147 deg ≤ hd(t) ≤ 153 deg
    then "adjust hmp(t+D(AD HMP))=190 deg"
        and perform ST "AC LM"

If 178 deg ≤ hd(t) ≤ 195 deg
    then "adjust hmp(t+D(AD HMP))= 90 deg"
        and perform ST "AD ERPM"

The task "monitor heading" (MO HD) occurs during certain heading changes. When the autopilot is controlling heading changes, the pilot probably uses his ability to make time estimates and predictions to determine when to again change the position of the heading marker. During this task he has (1) to evaluate whether the heading is in its tolerance between 185 deg and 263 deg, (2) to compare the actual value of the heading with its desired value of 190 deg, and (3) to estimate the time available until reaching 190 deg. If the heading is out of tolerance then a failure procedure is required; if the heading reaches the value of 190 deg then a new adjustment of the heading marker position has to be performed; if the time available is greater than 25 s the pilot can be idle for a while; if the time available is between 25 s and 10 s he will perform a cross-check starting with the task "check indicated air speed" (CH IAS); if the time available is less than 10 s he continues to monitor the heading. For describing this task, again a list of productions arranged sequentially is generated. The first matched production is the one used. With hdv denoting the variation rate of the heading hd these productions are:

If hd(t) < 185 deg or hd(t) > 263 deg
    then perform ST "failure procedure".

If hd(t)=190 deg
    then perform ST "AD HMP"

If (hd(t)-190)/hdv(t) > 25 s
    then perform ST "idle":

If 10 s < (hd(t)-190)/hdv(t) ≤ 25 sec
    then perform ST "CH IAS"

If (hd(t)-190)/hdv(t) ≤ 10 sec
    then perform ST "MO HD"

It is evident that each task can be described by a set of productions. This set constitutes a production subsystem (Winston, 1977). By considering each task as a production subsystem, the large amount of knowledge required by the pilot during the landing approach can be partitioned advantageously and structured in form of a task network. For describing each task, some productions characterize relations between pilot's inputs and outputs. Others describe transfers of attention to successor tasks. Thereby, the task duration required by the pilot is also considered. For each task, the duration was described by a normal distribution function characterized by mean and standard deviation.

MODEL IMPLEMENTATION

Once a representation of the problem, i.e., a conceptual model, has been developed, the next step with the digital computer simulation is the implementation of the model. With this step, the conceptual model is translated into a computerized model (Schlesinger and othe 1979). Although a general purpose language, e.g., FORTRAN, can be applied by this step there are distinct advantages to using a simulation language such as SLAM (Simulation Language for Alternative Modeling) (Pritsker and Pedgen, 1979). In addition to the savings in programming time, SLAM also assists in model formulation by providing a set of concepts for problem representation. Futhermore, because in most cases it is neither useful nor possible to store all data that are generated during simulation, SLAM offers certain basic statistical functions for data reduction, aggregation, and documentation such as calculation of means and standard deviations, and certain formats for presenting results such as histograms and certain sorts of plots.

The computerized model developed comprises a continuous part a network part, and an event-related part. With the continuous part, the flight processes are characterized by variables that are defined by state equations and assumed to change continuously with time. With the network part, discrete changes of variables characterizing pilot's task performance are modeled. There are interactions between the continuous and the network part, which are modeled by a separate event-related part.

For implementing the conceptual model of flight processes described previously, SLAM offers a special FORTRAN subroutine named STATE to define the dynamic equations for state variables such as heading, indicated air speed, altitude, etc. If difference equations are used in subroutine STATE, the executive routine will call STATE at periodic time intervals called steps unless an intervening event or information request is encountered. The values of state variables at the end of a step are computed from those equations that represent the conceptual model and that have to be written in FORTRAN by the user. In these equations, the SLAM variable SS(I) is used to represent state variable I. The state variable values at each immediately preceding step are maintained as the SLAM variable SSL(I). With these SLAM variables, it is possible to write a difference equation for state variable I in subroutine STATE in the following fashion:

$$SS(I)=SSL(I)+DTNOW*RATE(I)$$

where DTNOW is the interval size (t_n-t_{n-1})

and RATE(I) the rate of variation of state variable I during DTNOW. The value of RATE(I) can be determined as a function of other system variables. The executive routine calls subroutine STATE every time a step is to be made so that the updating of time is implicit in the equation given above for SS(I).

When programming process-oriented simulation that is based on a network, the flow of temporary elements, in our case the pilot, through processes, in our case the task network, is considered from their arrival to their departure. These temporary elements are called transactions or entities. For modeling that flow, SLAM provides a network structure consisting of specialized nodes and branches that are used to model resources, queues for resources, activities, and entity flow decisions. In our case, each pilot task represents a process component of the network. The pilot is regarded as an entity that flows from task to task. Each task consists of a combination of nodes and branches dependent on the type of task. For instance, the basic structure of an adjusting task is constituted by the following sequence of SLAM elements: ASSIGN node, EVENT node, task activity, EVENT node, and various branching activities (Fig. 2).

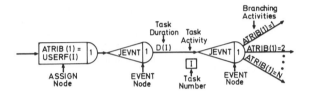

Fig. 2. Basic structure of adjusting tasks

The ASSIGN node is used to prescribe values to attributes of an entity passing through the node, here to attribute ATRIB(1) of the entity representing the pilot. The attribute values is generated with the user-written function USERF (I) in which specific productions are coded for each task I. The value of ATRIB(1) determines the branching to the successor task.

The EVENT node is included in the network model to interface the network portion of the model with a discrete event. The EVENT node causes a user-written subroutine named EVENT(JEVNT) to be called every time an entity arrives at an EVENT node. The value of JEVNT specifies the event code of the discrete event to be executed. In the case of an adjusting task, the first EVENT node starts the adjusting operation and the second EVENT node stops that operation. The desired value of the adjustment is determined in the subroutine USERF(I) as the action part of that production of which the situation part is matched. The modifications that happen according to the adjustment to the flight processes modeled in the subroutine STATE are coded in the subroutine EVENT(JEVNT). With the task "adjust heading marker position", for instance, the variation rate of the heading marker position is determined when the entity arrives at the first EVENT node. The variation rate of the heading marker position is set to zero when the entity arrives at the second EVENT node.

The task activity is modeled by the branch between first and second EVENT node. Each activity is given a number which corresponds to the task number I. The duration D(I) of the activity I is the time delay that an entity encounters as it flows through the branch representing the activity. That is, the activity duration represents the time which the pilot needs for performing the task. Task duration values

used for each performance of each task are determined randomly of a sample of a normal distribution.

The branching activities that are emanating from the second EVENT node are determining conditionally the flow of the entity to the successor task. The entity is following that branch of which the condition is true. The condition is prescribed by the value of ATRIB(1).

Because it is assumed that, in an activating task, the event becomes effective at the end of that task, the basic structure of an activating task differs from an adjusting task only in omitting the first EVENT node. In the structure of monitoring and checking tasks the first EVENT node is omitted and the second EVENT node is replaced by a GO ON node. The GO ON node is included as a continue type node. It is used for modeling the separation of sequential activities. Here, the GO ON node separates the task activity from branching activities.

The subroutine EVENT(JEVNT) mentioned is also used for interfacing the continuous part of the model with a discrete event. At such event, a discrete change in value may be made to a continuous variable. To model such events, the state-event feature of SLAM is used. A state-event happens when a specified state variable is crossing a prescribed threshold. For instance, the localizer capture event occurs when the course deviation variable cd is crossing the value of 1.9 dots. If a state-event I is detected, the executive routine calls subroutine EVENT(I). In this subroutine, the user has to code the changes to be made to the system when a particular state-event occurs. At the localizer capture event, the user has to specify that the heading changes now with a rate of turn of 3 deg. Additionally, the functional description of continuous variables may be changed at discrete time instants, e.g., when the autopilot mode of operation is changed by the pilot or by certain state-events.

SIMULATION OF THE ILS-APPROACH

When the computerized model of the ILS-approach has been implemented, the digital computer is used to simulate system activity. Generally, simulation means an exercising of the computerized model to generate a chronological succession of state descriptions, i.e., of values of relevant state variables describing the system behavior. Since the model has probabilistic aspects, e.g., with the duration of each task, the generation of simulation data is statistical in nature. Thus, the state trajectory must be replicated many times to obtain a sufficient number of samples of performance measures to be able to estimate average performance within reasonable confidence limits (Rouse, 1980). With the computerized model, 30 ILS-approaches were simulated.

MODEL VALIDATION

Experimental validation of a model involves
using the model to predict performance and
then empirically determining how close the
predictions are to actual occurrances (Rouse,
1980). The computerized process model when run
simulates the pilot, certain aircraft processes
and certain events. Pilot processes of the model would be validated if performance values
obtained closely matched the performance of a
real pilot in the flight simulator. Aircraft
processes and event time points would be validated if measures concerning these closely
match values of the realistic flight simulator.
Ten simulated ILS-approaches were performed in
the flight simulator. Pilot and simulator activities observable in the cockpit were recorded on a video-tape and later analyzed. Data
obtained during the approaches constituted the
basis for validating the model.

Techniques for model validation are discussed
in detail by Sargent (1979). From the various
validation techniques, we applied face validity, traces, internal validity, and event
validity. For evaluating the face validity of
the model, pilots knowledgeable about the ILS-
approach were asked whether the conceptual model was reasonable. Particularly, the logic of
the task networks, details of the flight processes described, and interrelations between
network and flight processes were discussed and
checked in this way. Traces were used to check
the computerized model, i.e., the behavior of
the pilot as represented by the network was
traced through the model to determine if the
model's logic and the computer program were
correct and if the necessary accuracy was obtained. The internal validity is determined by
the amount of stochastic variability in the model. To test this type of validity, several
stochastic simulation runs were made and the
variability of some variables, e.g., of glide
path interception altitude, glide path deviation, and course deviation were determined and
compared with the variability of experimental
data obtained by the flight simulator. To check
event validity, the occurance times of the 17
events mentioned above were used as performance measures. For all events, mean times of
occurance were determined both from the 10 pilot
flight simulator ILS-approaches and from 30

runs of the digital computer simulation. The
times were compared statistically by means of
the T-test and no significant differences were
obtained between flight simulator and computer
simulation model data. Fig. 3 shows the results of only the first occurring events with
their 99% confidence intervals.

ANALYSIS OF SIMULATION OUTPUT DATA

After model validation, the final step in our
simulation study was the experimental application of the model to generate simulation output data for later analysis. In this study,
information flow requirements were to be determined for the pilot-cockpit interface. Specified data requirements include those variables
to be controlled about which information is
transmitted to the pilot or which are subsequently affected by control outputs of the
pilot, and the mission required range, sequence, use frequency and duration of those
variables. (Döring, 1976a; Shackel, 1974). To
facilitate the determination of these data the
total ILS-approach was partitioned into an intermediate (IP) and a final phase (FP). The
indication range and use frequency of the variables were identified for each phase by analyzing output data of 30 simulation runs. For
this analysis, SLAM routines for data collection and statistical calculation were applied.

Fig. 4 shows, as examples, the resulting display ranges of heading, course deviation, and
indicated air speed for each phase. In Fig. 5
the total use duration of those variables in
each phase is depicted. It can be seen, that
heading is utilized for only about 60 s of the
IP with values between 90 and 263 deg. Although
course deviation is also used during the IP, in
the FP course deviation is used inplace of
heading, because it gives more accurate information about current aircraft position in relation to the final course. Course deviation
stay between 1.4 dots to the left and 0.25 dots
to the right of the final course during the IP
for about 8 s when intercepting that course.
During the FP course deviation ranges from 0 to
0.33 dots left and is used about 13 s for
checking aircraft position when approaching the

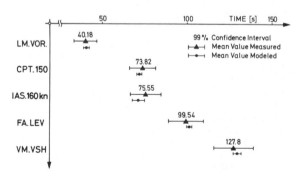

Fig. 3. Comparison of approach events
both measured and modeled

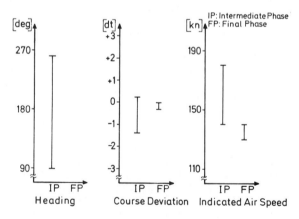

Fig. 4. Phase related range of variables

glide slope. Indicated air speed is verified every time that an instrument cross-check is performed by the pilot. This variable takes values between 180 and 140 kts during 23 s of the IP, and values between 140 and 130 kts during 18 s of the FP.

The sequence of tasks performed and of information used in their performance was obtained by plotting tasks and corresponding information used over time. In Fig. 6, a section of the information time line obtained is depicted. In this figure the time of utilization has been recorded for various variables.

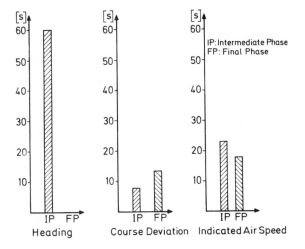

Fig. 5. Phase related duration of used variables

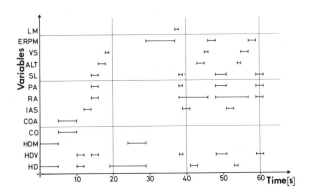

Fig. 6. Section from information time line

HD : heading;   HDV: heading variation;
HDM: heading marker;   CO : desired course;
COA: actual course ;   IAS: indicated airspeed;
RA : roll attitude ;   PA : pitch attitude;
SL : slip;   ALT: altitude;
VS : vertical speed;   ERPM: engines revol./min;
LM : lateral mode.

It is obvious that more than one variable is often used concurrently when the pilot performs a task, e.g., during the task "adjust heading marker position" (AD HMP), which starts at time 0 and lasts about 5 s, hd and hdv are used by the pilot.

Also, the average utilization time in percent

of total approach time of display and control components can be determined by analyzing output data of the simulated approaches. In Fig. 7 this utilization time has been drawn on the y-axis. The display and control components are listed on the x-axis. It can be seen, that, e.g., the most often used display components are the flight director indicator, variometer, course indicator, and altimeter. Because of the highly automated ILS-approach, control components are seldom used. The control component used most often is the vertical command control with which the vertical speed is adjusted when changing altitude to the required value for intercepting the glide slope.

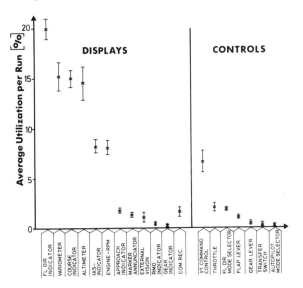

Fig. 7. Average utilization of displays and controls per run

CONCLUSIONS

With the normative approach used the task behavior of the pilot can be described in terms of the situation-action characteristic of production systems. This was done by identifying tasks and their inputs and outputs, randomizing task performance durations, and structuring the tasks in the form of a network. When a network technique was combined with production systems, a method was created which permits the modeling of both the sequencing of attention among different tasks as well as the performance of those tasks. Prerequisite for the application of that method is a comprehensive analysis of tasks that must be accomplished and the identification of the system processes affecting those tasks. The advantages of that method are that the model can be easily established in a relatively short time and later easily modified. Furthermore, because of model characteristics it can easily transformed into a simulation program using a high level simulation language such as SLAM in order to exercise the model dynamically with a digital computer. By analyzing dynamic simulation output data, information flow requirements which are necessary for performance of a mission and critically useful for cockpit design and evaluation can be determined. This method can be used in the

early development phase of man-machine system,
e.g., for determining information flow requirements of procedure oriented operator tasks.

REFERENCES

Chu, Y.Y., Rouse, W.B. (1979). Adaptive allocation of decision making responsibility between human and computer in multi-task situations. IEEE Transaction
on Systems, Man, and Cybernetics, SMC-9,
(No. 12), 769-778.

Collins Radio Company (1974). Pilot's Guide
AP-104/FD-109H/FD-109F, Flight Control
System Collins Radio Company, Cedar Rapids,
Iowa.

Davis, R., King, J. (1977). An overview of
production system. In Elcock, E.W., Mi-
chie, D. (Eds.), Macnine Intelligence 8,
Halsted Press: A division of John Wiley &
Sons Inc., New York, pp. 300-332.

Döring, B. (1976a). Analytical methods in
man-machine system development. In
Kraiss, K.F., Moraal, J. (Eds.), Introduction to Human Engineering, Verlag TÜV
Rheinland GmbH, pp. 293-350.

Döring, B. (1976b). Application of system human
engineering. In Kraiss, K.F., Moraal, J.
(Eds.), Introduction to Human Engineering,
Verlag TÜV Rheinland GmbH, pp. 384-415.

Goldstein, I.P., Grimson, E. (1977). Annotated production systems - A model for
skill aquisition. Proceedings of the
Fifth International Conference on Artificial Intelligence, MIT Cambridge MA,
MIT Artificial Intelligence Laboratory,
311-317.

Johannsen, G., Rouse, W.B., Hillmann, K.
(1981). Studies of Planning Behavior of
Aircraft Pilots in Normal, Abnormal, and
Emergency Situations. Forschungsinstitut
für Anthropotechnik, 5307 Wachtberg-
Werthhoven, Germany, Report No. 53.

Kok, J.J., Van Wijk, R.A. (1977). A model
of the human supervisor. Proceedings of the
13th Annual Conference on Manual Control,
MIT, Moffett Field, CA: NASA, 210-216.

Kraiss, K.F. (1981). A display design and eva-
luation study using task network models.
IEEE Transactions on Systems, Man, and
Cybernetics, SMC-11, (No. 5), 339-351.

Miller, K.H. (1976). Timeline Analysis Program (TLA-1). Final Report. Boeing Commercial Airplane Company, Seattle, Wash.,
NASA CR - 144942.

Muralidharan, R., Baron, S. (1979). DEMON: A
human operator model for decision making,
monitoring, and control. Proceedings
of the Fifteenth Annual Conference on
Manual Control, Wright-State University,
Wright-Patterson AFB, OH: Air Force Flight
Dynamics Laboratory.

Nilsson, N.J. (1980). Principles of Artificial
Intelligence, Tioga Publishing Company,
Palo Alto, Calif.

Pew, R.W., Baron, S., Feehrer, C.E., Miller
D.C. (1977). Critical Review and Analy-
sis of Performance Models Applicable to
Man-Machine Systems Evaluation. Bolt Be-
ranek and Newman Inc., Cambridge, Ma.,
BBN Report No. 3446.

Pritsker, A.A.B., Pedgen, D.D. (1979).
Introduction to Simulation and SLAM: John
Wiley and Sons, New York.

Rouse, W.B. (1977). Human computer interaction in multi-tasks situations. IEEE
Transactions on Systems, Man, and Cybernetics, SMC-7, (No. 5), 384-392.

Rouse, W.B. (1978). A model of human decision
making in a fault diagnosis task.
IEEE Transactions on Systems, Man, and
Cybernetics, SMC-8, (No. 5), 357-361.

Rouse, W.B. (1979). A model of human decision making in fault diagnosis tasks that
include feedback and redundancy. IEEE
Transactions on Systems, Man, and Cybernetics, SMC-9, (No. 4), 237-241.

Rouse, W.B. (1980). Systems Engineering Models of Human-Machine Interaction. North
Holland, New York.

Sargent, R.C. (1979. Validation of simula-
tion models. Proceedings of 1979
Winter Simulation Conference, San Diego,
Cal., Dec. 3-5, 496-503.

Schlesinger, S., Crosbie, R.E., Gagne, R.E.,
Innis, G.S., Lahvani, C.S., Loch, J.,
Sylvester, R.J., Wright, R.D., Kheir, N.
and Bartos, D., (1979). Terminology for
model credibility, Simulation, March,
103-104.

Sheridan, T.B. (1976). Preview of models of
the human monitor and supervisor. In
Sheridan, T.B., Johannsen, G. (Eds.),
Monitoring Behavior and Supervisory
Control, Plenum Press, New York, 1976,
pp. 175-180.

Sheridan, T.B., Johannsen, G. (Eds.), (1976).
Monitoring Behavior and Supervisory Control, Plenum Press, New York.

Walden, R.S., Rouse, W.B. (1978). A queueing model of pilot decision making in a
multi-task flight management situation.
IEEE Transactions on Systems Man, and
Cybernetics, SMC-8, (No. 12), 867-875.

Wesson, R.B. (1977). Planning in the world of
the air traffic controler. Proceedings of
the 5th International Conference on Artificial Intelligence, MIT, Cambridge, MA,
MIT Artificial Intelligence Lab., 473-479.

Winston, D.H. (1977). Artificial Intelligence,
Addison-Wesley Publishing Comp., London.

A FUZZY MODEL OF DRIVER BEHAVIOUR: COMPUTER SIMULATION AND EXPERIMENTAL RESULTS

U. Kramer and G. Rohr

Institute of Automotive Engineering, Technische Universität Berlin, Berlin, Federal Republic of Germany

Abstract. A model for driver-behaviour founded on fuzzy set theory will be presented here. Our results show that this method achieves a high degree of accordance between observed and simulated eye- and steering movements, and, accordingly, it is highly suitable for heuristic modelling of complex systems. In particular, a close connection between eye-movements and steering wheel turning has been established.

Keywords. Biocybernetics; electrophysiology; man-machine systems; driver behaviour; modelling; fuzzy set theory; pattern recognition.

INTRODUCTION

Visual pattern processing while guiding a vehicle is characterized by extracting such parameters from a complex visual scene in a finite time which can be transformed directly into motor commands, e.g. for turning the steering wheel or actuating the pedals. The human visual motor system performs these tasks so perfectly after a short training period that it becomes an interesting question how the processing of visual information works.

Recently, there has been some evidence supporting the viewpoint that the human visual system is a spatial frequency analyser which reacts specificly to certain frequency components under defined attentional conditions (Davis, 1981).

Kramer and Rohr (1981, 1982) have assumed parallel working spatial frequency filters which allow the driver to derive simultaneously parameters for lane keeping and speed control. Thereby, high- and low-frequency components of the actual visual pattern will be processed by a kind of spatial correlation (fuzzy correlation) with a corresponding internally represented reference pattern.

Furthermore, the authors assumed lane keeping to be a result of low frequency processing applied to the actual pattern. For this, it has also to be taken into account that, generally, hand movements towards.a specified target are not possible without fixating this target with the eyes (Prablanc et al., 1979).

Therefore, a first process is required which determines the primary area of eye fixations by extracting relevant low frequency features from the visual scene. On the basis of this first process, a second process is introduced which provides a motor command for the hand movements.

This part of the driver-behaviour will be presented more explicitly in this paper.

MODEL OF DRIVER-BEHAVIOUR

For the modelling of driver-behaviour at lane-keeping it is presumed that the selection of the low frequencies from the visual scene has already be done. The road course contained in this low frequency pattern is represented as a matrix, $\underline{M}_0$, (object matrix) of membership values, $\mu_0(x_i, y_j)$, which indicate the degree of membership of the interval (x_i, y_j) from the visual scene to the object 'road'.

This object matrix, $\underline{M}_0$, and a reference matrix, $\underline{M}_R$, mapping the vehicle's lane width requirements estimated by driver are used within feature extraction (Fig. 1) to perform the fuzzy horizontal correlation (Eq. (A9)):

$$\underline{M}_{OR} = \underline{M}_0 \overset{hor}{\circledast} \underline{M}_R \qquad (1)$$

which contains the information about the horizontal shift to get the maximum conformity between $\underline{M}_0$ and $\underline{M}_R$. Vertical correlation is not considered here because of its irrelevance related to lane keeping.

The interval, (x_i, y_j), of the visual scene indicating the maximum conformity corresponding to, $\underline{M}_{OR}$, is assumed to be the area of eye fixation (cf. Eq. (A10)):

$$\underline{\mu}_{F_h} = \text{proj}_Y [\underline{M}_{OR}] \text{ hor. fixation} \tag{2}$$

$$\underline{\mu}_{F_v} = \text{proj}_X [\underline{M}_{OR}] \text{ vert. fixation}$$

The rows of $\underline{M}_{OR}$, whose supremum , $\mu_{F_v}(x_i)$, exceeds a certain value, are processed as significant rows in the subsequent pattern recognition stage (Fig. 1). In this, the significant rows of $\underline{M}_{OR}$ are compared with the same rows of the matrix, $\underline{M}_{PR}$, resulting from

$$\underline{M}_{PR} = \underline{M}_P \overset{hor}{\circledast} \underline{M}_R \tag{3}$$

representing the feature extracted prototype matrix (prototype matrix, M_P: straight course with diminishing relative heading angle and lane deviation of the car). With

$$\underline{M}_{OR,PR} = \underline{M}_{OR} \overset{hor}{\circledast} \underline{M}_{PR} \tag{4}$$

one gets a fuzzy horizontal correlation of second order whose projection to Y generates the fuzzy steering command:

$$\underline{\mu}_\Lambda = \text{proj}_Y [M_{OR,PR}]. \tag{5}$$

DIGITAL SIMULATION

At first, the pattern of road course is computed as a perspective display with the inputs: position of the vehicle, road curvature along the visible course, relative heading angle of the vehicle, and relative lane deviation of the vehicle (related to the middle of the road).

This pattern is decomposed into an orthogonal raster consisting of 16 rows and 79 columns (Fig. 2), from which the object matrix, M_O, is derived. For the simulation of the driver-behaviour a binary (0-1)-distribution is assumed.

To the matrix arising from this the driver model is applicated which produces the fuzzy eye fixation areas according to Eq. (2) as well as the fuzzy steering commands according to Eq. (5).

EXPERIMENTAL PROCEDURE

Three adult subjects had to perform four driving tasks each 83 seconds long in a driving simulator. The subjects were sitting in a cabine, facing a TV projection screen (1.8 m x 2.4 m), and guided the simulated vehicle along the road generated by a special hardware and shown on the projection screen.

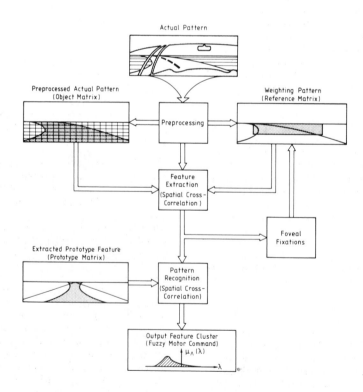

Fig. 1.   Conceptual model of pattern processing
in lane control (Kramer and Rohr, 1982)

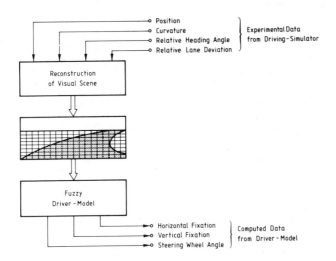

Fig. 2. Scheme of simulation of
driver-behaviour in lane control

The values of curvature, relative heading
angle, and relative lane deviation were
measured directly, the momentary position was
computed from the speed measurements. By
means of these values the picture sequences
of each driving task could be reconstructed
by the digital simulation programme (1000
frames per drive).

As person parameters served eye fixations
recorded by the method of electro-oculo-
graphy, EOG, (DC recording, vertical and
horizontal component) and the steering wheel
angle measured by a potentiometer mounted on
the axle of the steering wheel. To avoid
interferences between head movements and
EOG's the subject's head was fixated.

For EOG measurement a calibration referring
to the edges and the center of the screen
was conducted before and after each driving
task. The EOG data were then normalized
corresponding to these calibration points so
that drift and polarization effects were adjusted.

RESULTS

To generate the outputs of the driver model
the reference matrix, M_R, was determined in
an interactive procedure. It indicates
individually differing (0-1)-distributions
related to the subjects.

Using the best fitting reference matrix,
M_R, the computed variables, horizontal and
vertical eye fixations, steering commands,
are then characterized by ranges resembling
confidence intervals in which the corresponding membership values exceed the 0.8 possibility mark. Fig. 3 shows simulated and
empirical data of horizontal eye fixations
during the first 300 analyzed frames (about
25 seconds) of the first driving task.

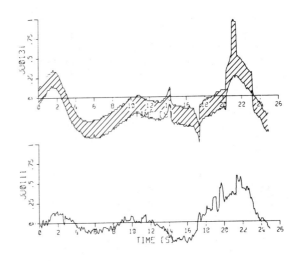

Fig. 3. Simulated (above) and empirical
(below) horizontal eye fixations
(y-axis: +1 ≅ right edge; -1 ≅
left edge of the screen)

For each variable the normalized cross-
correlation between simulated and empirical
data (complete driving task, 1000 frames)
was calculated. For this comparison the
arithmetic mean of upper and lower boundaries
of estimated range was taken for non-fuzzy
value. The maximum cross-correlation value
was tested by correlation analysis, the
results are specified in Table 1. This procedure seems to be justified to us because,
in the present state of the modelling, the
dynamical driver-behaviour is not the centre
of interest. In addition, the lag was in
most cases close to zero.

The simulation as well as the statistical
analysis was conducted by the DEC computer
PDP 11/55. The complete analysis of one
driving task took ten hours.

TABLE 1   Results of Correlation Analysis

	Eye fix hor	Eye fix vert	Steering Angle
Subject			
GA	0.79 **	0.19 *	0.88 **
JU	0.68 **	0.03 n.s.	0.58 **
ST	0.78 **	0.06 n.s.	0.86 **

** 1 p.c. level (N = 1000)
 * 5 p.c. level (N = 1000)

DISCUSSION

The correlation analysis indicates a good correspondence between predicted and empirical variables except for vertical eye fixations. This component seems to depend on other operations than horizontal correlation between actual and reference pattern.

As suggested by a first analysis the vertical eye fixations could play an essential role in estimating the velocity. Possibly, the inclusion of a vertical correlation, analogously to Eq. (A9), in the fuzzy driver model will explain the vertical fixation component.

Furthermore, a comparison between the absolute values of simulated and observed horizontal eye movements shows tendencies to overestimation of small eye movements and to underestimation of large eye movements. To avoid these effects, perhaps a dynamical version of the eye movement model could become necessary (Kramer and Rohr, 1982).

CONCLUSION

It can be stated that steering movements or corresponding motor commands are almost completely determined by the horizontal eye movements, which in turn depend essentially on the perspective road pattern. This result agrees with our former hypothesis, derived from the general findings of Prablanc et al. (1979) that the steering movement is a goal reaction in his sense.

REFERENCES

Davis, E.T. (1981). Allocation of attention: uncertainty effects when monitoring one or two visual gratings of non-contiguous spatial frequencies. *Perception and Psychophysics, 29,* pp. 618-622.

Dubois, D. and Prade, H. (1980). *Fuzzy Sets and Systems: Theory and Applications.* Academic Press, New York.

Kramer, U. and Rohr, G. (1981). Psycho-mathematical model of vehicular guidance based on fuzzy automata theory. In

H.G. Stassen and W. Thijs (Eds), *Proceedings of the European Annual Manual '81.* Delft University of Technology, Delft.

Kramer, U. and Rohr, G. (1982). A model of driver behaviour. *Ergonomics, 25,* (in press).

Prablanc, C., Echallier, J.F., Kommilis, E. and Jeannerod, M. (1979). Optimal response of eye and hand motor systems in pointing a visual target. *Biological Cybernetics, 35,* pp. 113-124, pp. 183-187.

APPENDIX

The fuzzy subset, A, of a finite universe, X,

$$A = \{(x_i, \mu_A(x_i)): x_i \in X \, (i = 1, \ldots, n),$$
$$\mu_A(x_i): A \to [0,1]\} \qquad (A1)$$

can be characterized by an n-dimensional vector, $\underline{\mu}_A$,

$$\underline{\mu}_A = [\mu_A(x_1), \ldots, \mu_A(x_n)]^T \qquad (A2)$$

whose elements indicate the degree of membership of $x_i \in X$ belonging to the fuzzy set A. For a closed interval $[a,b] \subset R$ and $x_i \in [a,b]$ $(i = 1, \ldots, n)$ as a complete partition of $[a,b]$,

$$\bigcup_{i=1}^{i=n} x_i = [a,b], x_i \cap x_j = \varnothing \, (i \neq j) \qquad (A3)$$

the vector, $\underline{\mu}_A$, represents a fuzzy sub-interval of $[a,b]$.

The similarity of the fuzzy subintervals A,B of $[a,b]$ is defined by

$$\sigma_{AB} = \sum_{i=1}^{i=n} \mu_{A \cap B}(x_i) / \sum_{i=1}^{i=n} \mu_{A \cup B}(x_i) \qquad (A4)$$

(Dubois and Prade, 1980), where $\sigma = 0$ iff $A \cap B = \varnothing$, and $\sigma = 1$ iff $A \equiv B$.

Introducing the shift-operator, $q^k, (k \in Z)$ such that

$$q^k \underline{\mu}_A = [\mu_A(x_{1-k}), \ldots, \mu_A(x_{n-k})]^T \qquad (A5)$$

($\mu_A(x_{i-k}) = 0$ for $n < i-k < 1$) and Eq. (A4) a measure of conformity

$$\underline{\sigma}_{AB}(k) = \sigma(\underline{\mu}_A, q^k \underline{\mu}_B) \qquad (A6)$$

can be defined in accordance with the conventional correlation function.

For odd n the vector, $\underline{\sigma}_{AB}$,

$$\underline{\sigma}_{AB} = [\sigma_{AB}(-(n-1)/2), \ldots, \sigma_{AB}((n-1)/2)]^T \qquad (A7)$$

can be interpreted as a fuzzy set on the interval $[a,b] \subset R$, which will be designated a fuzzy correlation function.

Applying these considerations to the cartesian product, $X \times Y$, of the two finite universes, $X = \{ x_1, \ldots, x_n \}$, $Y = \{ y_1, \ldots, y_m \}$, a fuzzy plane object, O, is, by the matrix, $\underline{M}_0$,

$$\underline{M}_0 = \begin{bmatrix} \underline{\mu}_{0_1}^T \\ \vdots \\ \underline{\mu}_{0_n}^T \end{bmatrix} = \begin{bmatrix} \mu_0(x_1,y_1) \cdots \mu_0(x_1,y_m) \\ \vdots \qquad\qquad \vdots \\ \mu_0(x_n,y_1) \cdots \mu_0(x_n,y_m) \end{bmatrix} \qquad (A8)$$

completely determined as a fuzzy set.

The fuzzy horizontal correlation of the matrices, $\underline{M}_0$ and $\underline{M}_R$, is defined by (m odd)

$$\underline{M}_{0R} = \underline{M}_0 \overset{hor}{\circledast} \underline{M}_R = \begin{bmatrix} \underline{\sigma}_{0_1 R_1}^T \\ \vdots \\ \underline{\sigma}_{0_n R_n}^T \end{bmatrix} \qquad (A9)$$

with the fuzzy correlation, $\underline{\sigma}_{0_i R_i}$, ($i = 1, \ldots, n$), of the i-th rows of $\underline{M}_0$ and $\underline{M}_R$ accordingly to Eq. (A7).

This matrix, M_{0R}, is also conceived as a fuzzy set on $\overline{X} \times \overline{Y}$.

The projection, $\text{proj}_X [\underline{M}_0]$, of $\underline{M}_0$ onto X is given by the vector

$$\underset{X}{\text{proj}} [\underline{M}_0] = \begin{bmatrix} \underset{j}{\sup} (\mu_0(x_1,y_j)) \\ \vdots \\ \underset{j}{\sup} (\mu_0(x_n,y_j)) \end{bmatrix} \qquad (A10)$$

Analogously, the projection vector, $\underset{Y}{\text{proj}} [\underline{M}_0]$, of $\underline{M}_0$ onto Y is defined by

$$\underset{Y}{\text{proj}} \ \underline{M}_0 = \begin{bmatrix} \underset{i}{\sup} (\mu_0(x_i,y_1)) \\ \vdots \\ \underset{i}{\sup} (\mu_0(x_i,y_m)) \end{bmatrix} \qquad (A11)$$

Both vectors are interpreted as fuzzy sets on the corresponding projection universes.

THE AIR TRAFFIC CONTROLLER'S PICTURE AS AN EXAMPLE OF A MENTAL MODEL

D. Whitfield* and A. Jackson**

*University of Aston, Birmingham B4 7ET, UK
**Royal Signals and Radar Establishment, Malvern, Worcs. WR14 3PS, UK

Abstract. Air traffic controllers refer to the 'picture': their
overall appreciation of the traffic situation for which they are
responsible. In particular, great importance is attached to the
controller's maintaining an appropriate picture when control and
decision making responsibilities are shared between man and computer.
We are making three initial approaches to elucidating the nature of
the picture. (1) Interviews with controllers have defined some major
characteristics. (2) Operational controllers have given 'verbal
protocol' accounts of establishing the picture before taking over a
sector position from a colleague. (3) Controllers working within a
real-time simulation later observe a complete replay of radar display,
flight plans, and speech communications for particular sessions,
during which replay they record verbal protocols and provide further
explanation. Initial results are presented, and the potentialities,
limitations, and future development of the techniques are discussed.

Keywords. Air traffic control, behavioural sciences, cognitive processes,
cognitive systems, computer applications, ergonomics, human factors,
man-machine systems, models, skills.

INTRODUCTION

Mental Models in Control Tasks

Various studies of control and management
skills have suggested that the human
operator depends on an internal
representation of the system he is
controlling. Various terms have been used:
'mental model', 'conceptual model',
'internal model'. The implication is that
the operator's decisions are based on his
understanding of the system, as an
internalised representation or set of rules
by which he may test possible actions.
Rasmussen and Rouse (1981) in their introductory summary of the conference on
"Human Detection and Diagnosis of System
Failures" remark:

"While .... this rather obvious notion is
well accepted, researchers' models of
internal models range from sets of
differential equations to functional
block diagrams to snapshots of physical
form. The current problem is not really one
of finding the "correct" internal model,
but instead trying to find where each
alternative applies and is useful....". (p.3)

It is envisaged that fuller knowledge of the
operator's mental models would lead to more
appropriate interface design, in the
selection and coding of system information
to accord with the models. If there are
indications of inappropriate operator
models, then training requirements may be
inferred. Individual differences in models
may suggest selection requirements, or the
need for flexibility in system planning and
design. Overall, the mental model is an
important element in operator skills.

The Air Traffic Controller's 'Picture'

Our present interest in this area derives
from system research for the U.K. Civil
Aviation Authority, in which the Royal
Signals and Radar Establishment (RSRE) has
been developing concepts for possible
computer assistance for the decision making
of air traffic controllers. The general
aim of the RSRE work is to develop computer
aids which will handle appropriate parts of
the controller's tasks, whilst still
preserving his ultimate judgement and
responsibility for action. We have reported
the ergonomics aspects of research on
"Interactive Conflict Resolution" (ICR), a
computer aid for the detection and
resolution of conflicts between aircraft
(Whitfield, Ball and Ord, 1980). There, the
use of the computer as a predictive aid was
shown to be successful, within a fairly

restricted simulation, and this added support to the emerging principles for this type of man-computer allocation of functions.

However, some important questions emerged from the ICR studies, in particular the issue of the controller's 'picture'. This problem of maintaining an overall appreciation of the traffic situation was mentioned spontaneously by the controllers in questionnaires, and was fully discussed after the simulation experiment. All of the subjects reported relying heavily on the ICR aids in order to deal with the traffic, particularly at the busy levels. There was a tendency for them to be driven by the computer to solve the next immediate problem, with the result that they felt not properly in charge of the situation. There were fears that an equipment failure might lead to their losing the overall picture. We concluded from these reports that, even though the computer aid extends the controller's capabilities in several res- pects, the requirement to preserve the picture might still set a fairly close limit on effective traffic handling capacity.

Thus, further investigation of the nature of the picture seemed important, as a back- ground to the application of this type of computer aiding, as another example of 'supervisory control'. In addition, it should enable us to understand further air traffic control skills, and it might provide some new approaches to questions of controller workload and admissible limits.

INTERVIEWS WITH TRAINING CONTROLLERS

We began by having individual discussions with ten training controllers, organised around topics such as describing the picture, its development, and the possibilities of losing it. Controllers found it difficult to describe the picture in verbal terms, and there were some con- flicting aspects of descriptions by diff- erent individuals. Nevertheless, there was broad agreement that the topic was worth investigating further, and the next section lists the major themes which did emerge from these interviews (Whitfield, 1979).

Reported Main Elements of the 'Picture'

Handing over of a sector from one controller to his relief was frequently mentioned in relation to the picture: the incoming controller has to establish his own picture, on the basis of observing and talking to the outgoing controller, until he is satisfied that he can assume responsibility.

Experienced and skilled controllers were

said to perceive problem situations more quickly and easily: there are obvious parallels with reports of other cognitive skills, where the more experienced operator seems able to handle larger units of information.

Selective aspects of the picture were emphasised. Controllers talked of parts of the overall traffic situation being in focus or in the foreground, as opposed to the rest out of focus or in the background. Never- theless, it is relatively easy to switch 'foreground' attention between different parts of the situation.

A plan for the development of the traffic is inherent in the picture, and the actual behaviour of individual aircraft is checked against this plan. This organisation of system information into a coherent pattern is similar to many other control tasks.

Predictive aspects were referred to often. Thinking ahead and planning ahead seem to be important issues, as is evidenced by the success of the predictive computer aid in ICR.

Further Studies of the 'Picture'

Having obtained these general accounts of the picture, techniques were sought for developing more detailed descriptions. Two approaches are discussed in the following sections. First, a study of the establish- ment of the picture, and second, a facility for obtaining evidence of the elaboration and development of the picture during simulator trials.

VERBAL PROTOCOLS FROM A STUDY
OF SECTOR HANDOVER

The verbal protocol technique has been used in attempts to analyse cognitive skills in several contexts (e.g. Bainbridge, 1974; Rasmussen and Jensen, 1974). The operator is asked to give a running commentary describing his sequences of thought and action, which is recorded and later analysed. The technique should ensure a detailed and accurate account of the operator's 'thought stream', without the rationalisations which might attend any 'off-line' description in an interview situation. However, there is the possibility that the extra task of providing the verbal description may modify, or interfere with, normal activities and modes of operation. Obtaining verbal protocols from working air traffic controll- ers seems unacceptable, because of the likely interference with the control task itself, and the inevitable conflict between R/T (radio-telephone) transmissions and the verbalisation required. However, noting the emphasis in the interviews on the picture in relation to handing over a sector, a slightly different approach emerged. We would use the technique to study the establishment of the picture by another controller in the state of preparing to take over.

Volunteer controllers in a civil operations room were asked to look over the shoulder of a working sector controller, observing the radar display and the board of flight progress strips (abbreviated versions of aircrafts' flight plans). The volunteer was required to pretend that he was about to take over the sector, and to tape record a verbalisation of his process of building up the picture. The flight progress strips were photographed at the beginning and end of each protocol, and a one minute time lapse camera recorded the radar display, to aid in subsequent analysis of the protocols.

Thirty protocols were obtained over a period of a few days, and 18 were fully analysed, as they represented the most interesting situations. The transcripts of the recordings were arranged as successions of separate phrases, annotated with information from the photographic records (e.g. fig. 1). These records were then analysed as frequency counts of defined elements and groups of elements, and in terms of the general structure of the protocols.

Frequency Counts of Protocol Elements

Figure 2 shows the percentages of protocol elements observed, arranged in four major categories. The data is presented as minimum/median/maximum across the 18 protocols analysed, demonstrating considerable variation. However, the relative frequencies of the major categories are of interest:

STRIPS: the predominance of elements referring to flight progress strips is obvious: controllers relied heavily on strip information in establishing the picture. Two evident reasons are that expected aircraft, either at airports or in adjacent sectors, are not yet on the radar display, and that the strips alone contain detailed information about each individual flight.

RADAR: by contrast, elements referring to the radar display are much less frequent. Moreover, there is still a strong link with the flight progress strips: the majority of occurrences represent a radar identification, where the controller reads a strip and locates the aircraft on the display. There are relatively few reports of using flight-level, call-sign, or squawk (ident- ification) data from the alpha-numeric aircraft labels on the radar display.

T, M, P, C, D: are "mental activities" inferred during the protocol analysis - Time check, Memory reference, Prediction, Calculation, Decision. The separate frequencies were so small, that they are merged in this initial analysis: the low frequencies could well be an artefact of this study.

SO, SS, SP, RR: are search activities prior to references to Strip or Radar data, again inferred during the protocol analysis, and appearing relatively infrequently.

Thus, fig. 2 illustrates the predominance of information derived from strips in the controller's verbal reports. It suggests also that further more detailed analysis within the categories would be useful.

So far, this further analysis has been applied only to the STRIP (S) category, and the results are shown in fig. 3. This category has been analysed first because of its obvious importance, and because it contains the most reliable data: the frequencies are substantial, and the elements are present in the protocol ab initio and not inferred. The data presented in fig. 3 gives a percentage breakdown, in the minimum/median/maximum form.

The following comments are appropriate:

(a) The high frequencies of SC (Call-sign) are partly a result of the request to the controllers to identify each aircraft to which they were referring at any stage of the exercise.

(b) In relation to the partially 'imposed' nature of SC, the comparable frequencies of SR (Route) are notable. Controllers appear to use route information (particularly departure and arrival airports, but also direction of travel) very much in thinking about aircraft. Obviously, route inform- ation will be of great importance in developing the general picture of the traffic situation.

(c) The next most important category is SL (Flight-level). Several controllers organise their thinking in terms of levels occupied.

(d) Both SB (Beacon time - estimated time at a navigational point) and SA (airborne time from an airport) are referred to frequently. The reliance on airborne times (or, fre- quently, the observation that the airborne time had not yet appeared) is due to the controllers' immediate responsibility for departing aircraft.

(e) The remaining categories of SN, ST and SO are less numerous, but each plays a significant role at times.

(f) This particular part of the analysis has been successful in tabulating the information used by the controllers, but less so with the clarification of the thought-links between sections of a protocol dealing with different aircraft. When the controller is moving systematically up the bank of strips, from earliest to latest, these transitions from one aircraft to another can be easily interpreted. However, there are several cases where the controllers did not adopt this procedure, and this behaviour needs further investigation. Some information results from the analyses in the next section.

Structures of the Protocols

This section attempts to describe some of the organisation or general pattern of the protocols obtained. Only broad, qualitative, indications are possible at present. Further analysis and discussion of the protocols, and the obtaining of more recordings, perhaps under different conditions, are necessary to develop this level of description. There are many variations between protocols, some due to the different controllers, but others due to the particular traffic conditions prevailing. For all of these reasons, numerical summaries, as in the frequency tabulation of the previous section, seem somewhat premature.

(a) The conventional procedure, when taking over a sector, is to examine each flight strip in time order, and relate it where possible to the radar indication. This procedure was observed on several occasions.

(b) There were also many cases not like (a), and it is difficult often to clarify the reasons for such patterns. Some examples are:

- detection of a potential conflict, and moving off to explore that;
- insertion of a new strip, such as an airborne departure, and the need to consider that;
- comment from the incumbent controller, or overhearing the R/T.

(c) Having safely considered an individual aircraft, either as in (a) or (b) above, a typical final comment is to say "no problem" or "I can forget that", etc., apparently mentally ticking off the aircraft before moving on to look at another. This aspect has been observed before, by others and by us: Sperandio (1974) shows that subsequent memory for non-conflict aircraft tends to be poorer than for aircraft which were considered to be in conflict. Nevertheless, working through the flight data would seem to be important to building up the picture, even if the controller then 'dismisses' the information.

(d) As alternatives to the conventional ordered strip approach, there are several examples of other means of organising the picture. Several controllers summarise the situation in terms of the inbound and outbound flows for the various airports, relying on their knowledge of the typical routes and procedures. Other controllers may organise their traffic in terms of levels, mentally listing aircraft as "my 6, 7, 8", for example. Several of these examples suggest a fairly complex three dimensional organisation of the flight data. Some specific examples include relating the paths of climbing and descending aircraft potentially in conflict.

(e) In general terms, the overall patterns of the protocols confirm the major reliance of controllers on the flight strip information.

(f) There are several examples of the picture being focussed on unusual or non-routine items. For example, one controller organised his appraisal around a low-level slow aircraft overflying the sector, and another had to consider the interaction of an easterly inbound to one airport, with predominantly N/S flows into and out of others. Another controller described aircraft in terms of such characteristics: "a joiner, ... at low level ....". Other 'negative' evidence of such approaches is given by comments such as "there's nothing descending, nothing climbing ... " and by a controller's not referring to standard flights, such as normal inbounds, in his report.

(g) There are examples - again, difficult to analyse in detail - of prediction and planning ahead. As expected, these depend largely on the controller's knowledge of typical procedures, such as the runway in use at a given airport.

(h) There are a few examples of the controller's thinking through potential conflicts, and of questioning the incumbent controller's decisions, for example on a restricted climb. Once again, this relies on knowledge of typical situations and procedures, but the detail of the thinking is difficult to establish. Sometimes, the tape recording picked up the subject controller's conversation with the incumbent, and also some R/T and intercom. conversations. This additional information is helpful, and might be analysed in further investigations.

REPLAY TECHNIQUE: THE INVESTIGATION OF ESTABLISHED PICTURE

We have referred to the difficulties of obtaining concurrent verbal protocols in the live ATC situation. One alternative is to collect retrospective protocols after the control process has been completed. Several French researchers have employed variations of this technique which can involve free verbalisation after the control process has been interrupted (in simulation) or verbal response to detailed questioning. Leplat and Hoc (1981) provide a discussion of this use of retrospective report. In a major contribution to the understanding of the use of verbal data, Ericsson and Simon (1980) have suggested that the ability to produce verbalisations bearing a direct relation to actual processing is limited by a number of factors, including the duration of short term working memory.

The replay technique attempts to overcome some of these difficulties in establishing the vocabulary or repertoire of objects and procedures employed by controllers. While the concept is being developed on a simulator at RSRE, the intention is to produce a technique which could be used to study the operation of picture in real ATC by means of detailed recordings of controller activity, its consequences and the production of subsequent protocols.

The technique requires complete and synchronised recordings of the data presented on radar displays (digitally on disc storage) and all verbal transactions involving the controller (on magnetic tape). In addition the controller annotated flight strip listings are retained. This data serves two purposes.

(a) It provides a very full description of the controller, and traffic, activity during the one hour session.

(b) It can be replayed to the controller and used as a continuous and accurate prompt for the production of a retrospective verbal protocol.

Since the radar information is recorded digitally, image quality on replay is identical to the original radar display. The subject, presented with this image, the auditory playback in real time and the annotated flight listings, re-experiences the information input from the original session.

The technique is still very much in the development stage. A number of pilot runs have been conducted using two Air Traffic Controllers, on a representative en-route sector, and even at this early stage several observations, confirming previous findings can be made.

(a) As well as familiarisation with the sector, the controllers require time to become used to the production of protocols. The volume and level of detail obtained increased substantially over the introductory sessions. There was also some

suggestion, from the fluency of the protocols, that the controllers were organising their verbalisations rather than producing them with complete spontaneity. Leplat and Hoc (1981) have discussed the tendency for subjects to adapt their verbalisation to their representation of the analyst's knowledge of the skill. In the present case, both controllers' experience as instructors almost certainly proved a contributory factor. The role of instructions in influencing the quality of the protocol is well known, and we are exploring modified instructions to alleviate this difficulty.

(b) There has also been some evidence that attending to the R/T Replay exerts an inhibiting effect on protocol production. Since the taped material is not under the subject's control, it can interrupt verbalisation. There is also a possibility that one's own voice constitutes a more potent distractor than the voices of others.

(c) Observation and discussion with the controllers have provided details of the way in which the simplified flight strip listings were employed, and these tend to support the findings of previous researchers. Most important was the observation that the manually updated flight strips are used primarily during planning phases, for 'looking ahead', while radar dominates during the implementation of plans. This is in keeping with the importance of flight-strip information in the establishment of picture as described above.

At a more detailed level, the two controllers showed slight differences in the way in which flight strips were employed, but in general their logic was similar to the 'flight level strategy' group described by Leplat and Bisseret (1965). The entry height of a new aircraft was compared with those aircraft up to 10 minutes ahead. Next the range between entry and exit heights was compared with the other aircraft to establish potential conflicts before checking to ensure that the potential conflicts were on different routes. The results of this process constrain the level allocated initially to an aircraft and also suggest levels for aircraft further down the list. The annotations of levels already allocated can be used to check back for levels still available. Although the radar display can also provide some of this information, it displays only present height and not allocated level.

A further use of flight strips involves checking ahead to predict R/T loading at future time. Control alternatives may be discarded if potential R/T load will be too high.

(d) The two controllers participating in our pilot sessions evolved rather different patterns of control. However, this difference may not reflect a fundamental difference in their strategies, since one of them had prior experience in a slightly different context. This might have served to impose a particular structure on his methods. It is worth noting that current methods of validating a controller on a sector would tend to transfer the control patterns of existing controllers onto the trainees. By providing a novel sector and allowing individual controllers the opportunity of developing their control strategies based on their own reservoir of experience, we may improve our chances of establishing the factors which contribute to that reservoir.

THE DEVELOPMENT OF THE REPLAY TECHNIQUE

The technique is still at an early stage of development, and future work will have to explore a number of issues, not least the relationship between verbalisations obtained and the processes employed in the ATC task. It should be stressed, however, that the verbal report component, while important, is only one aspect of the data collected. Data from any one source must be interpreted in the context provided by the others.

Future development includes the following:

(a) We need to discover more about the role of memory from the initial session, in the replay protocols. This might be investigated by comparing protocols from immediate replays with those obtained after a delay of days or weeks, perhaps with the interposition of interfering replay sessions. A pilot study, again using the same two controllers, is in progress but the data has not been processed at the time of writing. Indeed, the volume of material deriving from a single one hour session, and the amount of analysis, is substantial. We will be seeking qualitative differences in the specificity of the protocols produced. Generally, retrospective protocols are more global than those produced concurrently, possibly as a result of memorial processes. Since delayed replay cannot rely on memory to the same extent, and has the alternative of reprocessing the data presented, it may produce a more specific protocol. As always, caution would be required in exploiting this specificity if it were found. If reprocessing takes place, we have no proof that it is the same as they employed in the initial session. Rather we would wish to imply that there may be similarities. Examination of replay data can, at best, generate hypotheses which must be validated in the real control context.

(b) We have referred already to the difficulties of interrupting R/T replay in order to generate protocols. Two methods

which could alleviate the problems are being considered:

(1) to allow the controller to halt the replay process while verbalising. Such a facility already exists in the replay system.

(ii) To use replay sessions as the framework for a series of interviews. A variation on this method has been incorporated in our pilot work, but only after the controllers had generated an immediate replay. The experimenter used the recorded data to generate a description of factors underlying various controller decisions. This description was shown to the controller and then controller and experimenter stepped through the replay discussing the description and qualifying it. Columns 4 and 5 of the protocol sample in fig. 4 contain extracts from the description and additions.

Both these methods have some advantages over the production of spontaneous protocols but they lose many of the temporal aspects and pressures which replay retains.

(c) A third development relates to individual differences between controllers. We have mentioned that even with two controllers different approaches are evident. One advantage to running on a simulator is that different controllers may be presented with the same traffic samples, their data recorded and then replayed in synchronisation. It may be possible to arrive at some classification of the different strategies and identify choice points where strategies diverge. These choice points may represent significant aspects of picture. The fact that a choice can be made suggests that (a) individual controllers represent more than one alternative at that point, or (b) controllers' representations differ at that point.

REFERENCES

Bainbridge, L. (1974). Analysis of verbal protocols from a process control task. In E. Edwards and F. P. Lees (Eds), The Human Operator in Process Control, Taylor and Francis, London.

Ericsson, K. A. and Simon, H. A. (1980). Verbal reports as data. Psychological Review, 87, 215-251.

Leplat, J. and Bisseret, A. (1965). Analyse des processus de traitement de l'information chez le controleur de la navigation aérienne. Bulletin due Centre de'Etudes et de Recherches. Psychotechniques. 14, 51-67.

Leplat, J. and Hoc, J. M. (1981). Subsequent verbalisation in the study of cognitive processes. Ergonomics, 24. 743-755.

Rasmussen, J. and Jensen, A. (1974)
    Mental procedures in real-life tasks:
    A case study of electronic trouble-
    shooting. Ergonomics, 17, 293-308.
Rasmussen, J. and Rouse, W. B. (Eds) (1981).
    Human Detection and Diagnosis of System
    Failures NATO Conf. Series III
    Plenum, N.Y.
Sperandio, J. C. (1974). Complements à
    l'étude de la mémoire opérationelle
    des contrôleurs de navigation aérienne.
    Institut de Recherche d'Informatique
    et d'Automatique, Report C.O. 7403.

R42 (RSRE translation No. 518:
    Extensions to the study of the
    operational memory of air traffic
    controllers).
Whitfield, D. (1979). A preliminary study
    of the Air Traffic Controller's
    'picture'. Journal of the Canadian
    Air Traffic Controllers' Association.
    11(1), 19-28.
Whitfield, D., Ball, R. G. and Ord, G. (1980).
    Some human factors aspects of computer
    aiding for air traffic controllers.
    Human Factors. 22, 569-580.

LIC-1   I can see   (Epinal-Manchester)
1913   FI (GBFFI)
               level at 60
               He's put him on approach
               he's level at 50 now

LCE-3   One just cleared out of C/Don. (BD212)   (E,Mid-Liverpool)
1919+3   40 only
               so I am looking for the traffic
               (Operating controller:
               He's going up the way for starters, (BA5656)
               going all the way to 180
               He's traffic to these two at moment (BD212 and BD946)
               but he's coming down.) (BD946)

STF-1   One out of Birmingham (BA5656)   (BB-Glasgow)
1914 6   to go North

FIGURE 1:   Example of verbal protocol in establishing the picture

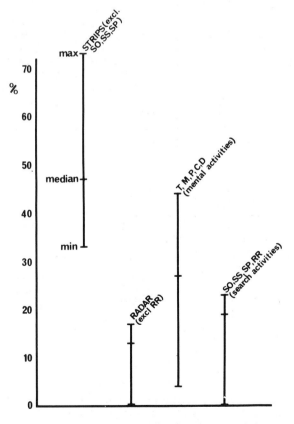

FIG 2:   PERCENTAGE OF PROTOCOL ELEMENTS

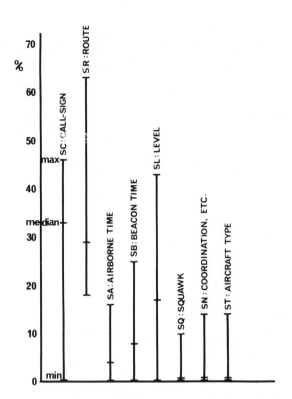

FIG 3: PERCENTAGE CONSTITUTION OF STRIP

CATEGORY, EXCLUDING SO, SS, SP

COL 1 Time	COL 2 RT Transaction	COL 3 Replay Protocol	COL 4 Description	COL 5 Comments on Description
05.10 (031)	A/C Springbok 2210 roger Scottish, turning right from 150 to 160 degrees over.	That should bring him back towards the centre of track.	Should bring 2210 on a course roughly parallel to the opposing 2414 but displaced about 12 miles.	It will keep him in the picture (not too far to edge of display) and also give earlier crossing on the Yugoslav (JG3642 on crossing course)
05.20 (032)	A/C OM1186 on your frequency at ... cruising at FL350 U25 for EW.	The Iranian 3846 is in conflict with the Yugoslav ... they are about 60 miles distant at this time		
05.30 (033)	1, over CONT 186 roger cleared present position, direct EW1. A/C OM1186 roger Scottish taking up heading now direct for EW1.		Taken OM1186 completely clear of everything else..	especially the AZ and avoids Dean's Cross

FIGURE 4 Example of verbal data using replay technique from a simulation experiment.
CONT = Controller, A/C = Aircraft, bracketed statements are explanatory
insertions by experimenter

THE DERIVATION OF HANDLING QUALITIES CRITERIA FROM PRECISION PILOT MODEL CHARACTERISTICS

K. J. Brauser

Anthropotechnik (Human Engineering Department) MBB-GmbH, Munich, Federal Republic of Germany

Abstract. Handling qualities criteria for several controlled elements are derived from a linear closed loop analysis including a quasilinear precision pilot model, which is adapted to the characteristics of the controlled elements. The analytic treatment of the linear closed loop transfer function results in a bandpass - filter like frequency response for arbitrary operating points within the performance envelope of the system. Handling qualities criteria are derived from the conditions,that the response of this constant-bandwidth- filter remains within limits which have been evaluated from a pilot - aircraft,single - axis system with well known handling qualities,from which the adaptability limits of the human pilot have been defined. The method also is extended to other controlled elements, for example to the "critical task" or to systems with characteristics of the type $F(s) = 1/s(s - \lambda)$.

Keywords. Handling qualities criteria; manual control; precision pilot model; closed loop analysis; adaptibility limits; bandpass- filter.

INTRODUCTION

Handling qualities criteria for manually controlled systems have been derived in the past preferably by the examination of the operator and his rating during system performance tests.The evaluations best known are those of flight tests and, as a result,the sampling of handling criteria for aircraft(e.g. Military Specifications).Analytical treatments of the human operator in the control loop have been published, in order to assess handling criteria of controlled systems theoretically, including a prediction of operator ratings for systems under development(Anderson 1971; Arnold 1973; Smith 1976) Some methods are ma - thematically perfect covering all possible model and system conditions(Kleinman, Baron 1973). The tribute which has to be payed for this perfection however is the complicated calculation to be performed with computer aid.

Recently, researchers tended to exclude the human operator from the theory of handling criteria by open loop analysis.This may be justified to a certain amount since differenc es between open loop measurements and theoretical calculations sometimes are very small.But, if the conventional treatment of control problems is preferred, an operator model again should be included in open and closed loop analysis, in the form of a "servo model", the characteristics of which had been defined parametrically.(McRuer and Magdaleno 1966; Bubb 1978) With the assumption that the characteristics of the human pilot

will vary with the operational conditions of the controlled system (Hess 1981),the characteristics of the closed loop which depend on these operational conditions, are to be determined analytically.This means that the roots of the frequency transfer function of the closed loop are to be evaluated numerically for all stationary operating points of interest. These closed loop responses must remain within the limits of good handling, the criteria of which have to be defined and validated by measured data.

The handling quality analysis described below has been based on known handling qualities criteria for aircraft which are naturally stable. They are extended by this method to other systems also, especially to unstable systems like the "critical tracking task" (Jex et al.1966).

CLOSED LOOP ANALYSIS

Closed loop modeling.

The closed loop frequency response of a system including the operator and the manually controlled element, is formally described by

$$\left[F_{CL}(s) \right]_{op} = \left[\frac{F_p(s) F_C(s)}{1 + F_p(s) F_C(s)} \right]_{op}$$

$$= \frac{K_e Z_p(s) Z_C(s)}{Z_p(s) Z_C(s) + N_p(s) N_C(s)} \quad (1)$$

$F_p(s)$ is the operator transfer-function,and

$F_C(s)$ is that of the controlled element in its reduced form, while Z_p and Z_C define the numerators, N_p and N_C the denominators of both. The arrangement is also shown in figure 1 in which the switch symbol defines feedback alternatives.

The analysis is based on the knowledge of the roots of the $Z(s)$ and $N(s)$ which have to be defined for both the operator and the controlled element parametrically.

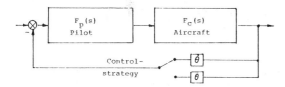

Fig. 1   Arrangement of the closed loop
         analysis

If this condition is true, the closed loop formula (1) can be developed into

$$\left[F_{CL}(s)\right]_{op} = \frac{\Pi_i(1 + a_is + b_is^2)}{\Pi_j(1 + a_js + b_js^2)}\ e^{-\tau_e s} \qquad (2)$$

which is an expression for an active and adaptive bandpass filter with delay. The index op defines the operational state, the indices i and j determine the order n of the filter. generally is $i < j$. The human controller is assumed to adapt his dynamic parameters to the momentary stationary characteristics of the controlled system, which are defined for the operating conditions, by means of his gain, lead, lag, and possibly his delay parameters. If the operating conditions change slowly, the operator may adapt to the new situation, remainig quasilinear. At rapid changes the closed loop will appear time-variant and possibly non-linear, this may contribute to the "remnant" which always is present according to the random character of input and output signals of systems in the real world.

The analysis is concerned primarily with the problem

$$\left[N_{CL}(s)\right]_{op} = \left|N_p(s)N_C(s) + K_e Z_p(s)Z_C(s)\right|_{op}$$
$$= \left|\Pi_j(\ 1 + a_js + b_js^2)\right|_{op} \qquad (3)$$

This problem is solved generally by the assumption that $Z_p(s)$ is adapted by the operator to $Z_C(s)$ to equalize the part $N_p(s)N_C(s)$ with the aid of K_e, if necessary.   Equalization means the generation of a denominator $N_{CL}(s)$ with the lowest order possible. It has been experienced that this equalization can be based on

$$\left[N_{CL}(s)\right]_{op} = N_p(s)N_C(s) +$$
$$+ \Pi_i(1 + a_i(s) + b_is^2)\ e^{-\tau_e s}$$

or, using the formulation of eq. (3) and (5) the general solution of the problem is the following

$$\left[N_{CL}(s)\right]_{op} = \qquad\qquad (4)$$
$$= N_p(s)N_C(s)(1 + T_es) +$$
$$+ \Pi_i(1 + a_is + b_is^2)(1 - T_es)$$

according to the first order Padé - substitution

$$e^{-\tau_e s} = \frac{1 - T_es}{1 + T_es}\ ,\quad T_e = \tau_e/2 \qquad (5)$$

In many cases the numerator of eq. (2) has the form

$$Z_{CL}(s) = \Pi_i(1 + a_is)\ ,\quad b_i = 0. \qquad (6)$$

The operator model, used in this analysis, is the model proposed by Bubb (1978), modified by Brauser(1980)

$$\left[F_p(s)\right]_{op} =$$
$$= \frac{(1 + T_As)(1 + T_Vs)\ e^{-\tau_p s}}{\overline{K}\ (1 + T_Ws)(1 + T_ms + T_m^{*2}s^2)} \qquad (7)$$

T_A = information input lead time const.

T_V = information processing lead time c.

T_W = neuromuscular lag time const.

$\left.\begin{array}{l} T_m = 2\zeta_m/\omega_m \\ T_m^{*2} = 1/\omega_m^2 \end{array}\right\}$ man - manipulator response characteristics

τ_p = operator delay time constant

$\overline{K}$ = $1 + K_k + K_EK_W$ the combined force & joint sensor feedback gain

For this analysis, it is required that $\omega_m > 10$ rad/sec which is in accordance with handling quality requirements (MIL - F - 8785 B, 1969). In this case, the man - manipulator system can be excluded from the analysis.

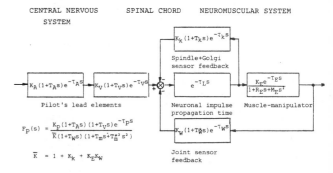

Fig. 2 The model of the human operator, from
        Bubb (1978)

The replacement of $F_p(s)$ in equation (2) by that of equation (7) and, setting in (2) the appropriate formula for the controlled element, results in an expression for the closed loop frequency response, the Bode diagram of which should remain within the limits drawn in figure 3 if "good" handling qualities are required. These limits have been extracted from the publication of Neal and Smith(1970) and Diederich(1980), which were concerned with fixed wing aircraft handling qualities.

Analysis for fixed wing aircraft.

For this analysis the naturally stable fixed wing aircraft (pitch axis short period) was selected:

$$\left[F_c(s)\right]_{op} = \frac{K_c K_\theta (1 + T_\theta s) \, e^{-\tau_c s}}{s \, (1 + T_c s + T_c^{*2} s^2)} = \frac{\theta}{F_s} \quad (8)$$

K_c = control stick force
K_θ = stick force to elevator deviation gain
T_θ = lift coefficient lead time const.
   = $1/T_\alpha$
T_c = $2\zeta_{nsp}/\omega_{nsp}$, pitch axis damping
T_c^{*2} = $1/\omega_{nsp}^2$, pitch axis resonance
θ = pitch angle
τ_c = system delay (preferably = 0)

The combination of eq.(7) and eq.(8) into eq. (2) has - with respect to eq. (4) - the result

$$\left[F_{CL}(s)\right]_{op} = \theta/\text{stick input} \quad (9)$$

$$= \frac{(1 + T_A s)(1 + T_V s)(1 + T_\theta s) e^{-\tau_e s}}{\overline{K}* \{1 + (2\zeta_k/\omega_k)s + s^2/\omega_k^2\}(1 + T_c s + T_c^{*2} s^2)}$$

$2\zeta_k/\omega_k$ = $(\overline{K}/\overline{K}*K_e) - T_e$

$1/\omega_k$ = $\overline{K}T_W/\overline{K}*K_e$

K_e = $K_p K_c K_\theta$

τ_e = $\tau_p + \tau_c$

It has been proved that the closed loop response of eq.(9) remains within the limits for good handling qualities drawn in figure 3 if the aircraft characteristics are defined according to MIL -F- 8785 B, and, if the pilot characteristics meet the following conditions:

T_A = T_e ⎫ (9a)
$T_V T_\theta$ = $1/\omega_{nsp}^2$ ⎰ lead equalization, (9b)

$$K*(s) = \frac{(1 + T_V s)(1 + T_\theta s)}{1 + (2\zeta_{nsp}/\omega_{nsp})s + s^2/\omega_{nsp}^2} \quad (9c)$$

$\overline{K}*$ = $C\left|\overline{K}*(s)\right|$, gain equalization. (9d)

One important result of the conditions given above is, that the adaptation of the characteristics of the pilot model is limited (Brauser,1981). These limits are also published elsewhere (Anderson 1971; Arnold 1973) and are compiled in table 1.
The equation (9) and its conditions (9a) through (9d) are representative for "compensatory tracking" tasks. In real cases the complex gain equalization K*(s), the Bode - diagram of which is shown in figure 4, is replaced by a constant gain $\overline{K}*$ without an average phase shift.

Large manoeuvre inputs of the pilot will alter the situation to a certain amount. While many authors handled this problem by open circuit analysis, it can be seen that actually the system remains closed, since the pilot has to keep his hands on the stick in manoeuvres!

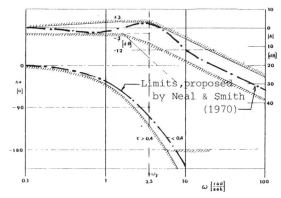

Fig. 3 : Limits for "good" handling qualities of the closed loop frequency response ($\sqrt{}$)

According to fig. 1 the feedback of the angle (θ) will be replaced by an angle-rate feedback ($\dot{\theta}$) during hard manoeuvres. This means that eq. (8) is multiplied by s, resulting in the following form of the controlled element:

$$\left[F_c(s)\right]_{op} = \frac{K_c K_\theta (1 + T_\theta s) \, e^{-\tau_c s}}{(1 + T_c s + T_c^{*2} s^2)} = \frac{\dot{\theta}}{F_s} \quad (10)$$

Closing the loop with eq. (7) yields

$$\left[F_{CL}(s)\right]_{op} = \dot{\theta}/\text{stick input} \quad (11)$$

$$= \frac{(1 + T_A s)(1 + T_V s)(1 + T_\theta s) e^{-\tau_e s}}{\overline{K}*\overline{K}_1^*(1 + \overline{K}/\overline{K}*\overline{K}_1^* K_e)(1 + T_W s)(1 + T_c s + T_c^{*2} s^2)}$$

which is nearly identical with the open loop formula $F_p(s) F_c(s)$. The adaptation conditions for the pilot are the same as for the solution (9), but there are some additional ones:

1. The term $\{1 + (2\zeta_k/\omega_k)s + s^2/\omega_k^2\}$ is replaced by the gain term of equation 11:
   $\overline{K}*\overline{K}_1^*(1 + \overline{K}/\overline{K}*\overline{K}_1^* K_e)$

2. The neuromuscular lag term $(1 + T_W s)$ is equalized by the complex gain

$$K_1^*(s) = \frac{1 - T_e s}{1 + T_W s}, \quad \overline{K}_1^* = C\left|K_1^*(s)\right| \quad (11a)$$

This adaptive gain factor may also be adjusted to be 1.0 , but contrarily to $\overline{K}*$, an average phase shift cannot be neglected, but which may be positive (figure 4).

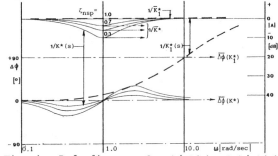

Fig. 4 : Bode diagrams for 1/K*(s) and 1/K$_1^*$(s) the adaptive gain factors for closed loop handling qualities

The additional gain $\overline{K}_1^*$ equals unity for the
case $T_W = T_e$, which means that the neuromuscular lag T_W is smaller than the delay τ_e.
The most interesting consequence, however,
is the fact that the closed loop response of
eq. 11 is almost identical with the open loop
response from theory. The difference may not
be detected during measurements.This fact
appears to be an explanation for the definition of measured data as "ōpen loop" data in
the literature.Calculated data show, that a
positive phase shift is the indicator for the
existence of the $\overline{K}_1^*$ - adaption of the pilot,
shown in figure 5.

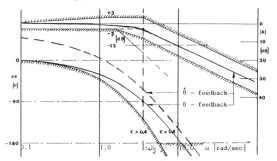

Fig. 5: Examples of the closed loop response
for θ- and $\dot{\theta}$- feedback.

Analysis of the "critical task".

The generality of the method described can be
demonstrated by the application to other controlled elements.
Since the hypothesis is proposed,that the human operator prefers to remain within the
adaptibility limits of table 1, the analysis
is applicated to the "critical task" (Jex et
al.,1966),the frequency transfer of which is
described by

$$\left[F_c(s)\right]_{op} = \frac{\lambda}{s - \lambda} = -\frac{1}{1 - s/\lambda} \qquad (12)$$

Here is λ the "operating point" characteristic. Division by λ inverts the sign of the
gain, which formally has the consequence that
the phase shift will start at $-180°$ for $s=j\omega$
near zero.(Landgraf & Schneider,1970).
The closed loop response is now

$$\left[F_{CL}(s)\right] = \frac{(1 + T_A s)(1 + T_V s)\,e^{-\tau_e s}}{\overline{K}_C^*(1 - \overline{K}/\overline{K}_C^* K_e)(1 + T_W s)(1 - s/\lambda)}$$

$$= \frac{(1 + T_e s)\,e^{-\tau_e s}}{\overline{K}_C^*(1 - \overline{K}/\overline{K}_C^* K_e)(1 - s/\lambda)} \qquad (13)$$

with the conditions

$$T_A = T_e = \tau_e/2 \qquad (13a)$$

$$T_V = T_W \qquad (13b)$$

$$K_C^*(s) = \frac{1 - T_e s}{1 - s/\lambda} \quad, \quad \overline{K}_C^* = C\left|K_C^*(s)\right| \qquad (13c)$$

This gain $K_C^*(s)$ which is described by its Bode diagram in figure 6, may be realized as

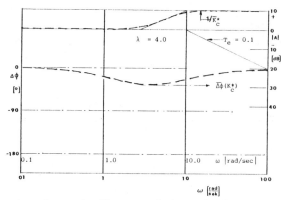

Fig. 6 : Bode diagram of the gain factor
$1/K_C^*(s)$

a constant average gain factor $1/\overline{K}_C^*$.Addition-
ally, a constant negative phase shift has to
be assumed, say $-45°$. The calculation for
$\lambda = 4$ results in a frequency response which
is comparable with measured data,published
by Stein (1975, figure 7).
The "handling qualities" are "poor" for this
closed loop. Phase shift is exceeding the
limits at low frequencies, the amplitude is
preferably above the 0 -dB line and does
not decrease with - 20 dB/decade for $\omega > \omega_g$.

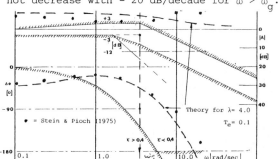

Fig. 7 : Bode diagram for the closed loop
with the controlled element
$-1/(1 - s/\lambda)$

Analysis of an instable "aircraft".

The method now is applicated, last not least,
to the "unstable aircraft" with one positive
real root:

$$\left[F_c(s)\right]_{op} = \frac{K_C}{s(s - \lambda)} = \frac{-K_C}{s(1 - s/\lambda)} \qquad (14)$$

The closed loop response is now described by

$$\left[F_{CL}(s)\right]_{op} = \qquad (15)$$

$$= \frac{(1 + T_A s)(1 + T_V s)e^{-\tau_e s}}{\overline{K}_C^*\{1 + (2\zeta_k/\omega_k)s + s^2/\omega_k^2\}(1 - s/\lambda)}$$

Again λ is the operating point of the system.
The factors ζ_k and ω_k are the same as appear
in the solution eq.(9)!Does this mean,that
the human operator prefers to activate a 2d
order resonance system when engaged in complicated manual tasks?
The conditions for the solution (15) are now

the following

$$T_A \quad = \quad T_e \quad = \quad \tau_e/2 \qquad (15a)$$

$$|T_V| \quad = \quad |1/\lambda| \qquad (15b)$$

$$K_C^{**}(s) \quad = \quad \frac{1 + T_V s}{1 - s/\lambda} \qquad (15c)$$

$$\overline{K}_C^{**} \quad = \quad C|K_C^{**}(s)| \qquad C \leq 1.0 \qquad (15d)$$

The human operator tries to compensate $1/\lambda$ by the lead term T_V and by the gain $\overline{K}_C^{**}$. But, it has to be mentioned that this solution is true only for the substitution

$$1/(s - \lambda) = +1/(1 - s/\lambda) \quad \text{for } \Delta\phi = -180°$$
$$\text{at } j\omega = 0$$

Since $1/K_C^{**}(s)$ also has a phase shift of $\Delta\phi = -90°$ at $j\omega = 0$, the phase shift of the closed loop response will start at ca. $-270°$. Figure 8 shows the Bode diagram of $K_C^{**}(s)$, and that of the solution (15) is shown in figure 9 for the value $\lambda = 1.0$, together with values measured by McRuer et al (1968).Pilot ratings published by McDonell (1968) were 8.5 of the Cooper scale, which means nearly uncontrollable,for the same system.

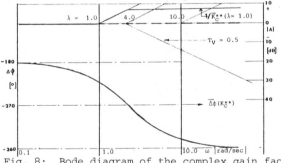

Fig. 8: Bode diagram of the complex gain factor $1/K_C^{**}(s)$

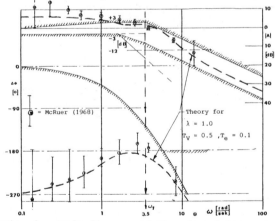

Fig. 9: Bode diagram of the closed loop solution equation (15).

The "handling qualities"of this system turn out to be extremely poor.Indication for this rating predicted, is the extreme negative phase shift exceeding the limits shown in fig. 9 as applied and extrapolated from the stable aircraft (fig. 3).
The result of the analysis for both the unstable controlled elements is,that $1/s(s - \lambda)$ is the pursuit tracking case while $1/(s - \lambda)$

represents the compensatory tracking case, when compared with the stable aircraft analysis.

CONCLUSIONS

The method of the closed loop analysis in combination with the parametrically specified operator model appears to have become a relatively simple tool in order to derive handling qualities criteria by means of the response limits of figure 3 gained from well established aircraft data.
The pilot adaptation limits are derived from these response requirements and have been proved to be in agreement with published data.
The analysis was based on the following hypotheses:

1. The human operator always closes the circuit during active control.
2. He always tries to reduce the order of the closed loop to the lowest possible, by means of lead compensation for lag elements.
3. In "pursuit tracking" tasks the operator tries to stabilize the system by means of gain in the form $\overline{K}^{*}(1 \pm \overline{K}/\overline{K}^{*}K_e)$
4. In "compensatory tracking" tasks the operator prefers to stabilize the system by means of lead and the 2d order control behaviour expressed by the term $\{1 + (2\zeta_k/\omega_k)s + s^2/\omega_k^2\}$
5. The closed loop "bandpass filter" response has a constant breakoff frequency $\omega_g = 3,5$ rad/second which is independent of the resonances of the controlled element.
6. The human operator tries to adapt his characteristics to stationary operating point characteristics of the controlled element. Rapid changes of these increase the "remnant".

Table 1 : Limits of adaptibility of the human operator for good handling qualities.

characteristic	symbol	adaptability limits
lead time constant	T_A ; T_V	T_A, T_V $T_A + T_V$ 2 - 3 sec
order of lead term (pilot only)	$i_p \leq i$	i_p ≤ 2
delay time	τ_p	$0.05 \leq \tau_p \leq 0.4$ sec
man - manipulator resonance	ω_m	$\omega_m \geq 3.5\omega_{nsp}$ $6 \leq \omega_m \leq 15$ rad/sec
man - manipulator damping	ζ_m	$1.0 \leq \zeta_m \leq 1.5$
operator control frequency	ω_k	$2 \leq \omega_k \leq 6$ rad/sec
operator control damping	ζ_k	$0.4 \leq \zeta_k \leq 1.5$ (ca.)
bandpass filter breakoff frequency	ω_g	3 - 4 rad/sec (set to 3.5 rad/sec)
neuromuscular lag time constant	T_W	$0.05 \leq T_W \leq 0.2$ sec

CONSEQUENCES

Handling qualities criteria for the manual
control of a system have been derived from
the single axis closed loop analysis which
has been demonstrated for 3 different controlled elements. The method has been kept relatively simple for the application in early
development phases of a system. Controlled
elements of higher order lag elements have
been treated effectively (Brauser, 1981)e.g.
aircraft system of 3d and 5th order lag,pub-
lished by R.E.Smith (1978)
The method explained has to be extended to
multiaxis problems, and adapted to higher
sophisticated methods. "Fraction of attention"
(Johannsen, 1978) or "control urgency" (Onstott & Faulkner,1978) may be leading signals for an effective extension.

REFERENCES

Anonymous (1969). Military Specification:
    Flying Qualities for Piloted Airplanes.
    MIL - F - 8785 B (ASG)
Anderson, R.O.(1971). Theoretical Pilot Ra-
    ting Predictions; AGARD - CP -106.
Arnold,J.D.(1973). An Improved Method of Pre-
    dicting Aircraft Longitudinal Handling
    Qualities Based on the Mimimum Pilot Ra-
    ting Concept ; Wright - Patterson AFB,
    MSc. Thesis CGC/MA/73 - 1 .
Brauser, K.J.(1980). Untersuchung über ein dy-
    namisches Modell des Zusammenwirkens von
    Pilot und Bedienelement mit Anwendung auf
    die Steuerbarkeit ; MBB München, TN - FE
    301 - 6/80
Brauser, K.J. (1981). Die Ableitung von Steu-
    erbarkeitskriterien für Kampfflugzeuge
    aus einem quasilinearen Präzisionsmodell
    des Menschen mit Hilfe der Filtertheorie;
    MBB München, FE 301/S/STY/33
Brauser, K.J. (1981). Theoretical Linear App-
    roach to the Combined Man - Manipulator
    System in Manual Control of an Aircraft;
    Rep. on the 17th Annual Conference on
    Manual Control,UCLA;Los Angeles, June 16
    to 18,1981.
Bubb, P.(1978). Untersuchung über den Einfluß
    stochastischer Rollschwingungen auf die
    Steuerleistung des Menschen bei Regel -
    strecken unterschiedlichen Ordnungsgra-
    des. Dissertation TU München,1978.
Diederich,L.(1980). Gütekriterien für die
    Anstellwinkelschwingung. MBB München, TN-
    FE 132 - 1 / 80

Hess, R.A. (1980). Structural Model of the
    Adaptive Human Pilot. J.of Guidance and
    Control 3,No. 5, 416.
Hess, R.A. (1978). Dual Loop Model of the Hu-
    Controller. J. of Guidance and Control 1,
    No. 4,254.
Hess, R.A.(1981). A Pilot Modeling Technique
    for Handling Qualities Research. From
    personal Communication to the author.
Jex, H.R., McDonnell,J.D.;Phatak,A.V. (1966).
    A Critical Tracking Task for Manual Con-
    trol Research. IEEE Trans. HFE -7, No. 4
    1966.
Johannsen,G.;Boller H.E.;Donges E.; Stein W.
    (1976).Lineare Modelle für den Menschen
    als Regler. Forschungsinstitut für Anthro-
    potechnik, Meckenheim. FAT-Bericht No. 24
Johannsen,G.(1978). Auslegung von Flugfüh -
    rungsanzeigen mit Hilfe des optimaltheo-
    retischen Modells für den Menschen als
    Regler. Deutscher Luft- und Raumfahrt-
    kongreß, Darmstadt 1978, Vortrag Nr.78-
    149.
Kleinman,D.L.;Baron,S.(1971). Manned Vehicle
    System Analysis by Means of Modern Control
    Theory. NASA - CR - 1753 (1971)
Magdaleno,R.E.;McRuer,D.T.(1971).Experimental
    Validation and Analytical Elaboration for
    Models of the Pilot's Neuromuscular Sub-
    system in Tracking Tasks. NASA CR -1757
    (1971)
McDonnell, J.D.(1968). Pilot Rating Techni -
    ques for the Estimation and Evaluation of
    Handling Qualities. AFFDL - TR - 68-76
McRuer, D.T., Krendel, E.S.(1974). Mathemati-
    cal Models of Human Pilot Behaviour.
    AGARDOgraph AG - 188, 1974.
Neal,P.T.;Smith, R.E.(1970). An Inflight In-
    vestigation to Develop Control System De-
    sign Criteria for Fighter Airplanes.
    AFFDL - 70 - 74.
Onstott, E.D.;Faulkner, W.H.(1978).Prediction,
    Evaluation, and Specification of Closed
    Loop and Multiaxis Flying Qualities.
    AFFDL - TR - 78- 3.
Smith, R.H.(1975). A Theory for Handling Qua-
    lities with Application to MIL -F-8785 B.
    AFFDL - TR -75 -119.
Smith, R.E.(1978). Effects of Control Systems
    Dynamics on Fighter Approach and Landing
    Longitudinal Flying Qualities,Vol. I.
    AFFDL - TR - 78 - 122.
Stein,W.;Pioch,E.(1975)Zur Anwendung   von
    Regelaufgaben bei ergonomischen Untersu-
    chungen . Forschungsinstitut für Anthro-
    potechnik, Meckenheim, FAT - Bericht Nr.
    23,1975.
McRuer,D.T. et al.(1968) New Approaches to
    Human Pilot/ Vehicle Dynamic Analysis.
    AFFDL - TR - 67 - 150

DISCUSSION OF SESSION 1

1.1 S PERSPECTIVES ON HUMAN PERFORMANCE MODELLING

Martin: In your presentation I miss the concept known as Action Theory. This is being developed, e.g., by VOLPERT, West Berlin, and ULICH, Zürich (references given in paper 3.5 I). We apply this successfully in designing improved manufacturing systems. The concept tries to also include such complex items as mental stress, motivation, and operator satisfaction. Social effects are missing in your paper completely.

Baron: You are correct in that we did not discuss Action Theory and for this the reader is referred to your references. Also, social effects are not discussed and/or included in a predictive manner in this model. It is possible to adjust some model parameters to reflect certain social interactions - e.g. the level of authority or the reliability of the source could be allowed to influence communication of information or action based on that information. Mental stress due to excessive workload is an aspect of the model In fact, such workload (time stress) is a prediction or output of the model and will lead to operator "errors". Finally, the modelling approach generally assumes that operators are well-motivated and trained; alternative assumptions may be examined by changing subjective goal structures and values as well as other operator related model parameters (e.g., operator noise levels).

1.2 T VISUAL SCAN AND PERFORMANCE INDICATORS OF MAN'S WORKLOAD

Henning: The method presented includes only the static workload. The transinformation measurement does not describe the dynamic system behavior. Extended information theory methods should be used, that include the dynamic system behavior. Similar investigation at the RWTH Aachen have shown, that especially the presentation of stability of the certain controlled axis is an important criteria for workload. Sometimes even the dynamic workload leads to more stress than the static workload.

Kimmel: The original purpose of the study was an investigation in how far the transinformation measure is a useful performance index for the discrimination between different display configurations in multi-loop control tasks. So, we were not concerned with aspects of dynamics. Some findings in conjunction with the insertion of a side-task showed that there should also be workload effects reflected by the transinformation measurement. The study of eye-movement sequence lengths turned out to be a useful method to measure workload, confirming our suggestion that our (static) information measure could also show such influences. This led us to the formation of a workload index which combined the measures. Your findings concerning extended information measures as a useful tool of investigating dynamic workload supports our findings and we agree with you that we should include this method in our future experiments.

1.4 T CONSTRUCTION AND APPLICATION OF A COMBINED NETWORK AND PRODUCTION SYSTEMS MODEL OF PILOT BEHAVIOR ON AN ILS-APPROACH

Mancini: a) Why have you chosen a normal distribution for task duration?

b) Have you considered in your simulation that some tasks may have been omitted or wrongly executed and recovered later on?

Döring: a) In our approach for verifying the model we used a normal distribution for describing pilot task performance times because the mean values and standard deviations were already available to us from the literature and from our measurements. Since then, we have learned, e.g. from G.L. Berry (1981; The Weibull Distribution as a Human Performance Descriptor; IEEE Transaction on Systems, Man and Cybernetics, SMC-11, no. 7, 501-504) that the Weibull distribution is a better fit to refine the model with this special type of distribution.

b) Since the objective of our efforts primarily was the development of a simulation supported analytic approach rather than the analysis of effects of incorrect pilot performance on system performance we used a normative approach because of simplicity.

Because of the normative point of view taken for establishing pilot tasks, only ILS-approaches that were completed without any failures were considered. In a real system development situation we, of course, would have considered the likelihood and consequences of pilot errors.

1.5 T A FUZZY MODEL OF DRIVER BEHAVIOR: COMPUTER SIMULATION AND EXPERIMENTAL RESULTS

Ekberg: I agree on the concept of differentiation between fuzzy set and probability but I strongly disagree with equalizing fuzziness with uncertainty! E.g.: in a pool you might have 100 whales. You say that a whale is 0.7 a fish. That does not mean that there will be 70 fishes in the pool as probability theory would say. It does not mean that you are 0.7 uncertain that a given animal in the pool is a fish. Therefore I do think that you are mistreating the definition of fuzzy set theory.

Kramer: Both cases outlined by you can be characterized by uncertainty, because the decision whether a given animal is a fish or a whale, cannot uniquely be predicted. In the first case the cause of the uncertainty is the incomplete pre-information about the occurancy of the event 'a given animal is a fish'. In the other case the cause of uncertainty is the incomplete information about the list of features which enables someone to classify a given animal as a fish or as a whale. In my opinion you are in both cases 0.7 uncertain before the decision has to be made. The difference between both models is that in the first case the a-posterion uncertainty becomes 1.0 (or 0.0), while it remains just the same value in the latter case.

1.6 I THE AIR TRAFFIC CONTROLLER'S PICTURE AS AN EXAMPLE OF A MENTAL MODEL

Baron: Do you infer two types of "images" of the situation corresponding to the two types of information presentation (scripts and radar)?

Whitfield: I think it is premature to talk of "images", as we have no real indication of the nature of the "picture". However, I suggest that our data show that the two types of information are merged together, though perhaps yielding varied types of picture with different controllers. The two types are different in their level of detail, and in their time span, and make complementary contributions to the picture. However, as reported, there does seem to be more weight attached to the flight strip information.

Bainbridge: In process control studies identifying the "picture" from verbal protocols, it is necessary to infer the cognitive processes from which the statements could have been generated, and infer from these to the stored data used. I would suggest that it will be difficult to identify the picture directly from the verbal statements.

Whitfield: I agree that such extended analysis is necessary. The present results can be regarded as a starting point.

Lenior: Do you have any clarification for the (commonly found) great individual differences in strategy? For instance: is there a relation with the training received by the operators?

Whitfield: We do not have enough additional data available, to draw such conclusions. It is sometimes suggested that older controllers, who received their initial training in a largely "procedural" system (essentially based on flight strips) will rely on strip information now, much more than their younger colleagues who were trained ab initio in a radar environment. There are doubtless more influences involved: in our study, one such young controller hardly referred to the radar display at all!

Brauser: Did you find a reference or correlation of technical aids given to the controller, to his capability to build up the picture of the control situation?

Whitfield: Our studies have been conducted only in the context of the present U.K. A.T.C. system: processed radar display including computer-generated labels (tags) for the aircraft, and computer-printed flight strips. Thus, we have no comparisons between the effects of different technical aids. However, considering controllers' comments about limits on their pictures in a simulated predictive aiding scheme (the original stimulus for this research), it is possible that such aids might not have much effect on the picture.

Tainsh: Do you consider your work on verbal protocols will help you identify the need for decision aids, and if so do you imagine these might involve expert systems?

Whitfield: This work is more likely to help make design decisions concerning display formats rather than decision aids. I do not think the need for expert systems can be inferred from verbal protocol analysis alone.

Brooke: Is the differentiation between different controllers' "pictures" very great, given that the "picture" is apparently handed over adequately at changeover?

Whitfield: "Handover" is basically a one way passive movement of information with the new controller placing his own interpretation on the data.

1.7 I THE DERIVATION OF HANDLING QUALITIES
      CRITERIA FROM PRECISION PILOT MODEL
      CHARACTERISTICS

Henning: The compensation of deadtime of the controlled element has been investigated 8 years ago (Henning, 1973; Kuchen, 1974). The result is an approximation of a Smith Predictor, that is the approximation of a positive feedback of the controlled element deadtime as an internal model of the controller.

Brauser: I like to learn, that there are measurements which seem to be background to my hypothetical findings!

Vogt: Would you please comment on how you obtained the resonance frequency of the arm-manipulator system.

Brauser: In an additional study published in 1979 and 1981 ("Theoretical linear approach to the combined Man-Manipulator System..."). Please refer to "REFERENCES".

ON THE MODELLING OF THE HUMAN PROCESS OPERATOR

L. Norros, J. Ranta and B. Wahlström

*Technical Research Centre of Finland, Electrical Engineering Laboratory, Otakaari 5 I,
02150 Espoo, Finland*

Abstract. The human operator in the control room of a complicated process
plant plays a crucial role influencing the safety, availability and economy
of the plant. An understanding of human behaviour and limitations is
therefore essential both for the assessment of the plant safety and for the
design of an optimal man-machine interface. In the paper different models
of the human operator and their applicability to process supervision are
discussed.

A qualitative framework for operator models of a complicated process such
as a nuclear power plant is discussed in more detail. The implications
of the model in the control room design, the design process itself and the
training of operators are indicated. Finally some general needs for new
research are discussed

Keywords. Man-machine systems, human factors, automation systems.

INTRODUCTION

The high level of automation, which is common
in complex process plants, has brought up the
situation in which the process operators have
to make decisions on a rather abstract level
using conceptions concerning plant safety,
availability and economy. The fast changing
technology has raised new questions and
problems in structuring and presentation of
process data and in aiding the process operator during difficult plant transients. It has
been realized in practice that the human
operator in the control room of a complicated
process plant plays a central role influencing
plant safety, economy and reliability.

The recent incidents in the process plants,
like TMI-2 incident, have focused attention
to the role of process operator. The investigations of the incidents have shown (IEEE
1979 a, b) that the human errors committed
both inside and outside the control room have
had a central role during the incident. It
was also shown that those human errors were
caused by deficiencies in the control room
design, operator training, control room
procedures and management procedures. Those
problems have been pointed out also earlier
(Seminara, 1976) and the criticism could be
intepreted in such a way that a major misfit
exists between the demands placed on and
the resources provided by the human operator.

The development of technology has offered
increasing flexibility and power for computing

devices, which allow new ways of
preprocessing, structuring and ordering
the basic information and changing data, e.g.
by using computer graphics and interactive
display design. This gives more freedom and
new possibilities to a system designer, but
there are also new risks of misfits and
unfeasible solutions in man-machine interface
design.

To reach a balanced man-machine interface
we must be able to evaluate different design
alternatives. A necessary precondition for
a successful design evaluation is an understanding of human behaviour both with respect
to abilities and limitations. The limiting
factors of the mental resources of a human
operator are then particularly important for
avoidane of demand-resource conflict. This
fact is of increasing importance, because the
new interactive systems emphasize the human
information processing activities and problem
solving strategies as a part of a man-machine
communication.

The above mentioned facts require a study and
model of human performance in the control
room environment. The modelling and understanding of human needs and behaviour allows
new design criteria for man-machine interface
design, which are the guiding factors of the
practical design process. The need for
theoretical consideration of man-machine
interface has also been elaborated elsewhere
(NKA/KRU 1981, Rasmussen 1980, Rasmussen and
Rouse 1981).

ADE-C*

ON THE MODELLING METHODOLOGIES

The motivation for constructing models of the human operator is the same as for modelling technical, sociological, biological etc. systems. The model developed could serve as an aid for thought, communication, training and instruction, prediction, experimentation etc. A model is an efficient way for structuring available information and formalizing loose hypotheses. The use of the models for prediction is often the main goal of the modelling work and the predictive power of the model is used as a goodness measure. If the model has a good predictive power it could be used to support or even substitute experiments.

The main difficulty in modelling an activity of human operator is his versatility and variability, which makes it extremely difficult to find any general explanations for his behaviour. This means that the different roles, functions and capabilities of the human operator have to be modelled separately. The different models will thus reflect the different needs, and the use of the model will be restricted to the situation studied.

The mathematical models are a class of the quantitave models, and they have also been applied to man-machine studies. Typical examples are models based on queing theory, optimal control, decision theory and fuzzy systems theory. Those models include a hypothesis on the human behaviour and can be used for the testing of different behavioral hypotheses (Rouse 1977, Rouse 1980, Timonen 1980, Timonen et. al 1980).

Instrument monitoring is a typical but isolated task in control room environment, which has been studied with mathematical models. As the operator must scan several instruments, and as we can assume that the human operator concentrates his attention at one item a time, he allocates resources between them and also prioritizes them. The concept of time sharing between different channels is obvious and the queuing theoretic approach is a natural way to build up the model of an isolated monitoring task. When failure detection and process state estimation are essential estimation theory has been applied. Also optimal control models and decision theoretic models have been presented to describe resource allocation between different tasks.

Another typical problem, which has been treated with mathematical models, is to describe the human operator as a part of a control loop. The models are used to estimate possible instability regions of the system. Usually linear models with a time delay have been accurate enough for the design purposes. The basic model can, of course, be extended to take into account also goal settings, weighting aspects, trajectory tracking and prewiewing aspects. This problem statement leads to different optimal control models. The man in the loop problem can also be found in the control room environment. However, usually it is not an isolated task, but a part of diagnostic and prognostic activities.

The main applications of the mathematical models are in the aerospace area. Usually, the modelled tasks have been well defined and structured so that the criterion for the human behaviour has been easy to find. Those models have also been tested and validated in simplified laboratory or experimental environment. The isolation of well defined tasks is difficult in a real process control room environment, where problem solving and group dynamic plays an essential role in human behaviour.

Yet, it seems that problem solving and strategy finding activities are difficult to put into a rigorous mathematical model. Of course, different decision theoretic and fuzzy system concepts can serve as thinking aid and help to test different hypotheses, but it is apparent that a stronger framework and qualitative descriptions are still needed. This is important, because the problem of diagnosis and prognosis, as discussed later on, are critical activities in control room environment.

There are also other problems related to the use of mathematical models (Bainbridge 1981, Timonen 1980). For instance, if we like to study human information processing activities the input-output similarity equivalence can be good between the behaviour of the human operator and our mathematical model, but the hypothesis and theories behind the model do not necessary reflect the internal problem solving strategies of human operator.

The qualitative models can offer a very strong conceptual frame-work to study man-machine interactions. Also the problem solving elements can be included in the model in such a way that the model can be used as a base for developing design criteria of a process control room.

Qualitative models are usually good aids in deciding the system concept, in designing dialogue control principles etc. Moreover, before we can build a good mathematical model with a good predictive power, we need conceptual framework and qualitative description of human behaviour with theories assumption behind the model.

Qualitative models are built up with help of cognitive psychology and usually they are based on theories about human problem solving and strategy finding activities. Practical experiences have been gathered from verbal protocols and real control-room (or simulator) experiments. Again interpretation difficulties rise with experimental results and with the equivalence between model assumptions and the output of verbal protocols (Bainbridge 1979).

SPECIAL FEACTURES OF MONITORING
AND SUPERVISING OF COMPLICATED
PROCESS PLANTS. - A FRAMEWORK
FOR MODELLING

Operator tasks and task demands

In designing the process information system
we have to remember that this system
is basically a tool for the
personal in the plant to supervise, control
and manage the process. Starting from this
point, the problem can be seen to have three
main aspects, as Fig 1 presents. First we
have the technical production process.
Basically the nature of the process sets the
main goals and conditions of plant operation
as related to safety, availability, economy,
and often puts also serious requirements on
operation principles. These goals and
principles are reflected, e.g. in the process
instrumentation and automation system. On
the other hand process instrumentation and
automation is a tool to realize those principles and also a tool to aid the human
operator in his task. Second we have the
human operator, who supervises and controls
the process with the help of the instrumentation and automation. The operator has
experience, training and thus his own view
about the process and its operation principles.
The operator has an ability to solve
problems and formulate strategies and his
functioning is influenced and modified by
psychophysiological factors. Third we have
the social context, where the plant is
operating. Each society has its own laws,
regulations, authorities, cultural background,
habits, attitudes, which all are influencing
at least indirectly the functioning of the
global system human operator - automation -
process.

Thus it is essential to see the role of
process operator in two ways. First we must
realize that the human operator is acting as
a system component and has a crucial role in
the overall system. From the point of view
of the overall goals it is essential that the
human operator function as they have been
specified during the design stage can be
performed accurately during the process operation. Second the control room is a work
environment for the operator. To reach the
overall goals of the process operation the
control room must provide an appropriate
environment for the human operator in giving
him necessary support and possibilities to
develop himself.

We can summarize the problems of supervising
and control of a complex process plant, like
a nuclear power plant, as follows,

- complexity, dynamic interactions, feedback
  paths in the process play an essential role
  in the understanding and recognizing of
  the process properties,

- slow responding dynamic, or long time
  constants are usual and cause difficulties;
  on the other hand fast responses are also
  some times required,

- rare events and multiple failure situations
  are difficult to supervise, because plant
  variables develop in an unpredicted way,

- low activity level during normal operations
  although operators highly trained into tasks
  of high intellectual qualities,

- manual take-over supposed, although little
  opportunities is given to train manual
  control.

To begin with it is worthful still to emphasize that which we can call critical demands
of activity. That is important, because during
critical demands drawbacks and misfits in man-
machine interface design become overt and are
emphasized; individual differences in operator
activities and training become apparent. The
activity in a normal situation and in a rare
event cannot be found very different. Thus
the concrete operator activities that carry
out these critical demands are not isolated
actions or cognitive functions, but rather
an integrated cognitive motivational whole.
Thus it is apparent that a well-balanced
control room in normal situations is a
precondition also for successful operation
during critical task demands.

The basic contradictions in the activity
demands arise from the nature of the production process and the required tasks:
most of the time the process calls for low
activity tasks or mere supervising, but
occasionally sudden active operations of
high intellectual qualities are demanded.
Because of a high grade of automation the
change in demands might be very dramatic,
and the risks in case of an unsuccessful
change are very high; e.g. in nuclear power
plant control. The contradictions at the
action level that originate from this objective task feature are of motivational and
cognitive character: In the long run it is
very difficult to be motivated in an under-
demanding task, or at least the activity
must be motivated through task extern goals.
Because of the low variance in the task
conditions, the formation of the skills
related to critical demands is hindered
during the work.

From the actual events the experienced
operator creates in his mind candidate
chains of events and tries to relate the
actual event to a familiar and experienced
event. Thus the increasing skills increases
the selfsconfidence and then it might be
difficult to begin to analyze and diagnose
the actual event as a new one, although
there would be a need for that.

A second task characteristic which makes
the activity internally contradictory is
the rising role of abstract information

as a basis of control actions. The cognitive motivational structure of a working activity is traditionally formed in working situations where the material effects of the decisions are close, immediate and affective. To be able to control the modern processes one needs abstractions that contradict this perceptual and concrete closeness.

This also means that in a conventional working process the orientation basis is formed through empirical generalizations on the basis of the perceptual information. In the control of modern processes this is not so apparent, as the orientation basis and the internal process model of the operator are to be formed through theoretical and conceptual means, and comprice both actual information and memory representations.

The role of orientation is perhaps clearest in the diagnostic demands, i.e. during plant state detection, hypothesis forming and testing. It is also evident that predictive abilities, i.e. evaluation of actions, planning and plan evaluation are important especially at the decision making and activity choosing stage. A particular problem in a predictive activity is that the operator has to be willing to put his efforts in such an activity, although the predictions can bery seldom be verified against the actual, real results.

Together with the diagnostic demands we can define the predictive demands as the critical operator demands.

As already stated above the critical demands can not be supported if the control room is not well-balanced during normal situations.

An important aspect, which must be added in that respect, is that the operators must, all the time, relate their decision to the overall goals of plant operation. The goals of the whole plant are generally defined in qualitative terms using different sub-goals.

In the sub-goals there is a priority which means that the plant safety is the primary concern and the plant economics should be optimized only when neither safety nor availability are endangered. The sub-goals could be used to derive more accurate conditions and objectives for the plant operation. The conditions could be given, e.g. as

- allowed plant variable limits,
- conditions on plant status,
- operational restrictions.

The required conditions could be assured by the man-machine system through different operational principles and through a proper task division between man and machine, i.e. through a proper task allocation between manual, semi-manual, and automatic operation. Thus we can state that the automation and information system is introduced to realize a proper task division between man and machine and also to aid human operators during

different operational task. From this starting point we must evaluate the task division with the overall goals of the plant.

Considering the multitudes of systems, tasks and situations, which the operators should manage, it is clear that no single model, description nor theory would cover them all. We need a larger framework, which should integrate different views and descriptions into a global theory and model. The construction of such a framework has been tackled in the Nordic Cooperation project (NKA/KRU 1981) and will shortly be described below.

Describing the operator in process supervision

The diagnosis of events, prediction of consequences and planning of activities include typically complex demands, which are not easily covered by models. However, to understand how these demands transfer into concrete activities, a concept and description of the control structure of the diagnosis, prognosis and planning activities are needed.

The activity has the character of fulfilling tasks, which is related to the nature of the working process in question. This means that the activity is characterized through goals that are organized in a hierarchical way, forming the hierarchical control structure of the activity. The hierarchy of a particular activity is not pre-defined. Because goals can be reached through different means and the same means can serve different goals, there exists degrees of freedom that are used in different ways depending on the individual and the situation. However, the fundamental role of goals holds for every activity. Fulfilling the goals and sub-goals can be described as a sequence of (conscious or unconscious) mental operations, which form the sequential structure of the activity.

The components which are defined in many information processing models are principally in agreement. On a rather general level we could say that the activity is comprised of goal formation, orientation, planning, decision and control. The orientation function is often analyzed in more detail, and several sub-components or functions are being formulated.

The hierarchical and sequential structure are integrated through a set of internal models, which are different kinds of memory representations. The models can be differentiated through the contents. They also include programs of varying levels, and they develop and their content can change, e.g. according to training and experience.

As mentioned the internal models or memory representations serve in the regulating of activity. Only by postulating the internal models can the fluency and flexibility of human activity be explained. It is usual to separate three levels of regulation, that are from the higher to the lower, intellectual, perceptive-conceptual, and senso-motor, or knowledge based, rule based and skill based, according to Rasmussen (1980).

Within the frame of Nordic projects, an operator activity model is developed that reflects the general principles of working activity just mentioned. This model has been adapted to the conditions of operator activities and serves as a common framework for human reliability control room design and training. A description of this model can be seen in Fig. 2. It conceptualizes the sequential structure of the operator activity in differentiating eight functions and corresponding states of knowledge. The regulative aspect to the activity is taken into account by postulating the possible "leaps" or "short cuts" in the complete structure. The highest level is knowledge based activity and it is represented in the model as going through the complete sequence of functions. Should the activity be reduced through associative leaps from the identification or observation function to the target state or other lower states of functions it can be defined to be rule-based. If the activating phase is immediately followed by executing pre-learned senso-motor patterns, the activity is defined as skill-based, for more details, see (NKA/KRU 1981).

As stated above the model has been proposed to serve as a framework for human reliability, control room design and training studies (NKA/KRU 1981, Rasmussen 1980). However, there still exist problems and the concept above must still be extended. We try to sketch a concept for the future elaborations of the model.

Concerning the model itself we could state that by emphasizing the sequential aspect we loose easily the dynamics of the activity carried out through the goal structure and represented in the internal models. Although an activity is actually controlled mainly by processes of a particular regulation level, it always includes also features from the lower levels crystallized in some concepts, and is also a dependent part of a more general process that is regulated through higher level processes. This is the only way you could understand the ability to exchange the actual regulating level to another when the conditions are changing. Following the frame of the Rasmussen model we had to postulate another additional principle or mechanism for the changing of the behavioral level.

There are many factors, which influence the exchange of the regulation level, like emotional factors: stress, boredom, etc., interest and attitudes of operators, motivational aspects etc. The change of regulating level is still an open question and it is also important from the design point of view. Moreover, what is the role of experience, and that of the theoretical knowledge, and how can the latter regulate operational activities.

A particular technological system is created by man, and there exists therefore the principle possibility of knowing its states and functions. But when the system is getting more complex, failures of different components become also complex; the unambigious definition of the system normal vs. abnormal state is not always possible. Thus the diagnosis of a technical system includes the ability to hadle uncertainty and fuzzy relations and probabilistic aspects of different failure mechanisms.

Experimental investigations of man's reasoning seem to suggest that he mainly relies on deterministic and linear hypothesis. The ability to handle uncertainty is particularly difficult to acquire even through extra learning. There are certainly different causes for that, but one is perhaps the most important: probability and fuzzines are not evident, they have to be inferred.

These examples are only suggestive. Their main message, however, should apply generally: The essential way in improving the diagnosis is to enhance the conceptual and theoretical level of the process knowledge. This, as such, is a research problem: How to make the necessary theoretical knowledge and conceptual information operative, that is, regulative?

In an optimal situation a process operator is working in a predictive manner. Problems arise when the complexity of the situation increases. If the understanding of the interactions of the process is not sufficient no basis exists for predictions, and the operator has to switch only to a feedback and reactive strategy. The major differences of the strategies concern the extension of the goals: in feedback strategies you can operate on the basis of rather isolated nearby goals, open loop and predictive mode is controlled through distant goals and goal hierarchies. Also the signals used as indicators of necessary actions differ: in feedback strategy the signals are limited and actual, in feedforward elaborate and more or less "warning" signals which are forecasting possible future events.

It is not the question which one of the working methods is better in most of the cases but rather how to keep the operator also on the predictive mode. It seems that operator aids could be quite helpful, like a predictive display system. The computer can operate on the basis of the interactions and count the values of relevant parameters, but in order to use this knowledge you still have to understand the interactions. Again we end up with the problem of what is the operatively adequate process knowledge and how it is, and should be represented in the mind of the operator; what is the complete orientation basis? And what are the implication on the automation system design?

A very crucial difficulty in predictive activities is motivational in character. A predictive working method presupposes an active mode in every activity; from active seeking of information to eagerness in making efforts in thinking. This aspect of the predictive demands should deserve much more attention in the future investigations. The solutions are presumably to be found of the organisational level: There have to be enough degrees of freedom in order to keep the operator's interest awake and his competence available.

TOWARDS A NEW CONTROL ROOM

The task given to the automation system in the control rooms in the near future will probably be quite similar to the present one with some additions. In addition, to those standard facilities different new operational aids could be designed to support the operator in his work.

The tasks of the operator will then be found in the following areas:

- monitor the process and the automation system,
- operate during non-automated state transitions,
- diagnose and operate in situations where the process or the automation system deadlocks in a situation not foreseen,
- serve as a back up for the situations where the automation system does not function properly
- repair and maintain the automation system,
- communicate with the environment,
- evaluate and optimize used operational practices and procedures.

The discussion in the previous sections has pointed out the fact that a good interface system is context- and taskdependent. This requires special flexibility of the interface system.

The flexibility means that the interface system has to aid the operator in performing different tasks at the different levels of abstractions, which means the data needed to present are different.

Supporting the formation of operation skills requires clear structuring of the basic process data, and on the other hand easy operativity. Maintenance personel and the plant management should have their own information center outside the main control room. Also the information presented in the main control room must be pre-prosessed as far as possible. This is especially important when the alarm system designed; different filtering, supression, routing and prioritizing, must be used. Also the display system must be hierarchically organized and flexible in changing and choosing the picture formats, and special attention has to be paid to dialogue control system (Ranta,et.al,1982).

The rule-based behaviour could be supported by a special aid system for procedure execution. A compute-based process state identification aid can be used to present different alarming sequences and to present how disturbances are proceeding. A dynamic real-time interrogation system can be realized by the computer, where precondition, interlockings, states of important components, and procedural aid, and interrogation and help systems have to be realized to protect the operators from memory slips and to ensure a rapid information retrieval. It is also apparent that the realization of an advanced procedural aid requires utilization of design data, which sets special demands on the design process.

As pointed out in the discussion of the previous section, the activities during rare events are critical and these activities are based on the knowledge of the process operation. This emphasizes possibilities to test diagnosing hypothesis and to validate strategy choices for predictive purposes. In a real plant complex interactions means that effective diagnosis, prognosis and hypothesis testing are not possible without advanced operation aids. Thus we can conclude that the knowledge-based behaviour requires information about function principles of process and automation system. A special design data base is needed so that a special interrogation and information retrieval system can be realized. The design data bases are needed also in the realization of different operation aids, disturbance analysis system, hypothesis testing system, fast mode simulation and scenario generation system.

A difficult problem to solve still exists. As discussed in the previous section, it is important that the information system can give cues so that the operational behaviour can be initiated at the right behavioral level. This empasizes what discussed above and it is a special challenge to design methods and to the flexible use of databases. A diagnosing aid system must utilize of necessity information about design principles also on the system requirement specification level. It is also evident that a good overall view of the process state is one of the best cues for the operator.

Besides supporting different data processing levels of an invidual operator, it is important that the display system can take into account the different experience back-ground of different operators and the need of different user groups. This requires that special attention must be paid to the dialogue control design and to the flexibility of system.

The problems discussed above require also a new approach to automation system design. The data needed in the realization of operation aid systems must be gathered during the design process. A special matheod must be developed to ensure this data need.

Computer-based design aids, like requirement specification, verification, function testing, and documentation systems, are quite natural candidates to facilitate the integration of design and operation.

The training of operators is mainly assosiated with means and methods to give the human operator an internal world model, which is relevant to his job. One other problem is to measure the knowledge and skill of the operator to get some estimation of the state of the internal world model. Considering the different types of internal models discussed earlier, it seems clear that a large variety of methods and aids are needed for the training and for the measurement of training results.

The models for functional meaning and abstract function could to a large extent be built using classroom training. The building of the functional structure models should be supported with observations and operations either in the plant or using a simulator. The models of physical function could not be acquired without operating on the plant or on a suitable simulator. The model of physical forms could again be acquired using classroom exercises, perhaps using also a mock-up of the control room.

The data processing activities on the different levels could be trained in simulated situations. On the higher level, a talk-through in the control room could provide a very cost-effective training. For the training of the dataprocessing on the intermediate level, a high fidelity simulator is perhaps the best training tool. The data processing on the lower level again could be trained with a part task trainer.

Considering the situation in the control room, it is evident that the emphasis in the training will be on the diagnosis of the very rare abnormal situations. Taking into account that by definition no available internal world model could provide an immediate solution to the diagnosis task, it is clear that the operator should execute the diagnose on the higher data processing level. He will, on the other hand, have the tendency to move the data processing down with increased experience. This means, e.g. that to ensure that the operator at an incident executes the diagnosing task on the higher level, we will have to restore the suspicion and confusion of the operator in order to prevent him from falling in the trap of associating the readings with some previously experienced incident.

Gained experience makes it possible for the operator to integrate the action performed in the control room during different maneovers into a larger entity. This abstraction ability makes it possible for the operator to intuitively comprehend a set of actions as tasks with defined goals, restrictions, initial state and terminal state. Training of this ability should be supported by presenting the process and the task in a top-down fashion.

From the training point of view if would be interesting to investigate which abilities makes up a good operator for the control room work. It could be possible that good operators have accuired general and specific rules of thumb to perform the different activities. Such rules could be diagnosing strategies, set of important parameters, goals and restrictions which are imperative etc.

CONCLUSIONS

The work in the human factors area, which has been done mainly in the aerospace field, provides a very good foundation for the human factors research in process supervision. Owing to differences between the two fields more research has, however, to be done to reach the same level of understanding and design practice. Thanks to the massive research effort, which has been spent, it seems, however, that an understanding is emerging. The challenge is then to draw the design implications of that understanding and work toward some standardized practice for the design of future control rooms.

It is also interesting to note the possibilities which are emerging with the introduction of new hardware and software. It is, however, a danger that system vendors continue to produce systems on an ad-hoc basis in response to market pressure. This makes it important to spread good practices in man-machine interface design to make both the vendors and the vendees aware of the requirements on a control room design.

REFERENCES

Bainbridge, L., (1979). Verbal reports as evidence of the process operator's knowledge. Int. J. Man-Machine Studies, 11, 411-363.

Bainbridge, L., (1981). Mathematical equations or process routines. Human detection and diagnosis of system failures Eds. J. Rasmussen, W.B. Rouse, New York 1981, Plenum Press, 259.

IEEE, (1979a). The Human. The key factor in nuclear safety. Myrtle Beach. Conference Record for 1979 IEEE standard workshop on human factors and nuclear safety.

IEEE, (1979b). Special issue on Three Mile Island and the future of nuclear power. IEEE Spectrum 16, No 11.

NKA/KRU-project on operator training, control room design and human reliability
- Summary report, NKA/KRU-(81)11, Joint Scandinavian Research Project, The Nordic Council of Ministers, June 1981.
- Technical Summary Report on Operator Training, NKA/KRU-(81)12, Joint Scandinavian Research Project, The Nordic Council of Ministers, June 1981.

- Technical Summary Report on Control
  Room Design and Human Reliability,
  NKA/KRU-(81)13, Joint Scandinavian
  Research Project, The Nordic Council
  of Ministers, June 1981.
- Publication List, NKA/KRU-(81)14, Joint
  Scandinavian Research Project, The
  Nordic Council of Ministers, June 1981.
- Guidelines for Operator Training, NKA/
  KRU-(81)15, Joint Scandinavian Research
  Project, The Nordic Council of Ministers,
  June 1981.
- Guidelines for Man-Machine Interface
  Design, NKA/KRU-(81)16, Joint Scandi-
  navian Research Project, The Nordic
  Council of Ministers, June 1981.

Ranta, J., Tuominen, L., Uusitalo, M.,
  Rantanen, J. (1982). Specifying
  man-computer dialogues - the use of
  guidelines for design of interactive
  systems. Preprints of IMEKO 8th Trien-
  nal World Congress, Berlin-West.

Rasmussen, J., (1980). Some trends in man-
  machine interface design for industrial
  process plants. Proc. IFAC/IFIP Symp.
  Assopo'80. Trondheim 1980, North-
  Holland, Amsterdam.

Rasmussen, J., Rouse, W.B., (ed).
  Human detection and diagnosis of system
  failures. New York 1981. Plenum Press.

Rouse, W.B. (ed), (1979). Special issue on
  applications of control theory in human
  factors. Human Factors 19(1977)4.

Rouse, W.B., (1980). Systems Engineering
  Models of Human Machine Interaction.
  North Holland, New York.

Seminara, J.L., (1976). Human factors
  review of nuclear power plant control
  room design. EPRI NP-309 S 4, Project
  501, Palo Alto.

Timonen, J,. (1980). On the control theore-
  tic modelling of human process operator.
  Technical Research Centre of Finland,
  Electrical Engineering Laboratory.
  Research Report 55, Espoo.

Timonen, J., Wahlström, B., Tuominen, L.
  (1980). On the modelling of the tasks
  of operator in automated process plant.
  Proc. IFAC/IFIP Symposium Assopo'80,
  Trondheim 1980, North-Holland, Amsterdam.

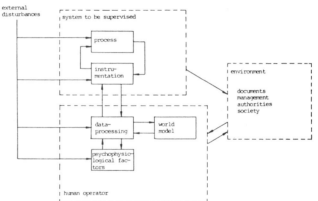

Fig. 1. A general model of the control room situation

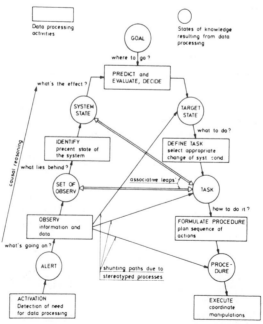

Fig. 2. Schematic model of the sequences
of operator's mental activities (NKA/KRU 1981)

AN INTEGRATED DISPLAY SET FOR PROCESS OPERATORS

L. P. Goodstein

Riso National Laboratory, DK 4000 Roskilde, Denmark

Abstract. In previous papers, the need to depart from traditional "one sensor - one indicator" approaches in order to take full advantage of the possibilities offered by the computer was pointed out and an approach based on the conception of the operator as a multi-mode information processor was advanced. This formulation of operator functioning as being categorisable into elements of skill-based, rule-based and knowledge-based behaviour gives, among other things, a structure for generating suitable forms for information presentation which is compatible with these elements and thus can counteract certain recognized tendencies for human malfunction arising from insufficient or inadequate displays of information.

These ideas have been developed further and incorporate now a multi-level representation of the process to be controlled together with the associated control task/information need spectrum. With this foundation, it is possible to structure the content of an appropriate set of displays for supporting operator strategies and internal system representations - particularly when dealing with plant disturbances.

The paper concludes with some preliminary examples of an application to a nuclear power plant to illustrate also a possible language for presenting the necessary informational content.

Keywords. Process operators; computer displays; information structure; models; complexity.

INTRODUCTION

The design of a suitable information retrieval and display system for operating staff has of necessity to be based on realistic and usable characterisations of the two sides of the man-machine interface - i.e., the human and the process respectively. In previous papers (Rasmussen, 1980; Goodstein & Rasmussen, 1980; Goodstein, 1981), the conception of the human as a multi-mode information processor has been advanced. This formulation of human functioning as being categorisable into elements of skill-, rule- and knowledge-based behaviour has, among other things, significance for information display and, as will be seen, support of the knowledge-based mode is of special interest.

Recent work (Rasmussen & Lind, 1981; Lind, 1982) has continued earlier efforts to also suitably characterize the process system - mostly in order to ease human problems in coping with the complexities of modern plant. Thus, for example, while operators do develop means for dealing with large quantities of data, as is the case in traditional one sensor - one indicator control

rooms, it would appear from the record that the rare unfamiliar situation can and often does lead to undesirable consequences because the usual techniques and tactics for coping turn out to be ineffective and inadequate. Thus the need exists for an interface design which assists in the daily routine but also enhances the building up of professionally sound habits for information handling and diagnostic strategies which then in a natural way support the operators in their use of the same interface to identify and deal competently with the unfamiliar and unexpected event.

SYSTEM REPRESENTATION

Operating in the knowledge-based mode implies the use of a set of mental representations or models of the system with which one is dealing. These serve as references of expected behaviour/functioning/structure against which actual conditions can be compared in order to identify state and to predict the effects of changes. As discussed in Rasmussen & Lind (1981), various "mental tools" are often employed to perform transformations on the system under consideration

63

to make it more amenable to such treatment. In general, these tools enable one either to aggregate/decompose (parts of) the system so as to change the resolution or level in the detail of attention with which one views the system or they permit a change in the level of abstraction with which the system is considered. In a process plant context, the various representations which are relevant can be listed as follows:

Abstraction levels
- Functional purpose
- Overall causal structure
- Generalized functions
- Physical functions
- Physical form

In a complementary manner, a decomposition at, e.g., the physical level would be equivalent to zooming in from a view of the total plant, to subsystem, to equipment, to parts, to nuts and bolts. An aggregation would of course occur in the reverse fashion. A consequence of this structuring process seems to be that the mental span of attention or "degree of zooming" usually becomes more restricted as attention moves down through the levels of abstraction from functional purpose to physical function and form. This seems compatible also with the diagnostic goal of "narrowing in" on a problem.

As far as information display is concerned, there are (at least) two important ideas to be gleaned from this ability of humans to shift their internal representations of the world according to needs:

- The data necessary to characterize the state of the system also has to be transformed to match the current representation (e.g., valve positional data are not compatible with speculations about mass balances).

- Moving between levels of abstraction is not just a matter of altering details or changing descriptors. Thus a shift downwards in abstraction level implies the imposing of a given set of requirements/specifications for the implementation at the lower levels. On the other hand, moving upwards carries with it information on means, capabilities and limitations in the actual implementation. Both of these are necessary ingredients in the operating staff's need for information. In more general terms, one can say that, with such a representational structure, reasons and rules for desired operation come from above while causes of actual behaviour (including malfunctions, and their effects) propagate upwards.

INFORMATION DISPLAY STRUCTURE

In developing these ideas further for information display purposes, it is useful to consider the total system as a hier-

archical collection of subprocesses coupled together in a coordinated fashion as the result of a specific "setting up" sequence - all of which is required in order to enable the main processes of the top level to be established and maintained in adherence with overall goals and constraints. Fig. 1 illustrates this. For each major operating mode of the main process, a given set of subprocesses has to made available (either actively or in a "ready" state) via a bottom-up progression in which the functions provided by each subprocess serve as necessary conditions[1] for the establishment of proper operation at the next higher level. The functions of this level then act as conditions for setting up proper operation at the next level - and so on.

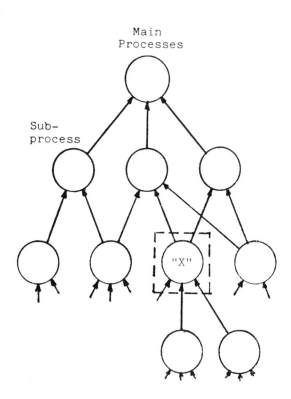

Fig.1 Overall Process Structure

As far as operating staff is concerned, they of course in principle have the need to be able to consider any (sub)process on any of the previously mentioned levels of abstraction - i.e., its purpose, causal structure, functional arrangement, physical properties. In a practical sense, one could say that the process operators' most pressing requirement is to have a convenient access to each (sub)process' "WHY", "WHAT" and "HOW" in order to support their various speculations about the (sub)process and its relations to the rest of the system. Expanding on this

---

[1] Conditions can include process-relevant support functions and supplies and/or reflect regulatory, safety-related or organisational requirements.

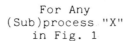

For Any
(Sub)process "X"
in Fig. 1

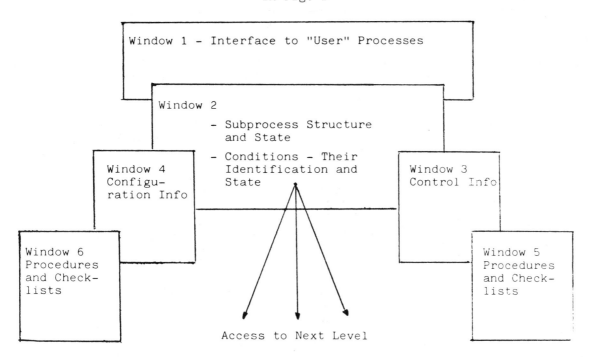

Fig. 2

Windows of Information

then, Fig. 2 sketches the important components for fulfilling this need. Thus, for any selected (sub)process, the following (nested) display "windows"[2] to information should be available:

Window 1 (WHY)

It is not customary to specifically include "why" information on control room displays. Perhaps it is assumed that if operators will just follow procedures, the "why's" will be taken care of implicitly. This, however, is not compatible with the notion of operators as competent knowledge-based problem solvers who are capable of recognizing the unexpected and dealing with it. It therefore follows directly that any planning for recovery from an identified disturbance must represent an acceptable compromise between the capabilities of the actual physical plant (possibly in a disturbed state) and the original purpose of / reasons for the design as reflected in the functional requirements and constraints.

Thus window 1 is an instance of an "interface" between the particular (sub)process and its "user" where (a) the adequacy of the delivered function can be judged against requirements coming down from above and (b) the effects in an upwards direction of deviations, alternate possibilities for recovery in the event of a disturbance can be evaluated from considerations of intention, span of tolerances, cofunctioning constraints, etc.

It is our feeling that the advent of computer-aided methods for design and documentation will make it feasible to record and store the bases for design decisions made regarding safety and availability. In this way, the design base will be available (years) later in connection with commissioning, training, procedure generation, dealing with disturbances, modifications, etc. and thus enhance, for example, the possibilities for incorporating Window 1 "WHY" displays.

Window 2 (Causal Structure, a WHAT component)

Window 2 is intended to give information on the causal structure of the particular (sub)process as well as its current operational state. What is needed here is a high

[2] My acknowledgements to Xerox Corporation's "Smalltalk" for this term (see, e.g. Testler, 1981).

level representation which is relatively
independent of implementation details and
yet can serve as an effective visualisation
of state and, hopefully, also of task. We at
Risø are proponents of using a flow modelling approach - be it for energy, mass,
information - and have employed a high level
symbolic flow language for depicting the
basic flows through the process. See Lind
(1981) and Rasmussen & Lind (1981).

Window 2 would thus supply operators with
the following types of information about
"WHAT":

- Topological flow diagram of the process
  reflecting actual paths, distribution of
  flow, deviations from target values.

- Quantitative information on the state of
  the "critical variables" which is the set
  sufficient to indicate that the given
  function is/is not realized.

- An identification of the set of support
  conditions which are required to be
  satisfied in order that the (sub)process
  can be operable, together with qualitative
  information on their state and interrelations.

Window 3 (Control System – A Special Condition)

In general, there are three major requisites
to proper (sub)process functioning - the
necessary support functions and supplies are
available, the appropriate flow paths are
activated and the degrees of freedom are
under control. Thus the control system is
equivalent to a kind of "finishing touch"
condition which overlays the physical system
with a web of interrelated constraints.
These then serve to restrict (sub)process
behaviour to an allowable - preferably
optimal - region compatible with higher
level requirements in spite of variations,
drifts, etc. Since operators can have
greater difficulty in understanding the
control system than other more process-related aspects, the need for appropriate
control information would seem to be quite
high.

However, it seems important that the possibilities for taking a control action first
be identified on the flow schema of the
(sub)process (i.e. Window 2) if these
actions can affect the overall flow through
the (sub)process. Identification of control
actions at this level would thus be more
likely to be compatible with operator
speculations about process behaviour including the possibilities for making corrections and adjustments. However, the additional window (Window 3) providing information such as the following would give a
better basis for selecting, performing, and
checking a control action.

- Control requirements
  . modes
  . transitions
  . limits
  . constraints (interlocks)

- Operation
  . control organs
  . actual vs target state
  . dynamics

- Backup/alternate forms for control

- Access to relevant procedures for manipulating the control systems

It is also at this window that implemen-
tation-dependent "HOW" details can begin to
arise - e.g., if the system actually
consists of two or more loops operating
simultaneously - each of which is under
separate control.

Window 4 (Configuration = HOW)

It is first at Window 4 that the more
familiar view of the plant at the mimic
diagram level appears. Configuration monitoring and control are tasks related to the
particulars of the given plant and require
support which can facilitate a search and
localisation within an equipment-based
structure. Information which is important at
this level relates to:

- Physical paths and connections - prescribed, actual and alternate.

- Actual vs target state in terms of the
  variables reflecting equipment operation.

- Equipment capability and limitations.

- Access to relevant procedures and checklists at the equipment level.

The capability of advanced display systems
to perform "windowing and panning" should be
particularly applicable here.

Windows 1-4 are thus intended to support
operators in knowledge-based operations of
observing, weighing, deciding and planning
while meeting requirements for maintaining
an overview, reducing memory requirements
and encouraging the use of top-down search
strategies through an orderly and easily
accessible display hierarchy.

Windows 5 and 6 (Procedural Support)

Windows 5 and 6 are available to support
procedural (rule-based) tasks as either
prescribed for operators or for automatic
sequential equipment for manipulating the
plant. As described elsewhere (Goodstein,
1979, 1982), these aids are composed of two
components:

- Support of memory - what is in the
  procedure.

- Support of the procedural steps themselves which consist of sequences of "check state, do action, check result".

The possibilities for combining computers and displays for this purpose are of course quite obvious.

In summary then, an integrated information structure based (a) on a multi-level representation of the process and (b) on the concept of abstractional shifts in human thinking in order to cope with complexity has been proposed. A generic family of "display windows" has been described which can be applied to any selected (sub)process to inform about its WHY, WHAT and HOW - i.e., its "interface" with higher level requirements, its flow structure, the relevant conditions necessary for satisfactory functioning, the associated control system and details about the physical implementation. This knowledge-based foundation is also supplemented by procedural and checklist windows to support rule-based plant maneouvering.

AN EXAMPLE

The remainder of the paper will deal with an illustration of some of the ideas presented, using a nuclear plant at power as the reference system. See also Lind (1982). Thus Fig. 3 indicates a simplified application of the structure of Fig. 1. In this case, however, convenient upper level descriptions will have to do with the dual goals of maintaining safe operation while maximizing the availability of the production of electricity. Hence segment (B) of Fig. 3 describes the plant as an energy conversion system, the goal of which is to distribute energy from the source to the designated sinks. To achieve this, the indicated support functions having significance for overall efficiency are required. A corresponding goal exists for safety and has to do with maintaining non-flows of radioactive materials through the protective barriers (see segment (A)) which, in turn, requires that the indicated conditions are satisfied. At the next level, (some of) the subprocesses required to provide the high level support functions are indicated. Thus maintaining the integrity of the primary system requires adequate heat removal so as to keep the temperature and pressure within limits. To produce electricity requires a chain of energy transport and conversion elements which, in turn, require that a given set of conditions be met. Not surprisingly, many of these are similar to those to ensure adequate heat removal for safety purposes. Thus support (sub)processes are not mutually exclusive; smaller or larger bits and pieces of process can be involved in more than one function. Displaywise this state of affairs calls for the ability to be able to selectively view the same elements in different contexts and possibly in varying degrees of decomposition/aggregation.

Continuing then to the third level, the subprocesses to provide such things as adequate inventory and circulation (i.e. energy transport requires a suitable medium), pressure, etc. appear. At the fourth level, details on pump functioning begin to become evident. At these component-oriented levels, in contrast to the more abstract, there exists a mutually exclusive set of physical elements which are connected in a topologically orderly fashion. Displaywise, this calls for facilities to "wander" about in the terrain in search of particular locations of interest.

When carrying out this breakdown, it is also possible to begin to get an idea of the kinds of control tasks operators will be confronted with and the concept of a generic set looks promising. Indeed, this has been implicit in the set of display windows suggested earlier. At any rate, the idea of being able to consider each level of a WHY, WHAT or HOW basis as one moves up and down the hierarchy should be kept in mind. For example, "Energy Production" and "Heat Removal" place requirements on "Circulation". For HOW this is done, we limit our attention and move down to "Circulation" but begin to consider it as a WHAT and so on. Moving up in response to a WHY inquiry acts in a similar way.

This structure has been used as the basis for some limited studies of display formats for depicting the information required within the various windows. For example, Figs. 4 and 5 illustrate possible implementations of WHAT windows (Window 2) for segments C and A of Fig. 3. These black and white photographs of colour originals use the flow symbology of Fig. 3 for indicating process structure. Thus, in accordance with the requirements for Window 2,

- The target and the actual production figures are given.

- The flow topologies (actual = solid and alternate = dashed) are shown and the amount of energy flow at various points is indicated by the degree to which the arrows are filled - hopefully giving an effective visualisation of energy transport through the main process.

- The support functions are identified with * (which also gives an addressable access to the appropriate Window 2 at the next lower level) and the states of the most critical of these are displayed qualitatively (via bar graphs, polar displays).

Fig. 5 attempts a similar representation of the overall situation regarding safety in terms of a display of the status of the multi-barrier configuration which is typical

of nuclear power plants - i.e., the fuel
cladding, the primary system (RCS) boundary,
the containment, the secondary system (SCS)
and so forth, which serve to contain and/or
impede the flow of radioactive materials to
the environment. Thus the (non) flow structure is indicated (see Fig. 3 again for a
definition of the symbols) as is the status
of the conditions which contribute to
maintaining the barriers. Any changes (leak-
age, collapse of barrier) would be reflected
on the display via appropriate changes in
the structure and flow.

A natural reaction to the proposed information display structure could be that the
total number of windows will become excessively high so that using the system will be
awkward and slow. There are several comments
to this potential criticism. The first is
that the relational links between levels
through the support function concept are
consistent as are the clusters of windows
for the (sub)processes. The second is that
the strategies for using the display set are
based on "top-down" searches through the
hierarchy in response to functionally-based
bottom-up directed warnings about disturbances. These searches essentially "zoom in"
on the relevant display window cluster(s) at
the level of abstraction where an appropriate recovery can be made. The strategies
are similar for daily routines and for infrequent incidents. However, to derive full
advantage of such an approach will require
adequate training - not only in the manipu-
lative details of information selection and
retrieval but, more importantly, in the
basic problem solving skills which are
prerequisite to a successful diagnostic
encounter in the control room.

CONCLUSION

It has been argued that a systematic
structuring of the process is necessary in
order to be compatible with the capabilities
and limitations of humans to cope with
problems of complexity. A hierarchical
structure which lends itself to multi-level
speculation about process WHY, WHAT and HOW
has led to an information display window
concept comprising six elements for satisfying this need. The ideas have been illustrated. However it should be emphasized that
Figs. 4 and 5 are examples of possible
implementations of a single element within
the suggested information cluster. An ultimate evaluation of the approach has to
consist of two phases - one which attempts
to verify whether the total concept is
viable and a second which aims at optimizing
the details of the display language utilized.

This work has been partially supported by
the Nordic Council of Ministers.

REFERENCES

Goodstein, L.P. (1979). Procedures for the operator - their role and support. IAEA NPPCI Specialists' Meeting on Procedures and Systems for Assisting an Operator During Normal and Anomalous Nuclear Power Plant Operation Situations, Munich.

Goodstein, L.P. (1981). Discriminative display support for process operators. In: J. Rasmussen & W.B. Rouse (Eds), Human Detection and Diagnosis of System Failures, Plenum Press, pp. 433-449.

Goodstein, L.P., and J. Rasmussen (1981). The use of man-machine system design criteria in computerized control rooms. In: A.B. Aune and J. Vlietstra (Eds), Automation for Safety in Shipping and Offshore Petroleum Operations, an IFIP/IFAC Symposium, North-Holland.

Lind, M. (1981). The use of flow models for automated plant diagnosis. In: Jens Rasmussen and W.B. Rouse (Eds), Human Detection and Diagnosis of System Failures, Plenum, pp. 411-432.

Lind, M. (1982). Multilevel flow modelling of process plants for diagnosis and control. International Meeting on Thermal Nuclear Power Safety, Aug. 29 - Sept. 1, 1982, Chicago, Ill.

Rasmussen, J. (1980). Some trends in man-machine interface design for industrial process plants. In: A.B. Aune and J. Vlietstra (Eds), Automation for Safety in Shipping and Offshore Petroleum Operations, an IFIP/IFAC Symposium, North-Holland.

Rasmussen, J., and M. Lind (1981). Coping with complexity. Risø-M-2293.

Testler, L. (1981). The Smalltalk environment. BYTE, 11, 8, pp. 90-147.

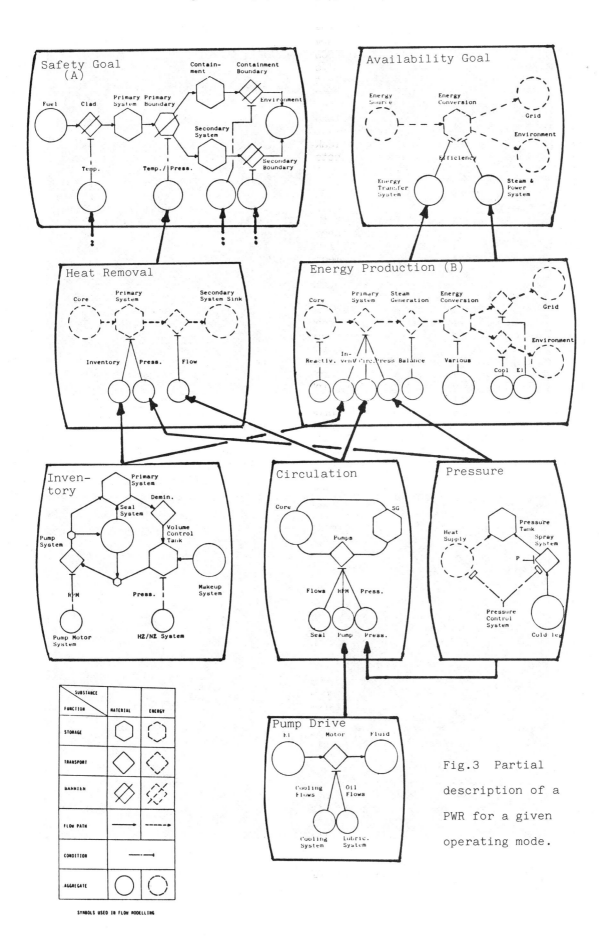

Fig.3 Partial description of a PWR for a given operating mode.

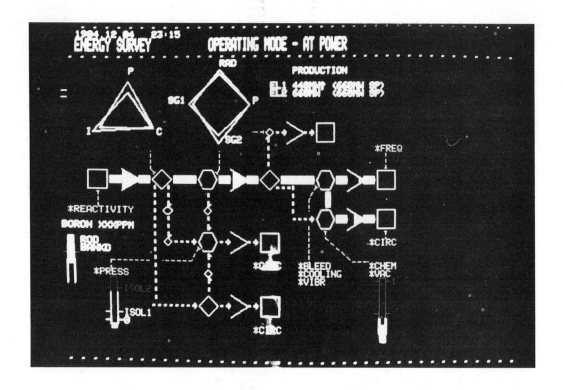

Fig.4   An example of a high level Window 2 display for production

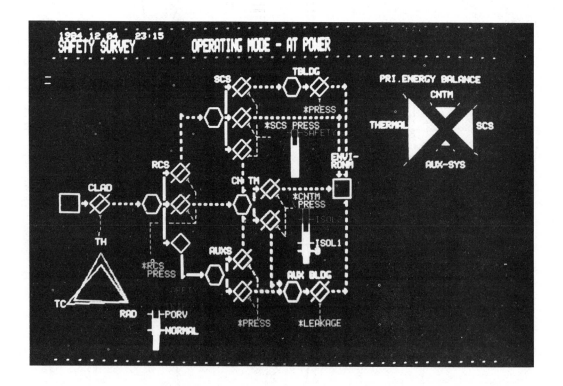

Fig.5   An example of a high level Window 2 display for safety

A MULTI-LEVEL ALARM INFORMATION PROCESSING SYSTEM APPLIED TO THERMAL POWER PLANT

K. Oxby*, R. Sykes*, P. Trimmer*, K. Kawai** and K. Kurihara**

*The Electricity Commission of New South Wales, Sydney, Australia
**TOSHIBA Corporation, Tokyo, Japan

Abstract. The purpose of this alarm system is to assist the operator during emergencies as well as during normal operating conditions by use of the process computer system to achieve safe and reliable power plant monitoring functions. During the normal operating conditions, the conventional computer alarm system provided sufficient information for the operator. During the abnormal plant conditions, however, too much information is produced due to momentary plant excursions above alarm limits. This flood of information impedes correct interpretation and correction of plant conditions by the operator.

In this paper, a new man-machine system is introduced, which is realized by the operator and process computer system. The facility to suppress the excess messages is provided to control the message output rate to the operator. The measure of the output messages is defined to be based on the "Alarm Message Activity". By adopting this new alarm system, the safe and reliable operation of the thermal power plant can be achieved.

Keywords. Alarm systems ; computer-aided instruction ; Human factors ; man-machine systems ; self-adjusting systems ; steam plants.

INTRODUCTION

The followings are the three major facilities, which are used to monitor the thermal power plant by the operator:-
(1) Plant Control Boards (Unit Control Board, Unit Subsidiary Board)
(2) Plant Control Panels (Analog Control Panel, Sequential Control Panel)
(3) Process Computer System (Alarm CRT, Annunciator Window).
The protective system which is also called the unit interlock, is also provided in the thermal power plant.

In the current operational philosophy, the full-scale automatic operation system without operator intervention is not possible due to the complexity of plant condition and alarm diagnosis. The role of the operator and his relationship with the control system shall be considered in detail in the current situations.
In Fig. 1 , the layout of plant control room is shown as an example of the operator's environment.

It is rather easy, paradoxically speaking, to achieve the power plant safety. That is to stop the plant as soon as possible on detection of the emergency. This is by no means desirable, however, because the stoppage of the plant will initiate disturbances to the power transmission system. Furthermore the power supply responsibility must be considered from the social effect point of view. In this sense, the alarm system must have the capability to avoid plant trips as much as possible.

On the other hand, it is generally said that the number of occurrences of accidents during emergencies is almost the same as that in the normal operating conditions, although the time period of the emergencies is much shorter than that of the normal stages.
A new alarm system, which is effective during emergencies as well as during normal operating conditions, is discussed in this paper.

71

OPERATOR'S CHARACTERISTICS

Errors made by the operator occur during of the following phases as shown in Table 1 :-
(1) Recognition - Of the fault condition
(2) Judgement - As to cause of fault
(3) Operation - Removal of fault or faulty plant
In this alarm system, the process computer is used to assist the operator, especially in the "Recognition" phase.

When the operator is considered as an element of the power plant monitoring system, such conditions as follows should be considered :- fatigue, responsibility, will, form of shift, age, nationality, education, and so forth. First of all, however, the processing speed of the operator to recognize correctly the situation is discussed.

The time lag of the judgement by the operator is estimated as follows (Hashimoto, 1980) :-
(1) Simple judgement: 0.1 sec
(2) Reading of the instruments : 0.5 sec
(3) Complicated judgement :   5 sec
Accordingly, there are no problems for the operator in absorbing conventional alarm system information, if the power plant is in the steady state.

In the non-steady state of the power plant, many process variables exceed their limits simultaneously. In the case of TMI accident (U.S. NRC, 1979), there were 52 alarms before its disturbance. In addition to them, more than 80 alarms were output to the operator for subsequent 30 seconds after its disturbance. The operator's ability to judge properly would have been degraded on account of the sudden flood of alarms, many of which were irrelevant.

The multi-level alarm information system in this paper is the computer aided monitoring system, which is effective in the non-steady state as well as in the steady state of the plant.

MULTI-LEVEL ALARM INFORMATION PROCESSING SYSTEM

Hardware System

The computer system hardware configuration (Kawai, 1982) is shown in Table 2. The man-machine peripherals for the operator are as follows :-
(1) Alarm CRT   1 set
(2) Information CRT   1 set
    (Man-Machine CRT)
(3) Utility CRT   2 sets
    (Output CRT)
(4) Printer   2 sets
    (Information/Historical)
(5) Console   2 sets
    (Communication panel/GSM panel)

In addition to the peripherals described above, such equipments as follows are also provided :-
(1) Computer annunciator
    windows :   100 sets
(2) Audible alarms   2 sets
    (WRS/BTR type)
(3) Annunciator Windows : 450 sets
    (hardwired)
The annunciator windows (hardwired) on USB panel are turned ON or OFF directly from the power plant monitor and control system as shown in Fig. 1.

Multi-level alarm system philosophy

In the conventional alarm system, the process variables are scanned periodically and judged whether they are in alarm or in normal state. Once a process variable enters into the alarm region, its corresponding alarm message is shown on the alarm CRT including the point(s) value and its alarm limits.

In this new alarm system, however, the alarm region is divided into more than 15 sub-regions by the unit called DELTA or significant change value which is assignable for every process variable as shown in Fig. 2(a). By using this configuration, the alarm tracking information can be output to the operator by such messages as WRS(Worse), BTR(Better), and so on. So, this new alarm system can call the operator's attention to the process value, while it is trending in the alarm region.

ALM(Alarm) and RTN(Return) messages have the same meaning as those in the conventional alarm system.

As shown in Fig. 2(a), Zone No. is defined to be equal to the degree of penetration of a process variable in the alarm region, where every point is normalized by this unit DELTA. This Zone No. can be regarded as the degree of urgency among the process variables in the non-dimensional unit.

If a point exceeds beyond Zone No. N, a message format change can occur to call the operator's attention. In this application the color may be changed from cyan to yellow.

In order to introduce the message suppression feature, the idea of SP pointer (Suppression Pointer) is shown in Fig. 2(b).
SP pointer is an integer-type variable which changes its value from 0 to N. In

the emergenices, the excess messages below this SP pointer are to be suppressed. SP pointer can be changed by any of the following events:-
(1) Plant emergency (disturbance)
(2) Heavy alarm message activity (AMA)
(3) Operator console manipulation

AMA (Alarm message activity) is defined to be the degree of alarm message output on the alarm CRT. The alarm message control block diagram is shown in Fig. 3. It is noted that some CRT messages are to be suppressed by relationship with the SP pointer, while all the messages are to be stored and output to printer.

CRT presentation

In Fig. 4. an alarm CRT presentation example is shown. Alarm Type can be classified as either of the following :-
(1) HI : High analog alarm
(2) LO : Low analog alarm
(3) DA : Digital alarm.

Alarm message type can be classified as follows :-
(1) ALM : New Alarm message
(2) WRS : Worse message
(3) BTR : Better message
(4) RTN : Return message

If an alarm occurs, alarm type and message type will start blinking until acknowledged by the operator. If the operator acknowledges an alarm, the blinking is stopped and the occurrence time and message type are deleted as a form of acknowledgement by the operator. A message which exceeds Zone No. N will change its color from cyan to yellow.

DESIGN AND IMPLEMENTATION

The alarm message suppression characteristics is so deeply related to the dynamic behavior of the SP pointer that the definition and adjustment of the SP pointer is of primary importance to the new alarm system. The dynamic behavior must have tuning capabilities built in to avoid unnecessary alarm message suppression. The suppression pointer processing block diagram is shown in Fig. 5. As shown in this block diagram, the system is so designed that the adjustment of the algorithm can be realized by changing the control parameters only. In Fig. 5, the following SP pointer calculation blocks, namely F_{AMA} , F_{EVE} and F_{OPE} and their corresponding SP pointer calculated results, namely ASP, ESP and OSP are introduced, respectively.
(1) F_{AMA} : Alarm Message activity SP pointer (ASP) calculation block

(2) F_{EVE} : Event initiated SP pointer (ESP)calculation block
(3) F_{OPE} : Operator manipulated SP pointer (OSP) calculation block

The TMR (Decaying Timer Processor) block is introduced to modify ASP, ESP or OSP with the decaying factor. The purpose of this TMR is to restrain the output rate of the once suppressed alarm messages when the SP pointer changes its value. The combination of modified ASP and ESP is fed to the optimizer block, where the optimum value between 0 to N is decided considering the current process status. The operater retains the right to change the SP pointer output from optimizer block by plus/minus one degree including the timer processing. The SP pointer so calculated is updated at a fixed interval to optimize its value constantly.

Alarm Message Activity SP Ptr.(ASP)

The definition of AMA (Alarm Message Activity) is stated in Eq.(1),

$$AMA^{(n)}=a_{AM}.M(AMA)^{(n)}+(1-a_{AM}).AMA^{(n-1)} \quad (1)$$

where $AMA^{(n)}$ is a value of alarm message activity at sample n, a_{AM} is a filtering constant, and $M(AMA)^{(n)}$ is a measured value (instantaneous value) of AMA at sample n.

The definition of $M(AMA)^{(n)}$ is obtained from the definition in Eq.(2).

$$M^{(n)}(AMA)=k_A.R^{(n)}(ALM)+k_W.R^{(n)}(WRS)$$
$$+k_B.R^{(n)}(BTR)+k_R.R^{(n)}(RTN)$$
$$+k_D.R^{(n)}(DIG) \quad (2)$$

where k_A,k_W,k_B,k_R and k_D are constants.
And $R^{(n)}(ALM)$ and so forth can be defined as follows:-

$$R^{(n)}(ALM)=a_A.M^{(n)}(ALM)+(1-a_A).R^{(n-1)}(ALM) \quad (3)$$
$$R^{(n)}(WRS)=a_W.M^{(n)}(WRS)+(1-a_W).R^{(n-1)}(WRS) \quad (4)$$
$$R^{(n)}(BTR)=a_B.M^{(n)}(BTR)+(1-a_B).R^{(n-1)}(BTR) \quad (5)$$
$$R^{(n)}(RTN)=a_R.M^{(n)}(RTN)+(1-a_R).R^{(n-1)}(RTN) \quad (6)$$
$$R^{(n)}(DIG)=a_D.M^{(n)}(DIG)+(1-a_D).R^{(n-1)}(DIG) \quad (7)$$

where a_A, a_W, a_B, a_R and a_D are filtering constants, and $M^{(n)}$(ALM) etc. are measured values of ALM messages etc. at sample n, respectively.

The $AMA^{(n)}$ so far obtained is related to the ASP as shown in Fig. 6.

Event Initiated SP Pointer (ESP)

As described so far, the ASP has the feedback features to control the excess alarm messages after they are generated upon the plant disturbance. Accordingly, ASP can be effective to suppress the excess alarm messages whose Zone No. is less than the value of SP pointer.
Actually, however, the initial message activity is so enormous that the feedback configuration might not be effective in such case as in TMI accident.

The ESP feature is introduced to overcome this burst of initial message activity. The plant disturbances are interfaced with the process computer system in a digital-type process inputs such as boiler trip (MFT : Master Fuel Trip) and turbine trip and so forth. Each event has the fixed level of ESP value and the fixed decaying time of its own. An ESP control example is shown in Fig. 7, where the two different inputs initiate this control logic.

Other considerations

(1) A message already displayed on the alarm CRT is not affected by the change of SP pointer. The increase of RTN messages are avoided by this feature when SP pointer increases its value.
(2) As a chronological record of all the plant inputs, all the process changes are output to printer, while there exist some suppressed alarms on the alarm CRT.
(3) In the conventional alarm system, a hysteresis or "dead band" is included to avoid frequent repetition of alarm occurrence and normal return. In this new alarm system, the additional dead band i.e. DELTA 1(for example, 1=0.5) is provided to avoid frequent BTR/WRS changeovers.
(4) If an analog type process input exceeds the Zone No. N in the alarm region, a format change will occur as an indication to the operator. This feature can be selected for each point. Thus the rank classification of alarms becomes feasible.
(5) As for the digitals, there are some computer systems which do not trend the digital alarms on the alarm CRT. In this new alarm system, the digital alarms are trended on the alarm CRT,

and the following two types of processing are provided :-
(a) A digital input which can not be suppressed in any situations, and always initiate the yellow message if it is in alarm regardless of the SP pointer value.
(b) A digital input which can be suppressed if the SP pointer exceeds the level M, which is a system constant for digital alarm filtering. When this kind of digitals are output, the color is always cyan.

EVALUATION

This new alarm system is currently applied and implemented to the actual power plant i.e. Eraring Power Station, Australia. Some features of this system proved to be valuable and effective from the operator's burden point of view. Considerable tuning and testing with actual plant dynamics are still to be carried out. The following are some examples of discussion toward this new alarm system :-
(1) A new alarm message with low Zone No. might be suppressed to the operator if the SP pointer is high. The operator must wait the alarm information until the alarm exceeds Zone No. N.
(2) Even if the process variables with strong mutual interaction exceed their alarm limits simultaneously, there exists a chance that only either of them are displayed on the alarm CRT.

These discussions can be classifed as the item for the plant diagnostic system described in Table 1.
The purpose of this new alarm system is to try to give positive aid to the opoerator by suppressing excess messages even if it might suppress legitimate plant error messages. During the normal operating conditions, much more stimuli are given to the operator by alarm tracking functions of WRS/BTR messages. In the emergencies, some of the alarm messages with low degree of urgency might be filtered and the alarm messages with high degree of urgency might change their format and be made to stand out conspicuously.

CONCLUSION

In this paper, a new alarm system, especially its design and implementation aspect is discussed. This system is intended to improve the plant operational safety from the standpoint of the recognition phase. Particularly, this system offers a much more positive role in plant monitoring compared with the conventional alarm system during major plant excursions.

Considering the general trend to adopt operation systems with many CRT's, the alarm filtering philosphy described herein appears to be valuable for inclusion in such system design considerations. Thorough evaluation of the actual field test results shall be done in order to verify the initial alarming philosophy of this system.

REFERENCES

Hashimoto, K. (1980). Human Characteristics and Error in Man-Machine Systems. J.Soc. Instrum. & Control Eng. , 19, 836-844.

Kawai K., K.Kurihara and Y.Yokota, (1982) Advanced Process Computer System Applied to Thermal Power Plants, Toshiba Review (International Ed.), 133, (To be published).

U.S. NRC. (1979). Investigation into the March 28, 1979 Three Mile Island Accident by Office of Inspection and Enforcement. NUREG-0600.

No.	Phase	Description	Mode of operator's error	Computer aided system
1	Perception	Information gathering and Acknowledgment (Alarm message acknowledge)	Misperception	Alarm system
2	Judgement	Judgement based on the information in step 1 (Cause of Trip judgement)	Misjudgement	Diagnostic system
3	Operation	Operation by the result of judgement in step 2 (Operation of control switch)	Misoperation	Automatic operation system

Table 1 Classification of operator's errors

(1) Central processing unit : 1 set
   -TOSBAC 7/40 (512 KB)
(2) Fixed head disk : 2 sets
   -FK-40 (8MB x 2)
(3) Magnetic tape unit : 1 set
(4) Color CRT display : 6 sets
   -CRT3200B (Alphanumeric/Graphic)
   Alarm display 1 set
   Information display 2 sets
   Utility display 3 sets
(5) Printer 3 sets
   -Versatec 1200A
(6) Typewriter unit 1 set
   -DEC LA36
(7) Hard copy unit 1 set
   -Tektronix 4632
(8) Communication panel 2 sets
(9) Group selection matrix 2 sets
(10) Process Input/Output 1 set
   -FCAI/UPC
   Analog inputs 1000 points
   Digital inputs 1300 points
   Pulse inputs 8 points
   Digital outputs 128 points
   Analog outputs 24 points

Table 2 Process computer hardware configuration

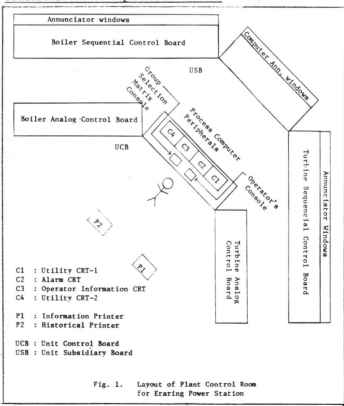

C1 : Utility CRT-1
C2 : Alarm CRT
C3 : Operator Information CRT
C4 : Utility CRT-2

P1 : Information Printer
P2 : Historical Printer

UCB : Unit Control Board
USB : Unit Subsidiary Board

Fig. 1. Layout of Plant Control Room for Eraring Power Station

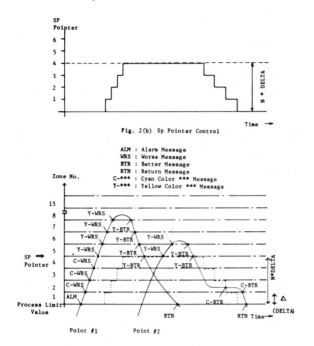

Fig. 2(b) Sp Pointer Control

ALM : Alarm Message
WRS : Worse Message
BTR : Better Message
RTN : Return Message
C-*** : Cyan Color *** Message
Y-*** : Yellow Color *** Message

Fig. 2(a) Alarm Message Output Control

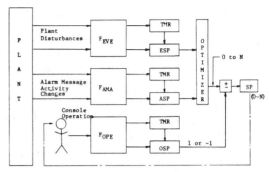

TMR ; SP(Suppression) Pointer Decaying timer processor
ESP ; Event Initiated SP Pointer
ASP ; Message Activity SP Pointer
OSP ; Operator's SP Pointer modification request

Fig. 5. Suppression Pointer Processing
Block Diagram

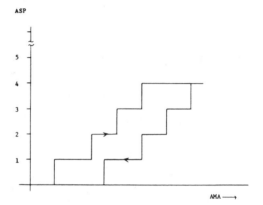

Fig. 6. AMA SP Pointer and its hysteresis

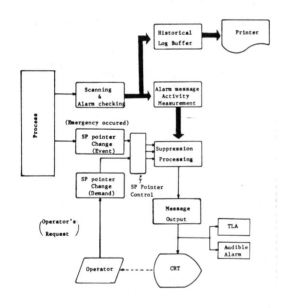

Fig. 3. Alarm Message Control Block Diagram

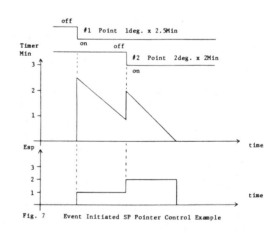

Fig. 7 Event Initiated SP Pointer Control Example

NO OF BACKPAGE	NO OF CURRENT ALM		NO OF BAD PTS		ALM LEVEL		
0	3		5		0		
TIME A-TYPE PID	DESCRIPTION	VALUE	Q UNIT	LIMIT	TYPE	ZONE	
20:35 HI Q30G11	H2 LEAKAGE APPROX	68.33	M3/HR >	1.50	ALM	9	Cyan to Yellow
20:33 LO A01F20	TOTAL AIR FLOW - 2	20.00	PERCT <	25.00	BTR	2	
20:30 DA A25P40	ID FAN A PWR OIL PRS	LOW			ALM		
20:28 A24S10	ID FAN A IB BRG VIB	2.47	MM/S		RTN		

Fig. 4 Alarm CRT Display Example

ERGONOMIC ASPECTS OF WORKING PLACES WITH VDU'S IN CONTROL ROOMS — COMPARISON TO WORKING PLACES WITH VDU'S IN OFFICES

R. Grimm, M. Syrbe and M. Rudolf

Fraunhofer Institute for Information and Dataprocessing, Sebastian-Kneipp-Straße 12-14, 7500 Karlsruhe, Federal Republic of Germany

Abstract. Man-machine communication in control rooms as well as in offices is performed more and more with the aid of computer driven video display units (VDU's). The differences in the system aspects are shown for the different working places such as data input, disposition and process control, giving examples for the workload determining features. General ergonomic aspects are described: information input and output are pointed out in connection with devices as well as information organisation and coding. The tasks for the working places referred to were defined individually for each working place and classified into seven categories. Five working places with VDU's were then examined using the "time budget method", including three control rooms. The results show the expected differences in the task distribution: 80 % of the time was spent for data input in an office working place, compared to 25 % in control rooms. During normal processing situations about 60 % of the time in the examined control rooms was taken up with inactivity, telephone calls and talks, thus necessary free capacity seems to be guaranteed to meet higher demands during process disturbances. The rest periods are also long enough not to put a strain on the operator due to the use of VDU's.

Keywords. Display systems; ergonomics; process control; human workload; man-machine systems.

INTRODUCTION

Video display units (VDU's) are added to work stations in many fields and partly seem to determine them. In the past and to some extent still today this led to - often very emotional - discussions about the "VDU work station" without specifying it precisely. If one looks at work stations which use VDU's and if one tries to classify them by their substance of work (Fig. 1) it can be shown that they extend from offices, starting with data input, via very different places such as programming or CAD or air traffic control to process control rooms. These again have different substances depending on the process state, e.g. steady production with little intervention or starting or ending situations with intensive manual control. Even though these differences are evident, the discussion is often still led on "the VDU-work station".

The goal of this work is to show the common features of the different work stations with VDU's, such as the devices, as well as to point out the differences such as the occupational attributes.

Office:

- Input of numerical data

- Input of alphanumerical data/texts from
  . Written original
  . Sound-track support

- Recall work station
  . Storekeeping
  . Information desk

- Disposition
  . Management information system
  . Expert system (including reservation)

Miscellaneous:

- Programming work station
- CAD/CAM
- Object protection
- Pseudopictures (e.g. ultrasonic)
- Radar (air traffic control)
- Education

Process Control:

- Steady production
- Process starting or ending
- Disturbances
- Training

Fig. 1. Work stations with VDU's

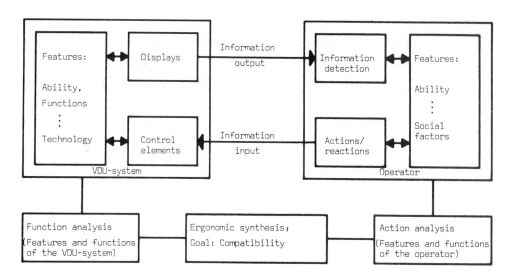

Fig. 2. Structure of the man - VDU interface

DESCRIPTION OF TYPICAL CLASSES OF WORK STATIONS WITH VDU'S

Survey

Figure 2 shows the components and relations between man and VDU within the man-machine system: Information is displayed via the VDU, on the other side man delivers information to the system via the input elements. The features of the VDU are determined e.g. through the functional extent, the ability and the technology used within the working system by a functional analysis.

On the one hand the operator detects information via his sensory organs, through his reactions on the other hand he gives information or instructions to the system. Information input and actions are determined by the kind, extent and sequence of the actions which have to be found out through action analysis. Boundary conditions are ability, aptitude, previous knowledge, motivation, but also social, economical and cultural background of the operator.

Goal of the ergonomic synthesis of the knowledge gained by the functional and action analysis is the adaptation of the VDU to the features and abilities of the operator to obtain a good compatibility of the working system.

Tasks and System Integration

Offices. Tasks and occupational attributes are given for two classes of work stations with VDU's in the office:

- Input of numerical data

  According to Fig. 3 the main actions for the solution of input tasks is the input of data and/or texts from an original into the VDU-system which above all carries out the

storage of these data. The screen is normally used only for the (rather seldom) control of these inputs.
The working scheme is therefore determined by the monotonous sequence

. look at the original,
. input of the data by hand,
. (rather seldom) control of the input data on the screen.

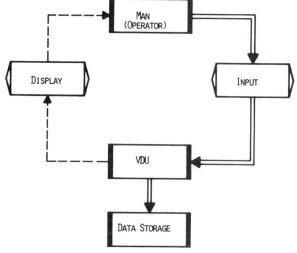

Fig. 3. Working diagram for an input action (in principle)

- Disposition

The task at the disposition working place (e.g. for management or expert systems ac-

cording to Fig. 1) in connection with the
VDU consists of information gathering and
evaluating as well as input of dispositional
information (e.g. booking).
According to Fig. 4 the working scheme is
determined by the interactive work with the
computer driven disposition system via the
VDU

. input,
. waiting for system reaction,
. evaluating the system answers.

During this time a fixation to the screen
and the keyboard can be given.

The aforementioned principles are found mainly
in work stations with VDU's in offices.

determined by

- occasional look at the display, including
  selection of process schemes,
- tracing of process states,
- conversion to new process states,
- detection of disturbances,
- initiation of steps for removal of
  disturbances.

Due to acoustic signalling for attention
when an alarm occurs there is normally no
necessity to watch the screen steadily or
for a longer period of time. Other monitoring
tasks can be examinations outside the control room.

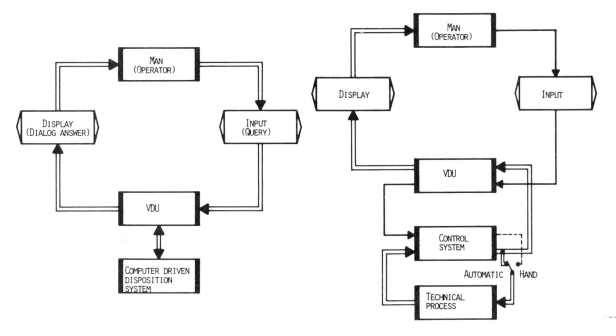

Fig. 4.   Working diagram for a disposition
          task (in principle)

Fig. 5.   Working diagram for a process
          control task (in principle),
          Display and Input are done via
          the VDU-system

Process control. The task in a process control
room is determined by the monitoring and control of a technical process. A closed loop of
action results from the manual control, whereby under normal process conditions the monitoring and control system is a distinct relief
for the operator. Within the whole system the
main task of the operator is the monitoring
of the technical process to detect deviations
from the given set points. Depending on the
degree of automation of the control system
manual inputs are carried out relatively rarely.
Other operator tasks can be protocolling, calculating of values for statistics and graphics,
communication, e.g. to arrange maintenance
work.

In principle Fig. 5 shows the tasks for process
control; the display of process data as well as
the input to the control system is carried out
completely via the VDU. The working scheme is

Interface Design

Within this section some aspects of the organisation and coding of information are
shown as well as different forms of user
guidance at work stations using VDU's.

The coding of information is carried out

- optically by alphanumerical symbols,
  graphical symbols, which can often be
  freely generated, time variant displays
  (curves, bars), colour, and/or
- acoustically by speech, signals.

Colour coding is used for

- separation of information classes,

- classification by importance,
- attention steering,
- ease and acceleration of searching within
  graphical or tabular displays,
- disposition of dialogue steps and clearance
  of dialogue sequences.

Through colour coding the naturalness of displays is greater, the motivation of the users and thus the acceptance of the VDU-systems can be raised. Therefore it should be welcomed that through the use of interactive videotex VDU-systems in offices use colour.

The organisation of information can have local, time and hierarchical criteria in input as well as in output direction. The information can be structured using tables, masks, diagrams, formats and graphical displays. The local organisation for the information display of technical or computer processes is done by graphics or in block structures. There are partly mixed displays, further video pictures (sometimes mixed with other kinds of information) and pseudo pictures.

By means of user guidance a better dialogue between man and technical system can be obtained concerning input as well as output of information. Some user guidance techniques for information input are shown in Table 1, with a list of their advantages as well as disadvantages.

In addition to the aforementioned activities features of work stations with VDU's can be classified as shown in Table 2.

METHOD OF INVESTIGATION

Service Model with Concurrent Customers

Service models are of special interest within man-machine systems. They can be applied to a relatively wide area of different situations and give information on the workload of operators. Figure 6 shows the scheme of a service system (Schumacher, 1978). The operator has to fulfil a series of tasks (customers of the system) which can arise to a certain extent simultaneously and have different priorities.

The operator in the service system can be represented by a model containing a waiting room (short-term memory) and service "mechanism". When the strain is too great the operator rejects customers; the service "mechanism" fills up more and more, the waiting room has no more spare capacity. Impatient customers (i.e. demands limited in time) leave the waiting room.

Service models like this can be applied to completely different situations. In process control or the driving of a vehicle for

TABLE 1   Types of Dialogue for Information Input

Dialogue Technique	Procedure	Advantages	Disadvantages
Query	Standardized queries	- simple, also for non-skilled user	- low flexibility - no alternatives
Key word	Input by key words	- close to natural language	- low flexibility - high possibility for mistakes
Coding (Programming style)	Technical abbreviations, abstract code	- short dialogue - powerful functional extent	- training necessary - high ability for abstraction and concentration necessary
Yes/No	Computer guided dialogue with decision trees	- very simple	- low flexibility - tedious
Instruction and response	Responses are given by the system	- very simple - alternatives possible	- low flexibility - tedious
Menu selection	Presentation of response menus	- clear dialogue guidance - easy to learn - alternatives possible	- slow, when many dialogue steps
Form filling	Fill-in-the-blanks	- simple - low possibility for mistakes	- low flexibility

TABLE 2   Some Features concerning Work Stations in Offices
          and in Control Rooms

Features	Offices	Control Rooms
Devices	- mostly monochromatical alphanumerical displays - mostly several degrees of brightness - alphanumerical keyboards - normally only few functional keys	- computer driven colour display systems - trend to input devices which allow direct pointing (e.g. light pen, touch panel) instead of alphanumerical or positioning input (e.g. cursor)
Information output	- mostly unstructured organisation of information, partly in tables - coding of information by alphanumerical symbols	- nearly always structured organisation of information by . process graphics . block structure (quick scan) . partly combinations hereof often hierarchical information display - coding of information . optic by alphanumerical and/or graphical symbols, bars, curves mostly combined, colours, blinking . acoustic by horn; speech output still seldom
Information input	- mostly by alphanumerical keyboards including few functional keys - input positioning with cursor - tabulator key	- structured by format control, masks or input lines using alphanumerical or graphical symbols
User guidance	- seldom or to a small extent (e.g. blinking of input position or input masks or formulas)	- mostly to a large extent, e.g. . search-aids . acoustic signalling (attention steering) . interactive input and feedback . automatic plausibility control - input support (e.g. input sequence, recipes for time and material driven automatic set point control)

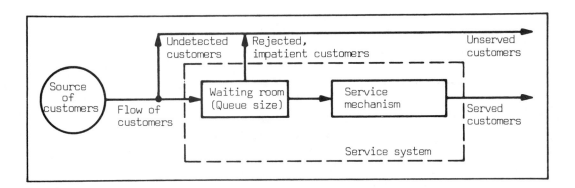

Fig. 6.   Basic structure of a service system (according to Schumacher)

example, each activity can be reflected in a
stream of demands, generated from a special
source. When only monitoring is required the
observation of one display means that the other
displays (which might show disturbances) cannot be monitored: The instruments are lined up
in a waiting queue and await monitoring by the
operator.

Since queuing theory is time oriented one can
consider the fraction of time devoted to each
task and the total fraction of time required
by all tasks (Rouse, 1980). The server occupancy R can then be determined as follows:
Let T_K be the mean time between arrivals and
t_K the mean service time for the K th task of
a set of N tasks, then $R_K = t_K/T_K$ is the fraction of time required by task K and
$R = R_1 + R_2 + ... + R_N$ is the fraction of time
required by all tasks. Thus, the server occupancy can be used as a measure of workload,
even though it is a rather gross measure.

Workload Evaluation using the Time Budget Method

Within a man-machine system the condition R <1
must be fulfiled to avoid losses in customers.
Computer simulations (Rouse, 1980) show that a
server occupancy of R ≥ 0.7 will result in a
steep rise in the average queue length and thus
in a high workload. This leads to rejected or
unserved customers. From experiences in the
assessment of workload and performance capacity
in air traffic control systems it could be
shown that the operators changed working strategies in order to save time for additional
loads when the workload approached 80 % of the
capacity limit. In communication e.g. the
operator started speaking with less redundancy
(Seifert, 1980). In Fig. 7 the performance
level is scaled in terms of percentage workload
(W = 100·R). Only up to about 75 % workload all
customers are fulfilled, above this threshold
losses occur.

The time budget summarized from the fraction of
times required by all tasks can therefore be
used as a first-order approximation for the

workload of the operator. That means the
workload (W) is given by the ratio of time
required to accomplish a task (no matter how
simple or complex it may be) to the time
available. One must take care however, that
the single activities are not defined too
globally to ensure that no workload elements
get lost (e.g. a task named "monitoring a
technical process" which contains the execution of different activities will not allow
statements on the workload). On the other
hand, when the single activities are too
short the measuring costs will be too high
compared to the consistency and accuracy
attained.

The time budget method can also be used to
analyse office work. If e.g. the data input
has to result time tapped, losses in the form
of unserved customers can occur if the number
of customers is too high. When the data input
is not time tapped notice is taken of all
customers (e.g. "forms to be written"). In
this case however, the error rate rises when
the input rate is too high, that means losses
occur in the form of input errors which show
an overload situation. The "operator" is
therefore unable to work at maximum speed
over a long period of time.

In the practice the time budget method means
a decomposition of the tasks and a time
rating similar to a REFA-analysis. The exemplified time budgets for the investigated
work stations are given in Table 3. The
demand for time has to be averaged over
different activities within the same class
as well as over the group of people working
there, who have differing abilities, motivations and temperaments. Therefore two means
of estimating T_i are used:

- estimation of the time budgets by a person
  carrying high responsibility (foreman,
  works manager),
- series of measurements under representative
  conditions.

Even though the estimation seems to result in
arbitrary fixations of T_i good statements over
the workload situation were to be found.

Analysis of some Work Stations

The described relations to work stations with
VDU's in offices and process control rooms
were supported by prototype investigations
in five work stations, two in offices and
three in control rooms:

- one work station for data input,

- one work station for disposition, here
  retrieval of literature in large data banks,

- one control room with conventional back-up,
  the VDU is used only as additional display,

- one control room with VDU-systems, alphanumerical and functional keyboards, without
  conventional back-up,

- one control room with VDU-system; the input
  of information to the process is done only

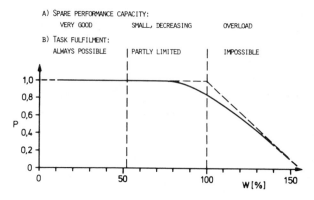

Fig. 7. Interrelation between task fulfilment
of the system P and workload W

TABLE 3 Classification of Activities at the Different Work Stations, $\sum_{i=1}^{7} T_i = 100\ \%$

Work stations Activities	Data input	Disposition	Control room
Operating (T_1)	- reading of forms - input of dates, control signals, corrections - reading at the screen (controlling)	- input of data, control signals, corrections	- selection of pictures, stations, elements - acknowledging of alarms - input of set points and thresholds - input of instructions, switches <u>simultaneously</u> - monitoring of displays (control panel, screen) - telephone calls directly related to the control task (e.g. when switching) - talks directly related to the control task
Monitoring (T_2)	- none (integrated in operating)	1. reading at the screen (monitoring, also during system reaction time delays) 2. reading of information in documents and notes	- monitoring of displays - control runs
Recording (T_3)	- short notes	- short notes	- recording in blank forms and logs - calculations, drawings, evaluations - reading of notes and directions <u>in connection with</u> - reading of displays - control runs
Telephone calls (T_4)	- not directly related to the operating task	- not directly related to the disposition task	- not directly related to the control task
Talks (T_5)	- not directly related to the operating task	- not directly related to the disposition task	- not directly related to the control task
Delay time (T_6)	- sorting of forms - system reaction time	- system reaction time (simultaneous to Monitoring 1.) - setting time	- system reaction time
Pauses (T_7)	- inactivity - absence	- inactivity - absence	- inactivity - absence (no control runs)

via the VDU-system using a virtual keyboard
and a light pen, without any conventional
back-up.

The investigations in the work stations were
carried out in three steps:

- Definition of the individual time budgets
  and measurement of the workload by summing
  the single activities - corresponds to R(t).

- Valuation of other load factors through observation and questioning of test persons and
  their superiors.

- Determination of subjective influences of the
  employees by the use of a questionnaire to
  estimate the strain.

The results of the first part of the investigations (measurement using the time budget method)
are given within this paper.

RESULTS

The activities were measured and classified according to Table 3. Figure 8 shows the normalized sums of the activities as a percentage of
the total time.

The results show

- high percentage of time needed for operation
  and monitoring at the office stations for
  data input and disposition during the measured period compared to rather low percentage of time for these activities in
  process control rooms under normal process
  conditions,
- much spare time in the process control rooms
  investigated (telephone calls and talks not
  directly related to the task and times of
  inactivity).

The percentage of the spare time was about 60 %
during the time of the investigations. The processes were in normal operation. This allows
the conclusion that on the one hand

- because of the low workload of 40 % d v ing
  normal operation there is a high additional,
  but necessary spare capacity to meet disturbances, and on the other hand
- the necessary pauses with a high degree of
  recovery (distribution of the time for pauses
  over the working hours in many small portions)
  are given in the investigates process control
  rooms, at least when the process is in normal
  operation.

ACKNOWLEDGEMENTS

The authors wish to thank Dr. Haller for his
interviews with the test persons and their
superiors and Dr. Carls as well as Mr. Thiel
for their contributions to the time measurements.

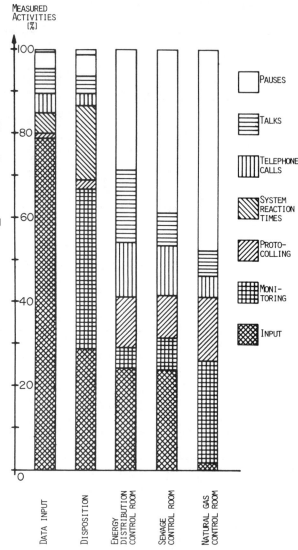

Fig. 8. Distribution of the time required
by each task as a percentage of
the total time

REFERENCES

Rouse, W. B. (1980). Systems engineering
    models of human-machine interaction.
    In A. P. Sage (Ed.), System Science and
    Engineering, Vol. 6. North Holland Series,
    New York.
Schumacher, W. (1978). Bedienungsstrategien
    des Menschen bei konkurrienden binären
    Forderungen in Mensch-Maschine-Systemen.
    Mitteilungen aus dem Institut für Infor-
    mationsverarbeitung in Technik und
    Biologie. FhG-Berichte 1/2-78, 27-33.
Seifert, R. (1980). Man machine systems per-
    formance as a function of the system
    demand/workload relationship. Conference
    on Manned System Design, New Methods and
    Equipment. Freiburg (W.-Germany). Pre-
    print, 160-180.

CONTROL AND SUPERVISION OF THE EURELIOS SOLAR POWER PLANT: DESIGN PHILOSOPHY AND OPERATION EXPERIENCE

C. Maffezzoni and M. Maini

ENEL, DSR-CRA (Italian Electricity Board, Research and Development Dept. Automatica Research Center), Via Valvassori Peroni, 77-20133 Milan, Italy

Abstract. Operating a solar power plant with limited personnel employment, while providing an operation flexibility adequate to the necessary experimental work, requires a non trivial coordination of the different subsystems realizing the complex energetic conversion and the adoption of a composite automatic control system keeping the plant components at their stipulated operating conditions. The paper describes the basic concept and the practical organization of the control and supervision functions adopted for the central receiver solar powered plant built near Adrano (Sicily) in the frame of the Eurelios Project founded by EEC. In particular, results of an accurate simulation study and experience of the first months of real plant operation are taken into account when discussing the most significant aspects of operator-plant interaction for which quite peculiar solutions have to be devised to comply with the continuously varying (dynamic) plant operation.

Keywords. Solar power plant; Man-machine Systems; Process Control; Supervision Systems; Modelling; Steam Plants.

INTRODUCTION

Exploiting solar energy for electric power generation is the objective of several pilot projects essentially aimed at establishing the feasibility, both technical and economic, of such an energy conversion by practically testing basic plant concepts proposed in a number of pioneering works. Restricting attention to the case in which the solar-electric energy conversion is realized through a more or less conventional thermal cycle, the concept of the underlying physical process can be described as in Fig. 1, where the basic transformations are indicated.

The process features the most relevant to the design of automation and man-machine interaction are the following:

i) The receiver has to work at high temperatures (about 500 °C), to obtain a reasonable efficiency of the thermal cycle, and with very high thermal power densities to limit its thermal losses.

ii) In view of item (i), the concentration subsystem is required to exibit very high performances, because the achievement of high concentration factors (in the range of 300 ÷ 1000) is generally possible only with excellent accuracy of the control equipments devoted to the sun tracking.

iii) The size of the plant (i.e. its nominal power output) is in general quite small, due to the distributed nature of the solar energy; for this reason, the economic application of such plants calls for the

employement of very few operating people (e.g. 1 or 2 operators).

Therefore, solar power plants are characterized by some critical components which require accurate control, monitoring and supervision, though only little employment of human work is admissible. These rather comflicting objectives can be achieved by a suitable application of computer and control technologies and, in particular, by designing effective man-machine interaction tools.

The paper is organized as follows:

- Section 2 describes the general concepts of the overall system organization;
- Section 3 described the main features and the basic operating criteria of the Eurelios[1] solar power plant (Hofmann and Gretz, 1978), together with the modelling approach adopted for the control design;
- Section 4 describes the automation and control functions which have been implemented on the Eurelios plant;
- Section 5 deals with the man-machine interaction problem, with particular regard to the realization of the plant manoeuvres and to the plant supervision;
- Finally, Section 6 draws some preliminary conclusions also relying on the first

----

[1] The aim of Eurelios Project, founded by the Economic European Community, is the realization of a 1 MW(e) solar power plant near Adrano, Sicily. Its first electric output to the grid has been obtained in May 1981.

months of plant operation experience.

OVERALL SYSTEM ORGANIZATION

The overall system under consideration can be thought as organized in three hierchical levels (as illustrated in Fig. 2 by the so called Process Information System (P.I.S.)):
i) the plant (P) level where all the equipments/components devoted to the realization of the energy transformation are included;
ii) the automation (A) level where the different subsystems allowing plant measurements and automatic control are suitably arranged;
iii) the supervision (S) level, which carries on the necessary interactions between the human operator and the plant.
The scheme of Fig. 2, whose details will be considered in the following sections, shows the essential information flows and actuating commands and emphasizes the modular organization of the A and S levels, which actually corresponds to the pratical realization of the different monitoring and controlling equipments. This is very important for the Eurelios plant, which, being totally experimental, could be subject to continuons modification/integration both at the P level and at the A and S levels, on the basis of the operating experience.

PLANT MODELLING AND OPERATION

The plant can be split into six different subsystems (see Fig. 2):
- the mirror field P3, which collects the solar energy and focuses it to a receiver located on a tower;
- the receiver P1, which essentially consists of a once through steam generator, made of two parallel tubes suitably disposed within a bell-shaped hollow (Castellazzi, 1979);
- the steam cycle P2, of which the essential elements are the turbine, the condenser, the feed water pumps, the turbine bypass and the necessary linking pipes endowed with a number of regulating or intercepting valves;
- the storage system P7 which, at complete filling, is able to supply 5 ÷ 10% of the nominal power for about half an hour;
- the alternator and the electrical sistem P6;
- the meteorological system (P4 and P5), getting the necessary information about the environmental conditions.
The detailed scheme of the thermoydraulic part of the plant is described in Fig. 3.
The nominal ratings of the plant (with full solar radiation at the equinox) are the following:
- generated electric power: 1 MW
- feed water mass flow rate: 4860 kg/h
- steam temperature at the receiver outlet: 510 °C
- steam pressure at the receiver outlet:64 bar

- steam pressure at the storage system outlet (when operating): 7 ÷ 19 bar.
Of fundamental importance to the design and optimization of the plant control and supervision is the knowledge of the dynamic behaviour of the plant, in particular during the most typical and critical transients (due for instance to cloud passages), where the variations of the main physical quantities (temperature and pressures) can compromise the operation continuity and, in the limit, the safety of the most stressed plant components (e.g. the solar receiver). To this end, a quite sophisticated dynamical model of the process (in particular of the solar receiver and of the thermal cycle) has been built according to the scheme of Fig. 3. Due to the length of the receiver tubes, transport phenomena are very important in the present case, so that the mathematical model has been built by using the computer code SICLE (Maxant and Perrin, 1979) conceived for the numerical integration of systems of partial differential equations describing thermoidraulic phenomena.
Details on this reference model are reported in (Maffezzoni and Parigi, 1982), where also the most significant plant transients are presented.
The complex dynamic behaviour of the process and the severity of the environemental disturbances (particularly in case of cloud passing) imply that non trivial solutions have to be found in order to obtain a satisfactory control of the plant. In particular, the following basic concepts have been applied:
- The steam pressure has to be made almost insentive to the solar radiation disturbances to avoid unacceptable motions of the evaporation zone.
- The position of the evaporation zone has to be kept as fixed as possible to prevent the receiver outlet temperature to go out its prescribed limits.
- The thermal balance in the preheating zone has to be very well controlled because possible unbalances strongly disturb the evaporation zone and can not be rapidily compensated by varying the feedwater flow because of the very long transport delay.
In addition, since the receiver operates at high temperatures and with high thermal fluxes, an important operation constraint is that of keeping the receiver metal temperature below a prescribed maximum value.
The above general control requirements can practically be met by equipping the process with a suitable measurement system and adequate control actions. Referring to Fig. 3 and to the general scheme of Fig. 2, the relevant manipulated variables are:
- the feedwater mass flow rates in both receiver tubes, by acting on the feedwater valves V1 and V2 and on the recirculation valve V3;
- the mass flow rate of the attemperating sprays of the evaporation zone, by acting on the spray valves VA1 and VA2;
- the mass flow rate of the attemperating sprays of the superheating zone by acting

on the spray valves VA3 and VA4;
- the position of the receiver isolation valve VIS and of the bypass valve VBY (devoted to the receiver pressure control).

Control and monitoring of the fluid temperature evolution, of the evaporating zone motions and of the receiver pressure require the following process measurements:
- fluid temperature at 4 equispaced points in the preheating zone;
- fluid temperature at the evaporator outlet, conventionally identified as the point where there is steam with 100 °C of superheating in nominal conditions;
- fluid temperature immediately upstream and downstream the superheater attemperating sprays;
- fluid temperature at the receiver outlet and at the turbine inlet;
- metal temperature at several points on the radiated surface of the receiver (in particular where the thermal flux is high);
- fluid pressure at the receiver outlet and at the turbine inlet.

Of course, all the above measurements are applied to both receiver tubes, since possible unbalances between the two have not to be allowed.

Special measure devices are devoted to the monitoring of the relevant environmental conditions and of the interactions between the mirror field and the solar receiver. More precisely, they are the following:
- direct solar radiation measurement obtained by some pyroeliometers located in suitable positions of the mirror field;
- meteorological measurements for the identification of not favourable or even dangerous environmental conditions;
- receiver thermal flux measurements realized by several fluxmeters suitably welded on the receiver tubes;
- mirror field state, identified by the condition of each heliostat.

The complexity of the man-machine interaction in the solar plant is mainly due to the fact that it has to operate very frequently in dynamic conditions, particularly because it has to be started each morning and because it can be frequently disturbed by solar radiation variations (passing of more or less dense clouds). Actually, almost every day the solar plant is expected to operate in the following different conditions:
a) normal operation, with or without solar disturbances, when the turbine is feeded by the receiver and the alternator is connected to the grid;
b) operation with the turbine feeded by the storage system and the alternator connected to the grid (this conditions occurs when the solar radiation is below a technical minimum);
c) cold and hot start-up;
d) shut down of the plant, when there is no sufficient thermal energy to feed the turbine;
e) fast starting with hot receiver, when the solar radiation failed for a limited

interval of time (5 ÷ 60 min).

The conflicting requirements of driving the solar plant to the condition of normal production in the minimum possible time and of ensuring sufficiently smooth temperature variations essentially determine the basic principles of the plant operation and make the plant manoeuvres quite critical to be performed without adequate control and operator guide tools. This is discussed in detail in Section 4 and 5.

It is finally worth mentioning that control strategies and operating procedures should be verified before to apply them to the real plant. This has been done by using a real time digital simulator (Maffezzoni and co-workers', 1981) based on a slightly simplified version of the basic reference model. Indeed, the real time simulator coupled to the real control system equipped with the main operator commands allowed a careful tuning of the operating procedures which were incorporated in the preliminary operating manual used by the operating personnel.

PLANT CONTROL AND AUTOMATION

As it is apparent from the general scheme of Fig. 2, the automatic control level (level A) includes the basic functions of regulation, manoeuvres automation and protection and can be split into the following interacting subsystems:
- A1 : Subsystem dedicated to the automatic control of the heliostat motions.
- A2 : Subsystem dedicated to the automatic control and protection of the receiver (steam generator), of the thermal cycle (turbine, condenser, etc.) and of the electric generator.
- A3 : Subsystem dedicated to processing of the solar radiation at the ground, of the thermal power entering the receiver and of the meteorological measurements.
- A4 : Subsystem dedicated to the automation of the operation manoeuvres.

The coordination of the different control actions has been achieved by conceiving the heliostat field control subsystem (A1) as hierarchically enslaved with respect to the couple A2 + A4. As a consequence the subsystems A2 and A4 receive information concerning the mirror field state and issue synthetic commands to A1, by which they are decoded and transformed into detailed commands to the single heliostats. The most relevant parameters to the receiver control is the total thermal power entering the receiver, whose estimation is performed by subsystem A3.

In fact, due to the particular nature of the solar energy source, the total radiation entering the receiver is affected by unwanted and practically not predictable variations (haze, cloud passing), so that the receiver control and its timely and correct interaction with the mirror field control can successfully be performed only if good estimate (Q_{er}) of the thermal power entering the receiver is available.

The interaction between (A2 + A4) in the opposite direction, i.e. the commands to the heliostat field, is handled by means of a pre-defined set of deviation/focusing sequences, applied both to emergency and to normal operating conditions. This has been realized by a special code device handling the interface protocol.

The interaction between A2 and A4 has been conceived in such a way that the manoeuvre automation subsystem sees the regulation subsystem as a slave system. In fact, commands issued by A4 generally consist of set point variation demands for suitable regulation loops belonging to A2. Moreover, to allow both automatic and semiautomatic operation, system A4 can be switched to manual operation; in this latter case, its role is assumed by the human operator who can manually perform the plant manoeuvres.

Heliostat control subsystem (A1)

The whole mirror field is subdivided in eight groups and managed by two digital control units (MCU1, MCU2) (four groups per unit) which assure the following basic functions:

a) to drive and then maintain the aforementioned groups in some well defined operational status like:
  a.1) stowage (i.e. safety position);
  a.2) stand by (i.e. sun tracking and focusing outside the boiler, ready for sending the flux into the receiver);
  a.3) receiver (i.e. sun tracking and focusing the flux into the boiler).
b) to deviate, in case of emergency, the whole mirror field, quickly driving all groups: to stand-by in case of emergency due to boiler and/or cycle; to stowage in case of emergency due to environmental conditions.
c) to test, diagnose and separately drive each single heliostat for supervision, check and maintenance purposes.
d) to handshake commands and status signals with receiver control subsystem (A2) and with operator's desk for allowing the automation and the supervision of the most significant mirrors manoeuvres.

The aforementioned features show that the heliostat control subsystem has been designed as a completely autonomous local device, which cleverly solves at its level all the problems concerning the heliostat field; moreover the digital interface protocol (just mentioned in d) allows its integration into a more complex hierarchic system, which provides a more efficient and safe control strategy.

Receiver control subsystem (A2)

Due to the severity of the solar disturbances, the most complex and critical part of subsystem A2 is that carrying on the receiver pressure and temperature regulation (see, Maffezzoni and Parigi, 1978 for a detailed description). A concise and simplified scheme of the receiver regulation functions is reported in Fig. 4, where the essential information pattern and the main elaboration blocks are indicated.

In particular, there appears the use of the thermal power estimate Qer, obtained by the special device SRES and exploited for the feedforward anticipation of the main receiver control actions when a solar disturbance occurs. There is also indicated the generation of the sequential commands to the mirror control system (A1) issueing heliostats deviation when this is required by the thermodynamic conditions of the steam generator.

The two principal regulations, i,e, that of the fluid temperature and that of the pressure, are, respectively, performed by the feedwater and sprays controllers (ESC, SSC and FWC) and by the pressure controller (PC) acting on the isolation and bypass valves (see Fig. 4).

For the sake of simplicity, the interaction between A2 and A4 is not explicitly indicated in the scheme; however, the regulation subsystem of Fig. 4 has been designed so as to receive command signals from the automation subsystem, which can replace the manual pressure set-point and the demands of feedwater and attemperating spray flows.

Meteo and radiation measurement subsystem (A3)

The main purposes of this subsystem are:
a) Estimating with sufficient accuracy the instaneous value of the net radiation incident upon the whole mirror field, which is an essential parameter for relevant operating procedures (like the start-up). To this purpose a special automatic sun tracking sensor (P4 in Fig. 2) has been employed, which gives an analog signal proportional to the net radiation value and displays the actual sun position and the possible tracking errors. Four of those devices installed at suitable corners of mirror field give an average value of actual incident radiation and allow the detection of partial mirror field shadowing due to irregular clouds.
b) Computing the best estimate of the current thermal power entering into the boiler: this measurement is essential for driving feedforward control actions, as explained before. To this purpose a special computing device has been developed which:
  - receives the measurements coming from twenty fluxmeter, suitably located on three different planes defined in the boiler body;
  - selects and takes into account only good fluxmeters, discarding the wrong ones;
  - computes, on the basis of the nominal map of the thermal flux, an estimate of the global thermal power (called Qer);
  - provides a complete monitoring of the

flux intermediate values and of the sensor availability.

c) Giving the operator on some chart recorders the status and the trend of the main meteorological parameters (atm. pressure, air temperature, humidity, etc.).

The just mentioned plant data (a, b, c) are the main operation parameters presented to the operator which are strictly typical of the solar plant.

The first experimental period shows that the operators become quickly familiar with actual energetic parameters (net incident radiation, global thermal power)but they learn more slowly to make use of meteo parameters and of meteo forecasts. In particular, it turned out that a specific training of the plant operators should be provided in order to improve their capability of using short and medium term weather forecasts, which mainly affect plant operation (e.g. kind of clouds and of wind).

Manoeuvres automation subsystem (A4)

In the case of the solar plant, manoeuvres are much more frequent then in conventional generating units and have to be performed in the minimum needed time to maximize energy production.

Moreover, since the dynamics of the plant is not much faster than the natural daily variations of the solar radiation, receiver heating and plant reloading have to be handled as large transients, in which the process dynamics plays a fundamental role (Maffezzoni and Parigi, 1982). Therefore, subsystem A4 can not be easily conceived as an open-loop sequence automation device, but rather as a digital control units (DCU) able to apply different control policies in dependence of the possible plant manoeuvres to be effected. The most important operating procedures to be dealt with are the morning start-up and the hot restart of the plant after a time interval in which the solar radiation was lacking and the receiver has been kept pressurised. Though the basic concept of the DCU has already been defined and its practical importance has been made evident by the first six months of plant operation, the optimization of the operating procedures and of the corresponding control algorithms is still matter of investigation. In this respect, the following critical points have to be overcome:

- the timely coordination between computer commands and manual operations to be locally performed on plant components.
- the optimal tradeoff between speed of receiver heating and procedure safety (in fact, when the heating of the receiver is too fast, it generally happens that water boiling starts in the central part of the receiver producing unberable pressure and flow variations of the receiver fluid).

MAN-MACHINE INTERACTION

As shown in Fig. 2 the first level (S1) of man-plant interaction has been designed around a conventional set of vertical panels and operator's desks, equipped with traditional instruments like: manipulators, pushbuttons, lamps, indicators, chart recorders and so on. This basic philosophy has been chosen also to facilitate the impact of the operators team with unconventional operational problems of solar plant. In fact, because there does not exist a well assessed training procedure or equipment (like a training simulator) for solar plant operators, we managed the plant during the first operational period (6 ÷ 8 months) mainly as a "training tool" for operation people who came from other kind of experience. During this first "special" phase the conventional (and therefore limited) supervision system (S1) was (paradoxically) in some way advantageous, because:

i) it allowed to focuse people attention on plant basic phenomena rather than on sophisticated supervision tools;

ii) it forced people to face the lacks and limitations of S1 system with their "human" attention, feeling and interpretation resources;

iii) it put in evidence to the design people, the most relevant points to be improved with the computerized supervision system (S2-S3).

Taking into account the aforementioned topics and the practical experience exploited during the first operation period, the most relevant results (as far as Man-Machine Interface (MMI) is concerned) can be sum marized as follows:

i) The solar plant, for its nature[1], is practically in a permanent (day by day, hour by hour) transient condition: this is quite different from traditional thermal or nuclear plants, where transients occur very seldom. The operation people, coming obviously from traditional plants, have found some difficulties and have been in some way upset by this new mode of operation. It this case, in fact, the traditional instrumentation ("dispersed" on panels and desks) forces the operator to a continuous "walk and look around", which appears as a potential stressing factor if it grows (for instance, for meteo reasons) over a certain limit.

ii) The complex dynamic behaviour of the plant (related to the energetic transformation chain, see Fig. 1) requires the operator to know (or to estimate) time by time the global energy and mass balance, or, more precisely, to exploit a kind of "mental model" of the process. Traditional supervision tools (S1) revealed themselves very limited in supporting those operator's needs, mainly because they have very poor computational resources built-in.

iii) The most typical operator's reactions we found, against the lacks of information or stressing factors just mentioned in i) and ii), where:

----

[1] The following considerations concern also wind power systems, photovoltaic plants etc.

- an increase of the number of erroneous manoeuvres and their recovery time;
- a reduction of operation speed, mainly during hot and cold start-up;
- an increase of the number of "no start-up conditions"

That obviously could reduce the plant global efficiency and safety.

iv) When a system engineer assists the operation people (as, for instance, during the first tests), we have found that a well calibrated set of advices can substantially help the operation people and also overcome most of just mentioned difficulties. Therefore we can conclude that this kind of plant needs a skilled set of real time "operator's guides", built into the computerized supervision system and extended to the most important plant components.

With reference to the last items i) - iv), the identification of the behavioural plant model, the operators should built in their mind and the real time operator guide should meet, is of primary interest both for operator training and for computer-man interaction design. In the present case it has to be based on the following fundamental facts:

- Up and down quick thermal power variations are very common even during the central part of the day, also because white clouds and irregular wind are typical for the geographical area where the plant is located.
- Fluid and metal temperatures of those parts of the boiler where the local thermal flux is high respond quite promptly to solar disturbances.
- The temperature of metal and fluid at points located downsteam high flux zones are subject to delayed variations, due to the transport time of the fluid temperature. However, the absolute values of these temperatures can also exceed those attained in the more heated parts.

Thus, manual control and supervision must be essentially predictive, in the sense that the operator intervention can be sufficiently prompt only if a successively improving estimate of the disturbance effect is made, first by looking at the Q_{er} signal which gives the thermal input variations, then by looking at the temperature measurements relative to the most radiated parts, finally by looking at the temperature at the receiver outlet, where the "long term" variations are larger. Actually, the sensor location in the receiver has been choosen on the basis of the described behavioural model, which has in this way been learned quite rapidly by the most skilled operators.

THE SUPERVISION SYSTEM

The MMI level (and therefore the plant performance) will be substantially improved by introducing the S2-S3 computerized system (in the following called DASS), which is presently in phase of advanced development and foreseen for activation during next December. The basic dual-computer structure of DASS is shown in Fig. 5 and its main features can be summarized as follows:

a) The dual computer structure decouples the task of acquiring and formatting plant data from that of processing them. This reduces the "risk" of software "dead lock" (which is a typical "ghostmare" of operators) and allows (by debug console) the operator to test "quality" of plant data as they enter the system.

b) The three levels of consoles (operator's, experimenter's, programmer's console) allow the operator to carry out the plant daily program, while the experimenter organizes and carries out a scientific analysis of plant data or evaluates in real time the results of his special programs. These consoles are also located in different rooms, to avoid interference between activities. Finally the third console (programmer's cons.) allows at any time to carry out programming, debugging and housekeeping standard activities.

c) The basic software features included into DASS allow:
   i) The presentation on colour and B/W CRT's the actual plant status, by means of a structured set of mimic diagrams, x-y diagrams, bar-graph istograms, alphanumeric pages.
   ii) Helping the operator in correlating and searching across the whole set of display pages, by means of a logic structure based on a physical subdivision of the plant in some well defined subsystems. This approach allows one to recall on CRT details of each subsystem by means of simple console commands.
   iii) Helping the operator in managing the plant by means of some skilled "operator's guides", presented on CRTs' in real time during the most critical manoeuvres and conceived according to the "behavioural model" illustrated at the end of the preceding section.

CONCLUDING REMARKS

Attaining competitive performances in electric power generation by solar energy calls for an integrated design of process, control and man-machine interface, in order to meet the quite peculiar features of such power plants. In particular, the following somewhat conflicting objectives have to be taken into account:

- the efficiency of the energy transformation requires the adoption of high temperature solar receivers and of radiation concentration equipments with high accuracy, i.e. of sophisticated processes.
- The distributed nature of the energy source essentially limits the size of the plant, so that very few operating people can be economically employed.
- The natural cycling and the very common irregularities (cloud passages) of the

energy source imply that the plant has to be operated in a much more variable way than the conventional power plants.

The above objectives can be complied with by equipping the plant with a sufficient degree of control and computer technology, whose features have to be well fitted to the process.

In the paper it has been shown how, based on an accurate mathematical model of the considered process, the concept of the overall process information system can be developed and how the different monitoring, control, automation and man-machine interface functions can be designed. In particular, the first months of plant operation have emphasized the importance of devising new modes for presenting information to the (possibly unique) operator of the plant, who has shown a certain degree of embarrassment in facing a continuously varying plant operation and its interaction with nonomogeneous environmental phenomena.

ACKNOWLEDGEMENTS

The participation of ENEL to EURELIOS Project, in the frame of which this work was developed, has been lead by Mr. G. Dinelli and Mr. G. Cefaratti of CRTN (Thermal and Nuclear Research Center) – Pisa. We wish to thank them for their cooperation and support.

We are also grateful to the European Economic Community, for the financial support of the study, to the engineering staff of AMN and Ansaldo Solar Energy Department for fruitful technical discussions and, in particular, to Prof. G.Francia who designed the solar receiver of the power plant. A special acknowledgement is also due to G.Pandini and F. Parigi of ENEL, who cooperate with us to this realization.

REFERENCES

Hofman,J. and Gretz, J. (1978). The concept of the 1MWel solar thermal power plant of the European Community. 2.International Sonnenforum, XII Zusammenkunft der Comples, Hamburg.

Castellazzi, P. (1979). High temperature thermodynamic conversion. Energia Domani ed. by CESEN, No. 10/11 (in Italian).

Maxant,M. and Perrin,M. (1979). Mathematical modelling of two phase flow applied to nuclear power plants. II Multiphase Flow and Heat Transfer Symposium, Miami.

Maffezzoni,C. and Parigi,F. (1982). Dynamic analysis and control of a solar power plant. Part 1: Dynamic analysis and operation criteria. Solar Energy, Vol. 28 n. 2, pp. 105-116.

Maffezzoni,C. and Parigi,F. (1982). Dynamic analysis and control of a solar power plant. Part 2: Control design and simulation. Solar Energy, Vol. 28 n. 2, pp. 117-128.

Maffezzoni,C. and co-workers' (1981). A real time digital simulator for testing the control system of the solar power plant Rivista di Informatica, Vol. XI n. 2 pp. 211- 216.

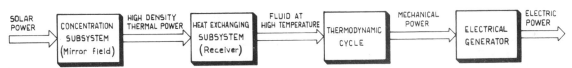

Fig. 1. Basic energy transformation on solar plant.

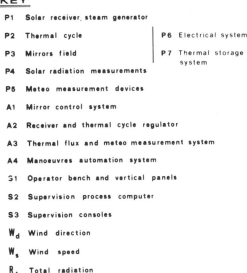

KEY

P1 Solar receiver, steam generator

P2 Thermal cycle

P3 Mirrors field

P4 Solar radiation measurements

P5 Meteo measurement devices

P6 Electrical system

P7 Thermal storage system

A1 Mirror control system

A2 Receiver and thermal cycle regulator

A3 Thermal flux and meteo measurement system

A4 Manoeuvres automation system

S1 Operator bench and vertical panels

S2 Supervision process computer

S3 Supervision consoles

W_d Wind direction

W_s Wind speed

R_t Total radiation

P Pressure

T Temperature

H Humidity

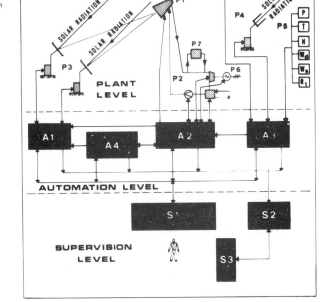

Fig. 2. Overall view of solar plant.

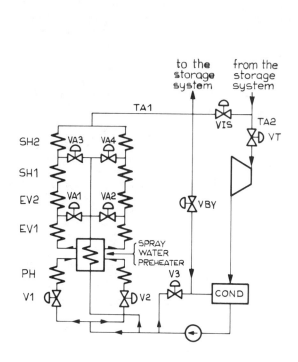

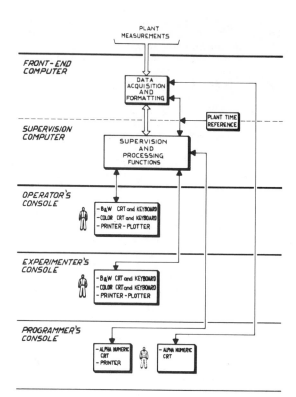

Fig. 3.   Scheme of the receiver and of the
          steam cycle.

Fig. 5.   The supervision system.

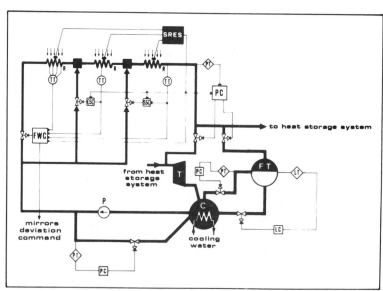

Fig. 4.   Regulation scheme.

COMPUTER AIDED CONTROL STATION WITH COLOURED DISPLAY FOR PRODUCTION CONTROL

H.-J. Warnecke, R. Dauser and L. Aldinger

Fraunhofer Institute of Production Engineering and Automation, Stuttgart, Federal Republic of Germany

Abstract. In the chemical industry and in power plants coloured graphic systems are used to make supervision and control activities more comfortable. Coloured graphic representations can replace all known optical displays. Physical and psychic strain can be reduced by appropriate forms of representation, choice of colour and image arrangement in order to obtain better working conditions. With better working conditions and new possibilities of representing information it is possible to increase the efficiency of supervision and process control
In the manufacturing industry only planning panels and simple display units for production control are realized. The efficiency is low, but with the help of computer aided information and planning systems the efficiency can be increased. For this purpose it was necessary to develop a new system for the plant control station, because the research report revealed that graphic systems could increase the efficiency of supervision tasks. Graphic systems enable us to realize new systems of man-machine interaction.

Keywords. Man-machine systems, manufacturing processes, computer-graphics, production control, chemical industry.

INTRODUCTION

In the chemical industry and in power plants similar problems of supervision arise as in the manufacturing industry. In the chemical industry process computer controlled systems are in use for these tasks (Dreher 1981). However, in the manufacturing industry computer-aided production supervision systems are not used yet. Therefore in the short term production control in job shops often planning panels and several systems of card files are used today. These aids make it possible to have more transparency of the production process than lists for instance (Bendeich, 1974; Bendeich and Dauser, 1977).

TASKS OF THE CONTROL STATION

In spite if different forms of organizational and a plant-specific distribution of the jobs, some typical jobs of production control can be established. These jobs constitute the basis for the planning, controlling and the supervising functions of the control station.
They are in particular:
  - data management
  - (plant) data collection
  — supervision of machines and jobs
  - job allocation
and the connection of these seperate jobs respectively the resulting activities into a feed-back control system for directed failure reaction (fig. 1).

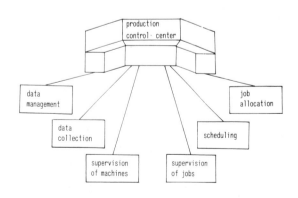

Fig. 1. Tasks of a Production Control Station

DATA MANAGEMENT

In conventional control stations data
management is carried out with the
aid of card-files. These include among
others:
- tool file
- work schedule file
- machine file
- finishing order file

It is usually accepted that the data of
these files show a certain redundancy.
This redundancy has to be put up with
in order to make working with these
files efficient; otherwise a certain
amount of search work would become
necessary in several files.

The handling of card-files in a conventional control station and the
manual registration of data in the
storage media cause considerable expense. Some data which carry paricu-
lar weight here because of their frequency are for example:
- order data (e.g. order numbers,
  part numbers, numbers of job procedures, part master data)
- order reference data (e.g. reference beginning date)
- actual order data (e.g. actual
  finishing date)

Apart from these, other data like tool
numbers, numbers of appliances, work
schedules etc. must be managed. Without up-to-date data files it is not
possible to build up a sufficiently
functioning production control system.
It is also part of the tasks of data
management to produce material cards,
wage sheets and further accompanying
papers for cut requirements, or at
least assort them and give them out
respectively pass them on in time.

TASKS IN CONNECTION WITH PLANT DATA
COLLECTION

For the tasks of data management it is
necessary to collect production data
(e.g. actual order data) and prepare
them for further processing. We can
choose, for this purpose, from a number of collection procedures. If EDP
is used for data management in the control station the data must be available either ready for entry station
or simply fixed by hand. The extent to
which the tasks of plant data collection must be carried out here depends
on the organization of the plant. Even
if the control station does not carry
out any further overlapping tasks
with EDP (e.g. postcalculation) the
data can partly be prepared here, in
connection with plant data collection,
in such a way that the data is increased. Therefore it is useful to
combine plant data collection with the
collection of data which is of impor-

tance for other departments within the
plant. In this sense the control station represents a switch in the flow
of information, increasing the data
density and helping to relieve other
departments.

PERFORMANCE OF MACHINE STATUS AND PRO-
CESS SUPERVISION

Performance of machine status (occupied, failure, set-up, in process
etc.) is, to a certain extent, done
with the aid of display units and data registrators.

Process supervision comprises the supervision of jobs which are in process
or which are on the point of being
processed, and of the job inventory
consisting of newly planned jobs and
of jobs which are not to be started in the near future.

Job performance itself is supported by
planning panels which serve as pre-
distributors. Jobs are represented by
planning cards which state the processing time on the machine (in form
of a time scale).

Through performance of machine status
failures can immediately be indicated
to the control station. Process supervision can provide short-term information on when a certain machine will
change its job and which job
will be next. It can also provide
medium-term information on when a load
peak will approach and which jobs are
causing it.

MACHINE LOADING SCHEDULE AND CAPACITY
PLANNING

While the machine loading schedule
deals with the short-term coordination of orders and machines (possibly
according to priorities) in the control station, job begin planning is
carried out in the long-term and
medium-term field of production control. In order to find out, in the
shortest possible time, whether a job
can be finished by the desired date,
the capacity layout in the control
station must be questioned.

With short-term planning activities
of this kind exact manual forward
planning (with regard to operating
sequences) is, though desirable, in
general not possible any more for a
great number of following operating
sequences.

In order to decide on the sequence of
jobs queueing at a machine, we must
make calculations considering marginal conditions and demands such as:
- technical conditions like set-up
  time

- minimum mean flow-time
- maximum capacity usage
- performance to schedule

Here we will meet with difficulties,
however, as the algorithms necessary
for heuristic or exact approaches are
often too extensive for manual calculation.

TASK DISPATCH

Basing on the results of the machine
loading schedule for the planned operating sequence, and after positive
checking of the availability resp.
operating condition, the control
station arranges for the transport
to the production center.

Deviations from the original schedule
(e.g. through machine noises, urgent
orders)can still be taken into
account here. If one capacity fails,
for example, it is possible to look
for back-up machines, to delay orders
or to increase capacities.

A high flexibility and a great transparency of the production process is
achieved with these systems, but
there are some weak points of these
aids which are shown in the next
chapter.

SUMMARY OF WEAK POINTS ANALYSES

The weak point analyses shows that
many routine activities with a considerable amount of manual calculation are needed. Among these routine
activities are:
- searching activities
- sorting activities
- writing of cards
- updating of lists
Activities of this kind, which depend
strongly on the individuals' efficiency, can to a certain extent be
carried out by computers and thus
make the operation of the control
system more efficient. Therefore EDP-
support is needed. But in the realized EDP-systems for production control lists are used. Lists are unsui-
table for supervision and short term
planning in manufacturing systems,
because they do not have the transparency of planning panels and other
aids used in the known production control stations. Colin (1980) has shown
that with computer graphic systems it
is possible to increase the efficiency of supervision and planning tasks
(fig. 2).

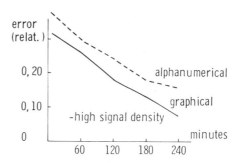

Fig. 2. Relation between Error Rate
        and Supervision Time.

The new aid opens new ways of short
term production control for the tasks
of supervision of machines and jobs.
The high expense in the case of
supervision and control tasks decreases and the disponent is free for additional tasks. These additional tasks
can be in the short term planning
area. Buth these new tasks mut be integrated in a EDP-supported system
for production control.

EDP-SUPPORTED DEVELOPING STEPS

Since EDP can only be applied under
certain organizational conditions,
and software modules can hardly be
used independently of eachother, a
directed step-by-step development of
EDP usage is advisable. For this purpose we have defined 5 developing
steps with the realization of a pre-
ceeding step being necessary for the
realization of the following step
(fig. 3).

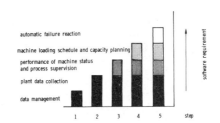

Fig. 3. EDP-Supported Developing
        Steps.

Step 1: Computer-supported data
        management

In this step card files and the manual activities attached to them are
replaced by data files and EDP-spe-
cific input activities. Manually written lists are replaced by computer
prints. The computer takes over sear-

ching and sorting activities. The advantages of this step is a reduction of many routine activities such as searching, sorting and writing of cards.

Step 2: Plant data collection

In this step the control station is equipped with its own process computer which is used for certain tasks of plant data collection. Data processing comprises arithmetical operations, writing and reading procedures in mass storage and input/output activities. One great problem of data collection is the correctness of the data. Therefore great efforts for plausibility checks are necessary. With the aid of graphic systems it is possible to reduce the effort for the plausibility checks, because the critical data can be personally controlled in an easy way. A system of this kind preserves the flexibility of a manual system on the one hand and makes sure on the other hand that the high speed of computer supported plant data collection systems is available for standard cases. But for the personal control of critical data, the data must be in a special form. For this purpose coloured graphic systems are of advantage but not necessary.

Step 3: Performance of machine status and process supervision

After step 1 and 2 the data necessary for the presentation of processing situations (machine status, job status, etc.) are available in the computer. The data in their alpha-numeric coding, however, can only be put out in badly arranged lists and tables. Accentuation and weighting of data is only possible to a limited extent. These data stores had to be transformed so far with the aid of optical displays and planning panels into clearly arranged representations giving both detailed information and a survey of precessing state and processing activities. In step 3 this transformation of the data into graphic and analog display images, which can also be accompanied by alpha-numeric parts, is carried out by the computer. Possible output media of the process computer are various optical displays. Since coloured graphic screens can replace all known optical displays and colour can represent additional information turning a two-dimensional image into a more-dimensional image, coloured screens are to be prefered for this purpose. On those coloured screens it is possible to show the layout of a factory with all machines. The colour of the machine

indicates when there is a machine failure or that the machine is idle or busy, If the coloured screen is sensitive for light pen touching it is possible to activate special information tasks with the help of a light pen and the coloured screen, which now must be controlled by a process computer. A coloured graphic system which is controlled by a process computer is the heart of the production control center in this step. The process computer for the graphical system can be the same as the process computer for data collection or it can be a seperate one only for controlling the graphic system.

With this system it is possible to build a very efficient supervision system. If a person touches the machine on the screen with a light pen, the person obtains new information of this machine. For example it is possible to fade in the number of the job which is on the machine and the numbers of the jobs which are waiting in a queue before this machine and waite for the manufacturing start. By additional touching of the screen with the light pen more information of the jobs (due-date, number of pieces etc.) can be faded in. Additional touching with the light pen can change the picture to show other production areas.

Step 4: Short-term planning tasks

In accordance with the possibilities of step 3 for representing order queues and job status, planning results can be arranged similarly in order to enable the assessment of both, individual data and the total result. In the case of job sequence, for example, a previous set up and calculated job sequence is changed interactively by calling the jobs that have to be changed onto the monitor with the aid of a light pen. The pictures for these operations are the well known Gantt-charts. For each machine there is a line on which several bars can be shown on the screen. Each card represents a job allocated to this machine. The length of the bar relates to the time the machine is busy. If the light pen is set on a bar the job is activated for new planning. Some of the important dates of the job are faded in on a special part of the screen. If the light pen is set again on a new blanket field in a machine-line, the job gets a new place in the job sequence of this machine (fig. 4).

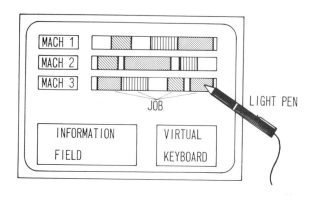

LIGHT PEN

Fig. 4. GANTT-Charts on a Screen
         with a Light Pen.

At the same time as the changed sequence is represented on the monitor the new beginning dates are
worked out and the new reference
data of the replanned jobs and of
the following operating sequence are
written into the data files.
With this aid it is possible to generate several job sequences very
quickly according to some production aims (reduction of average job
tardiness, reduction of average flow
time of the jobs). The generated job
sequences can be compared with
another, so that the best can be
chosen to be executed.

Step 5: Automatic disturbance
        reaction - job allocation

Despite the possibilities of representing situations and planning results on the monitor by the aid of
the computer, job allocation remains
to be done by the person in the production control station. But in the
case of machine failure the computer
can carry out automatical strategies
to react. A system of this kind is
equivalent to a feedback control
system which only has to be interrupted manually if there is a failure situation for which no failure
strategy has been prepared. In such a
highly automated system the graphic
screens merely function as supervising stations, only needed in exceptional cases. In these cases, however,
it becomes highly important to determine machine status and order progress
quickly and completely.

REALIZATION

The way of introducing a production
control station depends on the EDP-
support step realized and on the
already existing system for long
term production control.

In most cases long term production
control systems exist and therefore
only additional functions of short
term production control have to be
added. These functions of short term
production control are among others:

- to build up a graphical display
  for the data of the jobs
- actualization of graphical displays according to job process
  and machine status
- showing dependencies of the jobs
  in the production area
- interactive planning to make a
  new job schedule

If these additional functions are
realized, two ways to realize have
to be distinguished. We assume that
for both ways a semi-graphical system
is available.

First way to realize:
If there is already a production
control system for long-term tasks,
it can be used as a kind of heart for
the short term production control if
additional functions for graphic and
short term production control tasks
are added. The graphical system itself is only used to build up
coloured pictures according to the
results of the superior production
control computer with its additional
functions.

Second way to realize:
If a stand-alone-system for production control is to be developed,
a process computer which can do the
graphic tasks and at the same time

- data collection tasks,
- data processing tasks,
- supervision tasks and
- planning tasks.

This stand-alone-system for example
needs a connection with a superior
production control computer to exchange data only once a week.

In this exchange are included due
dates for jobs from the superior computer and the true date for the completion of a job. With this exchange
a connection between several stand-
alone production control centers can
be done.

In our institute in Stuttgart we have
developed a test system that can realize both ways. Fig. 5 shows the configuration.

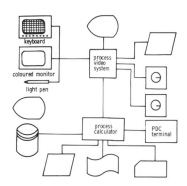

Fig. 5. Realized Configuration of a
       Production Control Station.

With this test system the most important functions which are fundamental
for short term production control
can be used for tests.

In the first step the well known
functions of the old production control centers have to be transformed
into new graphical functions. In the
next step the number of functions
were extended. In this extension the
possibilities for testing new
schedules are included.

To realize this step it was neccesa-
ry to develop a software-system and
to develop new graphics for the functions of production control. A part
of these developments was the determination of a changed way to control
the production.

Examples
For the supervision of machines and
jobs the layout of the production
area is used. On the screen the production area with the position of
machines and storages is visible.
Each machine has its own symbol. This
symbol shows if the machine is idle,
busy or if there is a failure. If
there is a failure the symbol for the
machine changes the colour and blinks.
It is also possible that a new symbol
for the machine can be faded in. If
the failure is very important it is
possible to make the whole screen
blink. The blinking screen attracts
the attention of the disponent. In
this case the disponent can see under
the symbol for the machine, which job
is in the machine or which job is next
in the queue for the machine.

For simple planning tasks the well
known Gantt-charts are used. For this
purpose each machine has a long
stretched area on which bars represent jobs for the shown machine. If
a bar is touched with a light pen
and afterwards a free place on the
line, the job is given a new date
(maybe even a new machine) (fig. 4).

The old bar has disappeared and at the
new place a new bar becomes visible.
The date of the start of the job is
calculated from the new place of the
visible bar. The computer also calculates the new data of the following
jobs on these machines. If it is possible to reduce idle time on a machine,
the following jobs get earlier beginning dates.

CONCLUSION

With the aid of graphical computer
systems it is possible to reduce the
number of routine activities and the
error rate. A new form of operational
guidance makes the system very efficient to use and allows employment
of persons without experience or
knowledge of EDP. This system allows
a new human-machine interaction for
the tasks of short term production
control.

The stepwiese introduction of a production control station with graphical coloured interactive screens
allows, that in relation to the
already existing production control
system the new system can be adapted easily or built up completely
new.
The advantages of such a system are:

- reduction of routine activities
- high transparency about the production process
- high up-to-date dates
- integrated information flow
- computer aided supervision of
  machine and job
- computer aided short term planning

With this system it is possible to
realize a closed information system
for production control.

REFERENCES

Bendeich, E. (1974). Fertigungs-
    steuerung mit Systemen der zen-
    tralen Arbeitsverteilung (ZAV).
    AV11, 6, 168-182.
Bendeich, E. and Dauser, R. (1977).
    Organisatorische Möglichkeiten
    der Arbeitsverteilung.
    FB/IE26, 3, 169-176.
Colin, I. (1980). Der Einfluß der
    Informationsgestaltung auf die
    Zuverlässigkeit von Wahrnehmungs-
    und Verarbeitungsleistungen bei
    Überwachungstätigkeiten.
    Diss. 1980, Frankfurt.
Dreher, P. and Grimm, R. (1981).
    Rechnerunterstützte Warten- und
    Automatisierungstechnik mit den
    Systemen SIMATC und EAF.
    Siemens-Energietechnik 3, 5,
    170-173.

STAR* — A CONCEPT FOR THE ORTHOGONAL DESIGN OF MAN-MACHINE INTERFACES WITH APPLICATION TO NUCLEAR POWER PLANTS

W.-E. Büttner, L. Felkel, R. Manikarnika and A. Zapp

*Gesellschaft für Reaktorsicherheit (GRS) mbH, D-8046 Garching,
Federal Republic of Germany*

Abstract. The paper deals with the development of computer based systems assisting the operators in nuclear power plants. The gap between describing process dynamics and information requirements resulting from human factors engineerng studies is shown to be narrowed using formal languages to encompass both these areas and at the same time designing actual systems realizing the goals for optimization of man-machine interfaces. As a result systems for disturbance analysis, post trip analysis, alarm reduction etc. are immediately obtained. Practical experiences from laboratory as well as pilot-installations are also reported.

Keywords. Optimization of man-machine-interfaces, design of computer-based operator aid's, information integration description of process dynamics, human factors engineering, formal languages, on-line process computers.

INTRODUCTION

The optimization of man-machine-interfaces of large technical processes requires a multidisciplinary procedure. First of all the dynamics of the process have to be described. This is mainly done by process engineers and physicists who are responsible for the development, construction, licensing and operation of the process. The important discipline therefore involved is process engineering.

On the other side the operator is to supervise and control the plant. Modern control rooms should help him to do this job displaying many measurements and controls. These of course have to meet ergonomic requirements. The second discipline as a consequence is human factors engineering. It turned out, however, that with forthcoming sophisticated electronic equipment replacing manual and electromechanical devices and with extraordinary enlargement of sensors and actuators, information integration has become both possible and necessary. The gap between process engineers and ergonomists has widened. Since computerbased solutions become more and more feasible computer science plays also an important role.

The paper will show the main problems on either side of the man-machine-interface and how information integration can be orthogonally designed using formal languages.

PROBLEM DEFINITION

To improve the man-machine-interface of large processes there are three main requirements to be fulfilled.

First the process behaviour has to be described in such a detail, that plant state as well as abnormal conditions and disturbances can be identified on-line. The causes and the consequences of actual plant situations or disturbances should be determined and additional information about the state compiled and analysed.

Second the operator's diagnostic stategies corresponding to specific plant situations have to be predefined as far as possible and according to the categories of levels as given by Rasmussen (1980) shown in Fig.1. At each level a different set of plant information is necessary and the degree of deterministic propositions about the plant situations decreases with increasing level.

Third, at each of the levels a different organization of the required information to operator and for plant engineers or specialists make a tailor-made display system a necessity. Although in principle general control room annunciator displays may be only involved colour CRT displays are considered here.

As mentioned in the introduction all three problem areas are dealt with by different disciplines which makes it especially difficult to design computer-based systems to provide the important information at each level and plant situation. An integrative tool has therefore to be designed to formally describe all requirements arising from the three problem areas which also allows the automatic generation of the computer-based system itself.

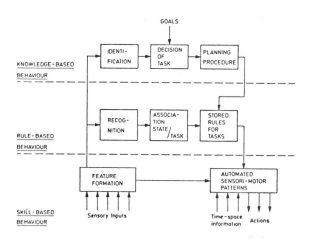

Fig. 1   Levels of human diagnostics

This can be achieved by means of formal languages which contain all essential aspects (semantics) of the problem areas and at the same time
facilitate the automatic construction of data
base required by on-line analysis algorithms.
Furthermore these languages allow better and
concise communication between the different disciplines involved.

THE LANGUAGE

It was found that for the purposes described in
the previous chapter formal languages are feasible. Formal languages are structured by formal
grammars. Although there exist a wide variety of
types of grammars those which are usually used
defining computer programming languages are especially applicable. The type of grammar used
here is the so-called context free grammar a
definition of which is given in sequel.

A context-free grammar is a quadruple (V,T,Π,S).
V is a set of symbols, T is the set of terminal
symbols, V-T is the set of syntactic variables.
V* is the free semigroup on the set of symbols
under concatenation. The string $e \varepsilon V^*$ is called
the empty string, i.e. the string which consists
of no symbols at all. $V^+ = V^* - \{e\}$. $\Pi \subseteq (V-T) x V^*$ is
the set of production rules. $\pi \varepsilon \Pi$ is usually writ-
ten as (a::=b), where $a \varepsilon (V-T)$ and $b \varepsilon V^*$. $S \varepsilon (V-T)$
is the start symbol. The relation $\rightarrow (V-T)^+ x V^*$ is
called direct derivation, written $\alpha \rightarrow \beta$ where α
$= \alpha_1 a \alpha_2$ and $\beta = \beta_1 b \beta_2$ and $(a::=b) \varepsilon \Pi$. The relation
$\rightarrow^* (V-T)^+ x V^*$ is the transitive closure of $\rightarrow$,
written $\alpha \rightarrow^* \beta$, where $\alpha = \alpha_1 \rightarrow \alpha_2 \rightarrow ... \rightarrow \alpha_n = \beta$.
The language L generated by grammar G is the set
of all terminal strings which can be derived from
the start symbol, i.e. $L(G) = \{w \varepsilon T^* \quad S \rightarrow^* w\}$.

For these types of grammars automatic systems
for the contruction of syntaxanalysers exist
which makes it very easy to design and experiment with the generated languages. As shown in
definition of the grammar it consists of a set
of input symbols (terminals) a set of syntactic
(structoral) variables and a set of grammatical

production rules. The syntax of the language
therefore is defined completely by giving the
the three sets. However, the meaning of each
language construct (rule) has to be given
according to what actions should be performed
when the construct is encountered. The meaning (semantics) given to each rule produces a data base which contains all information given by a special string of symbols put
together and satisfying the grammatical rules.

An example is given below:

```
1   <DESCRIPTION>  ::=  <MODULES>
2   <MODULES>      ::=  <MODULES>      <MODULE>
3   <MODULES>      ::=  <MODULE>
4   <MODULE>       ::=  SYSTEM : TEXT <NODES>
5   <NODES>        ::=  <NODES>        <NODE>
6   <NODES>        ::=  <NODE>
7   <NODE>         ::=  EVENT     <SUCCESSORS>
8   <SUCCESSORS>   ::=  <SUCCESSORS> <SUCCESSOR>
9   <SUCCESSORS>   ::=  <SUCCESSOR>
10  <SUCCESSOR>    ::=  EVENT
```

Here lines 1-10 denote the grammatical rules.
The set of syntactical variables are all
those elements of the rules enclosed in
square brackets < >. The goal symbol is
<DESCRIPTION>. All other symbols are ter-
minal symbols which are the actual elements for some description. Those using the
language for communication are only doing
this via the terminal symbols. They are
(at least should be) aware of the syntactic
variables denoting the structure of the description communicated.
To know the meaning of some description
each rule has to be given some informal
indication about it. For example the meaning of rule 1 is that each description of
the system to aid the operator consists of
a set of modules. In order to produce an
arbitrary number of modules repeated application of rule number 2 is necessary. To
terminate the production of modules rule 3
is used. The structure of each module is given in rule 4. By means of the terminal symbols SYSTEM : TEXT a system description can
be supplied the text possibly being used for
display on a CRT when the information given
by an on-line system (see next chapter). However, for each rule information has to be
generated similar to the one about TEXT in
rule 4. It should also be noted that whatever
SYSTEM : TEXT should mean is freely definable
and depends solely on the goals defined by
the human factors and process engineers. So,
for example, one could use the same rule for
taking the TEXT as an entry to a data bank
containing information about a specific plant
subsystem.

The rest of the rules define the syntactic
structure for the definition of so-called
nodes which eventually are parts of a cause-
consequence diagramm as shown in fig. 2.
Again, what should be done with this description, depends on the functions to be
carried out on-line. However, if the diagram
in fig. 2 is to be described some more rules
have to be added. For example information has
to be included for the logical combination of

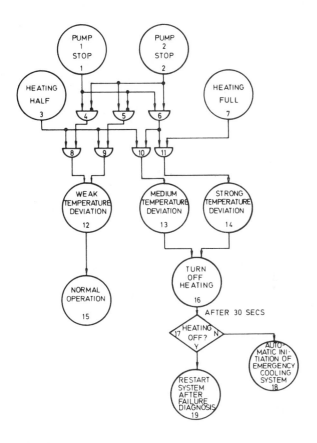

Fig. 2   Cause-consequence diagram

nodes. Semantically the nodes are events arising
during operation of the process given in fig. 3.

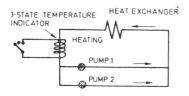

Fig. 3   A sample process

In this example the description is used for
implemantation of a data base which gets on-
line information about the occurence of an
event and the associated structure allows for
determination of plant state, disturbances, etc.

With a different meaning the same description
can be used for other functions.

In the same way as shown in the example where
the process behaviour can be described it is
possible to configure complete CRT-based display
systems.

As a consequence a complete computer-based operator aid can be designed and implemented. The
topdown (orthogonal) approach requires all disciplines involved to communicate in an artificial language as described in this paper. Furthermore, due to the formality of the language

automatic generation of associated data bases
can be made.

ON-LINE ANALYSIS

The task of extracting the status information
during the plant operation and giving the operator the necessary information is done in the
on-line analysis part of the system. The tables generated by the off-line model translator
controlled by the grammatical structure now is
interfaced with the actual process values in
accordance and the information produced now is
passed on to the display interface of the system. Since this is done in real time the on-
line analyser must be efficient. This is achieved to a certain extent by the translator
which produces preprocessed tables. These tables are prepared in such a way that operations
like logical combination, sorting procedures
etc. are carried out on a table-lookup basis.

The analysis task can be grouped into different sections depending on which level, Rasmussen (1980) (fig. 1), the information is
required by the operator in case of a disturbance.
A sample set of functions are given below.

POST TRIP ANALYSIS

The operator information required in a post
trip analysis can be considered to fall under
rule based level.

If the occurrence of alarms cannot be avoided
due to a disturbance which develops very fast
and propagates rather rapidly the operator
must be relieved from the very time consuming
task of gathering all the additional information necessary for the detection of the cause
of the disturbance. He must be able to trace
back the propagation of the disturbance from
the present status of the initiating event.
This can very easily be achieved by modelling
the trip sequences with cause-consequences
diagrams. The trip signal can be traced back
to its origins and at each step it may be
checked whether all required protection functions have been initiated properly. This
allows fast restart of the reactor if, as
in most cases, the trip was due to minor
causes.

ALARM HANDLING

In disturbance situations a considerable
amount of alarms occur in which a large number are activated due to system interactions.
The operator uses considerable skill to extract the source of the disturbance in such a
situation. The on-line analyser can help the
operator by fittering out unimportant alarms
and thus help the operator in his taks.

MASS, ENERGY AND MOMENTUM BALANCE
EQUATIONS

Some times the operator may find himself in
situations where a detailed process knowledge
becomes necessary.
To detect and identify disturbed situations

clearly and unambiguously even under difficult
conditions and with equipment degradation detailed information must be provided as to the
normal behaviour of the process. The information might be given by mass, energy and momen-
tum balance equations and evaluated on-line
using analog measurements.

Based on this information the operator can make
his own decisions. These are the typical tasks
carried out in the on-line analysis part of the
system.

CONCLUSIONS

It has been shown how advanced methods of computer science can successfully be used for providing a common language to communicate requirements for systems to aid the operators in technical processes. The advantages of this approach
are:

- Top-down design
- Common Language for both process engineers
  and human factors engineers
- information integration since all disciplines are involved
- automatic generation and implementation of
  the necessary data base
- universality of the concept.

The method described have been applied in the
design of the disturbance analysis system STAR[1]
which is implemented as a pilot installation in
the German Grafenrheinfeld nuclear power plant.
The grammar used for this application comprises
154 rules. Even larger grammars are presently
being designed for more integrative applications
in German plants in the near future.

REFERENCES

Bastl, W., Felkel, L. (1980) STAR-A distur-
      bance analysis system and its application
      to a PWR. ANS Meeting on Thermal reactor
      safety, Knoxville, Tenn.
Bastl, W., Felkel, L., (1980) Disturbance
      Analysis Systems state of the Art. NATO
      Symp. on Human detection and Diagnosis of
      system failures. Roskilde, Denmark.
Hopcroft, J.E., Ullman, J.D., (1969) Formal
      Languages and their relation to Automata
      Addison weseley
Lalonde, W.R., (1971) An efficient LALR parser
      generator, Tech. Rep. CSRG-2, University
      of Toronto.
Rasmussen, J., (1980) Some trends in Man-
      Machine Interface Design for Industrial
      Process Plants. RISØ-M-2228

---

[1]   Abbrev. from Störungsanalyserechner
            (disturbance analysis computer)

DYNAMIC AND STATIC MODELS FOR NUCLEAR REACTOR OPERATORS — NEEDS AND APPLICATION EXAMPLES

A. Amendola*, G. Mancini*, A. Poucet** and G. Reina***

*Commission of the European Communities, Joint Research Centre - Ispra Establishment,
I-21020 Ispra (Va), Italy
**Departement Metaalkunde, Katholieke Universiteit Leuven, De Croglaan 2,
B-3030 Heverlee, Belgium
***Studio MESA, Via Carnaghi 15, Milan, Italy

Abstract. This paper deals with the important problem of the impact of human errors on nuclear reactor operation. Human malfunctions are classified according to a taxonomy developed from a consideration of real records of incidents. Two main roles have been ascertained for the operator actions: controlling incident development, and carrying out test and maintenance procedures. Two different models have therefore been set up: a dynamic model which allows identification of critical timings of operator intervention and recovery, and a static model which takes care of the dependence structure of the various steps in maintenance and test procedures. These models may be used in "a priori" risk assessment, and also for optimization of procedures and the man-machine interface.

Keywords. Ergonomics; fault tree; human factors; man-machine systems; nuclear reactors; probabilistic risk assessment; reliability; safety analysis.

INTRODUCTION

One of the main sources of uncertainty when dealing with quantitative or qualitative assessments of the safety of complex plants, such as nuclear reactors, is the difficulty of correctly describing human interferences with the system. Human performance is difficult to predict because of its variability and because of the many different functions the "operator", in general, is asked to fulfil, leading to a multiplicity of inputs and outputs and of interfaces with the system.

Ignorance of these aspects in risk assessments of complex systems leads inevitably to incompleteness of the estimates. This is dramatically demonstrated when the analysts then try to force important occurrences into "a priori" risk studies. This again leads to great uncertainties in assessing the general credibility of risk estimates and certainly constitutes one of the main sources of difficulty in the nuclear debate today.

Indeed, in recent years there has been increased awareness of this problem, especially after the TMI incident. Several human models have thus been proposed for use in risk studies of which we mention here only the important contribution by Swain (1980).

The Joint Research Centre of the Commission of the European Communities has also approached this problem, with the help of other experienced national organizations in the field, as part of the "Risk and Reliability" Project of the Nuclear Safety Programme. In this project we are setting up a wide-ranging information system on failures and incidents of nuclear reactors in Europe (European Reliability Data System - ERDS); in addition basic reliability models are developed for risk analysis.

The possibility of exploiting real data on human errors coming from various sources and the need to develop adequate tools for risk analysis have characterized our approach. First we tried to analyze operation records of incidents with the aim of defining those items which uniquely identify the

various modes of interaction of the "operator" with the system, and we then implemented some of these items in the models.
These reliability models refer mostly to the
effect of human operations on the system and
are not models of human behaviour as such,
although they may also be of help to the analyst interested in the cognitive process of
human performance.

The analysis of real occurrences led to the
setting up of a wide-ranging human malfunction taxonomy (Mancini, 1981) that goes
beyond its use in reliability models to encompass some aspects of human performance
ance that are more appropriate to human
behavioural models (causes of human error
and its mechanisms, internal human malfunction, etc.). Eventually, it can be said
that such a classification, beside having been

at the basis of the model developments, as
will be shown later, deserves the more general function of organizing the information
in well defined categories, thus allowing the
analyst to identify recurrent patterns of
human errors and the collection and quantification of human reliability data.

In Table 1 the organization of the taxonomy
is sketched.
As far as the reliability models are concerned, two important different roles have
been ascertained for the "operator", namely
control of the reactor and of incident development and execution of test and maintenance
procedures, and so two different models have
been developed. The difference lies in the
time characteristics of the models: in the
first the dynamics of operator actions and
their interaction with the incident develop-

TABLE 1   Human Malfunction Taxonomy

PLANT

A Plant Identification
B Data System Identification

EVENT ANALYSIS

C Free Text Event Description
D Event Detection
E Plant State
   Systems (F) and Components (H)
   Affected
G Consequences of the Event
U Recovery Situation

H COMPONENT RELIABILITY
   DATA SYSTEM

HM Modes of Failure
HC Causes of Failure
HA Actions Taken

FILLING-IN
PREPRINTED FORMS

DATA COLLECTION BY

SPECIALISTS' ANALYSIS,
IN-PLANT INTERVIEWS, etc.

SPECIALISTS' ANALYSIS,
PRESELECTED TASK TYPES

HUMAN FACTORS DATA

Human System:

J Personnel Identification
K Personnel Location
L Personnel Task
M External Mode of Malfunction
N Potential for Self-Correction
P Situation Factors
HA Actions Taken
   Recommendations and Comments

HF Specialists' Analysis:

Q Internal Human Malfunction
R Causes of Human Malfunction
S Mechanisms of Human Malfunction
T Performance Shaping Factors
HA Actions Taken
   Recommendations and Comments

Quantification

ment are essential aspects. In the second
case, on the other hand, more emphasis is
placed on the dependence structure of the
various human acts.
Both models only refer to the interface of the
man with the system and do not take into full
account the interface of the man with other
human beings, so that communication aspects
(for instance between control and maintenance activities) are neglected together with
the modelization of checking and supervising
activities and of decision processes by more
operators.

The use of these models, as will later be
described in detail, is not restricted to risk
assessment studies but is also of interest in
all those decision processes, generally called
"risk management" activities.

DYNAMIC MODEL

Static analysis methods cannot be used to analyze an incident development in which an important role is played by the man-system interaction.
Usual safety analysis techniques, such as
fault-tree and event-tree techniques (or probability trees for human intervention, as
illustrated by Swain (1980)), separate the
logic and probabilistic system analysis from
the evolution of the physical process variables
which are the subject of a different evaluation
procedure. Cause-consequence diagrams are
more suitable to describe operator actions,
but they also require a set of separate physical calculations and are impractical for
modelling timings and delays, since each
initiating event is very soon split into a forest
of branches which cannot be evaluated in most
practical cases. Recent developments for so-
called continuous event-trees still separate
the process physics from the event occurrence evaluations.

Now, when dealing with man-machine interaction, not only the failures but also the
timings of operator interventions and possible
recovery are of the utmost importance. Recovery capability is a very dynamic process
which depends upon logical and intuitive mental processes, guided mostly by written procedures and interacting with signals changing
during the transient according to the physical
evolution of the system. The correctness and
timing of operator responses are also strongly
affected by stress factors, which in turn depend on the diagnosis made, e. g. recognition
of the possibility that an incident may rapidly
evolve into a serious accident.

These considerations demonstrate the need
for a new analysis methodology, able to
process and synthesize the information on
the physics of the process, on the component
fault analysis, on human behaviour under
normal and stress conditions, in a dynamic
modelling of the man-system interaction,
which may allow the analyst to
- perform overall a priori risk analysis,
- determine critical operator response
  timings,
- correctly understand the mechanisms of
  operator intervention, whether by retrieving actual incidents or from experiments
  on simulators,
- optimize the information which must be
  given to the operator to allow early diagnosis and recovery; and
- optimize incident procedures.

To this end we investigated by a study case
the adequacy of the dynamic ESCS technique
(Event Sequence and Consequence Spectrum)
to deal with such problems.

The ESCS Technique

The basic outline of the methodology, extensively described by Amendola (1981), can be
summarized as follows.

A system is described by a set of component
models obtained by a quantitative failure
mode and effect analysis resulting in physical
relationships characteristic of each component state. In these models step functions can
be used to represent responses of on-off
components (switches, alarms, etc.).
The resulting component models are constituted of "event variables" which define the
logical states (nominal, failed and degraded),
and of a set of analytical relationships, each
one describing the corresponding behaviour.

Thus, the whole system is synthetically represented in all its possible conditions by a
set of equations which contain both logical
and physical information. The appropriate
relations are selected by parametric operators, according to the event sequences generated by the analysis procedure.

In principle, by means of an automatic generation of the complete set of logical event
variable combinations, all incidental occurrences consistent with the assumed component models can be identified without limiting
the analysis to preselected abnormal transients or accidents. However, the generation
is performed in a controlled way according
to dials assigned to restrict the analysis to

the desired resolution.
To obtain a simple physical description of complex components, such as a reactor core, response surfaces may possibly replace outcomes of computer codes.

The physical consequences of a generated event sequence are obtained by the numerical solution of the corresponding equation set. Only those sequences which satisfy a pre-assigned quantified TOP condition are selected and logically analyzed to extract the minimal ones, so that, in some way, minimal TOP sequences correspond to the cut sets of the usual fault-tree approach. As appropriate to the dynamical nature of the problem, minimal TOP sequences are determined with respect to the complete time pattern of the variables considered.

According to this method the TOP event is no longer a simple logical event such as availability or not of a system, but is the set of values assumed by the physical variables investigated (e. g. the occurrence of a pressure above a certain value or within a certain interval, together with certain specified temperature values.

All algorithms have been implemented into the DYLAM computer code (Reina, 1981), which also computes the probability distribution of the TOP conditions investigated as a function of the transient time. Functional failure dependences can be taken into account, because of step-by-step evaluation of each physical variable.

Operator Intervention Study Case

In the same way as the system components, in the ESCS technique, we modelled the different tasks which an operator must accomplish during an incident, according to the diagram in Fig. 1.

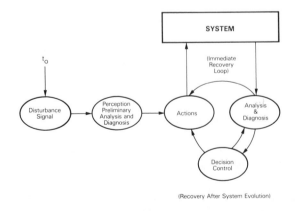

Fig. 1. Dynamic Operator Modelling Scheme

In this scheme the moment of incident initiation and the first analysis process have been represented as the trigger of the successive loops: actions ⇌ system evolution. A first action of the operator on the system will be followed by a system reaction which is monitored on the man-machine interface. According to the nature of the events a prompt recognition of an error may be made possible, and therefore a prompt recovery is possible.
In other cases both the diagnosis performed and the system responses might be subjected to "decision control" processes, which can result in the decision to perform the next action envisaged by the procedure, or to correct the first diagnosis and actuate a new procedure.
In each task all possible errors have been considered (i. e. omission, commission, etc. as described in the "external mode of malfunctions" of the taxonomy established (Mancini, 1981)); and also discretized distributions for reaction times have been included.

The next steps in the development of the method and of the DYLAM code will be the integration of the actual technique with Montecarlo procedures; in this way continuous distributions, possibly conditioned by stress factors, can be assumed for all operator failed or delayed responses.

The study case is derived from the analysis of a particular LOCA (Loss of Coolant Accident) in a PWR (von Hermann, 1980), in which a check valve failure provokes a break in the low-pressure injection system, outside the primary containment.

In our example, see Fig. 2, to obtain a much simpler equation set, we replaced the core by a tank, whose voiding timings are of the same order of magnitude as the real core melting.

Since in such an occurrence no water is available in the recirculation sump, the operator is asked to avoid or at least to delay as far as possible melting of the core by carrying out the following sequential procedure:

1. Manual activation of the high-pressure injection system (HPIS) at the minimum flow (only 1 pump) so that the available water in the emergency tank is injected as slowly as possible.

2. Switch off the low-pressure injection system (LPIS), which would be automatically

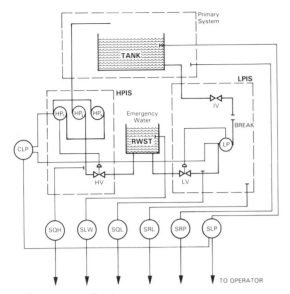

SLW = RWST level sensor
SQL = LPIS flow-rate sensor
SRL = LPIS radioactivity sensor
SRP = Primary radioactivity sensor
SLP = Tank level sensor
CLP = Tank level controller
SQH = HPIS flow-rate sensor

Fig. 2.   Operator Response Study Case

be called upon to operate as the pressure
decreases below a certain threshold, and
in this case would only waste water.

3. Attempt to close the isolation valve upstream of the LPIS.

In the case of LOCA towards the primary containment (much more probable event) the incident control is governed by the automatic
systems.

From the previous description it can be seen
how the timings of the operator interventions
can dominate the incident development.
Each error or delay, indeed, may provoke a
waste of the water available in the emergency
tank and recovery is possible only if allowed
by residual water in this tank.

The modelling logic is constructed as follows:

- at time t_o the alarm appears (primary level
  below a threshold; described by a step function);
- at time $t_1 = t_o + \tau_{al}$, the alarm situation is
  recognized by the operator (a new step function in which τ_{al} is a distributed operator
  perception delay. If τ_{al} is ∞, then the
  operator has not perceived the alarm at
  all);
- at time $t_1 + \tau_{syst} + \tau_d$ a first diagnosis is

made by the operator.
τ_{syst} is a time, characteristic of the system evolution, which can allow the diagnosis (e.g. fixed by a procedure, or the
time necessary until radioactivity increase
in LPIS local can be measured), τ_d is a
characteristic of the operator´s readiness
(τ_d = ∞ operator confused, not able to
formulate a diagnosis);
- the diagnosis results in the decision to
  perform a certain number of actions (e.g.
  actions 1 to 3 in the procedure mentioned
  above). An action timing delay τ_i is associated with each of these actions.
  The distributions assumed for τ_i can simulate either the non-performance of some
  of the actions envisaged, or the inversion
  of the order prescribed by the procedure;
- check points are identified where the operator can check his diagnosis against the development of the signals and, therefore,
  can reverse his action with a new distributed delay.

To show the influence of the timings, Fig. 3
shows the behaviour in time of the water
level in the vessel after a LOCA towards
LPIS in the nominal case and in the case
when the operator performs a correct diagnosis but actuates 2 HPIS pumps instead of
only one and recovers after 900 s. This delay is amplified by a factor 3 for the LOCA
being successfully isolated according to the
assumed procedures which envisage a diagnosis check at the trigger point of the LPIS,
before continuing with the prescribed actions.

This example alone validates the quality of
the approach for evaluating operator critical
timings, procedure optimization and identification of the signals necessary to the operator to reduce diagnosis delays.
But this example is only one out of several
sequences automatically generated by the
DYLAM code. Table 2 shows the number of
sequences which can lead to certain "low
level conditions in the primary tank" at a
given time.

Once the models are constructed, the same
system can be analyzed with respect to a
wide spectrum of consequences, for instance
identification of the sequences leading to
complete voiding, to specified low levels for
a specified time in the vessel, or to specified
residual water volumes in the emergency
tank.

Even if the limits of this paper do not allow
a more extensive discussion of methods and

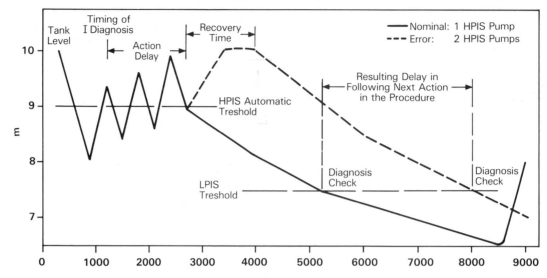

Fig. 3. Two Selected Time Sequences

TABLE 2 Number of Sequences (up to the IInd order) which Lead to Tank Levels:

Below m	9	7	5	3
at times (s)				
900	170	4	0	0
1800	95	55	9	5
2700	178	75	55	9
3600	158	85	67	55
4500	123	90	80	65
5400	132	90	79	72
6300	171	92	76	70
7200	122	107	70	68
8100	116	95	69	64
9000	144	92	64	64
9900	99	91	64	64
10800	91	86	64	64

results, the potential of this kind of analysis appears obvious.

STATIC MODEL

The model presented in this section does not deal in particular with the "time" aspect of operator behaviour and it is hence referred to as "the Static Model". It is based on the fault tree technique and its field of application is the same as that of the event (probability) tree developed by Swain (1980) with the basic exception that the fault tree technique, being a deductive type of analysis, selects only those events relevant to the TOP condition to be examined.
Furthermore, we have attempted to account more precisely for the various dependences and recovery actions implicit in each human task, as will be shown in the following.

The "static model" is used in the analysis of those tasks where the time dimension as such is not of paramount importance, i.e. in the analysis of the unavailability of systems due to previous maintenance or testing or in the analysis of single operation actions with no practical feedback to the system development. In all these types of analysis the main issue will be to model the dependence structure of the various actions properly.

The model is set up to be used in an interactive computerized method for fault tree construction and evaluation (Poucet, 1981) and it is integrated with the component models developed for the method mentioned above.

Model Description (Fig. 4)

In the model the different paths which lead to the undesired event are developed: the undesired event is the wrong state (unavailability) of a component as a result of operations performed on the component during testing or maintenance or during operation tasks: the task is thus subdivided into steps related to actions on components.

Failures leading to the undesired event can be divided into three types:

1. The combination of a correct action on the component with a failure of the latter (e.g. valve correctly commanded to open by remote control, failing because of a fault in the actuator). If the failed state can be observed by an indicator, a new action (new fault tree) is initiated by the operator; he must perceive the signal,

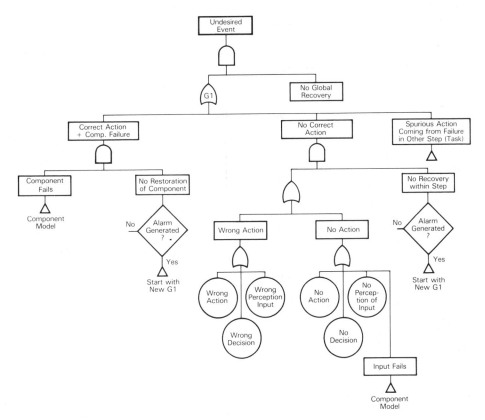

Fig. 4. Static Model Fault Tree

make a decision and restore the component.

2. An incorrect action or no action without recovery. In this case also an indicator may signal the fault and lead to recovery.

3. A spurious action on the component originated by a correct action on a "wrong" component in the same or in a different task. This event represents a link between actions performed on different components.

Recovery actions are also of three types:

1. Immediate:
   originated by the existence of indicators and already described above.

2. Global:
   these recovery actions are still performed at the component level and may be triggered by the carry-over of another action on the same component, but in a different step of the procedure and by the carry-over of an action on another component which reveals the incorrect state of the component under consideration. In the first case the incorrect state of the component will be described by an OR

gate of a series of AND gates, each of these describing the failure on the component in one step of the procedure and the lack of recovery in the other relevant steps.

In the second case a previously performed failure mode and effect analysis (FMEA) will help in the identification of the dependences.

3. General:
   this recovery refers to actions at the task level by which the full operation of the system(s) or of the subsystem(s) is tested.

Once the various steps have been modeled by the fault trees described above, these are put together in a general fault tree according to the top condition and task structure. Although the complete logical analysis of the general fault tree is possible with already existing computer codes, problems may rise in the numerical evaluation of the system unavailability if the dependence structure of the tree (conditional probabilities) is too complex. The use of Montecarlo techniques may ease the solution of these problems.

Amesz (1981) gives an example of the application of similar "static models" to the analysis of the unavailability of the Emergency Core Cooling System of a power reactor, caused by errors occurring during the regular testing of the system. The interesting results achieved in this analysis seem to encourage a more systematic application of the techniques developed.

CONCLUSIONS

Although the methods discussed in this paper are very appropriate for the probabilistic quantification of the risk induced by human failures, we think that data available are very poor, because of both the limited number of records and the inadequate way in which the data have been collected.

These methods should be seen as the way to analyze the quality of the man-machine interface design with respect to the completeness of the signals provided to the operator, the ergonomic features of the interface and the efficiency of procedures both for testing and maintenance and for incident control.

They also offer a correct framework for determining the data to be collected for reliable behaviour predictions: each psychological human failure investigation should be aware of the uncertainties arising from the physics and the man-machine interface, with which an operator is confronted, so that there should be a greater integration between the various disciplines mentioned above to solve the complex problem of man-machine interaction.

REFERENCES

Amendola, A., and G. Reina (1981). Event Sequence and Consequence Spectrum: A Methodology for Probabilistic Transient Analysis. Nucl. Sci. Eng., 77, 297-315.

Amesz, J., F. Francocci, and G. Mancini (1981). The Influence of Operator Error in Routine Testing. CEC-JRC Internal Report, available on request.

Mancini, G., J. Rasmussen, O. M. Pedersen, A. Carnino, M. Griffon, and P. Gagnolet (1981). Classification System for Reporting Events Involving Human Malfunctions. EUR 7444 EN.

Poucet, A., and P. De Meester (1981). Modular Fault Tree Synthesis. A New Method for Computer Aided Fault Tree Construction. Reliability Engineering, 2, 65-76.

Reina, G., and A. Amendola (1981). DYLAM: A Computer Code for Event Sequences and Consequence Spectrum Analysis. International ANS/ENS Topical Meeting on Probabilistic Risk Assessment, Port Chester, N. Y.

Swain, A. D., and H. E. Guttmann (1980). Handbook of Human Reliability Analysis with Emphasis on Nuclear Power Plant Applications. NUREG/CR-1278.

von Hermann, J., R. Brown, and A. Tome (1980). Light Water Reactor Status Monitoring During Accident Conditions. NUREG/CR-1440.

MAN–MACHINE INTERFACE IN THE DISTURBANCE ANALYSIS SYSTEM SAAP-2

F. Baldeweg, U. Fiedler and A. Lindner

*Zentralinstitut für Kernforschung Rossendorf, P.O. Box 19, DDR-8051 Dresden,
German Democratic Republic*

Abstract. In the paper presented, the surveillance partial
system for disturbance analysis SAAP-2 will be described.
SAAP-2 can be considered as an essential part of the digital
hierarchical decentralized informational system (IS) of the
Rossendorf Nuclear Research Reactor (RRR), which is being developed comprising the functions: surveillance, process analysis, optimal power control, start-up and shut-down.
SAAP-2 will be described from the point of view of assuring
high decision capability of the IS in cases of plant disturbances, i.e. among others - realizing an effective man-machine
interface. With special attention the dialogue system for model
manipulation, interactive diagnosis and diagnosis evaluation
will be considered. SAAP-2 comprises the tasks: monitoring of
disturbances, diagnosis and communication. It is based on a
discrete model, an event graph $G = [E, R, \emptyset]$, which represents the
disturbances as nodes and the technological relations as arcs.

Keywords. Man-machine systems; nuclear reactors; on-line
operation; graph theory; microprocessors; process diagnosis.

INTRODUCTION

In a paper given elsewhere (Baldeweg,
1982a) an informational system (IS)
for surveillance and control of the
Rossendorf Nuclear Research Reactor
System (RRR) was introduced, see
Fig. 1. This system comprises the
functions of surveillance, process
analysis, optimal power control,
start-up and shut-down.

In the paper presented, especially
the surveillance component of the
disturbance analysis SAAP-2 (Stö-
rungsanalyse-Applikationsprogramm)
will be described from the point of
view of man-machine-communication.
SAAP-2 represents a developed version
of SAAP-1 which was introduced in an
earlier paper (Baldeweg, 1980).

Disturbance analysis for sometime
being developed in several countries
especially for Nuclear Power Plant
surveillance becomes recently more
and more a matter of interest due to
TMI-accident (Meijers, 1980; Felkel,
1980).

Disturbance analysis must be seen an
informational process of real time
diagnosis which comprises in general
the tasks of monitoring disturbances,
finding out causes and predicting
consequences.

As a very important part the man-
machine-communication (MMC) must be
considered. MMC can be seen from a
technical (hardware, software), a
functional (dialogue oriented diagnosis) and an ergonomic aspect
(human factors).

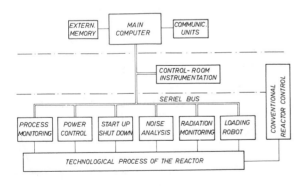

Fig. 1. Scheme of the informational
system at the RRR.

SAAP-2, which is based on a discrete
model of the technological process to
be investigated, will be more discussed from the point of view of
functional properties of MMC. In that
connection dialogue plays an important
role. It comprises functions for

- modification of the diagnosis model
  in dependence on the actual knowledge

- representation of automatically
  achieved diagnosis results and

- interactive diagnosis.

Dialogue must especially be seen from
the aspects

- that information, which is necessary
  for solving a complex decision problem can only be estimated step by
  step

- machine is not able to decide in
  all cases, decision will partially
  be made in dialogue with the operator.

THE DISTURBANCE ANALYSIS SYSTEM SAAP-2

SAAP-2 is a real time diagnosis
system for complex technological
systems (Baldeweg, 1982b).

By diagnosis we understand that set
of informational processes which
allows the estimation of the elapse
of consecutive disturbances of the
basic system (BS) and by the help of
which a set of primary causes and
possible consequences of the disturbed system states can be found out.
Diagnosis will be carried out by the
IS which is coupled to the BS by sensors and effectors at the one side
and by MMC to the environment/opera-
tor. SAAP-2 can be classified as a
system with the properties: guarantee-
ing system, on-line open loop, real-
time, deterministic, predictive,
dialogue oriented.

The rough scheme is drawn in Fig. 2.
According to this, it essentially
contains the preprocessing system,
the analysis system, the operator-
interface system.

From this the following can be seen:
after data acquisition and buffering,
first step analysis will be realized
using knowledge from the dynamic data
base and actualizing that as a result
of the first step analysis. Afterwards the diagnosis will be activated
on the base of the model (model data
base) and the actual process state
(dynamic data base). Presentation of
the diagnosis results, the model
manipulation and an interactive
diagnosis will be carried out in
dialogue, see also Fig. 2.

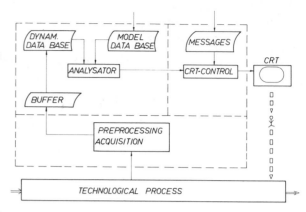

Fig. 2. Functional scheme of SAAP-2.

THE MODEL

The model of the technological process is noted as an event graph
$G = [E,R,\emptyset]$, with $E = \{e_i\}$ - set of
events/nodes; R - set of relations/
arcs; $\emptyset$ - incident function, it
corresponds to the mapping $\emptyset$:
$R \rightarrow E \times E$.

Furthermore there is $e_i \in E \subseteq N \times B \times D \times$
$W \times C$, with N - set of descriptions
(e.g. "heat exchanger"); B - set of
valuations (e.g. $B = \{0,1,2,3\}$ with
0 - state normal, 1 - state disturbed,
2 - state dangerous, 3 - trip state);
D - set of certain time properties
(e.g. reaction time, delay time);
W - set of probabilities (e.g. for
certain transitions); C - set of
further conditions.

The relation between e_i and the state
component z_i can be understood to be
$z_i = (e_i,t) \in Z \subseteq E \times T$, the state vector
$z(t) = (z_i)_k \in \Pi Z^k$.

That means, $z(t)$ corresponds to the
actual event pattern in the event
graph G.

Diagnosis must be therefore understood as follows: classification of
the event pattern, i.e. comparison
of $z(t)$ with those of the classes
"normal" and "abnormal", and search
for the graph to find the cause and
consequence partial graph.

DATA BASE

In general the data base should comprise the whole knowledge about the
technological and the informational
processes. In essential it has to
fulfil the following requirements:

it should contain the model and the actual process data; it should be actualizable, storage effective; it should support an effective dialogue and allow a very fast storage access.

According to Fig. 2 it consists of three non disjunctive partial sets:

- the set of process data and data for acquisition and preprocessing (margins of oscillation, priorities, event pattern, time marking)
- the set of model data for diagnosis (event pattern, time marking, causes, consequences pointer, operation instruction, delay times, theoretical calculation of nodes)
- the set of data for communication/ dialogue; it uses the same data as diagnosis does, excepting event pattern, but additionally messages for the user.

There can furthermore be distinguished between fixed and variable structure data.

Fixed structure data (table) contain pointers to variable structure data (graph), to data files, to theoretical valuations, to event pattern, to time marking. For each measuring point there exists a row in that table.

The graph contains cause-consequence pointers and corresponding specifications.
Each node will be implemented as a list element which can be written in the EXTENDED BACKUS NAUR FORM as
⟨list element⟩::=⟨node⟩⟨node valuation⟩

$$\left\{ \underset{o}{\{⟨\text{cause pointer}⟩\text{ OR}⟩\}^m} \atop \underset{o}{\{⟨\text{mark for AND}⟩ \atop \{⟨\text{cause pointer AND}⟩\}^{a_j}_2\}^n} \right.$$

⟨mark for consequence pointer⟩

$$\underset{o}{\{⟨\text{consequence pointer OR}⟩\}^q}$$

$$\underset{1}{\{⟨\text{mark for AND}⟩ \atop \{⟨\text{consequence pointer in AND}⟩\}^s_1\}^1_o},$$

with $m, n, q, s, a_j, j \in N$ (set of natural numbers),

$m + \sum\limits_{j}^{n} a_j$ – number of cause nodes

n – number of AND-tupels of cause nodes

$q+s$ – number of consequence nodes.

PREPROCESSING AND DIAGNOSIS

According to Fig. 2 there are in essential three informational sub-processes: preprocessing, diagnosis and MMC.

Preprocessing carries the interface between instrumentation and diagnosis. T_he process structures of both partial processes are shown in Fig. 3.

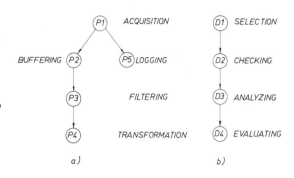

Fig. 3. Process structures for pre-processing a) and diagnosis b).

Diagnosis should be carried out with the properties: real time work, interruption, variable activation of the analysing process, simultaneous output of important results to the operator, MMC for diagnosis.

MAN-MACHINE-COMMUNICATION, MMC

MMC of SAAP-2 must be considered a partial system of the whole MMC system of the RRR, which in essential must be seen to be localized at the control room. It has been designed functionally and ergonomically to meet the following global requirements:

- the operator should not be overloaded by information
- information should be as compact as possible on the one side and as necessary on the other side
- certain information should be accessible in a certain manipulation time.

MMC of SAAP-2 comprises in essential three partial processes (TABLE 1):

1. Diagnosis evaluation and representation; real time diagnosis is able to yield a high quantity of information. But it is not always useful to present it completely; selection and representation on the colour CRT screen can be made by dialogue.

2. Interactive diagnosis; will be activated in cases of disturbed on-line diagnosis, e.g. disturbed data acquisition, contradictions in real time diagnosis, reaction in anomalous situations.

TABLE 1   Dialogue oriented Partial
          Processes

DIAGNOSIS EVALUATION	INTERACTIVE DIAGNOSIS	MODEL MANIPULATION

· FAILURE GRAPH
· CAUSE AND CONSE-
  QUENCE GRAPH
· PROPHYLAXIS
· THERAPY
· SEARCH OF
  PRIMARY NODES
	· HARDWARE
	  DEFECTS
	· CONTRADICTABLE
	  INFORMATION
	· OFF-MODEL EVENTS
		· IMPLEMENTATION
		· MAINTENANCE
		· ACTUALIZING,
		  ERASING,
		  ADDING OF NODES,
		  ARCS

3. Model manipulation; the event
graph must be handled as a very
complex tool. Therefore model implementation and maintenance are decided
to be made in dialogue.

Structures of these three partial
processes are recognized to be isomorphic and visible from Fig. 4 and
interpreted in TABLE 2.

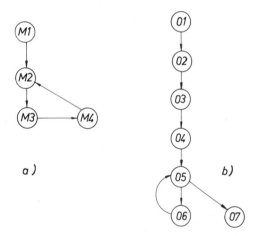

Fig. 4.   Structure of monitor, a)
          and dialogue oriented
          partial processes, b).

TABLE 2   Structure of MMC processes

subprocess	remarks
01 interruption	by user or system
02 selection	with a corresponding specification
03 loading	certain routine into main store
04 start	
05 synchro- nization	master function; synchronizing of calculations, dia- logue functions, in- put surveillance

subprocesses	remarks
06 communi- cation	input/output work
07 final operation	dialogue finished; interruption to the monitor

Concerning 02 for the subprocess
model maintenance the following
interface operation should be taken
into account:

– actualizing:
  at the given structure of the event
  graph node/arc evaluation should be
  actualized; input: node/arc;
  valuation element; new value

– adding:
  of nodes/arcs to a given event
  graph; input: e.g. node, position,
  node valuation

– erasing:
  nodes/arcs from a given event graph;
  input: node/node pair.

The function of dialogue in connection
with diagnosis can be seen from Fig.
5. As the control part of that system
the monitor works, whose operation
can be mapped by the process structure visible in Fig. 4 and TABLE 3.

TABLE 3   Monitor Operation

subprocess	remarks
M1 generation	implementation of process model
M2 start	activation of tasks by operator
M3 operation normal	on-line diagnosis
M4 operation disturbed	interactive diagnosis/ model maintenance

Concerning the ergonomic aspect of
MMC one has to start with the
following operator tasks which must
be understood to be embedded in the
set of interface measures mentioned
above, see TABLE 1.

– learning: model manipulation
– monitoring: surveillance of
               indication
– intervening: interruption of auto-
               matic procedures,
               manual control
– teaching: generation of commands
– planning: therapy measures

For this special properties of operator acceptance have to be taken into account, (Sublett, 1980). To reduce difficulties in operator acceptance, the following conditions for CRT oriented dialogue and representation have been considered:

- analogue information is suitable for dynamic variables and groups of variables

- pictures are very compact, but not suitable for detailed information

- it is not always favourable to use too much colour and/or blinking.

REALIZATION

Realization of the complete system at the RRR can schematically be seen from Fig. 1. The special part of the IS for diagnosis and dialogue is drawn in Fig. 5.

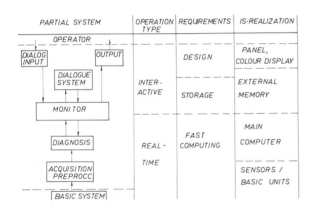

Fig. 5. System structure of diagnosis and dialogue system.

Procedures of analysis and presentation of graphs work recursive and use common working stacks. Each stack element contains pointer to list elements of the stacked node and a mark of the actual logic relation.

Generality of SAAP-2 can be supported, using parameters like: number of measuring points, number of nodes, length of node notation, maximum recursion depth, stack length.

Data structures comprise

- implementation of the event graph in list technique

- implementation of the relation between node notification and address of the list element and

- construction of stacks (handling of failure graph and recursive procedures).

Each node is stored as a list element. The valuation of nodes comprises the properties: event occurred/not occurred and the event class.

Properties of measuring points are mapped in a table structure; it contains: interval limits, measuring value class, address of the list element, other information.

Such problems can in an effective manner only be solved by using high level problem oriented languages. LISP has been used at a first stage of realization on cause of the following reasons

- data base exists as lists and the data structure should hold to be dynamic

- LISP seems to be dialogue effective

- LISP allows logical connections to be described very effectively

- LISP uses a dynamic management of the storage and therefore omits to touch the storage borders.

But there are disadvantages: LISP implemented on the microcomputer technique available currently works very slowly; for that reason it will be useful to gain a LISP subset, that would allow the time requirements to be met.

CONCLUSIONS

SAAP-2 must be considered under development. Disregarding technical properties, functional and ergonomic properties are a matter of steady further investigation.

Concerning the first aspect for diagnosis there are problems like: what should be overtaken by the IS in case of varying event graphs, recognition of technological relations processing of certain simple event pattern (to avoid expending graph analysis), self diagnosis of the IS.

Interesting problems arise with the task of taking into account automatic and dialogue oriented therapy control. This must also be seen to be tied up with following questions:

- what should be the border between automatic and dialogue oriented diagnosis and therapy

- what should be used from ARTIFICIAL INTELLIGENCE to allow an increasing decision capability for the operator and what should be done to shift that border in direction to machine.

REFERENCES

Baldeweg, F. (1980). On-line alarm
    analysis using decision table
    technique. Proc. IAEA/NPPC
    Specialist's Meeting, Munich, FRG,
    Dec. 5-7 (1979), pp. 301-319.
Baldeweg, F., W. Enkelmann, and
    J. Klebau (1982a). Überwachung
    und Kontrolle des Rossendorfer
    Forschungsreaktors mit einem
    mikrorechner-orientierten Auto-
    matisierungssystem. To be
    published in Kernenergie.
Baldeweg, F., and U. Fiedler (1982b).
    Dialog-orientierte Analyse von
    Ereignisgraphen - ein Verfahren
    zur Diagnose komplexer technolo-
    gischer Anlagen. To be published
    in Messen Steuern Regeln.
Felkel, L., R. Grumbach, A. Zapp,
    F. Øwre, and I. K. Trengereid
    (1980). Analytical methods and
    performance evaluation of the
    STAR application in the Grafen-
    rheinfeld NPP. Proc. IAEA/NPPC
    Specialist's Meeting, Munich, FRG,
    Dec. 5-7 (1979), pp. 283-300.
Meijers, C. H., B. Frogner, and
    A. B. Long (1980). A disturbance
    analysis system for on-line NPP
    surveillance and diagnosis.
    Proc. IAEA/NPP Specialist's
    Meeting, Munich, FRG, Dec. 5-7
    (1979), pp. 213-231
Sublett, I. G. (1980). The human
    component in man-machine systems -
    a balancing act. Control Eng.,
    Vol. 27, No. 12, pp. 62-64.

DISCUSSION OF SESSION 2

2.2 T AN INTEGRATED DISPLAY SET FOR PROCESS OPERATORS

Tanish: You gave some interesting examples of applications of multi-level flow diagrams or mimic displays. Have you considered the possibility of knowledge exchanges, or even rule changes agreed between the operator and the computer system?

Goodstein: Knowledge exchanges particularly concerned with why questions have been considered, and these types of possibilities are being pursued for the future.

Nzeako: Form your abstraction levels of the process (or system) by the human operator (in the upward direction) a static or a dynamic model? In other words: does it consider the points in time when the human operator moves from one level to the other and the fact that the operator can be promoted out of one, two or more levels of abstraction, or even out of the process control line, e.g., to supervisory or management positions?

Goodstein: Within the context of the control room and the operator's interaction with the plant, levels of abstraction refer to an important set of different representations of the plant within which the operator can base his thinking. As described in the paper, these range, for example, from a consideration of nuts and bolts to water circulation to overall production and safety goals. While on shift and dealing with the wide variety of tasks/situations/disturbances which can arise, the operator will/should move up and down among several of these levels of abstraction in order to detect, identify, evaluate, decide and act. Thus, in our view, operators should be trained and supported (via appropriate displays, etc.) in order to function in this way.
However, it is likely that a new junior operator or trainee will concentrate his thinking on the physical and physical-functional aspects while a supervisor will generally be more concerned with production goals and safety requirements and with the efficiency of the energy conversion processes and with material balances and inventory. However, this is not equivalent to stating that each level of employment corresponds directly to one level of abstraction since,

in dealing with the plant, answers about why, what and how will be required and these involve several levels of abstraction - possibly at different levels of plant detail for people of different rank/responsibility.

2.4 T A MULTI-LEVEL ALARM INFORMATION PROCESSING SYSTEM APPLIED TO THERMAL POWER PLANT

Hartfiel: Because we're also talking about ergonomics, that means, trying to reduce the workload of the operator, why do you output the alarms as messages and not as curves, since you also want to show tendencies?

Kawai: On the plant described, the number of process inputs which are checked for alarm exceeds 600. Further, at any given time up to 100 such points may be in alarm for a variety of reason (normally 5 or so). Current VDU technology for semi-graphic colour CRT's limits the number of graphic (trend curve) displays to about 5 per screen whilst the alphanumeric alarm message displays described can accomodate 20 alarms. The alarm display described is a dedicated peripheral. However, the curve type information can be requested by the operator on the VDU's on the control board if he wishes to observe a trend in more detail.
A further feature is that mimic diagram graphics are also provided which show a different colour when a process point is in alarm.

Miller: Re question by Hartfiel re outputs, messages, not curves, why?

Kawai: Trend analysis involving VDU color displays of alpha numerics, process mimics with color bars and alphanumerics, and color bar comparisons all have their place. Color bar comparisons are especially effective when operators must make quick judgemental decisions in a matter of seconds as on a hot steel strip rolling mill as the front end of the strip progresses through the six or seven finishing stands. The operator must quickly compare actual with predicted (calculated) stand rolling pressures. Higher pressures can result if the strip is colder than predicted, or the chemistry is incorrect (beyond expected

117

limits). The operator must decide whether or
not a set up modification is required to
avoid a mill cobble, or strip breakage. Generally it is not. It is important that users
be familiar with effective VDU displays
available for his process, or processes and
specify his preferences. VDU displays provided by vendors differ significantly in scope
and effectiveness. Software development is
expensive and displays new to a vendor are
unlikely to be provided if not specified by
the inquiry or the contract.

Nzeako: Considering Figure 2(a) "Alarm
message output control" of your model, can
you, please, explain whether the change over
from "worse" (y-wrs) to "better" (y-btr)
alarm is a function of the response of the
human operator, or that of a changed state
of the controlled process (or system), or
the interaction between the human operator
and the computer, or of a combination of any
two or all of the three functions (or agents).

Kawai: When an operator observes a "worse"
alarm he will attempt to correct this
situation by changes to his control of the
process.
The computer is used only for the checking
of a process input or calculation value
against its pre-assigned alarm limit and
delta and consequently indicates "alarm,
worse, better, return" based purely on this
check and has no control of the plant.

2.8 I CONTROL AND SUPERVISION OF THE
EURELIOS SOLAR POWER PLANT: DESIGN
PHILOSOPHY AND OPERATION EXPERIENCE

Pitrella: If the automatic system is based
on a mathematical model which has been
supported by experience, what is the operator's task: to make sure that the automatic
system is working allright or does he have
specific control duties?

Maini: Noting that the control system has
not been designed to control the plant during
the start-up or the restart of the plant
after a sufficiently long lack of solar
energy, the main tasks of the operator are:
- to guarantee plant safe operation also
  during the most demanding disturbances or
  component malfunctions, beyond the control
  system limits.
- to manage the (up to now) non-automatic
  manoeuvres like start-up and shut-down.
- to decide the operation mode of the plant
  taking into account the actual and foreseen meteo conditions and the amount of
  stored energy.

Pitrella: Exactly why are operators stressed
by meteorological changes, considering that
the automatic system works fairly well?

Maini: Taking into account the answers given
to the preceding question of mr. Pitrella,
it is possible to claim that the frequency
and the sudden changes of meteo conditions

do stimulate the operator more than in conventional plants. Furthermore, the always
incumbent time constraint (that is the unavoidable sun daily cycle) given another
subtle stress factor.

Lenior: You have an actual system with conventional displays and problems in operating
it. You have a CRT-instrumented simulator
for developing engineering characteristics
of the system; and intend to give the operators such sort of instrumentation systems
in the future. Are you already training the
operators from now on by the simulator to
prepare him for the future events?

Maini: Actually the simulator is not installed on plant but it is operated in laboratory
for preparing the plant dynamical test program, which will be carried out during next
1983 spring/summer. At that time we foresee
to use the simulator also on plant for
training the operators mainly in the face of
the most relevant transient.

2.9 I COMPUTER AIDED CONTROL STATION WITH
COLOURED DISPLAY FOR PRODUCTION
CONTROL

Klammer: Is the job description and dependencies among jobs entered into the system
interactively or by an 'off-line' method?

Aldinger: There are two ways, and it is
necessary to have both, because often the
dependencies are not all known or they can
change (e.g.: a component is used in several
products, but one product gets a new component instead of the old one).

Berker: Relating to the title of the conference, I want to ask, if there was made
something like a task analysis of the earlier
form of doing the job, as some kind of design
aid for the development of the system, e.g.
for the selection or the format of displayed
pictures?

Aldinger: There have been task analysis of
the earlier form of doing the job and there
are also results obtained by psychologists,
which show what are the best ways to inform
an operator. If you combine both approaches
you can build up a system, which has the
same possibilities for the operator and has
a very high level of comfort and performance.

Engelhardt: a) Are PERT or CPM used for
interlinking of all actions?
            b) What is the number of people
and the time of operation of the system?

Aldinger: a) There is no usage of PERT or CPM
in the short term production control. We
consider the usage of PERT or CPM in the
medium term or long term production control.
    b) The number of people in the
system depends on the content of each task.
(In some cases the data collection can be
done without the control station; in some,

many activities have to be done by the people
in the control station).

Arndt: Concerning your Figure 4, I assume
that the length of the bars is proportional
to true. Can the length/content of these bars
be scaled up and down in time, and does the
software allow an overall optimization of the
schedules for all machines? Or is this left
to the experience and creativity of the
operator?

Aldinger: The length/content of these bars
can be scaled up and down. There is no overall optimization for all machines. This is
left to the experience and creativity of the
operator. But the operator has the aid of
the computer to do new planning.

Arndt: How will this package now be transferred to industry? Can industry buy the
system?

Aldinger: Each shop has its own problems, and
therefore every solution needs adaptation.
That means, that our system has to be adapted
with the help of the IPA. Afterwards it can
be transferred to the shop.

Prutz: It seems better not to put the production planning on the process-computer due to
timing-problems.

Aldinger: The development gives us new pro-
cess-computers which are very powerful. And
therefore it is possible to do the data
management and the data collection with the
same process-computer.

Baldeweg: Would you like to tell more about
the automatic failure reaction system?

Aldinger: The automatic failure reaction
system is not included in the existing system. Perhaps there will be a system like
that in the future, but the effort for the
software will increase very strongly.

2.10 I   STAR-A CONCEPT FOR THE ORTHOGONAL
         DESIGN OF MAN-MACHINE INTERFACES
         WITH APPLICATION TO NUCLEAR POWER
         PLANTS

2.12 I   MAN-MACHINE INTERFACE IN THE DISTUR-
         BANCE ANALYSIS SYSTEM SAAP-2

Chairman's Discussion Summary: Papers 2.10 I
and 2.12 I will be treated together since
they share a common modelling approach based
on an a-priori appraisal of the spectrum of
possible disturbances which their systems
are to identify and diagnose as to cause,
consequence and (in some cases) indicate a
suggested recovery. Paper 2.10 I on STAR
dealt with a formal language which is intended to enhance the cooperation between
process engineers and ergonomists by serving
as a common medium for expressing the relevant information from both parties. This
essentially rule-formulated approach can,
according to the authors, in principle deal

with all conceivable mappings from process-
related conditions to operator-support
details. Among the support facilities mentioned were procedural guidance, post-trip
analysis, disturbance analysis, alarm handling, etc. During the discussion, it was
reported that the STAR system has been installed -using about 250 rules- at a nuclear
reactor site. The control room operators are
not directly involved in its use. Experience
to date has included several situations where
STAR has diagnosed a disturbance in advance
of the operators. Paper 2.12 I dealt with
the SAAP2 system which provided a disturbance
analysis function based on event graphs.
During the discussion the speaker indicated
that the size of the present system corresponded to approximately 120 nodes -a relatively modest system. Indeed SAAP2 is implemented as a subprocess of a subsystem of the
total information system. No analogue information is used. The speaker indicated that
this is possible but that there arise dialogue problems when these have to be modified.

Both systems have had to deal with the problem of coping with unforeseen disturbances.
Both require and utilize to a certain degree
an "interactive diagnosis" mode involving
operator participation when an automatic
analysis is not possible. For example, the
criterion for switching SAAP2 to an interactive diagnostic mode is signal-based.
However, according to the speaker, this is
not completely satisfactory. Thus the im-
pression was given by both speakers of a
striving after the best of the two "worlds"
of (a) a-priori "completeness" and (b)
enhanced "on-line" diagnostic aiding for
the remaining unforeseen disturbances,
while admitting serious problems with the
transition from one to the other -problems
which, from this correspondant's view, must
include the operator's role and involvement.

2.11 I   DYNAMIC AND STATIC MODELS FOR
         NUCLEAR REACTOR OPERATORS - NEEDS
         AND APPLICATION EXAMPLES

Chairman's Discussion Summary: The presentation of 2.11 I dealt with examples of two
forms for modelling the behavior of operators
interacting with the system. One was static
for situations where the time dimension is
not important and the other dynamic for
situations where the process can pace the
operational evolution. In the discussion,
the speaker emphasized that no behavioral
models had yet been incorporated within the
described modelling framework but that information from event reports, the literature,
etc. had influenced both the situations
chosen as well as the parameters employed.
All parameters (e.g. probability distribution
functions, etc.) could be altered. Other
models could also be incorporated if desired
since the structure was very modular. The
discussion touched on the interactive nature
of the dynamic modelling system which permitted the user to interatively constrain the
analysis' extent via several levels of inquiry.

MAN'S ROLE IN MAN-MACHINE SYSTEMS

F. Margulies and H. Zemanek

University of Technology Vienna, Vienna, Austria

Problems of man-machine relationship can be traced back to the very beginning of human history, when the conscious production and application of tools marked the transition of animals into the predecessors of human beings.

Man, the "toolmaking animal" is the only creature who with the power of his intellect transformed the permanent confrontation with nature, which every living organism has to experience, into a work process.

Life in every form is a continuous dialectical struggle with an evironment which provides the general conditions for life while at the same time threatening and destroying its individual existence. Plants and animals, directly confronted with this environment, cannot but react passively to it. Their weapons in this struggle are of a biological nature, and the character of this struggle only changes as a consequence of biological changes, i.e. over periods measured in thousands or millions of years. Their chance of survival lies not in changing nature, but in adapting themselves.

Only man confronts nature with weapons not developed by nature, but with artificial means created by himself. These means are tools, machines, the ability of man to put them to use, the organization of work, the social relations and the structure of society. By his own creativity and activity, by exploring, understanding and applying for his own benefit the laws of nature, man "made the earth his servant".

But if work - the employment of intellectual capacities and the conscious production and use of tools - constitutes the dividing lines between animal and man, this means that work is of paramount importance also for the individual of today and for the future of mankind. Work is not only a concept of physics, known by heart to every high school student. It is much more than that: work is a philosophical, a sociological, a psychological, or to put it in a better and more comprehensive way, a *humane category*. There is no work, however automated it may be, which does not involve man at some stage, there is no product which can be made completely without man. All the wealth of this world is the

product of human work, and vice versa, mankind could not exist without work. By his unceasing endeavour to make work better and more effective, man has developed new means of production as well as new ways of human cooperation and social organization complementary to these new tools.

For thousands of years throughout the history of mankind work was organized by the producer himself applying natural resources, traditional experience and personal knowledge at his disposal as well as experimenting with new ideas. This was feasible as long as the production process remained an entity of physical and mental work, as long as planning, production and control were all carried out by one person - the producer.

Industrialized mass production led to the division between mental and physical work. Distinct groups of people became responsible for planning, production and control respectively. Even with each of these groups, in particular in the production sphere, further division of work took place. This is to this day symbolised by the name of Frederick W. Taylor, who at the beginning of this century shaped his philosophy of company targets and his perception of human beings employed as workers into what he called "Scientific Management". This has been the mould for work organization ever since [1].

This kind of specialization of work and segmentation of life was accompanied by an elaborate hierarchical structure in which nobody would get more than his share of thinking and desicion-making. Human choice was restricted to what was left over, after higher levels had had their choice.

Thus man was degraded to a machine, to an automaton. In the process of work he cannot make any decisions, every move is regulated by instructions; the machine, his master, does not belong to him; the product, in whose production he participates, belongs to others. This way the worker has lost his personality, the personal relationship to his work.

All this went fine for some 150 years after the advent of the steam engine. The industrial revolution provided the tools, Frederick

W. Taylor supplied the ideology and the human
beings did the jobs which machines had not
yet taken over.

Much has changed, however, since Taylor's time.
Whether or not his 'Scientific Management'
was justified at his time, his principles do
not fit into today's situation.

Beginning with the end of the Second World War
the industrial nations entered a new stage of
technological development – the stage of
scientific technological revolution. It is
characterised by a comprehensive development
of *all* factors of production – automated
methods of production, new organization techniques, new materials, new sources of energy.
It is furthermore characterized by the transformation of science into a direct productive
force, by the application of mathematical
methods to many branches of science which
formerly were not amenable to them, by the
application of systems theory and of modelling
and simulation in experimental research, by
rendering the production 'scientific', by
an ever closer liaison between theory and
practice. This stage is – last but not least –
characterised by the growing importance of
man, in spite of automation and computerisa-
tion. Division of labour, man as an appendage
of the machine, these principles are no longer
in accord with the new technology. The more
complex a system is, the more it is dependent
on intellect, creativity and the human ability
to make decisions.

Equally important changes have taken place
since Taylor's times with respect to the work
force. Workers and employees have become the
majority of the population in all industrialized countries; their organizations have
become more powerful and influential, labour
unions are as a rule not outlawed and perse-
cuted any more, as they were 100 years ago,
but have become an integral part of political
and economic life. Workers and employees
have raised their level of education, they are
more self-confident, they are aware of their
personal, professional and economic value.
Thus resistance is growing against inhuman
working conditions. Where material need has
to some degree and for some period of time
been overcome, mental need as expressed in the
lack of challenge and content of work is all
the more present.

Thus technological needs and human aspiration
meet in the search for new patterns of work
organization and for new relationships between
man and machine.

Last not least, control engineers and computer
scientists have had to realize that automated
systems cannot and will not work without man.
The dream of the machine doing everything by
itself just cannot come true.

Yet some people still continue to dream the impossible dream. They may be hoping to protect
their machines against disturbances caused by
human operators or they may be scared by

stories of factories without men, offices
without men, a world without men. x)

Any serious analysis of these ideas proves
their lack of viability. Any attempt to replace human work even to a considerable degree,
let alone completely, poses unsurmountable
programming difficulties. A world without
work would require a universal algorithm which ·
as Goedel demonstrated long ago – does not
exist even for pure mathematics.

We better face the fact then: automated system:
will remain man-machine systems, they will increasingly require man to play his role as knowledgeable, educated and motivated master of the
machine, and they will require new forms of
work organization which allow for this kind
of human choice and human activities.

Norbert Wiener, the father of cybernetics,
wrote in his book "The Human Use of Human
Beings" more than 30 years ago [3]:

"In my opinion, any utilisation of a human
being is degradation and waste if it asks for
less and expects from him less than his full
capacities. It is a degradation to chain man
to a thwart and use him as a source of energy; but it is almost as great a degradation
to set him in a factory to a perpetually re-
iterating task which claims less than a
millionth of the capabilities of his brain."

Yet exactly this is still being done in our
factories, in our offices, in our data processing systems. It not only spells degradation of man, but also wastes one of the most
precious resources of our economy and our
society: human brain power. The waste is a
twofold one because we not only make use of
the resources available, but we also let them
perish and dwindle. Medicine has been aware
of the phenomenon of atrophy for a long time.
It denotes the shrinking of organs not in use,
such as muscles in plaster. More recent research of social scientists supports the hypothesis that atrophy will also apply to mental
functions and abilities.

It is interesting that this relationship between what the job demands and the development
of one's personality was mentioned by Adam
Smith, the British economist, 200 years ago,
without the background of medical knowledge
and research in this field. In 1776 he wrote
in "The Wealth of Nations":

"The understandings of the greatest part of
men are necessarily formed by their ordinary
employments. The man whose life is spent in
performing a few simple operations every day,
which will always produce the same result,

---

x) Scenarios of that kind have recently been
   presented at the Club of Rome meeting in
   Salzburg in February 1982, dealing with
   the topic "For Better or for Worse –
   Microelectronics and Society" [2].

has no occasion to exert his understanding. Since there are no obstacles occurring, he need not think of how to overcome them. He quite normally then forgets how to use his intellect and becomes as stupid and ignorant as a human creature can ever become."

To escape this vicious circle of designing stupid jobs to fit stupid people, thus making people stupid to cope with the stupidity of their jobs, we obviously have to start with the jobs. Modern technology offers a great opportunity to restructure our jobs, to give the stupid parts to the machines and to leave the intelligent parts to intelligent people as a chance, a challenge and a motivation to increase their intelligence.

Not by fighting technology but by putting it to a human use we shall find our way, not by returning to Adam Smith, but by translating his perception into our present and future. Let us repeat that Adam Smith was an economist who saw the economic implications in this situation, while some economists of today still maintain that Taylorism has to be applied for economic reasons, even if one regrets the inhuman effects. Alan Gladstone of the International Institute for Labour Studies, Geneva, has a different story to tell. In a framework paper which he presented at an international symposium in 1977 [4], he said:

"As the increasing interest in new forms of work organization seems to demonstrate, the economic advantages attributed to Tayloristic production systems no longer seem to hold true in the context of the present socio-economic situation. The more recent recognition of the economic disadvantages of its social effects has played an important role in the search for alternative forms of work organization which might simultaneously improve productivity, work satisfaction and motivation."

Similar opinions are expressed in a great number of publications by social scientists, trade unions, governments and enterprise managements. They all strike the note that 'Humanisation of Work' and 'Productivity' or 'Economic Viability' of alternative work organizations need not be contradictory or mutually exclusive. And in fact over the last fifteen years we have witnessed an increasing number of experiments and research concerning alternative work organizations.

They have been initiated by the discrepancies between Taylorism on the one hand and the changes in technology, educational level and consciousness on the other, as mentioned above.

In fact, every man/machine system links two extremely contradictory structures. Take, for example, the computer. It is an artificially created, formalized system with stable features. Once having been designed and built, it can fail, but it can never change. Man, on the other hand, is a natural, non-

formalized system. He is never stable, he changes as long as he lives - even if he does not fail.

A computer can be described completely by drawings, figures and wiring diagrams. Any expert in possession of such a description can reproduce this computer as often as he wants to, and it will always possess the same, previously determined qualities and parameters. Should you try to describe a human being the same way, you will soon find out that the more copious and detailed the description becomes, the less it will provide a picture of the personality. However successful one might be in producing human beings, production of a given pattern has never been accomplished.

Any machine offers a certain amount of perfectionism in its particular field but because of that quality is incapable of creativity. Man, on the other hand, is by definition not perfect, faulty, makes mistakes, but this is the supposition to his creativity. The perfection of a machine is restricted to the built-in formalism of its performance, while contents and meaning of the work orginate from and terminate in the human being. Just as the best quality of a TV set has no influence on the quality of the programme, just as perfect grammar and syntax cannot prevent a speech from being empty and meaningless, the same will happen even with the perfect computer: only permanent control and correction will prevent the contents from rotting and the sense from being distorted.

The classic computer is best described as a character storage and replacement device: conforming to strictly logical rules one set or sequence of characters is replaced by another one; the process can increase or decrease the total amount of characters. It is easy to show that a logical device of this kind can transform, select and destroy information but never generate information. Operating strictly by logical rules, it will amplify human intelligence, but it will equally amplify human un-intelligence [5].

While the machine will render best results if it is adept at applying a strict and well-defined routine, man has to be given space for applying his own discretion wherever this is possible.

If information processing as a technology and profession has already so many and important human aspects, its impact on society must correspondingly have many and important peculiarities. There is no non-trivial engineering innovation without an impact on society. Many political events of the last hundred years were, directly or indirectly, triggered by the progress of technology, although not as many as certain mechanistical political theories would like to have.

Until very recently, most technological innovations could be seen as a singular solution, improvement and phenomenon. But now we have reached such a density of excellent solutions and effective improvements that they hamper each other and produce chaos. The outstanding example for this effect is road traffic. Never in history has mankind had such excellent roads, so many traffic signals and such powerful, fast and well-designed vehicles. But the sum of all these excellent solutions and effective improvements are king-size traffic jams. We are much better in singular solutions than in systems design, not only in traffic - this is true all throughout technology and there is no risk in predicting that information processing will see itself soon in the same trouble. Even if the single, centralized big computer will remain a tool for order the mushrooming networks and mini-processors will lead to a chaos of equal or even bigger size.

A big computing center with its multitude of processors, storage units and peripheral equipment is a system which raises many problems of efficiency and cost/performance, and it must be admitted that there is not yet much of a useful systems theory to support its design or configuration. The advance of digital communications and the growing need for distributed information processing and the reduction of line cost makes it possible to distribute input, processing, storage and output over cities, countries, continents and the whole globe. If the computer started as a centralizing power, it has meanwhile developed into a tool for decentralized centralization or centralized decentralization. If properly conceived and designed, automatic information processing can provide a maximum of information, liberty of decision and human choice to peripheral subcenters without reducing the efficiency of central policy and overall success. Wherever a system suffers from bureaucratic inflexibility, slowness and inefficiency, the cure can be well-designed distributed computer assistance which re-establishes the flexibility, speed and success possible in a small structure. This is by no means easy to achieve, but there is a chance and much emphasis must and will be laid on the development in this direction.

It is the purpose of technology to either

- perform conventional tasks with less effort and at reduced costs, or

- perform new tasks which have been physically or economically impossible without the new technology

In the first case, technology obviously is a job killer - it reduces the required manpower and it tries to save money - and with the restricted pattern of thinking in our times this is done primarily on the salary sector.

In the second case, technology is a job creator, and nobody can question that applying modern technology is the only way to keep the present population of a continent like Europe fed and working. A computer-near example for killing and creating jobs is the typewriter. It certainly was a job-killer for scribes, but their number was in no way comparable to the number of secretaries and typists of today. We would laugh full-heartedly if we read a statement of a sociological pessimist who has just seen the first typewriter and in which he objects to the job-killing dangers of the new invention.

The analogy to the conventional computer as well as to the mini-computer is obvious. And much more than the typewriter, the computer initiates an avalanche effect: one computer put into operation prepares the need for further computers. Information processing technology is a job-creating invention. But this is not the full story.

A man whose qualification was the production of calligraphic documents will feel little consolation on hearing of many openings in typewriting jobs. The problem is one of transition and ageing. As long as technology develops at such a speed that obsolete jobs die out with the speed by which people retire, a modest educational opportunity will be sufficient to maintain the professional balance. In our days some job requirements change faster than our best educational measures can compensate for. There is a social problem and it is reasonable to sit down with all those concerned and responsible and discuss what can be done.

There is a second effect which aggravates the first, and that is the dehumanization which abstraction and labour division bring to the working place. A medieval shoemaker like Hans Sachs produced a pair of shoes out of raw materials he prepared either himself or bought from people he knew very well, and he sold the majority of his shoes to people he knew very well.

The distance between suppliers, producers and users in modern technology is steadily increasing and automatic information processing does more to separate them further than to bring them together again. The single working place tends to be a narrow slot through which it is almost impossible to see the complete product. In a big information processing system no single person knows all the details.

All of this has very many facets; some features are a necessity, others occur stochastically because nobody cares [6].

This was also proved by the report of Oliver Tynan, head of the Work Research Unit of the British Government, which was presented at a conference in Toronto last year [7]. His Unit sent a group of experts to a large trade show of robots and other new types of equipment and asked every exhibitor about the consideration given to human factors in designing the equipment - not a single exhibitor gave a positive response.

An empirical study on "Automation in the Steel Industry" showed much the same result. This cross-national research, in which ten countries participated, was published recently in Vienna [8]. It states in the conclusion:

"There was no indication that human factors had anywhere been considered consciously. Automation opens up new perspectives, offers new options to both users and designers, but these options can only be realized if everyone affected by technological change is given full information and opportunity to participate in the design process and in all relevant decisions and if this design and decision process includes the selection of alternative forms of work organization and of hierarchical structures to fit the human needs and to make the best use of technological opportunities."

Consideration in this direction should also be adopted by this "Conference on Analysis, Design and Evaluation of Man-Machine Systems". It is most encouraging that this topic has been chosen by the organizers. This meets with the intentions of the sponsoring federations. IFAC's Committee on Social Effects of Automation, which has been in existence for more than ten years, is itself the main sponsor of this conference, IFIP's TC 9 (Computers and Society) has expressed high interest in this event, having organized two conferences on "Human Choice and Computers" in 1974 and 1979 [9,10].

Recollecting the development over these last ten years one has to admit remarkable progress in the approach to and consideration of human factors when designing man-machine systems. Though ten years ago the idea of machines totally replacing human beings had to be dismissed as unrealistic and withdrew into science fiction and journalistic excursions, the prevailing approach of designers towards man was that of "unavoidable evil". Systems were designed to reduce human activity and even more so human responsibility. Man was confined to do the remaining auxiliary work which thus prevented him from disturbing the system. Experience has disproved this approach, too, it being detrimental to qualitative and quantitative efficiency of work. It was replaced by a more realistic one.

Man's role in man-machine systems was admitted to be of increasing importance in spite of or even because of sophisticated automation; consequently, to make sure that the new equipment was being adopted, man has to be treated, maintained and looked after like any other part of the system. Thus ergonomics came into the picture: user oriented design, interface comforts, reduction of stress, adjustable tables and chairs, noise reduction and many other things of that kind. A lot of research has been done in this field, a lot of recommendations have been issued and wherever these were implemented, certain hardships of the job concerned were reduced and, no doubt, improvements

achieved.

Much remains to be done, however, even in this limited field of ergonomic consideration. The great majority of work places still show severe shortcomings which are causing hazards to health, while physical improvements are being made up by increasing monotony, time pressure and mental strain. This is one reason why dissatisfaction with and resentment against new technology is increasing rather than being reduced.

In addition, it has to be realized that ergonomics, even if implemented to the best possible degree, will solve only part of the problem, while the work organization which will provide real satisfaction has yet to be found.

This will mean to rediscover work as an essential part of human life, as a humane category, as was mentioned in the first part of this paper, it will mean to overcome Taylor's principle of "division of work" and the separation between "thinking" and "doing". The worker should not continue to be considered the object of decisions taken by other people like designers, organizers, technologists, he should rather become the subject of work organization, participating in the design and in all decisions to be taken. The options provided by modern technology allow for more autonomy of the individual worker or of small groups of workers, they allow for work structures supporting the development of one's personality, they allow for identification with one's work.

So far, however, there are very few cases where these options are being made use of. Designing of jobs is usually aimed at *compensating for* former shortcomings, sometimes it is also intended *to prevent* future disadvantages. Both targets are valuable and important. But a third, equally important component of job design has been neglected widely up to now, i.e. what Ulich (1980) calls the *prospective design*, the conscious implementation of possibilities to apply *human choice* when using automated equipment, when structuring one's own job, when deciding on work contents, on work sequence, etc., thus allowing for the development of one's own personality [11].

These are not just romantic ideas proposed in favour of workers only. Renowned management advisers like Diebold Management Germany [12] have emphasized that productivity is not to be measured by machine performance alone, but always by the sum of machine capacity plus efficiency of the user which in turn depends on his motivation.

Social science researchers, on the other hand, like the American sociologists Maslow and Herzberg - to name only two of the most important ones - have proved that pressure and incentives lead to alienation rather than to motivation. A far better way to motivate people is to enrich the content of their work, to enhance their identification with the job, to give them the chance to develop their qualifications and their own personality.

This can certainly not be achieved by replacing one rigid work system by another one. What is needed instead is the replacement of rigidity by flexibility, the offer of new options and of a wider range of activities and decisions to everybody in accordance with new technologies.

Experiments with and implementation of alternative work organizations carried out over the last ten or fifteen years have proved that flexibility may be increased, options may be extended, job satisfaction may be enhanced, while technological needs and human aspirations need not be contradictory to economic considerations.

Some practical examples shall illustrate this.

A German insurance company found their work organization highly unsatisfactory. According to the description given by one of the directors it was a strictly Tayloristic structure [13].

"We had, for example, groups that were only responsible for filling out contracts. Other groups dealt with correspondence, reminders and alterations, and even here there were specialized jobs. That is, there were employees who only dealt with changes of address, changes of ownership or policy cancellations. As a result of this specialization it was possible for the individual employee to acquire very quickly the specialized routine knowledge which enabled him to deal with each case quickly and with very few errors.

On the other hand, we established that the fluctuation within these specialized groups was very high. Asked why they were giving notice, employees said that time and again that they wanted to leave in order to be able to do a more varied and interesting job in a small or medium-sized insurance company. Thus we had to pay for our job organization, which was certainly designed in full accordance with Taylor's basic principles, with the fluctuation and discontent of our employees.

We therefore started to consider how the structure of the departments could be altered in future. Instead of specialization we first considered dividing up the work according to file numbers. Thus a group of 8 to 12 employees assumed responsibility for all files with the end numbers oo-o9, 1o-19, and so on. However, this type of work distribution presupposed that if possible all the members of the group were familiar with the work which the processing and administration of contracts involved. Under the previous system this had not been the case. Our staff consisted mainly of specialists. By means of comprehensive

training and continuation courses, we had to put our employees in a position to carry out comprehensive tasks."

However, another very important step was taken. Through terminals the clerks were able to access data banks containing not only information about the customer, but also about the steps necessary to cater for customers' wishes. The computer thus became not only an archive without paper, but also a reference work and an aid to memory which enabled the employees to carry out comprehensive and diversified tasks.

The results of this change-over were reported as more job satsifaction, greater interest in work, easier substitution in the event of illness or holiday, and over and above all this, a productivity increase of 20-25%.

Management was so impressed by these results that they decided to extend the system to the other departments of the company. The related orders were carried out, but this time without the participation of the employees. The people responsible for the transition thought that copying the pilot department would produce a copy of the pilot results, too. They failed to recognize that satisfaction and identification can never be ordered and thus failed to achieve anything resembling the original positive results.

Another example is the use of word processors [14, 15]. We often come across a tendency to analyze the work of the secretary and to divide her activities by concentrating e.g. administration in one specialized administrative department, while typing is shifted to another central pool which is reinforced by word processors. The results have shown to cause serious physical strain to the staff, resentments against this kind of job in general and against the word processor in particular and a rather disappointing development with respect to productivity.

But in some cases word processors are used in an entirely different manner. Rather than intensifying fragmentation of work and specialization by establishing a new category of unskilled workers called, e.g. "operators of word processors", these machines are being implemented as decentralized auxiliary instruments to assist the secretary. Every major department gets its own word processor and every department has a sufficient number of personnel who have learned to operate it. The secretary who is charged with doing a certain job will use her own discretion to decide when and how she wants to use the machine.

The consequences of such an application are to be found in a reduction of routine work, in the possibility to deal with more responsible tasks, thus qualifying or upgrading the possibility to change the work place more often - nobody has to operate the word processor all day. Typing output is improved both in terms of quality and quantity while monotony and

drudgery are transferred to the machine. Last
not least work organization becomes more flexible with more people being capable of doing
different jobs.

Similar considerations were applied when the
Swedish car producer Volvo made its already
historic change away from the conveyor belt
and from Taylorism [16]. This change was
triggered by an increasing resentment against
the assembly line on the part of the workers.
Fluctuation had risen to 35% per year, absenteeism reached an average of 30%, and investigations showed that the social prestige of the
assembly line workers was the lowest on a
long list of different occupations. This
development was intensified to a considerable
degree by the reform of the Swedish school
system raising the educational level of young
people leaving school to a considerably higher
standard.

Thus, when a new production site was due to be
established completely new ideas were implemented. No more conveyor belts, no specialization to carry out minutely structured and
fragmented parts of the job, but autonomous
groups consisting of between 10 to 20 persons,
together responsible for a complete part of
the work, such as electrical equipment, engine
assembly, etc. each member of the group being
sufficiently skilled to carry out the whole
range of jobs belonging to this work section.

Finally, an impressive example of participative
design should be mentioned which was applied
in some British plants. In cooperation between management, social scientists and trade
unions the experts were asked to develop and
explain several solutions for a new computer
system. These solutions were discussed with
the employees, modified by their advice and
the final selection left to their decision.

These few examples should serve to prove that
there are always feasible alternatives available which will result in considerable improvements and even in the creation of new work
places if decisions are taken *not for but with*
the people concerned.

The philosophy behind these questions is based
on two assumptions. One maintains that *technology is not deterministic,* that its direction,
its application and its consequences for man,
for the economy and for society are not pre-
destined. For every technical or organizational problem there is a multitude of alternative solutions, each one having its merits
with respect to different criteria. Thus
technology rather than restricting human choice
to applying the "one best way" the experts
boast of having found, opens up a wide range
of new options to both designers and users.

The second assumption implies that every
*technological change requires corresponding
changes* in work organization, in working conditions, in the hierarchy, in the flow of information, in decision-making and in many
other things. Where this is neglected, the

new technology fails to live up to the expectations; the blame is then put on the technology while it really is to be found with the
conservative spirit of the respective user
or of our economic and social system as a
whole.

It is not sufficient to accept man as part of
a man-machine system just because all attempts
have failed to design systems which will work
without being "disturbed" by human operators.
It is not sufficient to include the operator
into the maintenance scheme equal to any other
part which has to be maintained to keep the
system going.

We ought to realize that in our systems, our
plants, our companies and our economy, each
and every progress in science and technology
was incurred by man in order to serve man.
It must never become an end in itself lest
people might ask the question, as they indeed
increasingly do, why do we have to have all
this? Why should we be dominated by forces
over which we obviously have lost control?
Let's stop this vicious circle and get off it
as quickly as possible.

This loss of faith in science and technology
may be witnessed in many countries; it spells
danger to our civilization. To combat it, we
have to prove in everyday life that there are
alternative ways to apply technology which
will not only help to avoid hazards and disadvantages such as unemployment, loss of
qualification, drudgery, monotony, stress,
etc., but will go far beyond that by improving
the material and the spiritual quality of
life for all of us.

If work design is restricted to ergonomic
criteria, we facilitate the escape from work;
if work satisfaction and self-realization
become the primary targets, as against economical and technological particularism, if
participation becomes the method as against
"happiness" prescribed from above, then we
shall facilitate identification with one's
work achieving beneficial results for an
overall economy and for society as a whole.

If technology and society do not match, it
is completely wrong to simply place the blame
on technology - though this seems presently
very much "en vogue". It is much more
appropriate to adapt our society to what this
new technology demands, e.g.

- a non-Tayloristic work organization
- a selective influence on technology
  (not everything which is possible is
  desirable)
- institutions to allow for participation of
  users in all decisions
- the priority of national economy over
  micro-economic considerations.

Above all we have to perceive man's role in
the man-machine system as that of the master,
with the machine designed, selected and
implemented to serve man. This has become

the real challenge of our time. Scientists and engineers should consider it their privilege to include into their work consideration about man and society.

The algorithm thinking improved and cultivated by the use of the computer forces us to improve planning and forecasting. But the non-mechanical nature of the world that counts becomes more and more apparent, and this will undoubtedly have its impact on society. The necessity to plan and predict the future in spite of many uncertainties will teach us that judgement beyond scientific confirmation is as important in many instances today as it was in the dark ages of the past. The only way to meet the uncertain elements in planning and forecasting is to strive for the maximum flexibility of our creation so that they can be easily adapted to the unexpected, unpredictable events and trends.

And this not only refers to the system as such but even more so to the unexpected and unpredictable properties of man within this system.

Science and technology of our times, based on the highest properties of the human mind, can in the long run have only one general impact: the return from the belief in mechanics to the belief in the humanities, the return of the power of decision-making from a few privileged to a computer-assisted democratic system.

Automation is a great instrument for better *and* for worse. It is being used as an excuse for failure and as an explanation for success; it is used to centralize or to decentralize work organization, to make it more rigid or more flexible, the computer may become Orwell's "Big Brother" watching everyone or a means for more democracy and freedom for everyone.

The choice is with each of us - let's hope we make good use of it!

REFERENCES

[1] Taylor, F.W. (1967). *The Principle of Scientific Management*. The Norton Library, New York.

[2] Friederichs, G., A. Schaff (Eds.)(1982) *For Better or for Worse - Microelectronics and Society*. Pergamon Press.

[3] Wiener, N. (1949). *The Human Use of Human Beings*. (German edition by Alfred Metzner Verlag, Frankfurt/Main, 1952)

[4] Gladstone, A. et al (1979). Framework Paper. In *Social Aspects of Work Organization, Research Studies, 33*

International Institute for Labour Studies, Geneva.

[5] Zemanek, H. (1980). Hoffnungen und Grenzen der Informationsverarbeitung. In *Proceedings of the 6th International Congress "Data Processing in Europe"*, Vienna.

[6] Zemanek, H. (1979). The Impact of Automated Information Processing on Society. Keynote address at *"Discoveries" Symposium*, Honda Foundation, Stockholm.

[7] Margulies, F. (Ed.) (1982). Automation and Industrial Workers - Optional Study Steel Industry. *European Coordination Centre for Research and Documentation in Social Sciences*, Vienna.

[8] International Conference (August 30 - September 3, 1981). *Quality of Working Life and the '80's*. Sponsored by the Ontario Ministry of Labour and Ontario Quality of Working Life Centre. Toronto. 1700 participants, including government officials, managers and union representatives.

[9] Mumford, E., H. Sackmann (Eds.) (1974). *Human Choice and Computers*. Proceedings of the IFIP Conference Vienna 1974, North-Holland Publ. Co.

[10] Mowshowitz, A. (Ed.) (1979). *Human Choice and Computers 2*. Proceedings of the Second IFIP Conference, Baden, Austria, 1979. North-Holland Publ. Co.

[11] Ulich, E. (1980). "Psychologische Aspekte der Arbeit mit elektronischen Datenverarbeitungssystemen. *Schweizerische Technische Zeitschrift, 75*.

[12] *Diebold Management Report* (1980).

[13] Ladner, O. (1974). Job Enlargement im Bürobereich. *Industrielle Organisation*.

[14] Margulies, F. (1979). Textverarbeitung und Angestellte. *Der Privatangestellte, 5*.

[15] Margulies, F. (1980). Trade Union Experiences with Work on VDU's. In *Proceedings of the International Workshop, Milan, 1980*; Taylor & Francis Ltd, London.

[16] Gyllenhammar, P.G. (1977). *People at Work*; Addison-Wesley Publ. Co.

IRONIES OF AUTOMATION

L. Bainbridge

Department of Psychology, University College London, London WC1E 6BT, UK

Abstract. This paper discusses the ways in which automation of industrial
processes may expand rather than eliminate problems with the human operator.
Some comments will be made on methods of alleviating these problems
within the 'classic' approach of leaving the operator with responsibility
for abnormal conditions, and on the potential for continued use of the human
operator for on-line decision-making within human-computer collaboration.

Keywords. Control engineering computer applications; man-machine systems;
on-line operation; process control; system failure and recovery.

'Irony': combination of circumstances, the
result of which is the direct opposite of
what might be expected.
'Paradox': seemingly absurd though perhaps
really well founded statement.

The classic aim of automation is to replace
human manual control, planning and problem
solving by automatic devices and computers.
However, as Bibby and colleagues (1975)
point out: 'even highly automated systems,
such as electric power networks, need human
beings for supervision, adjustment, maintenance, expansion and improvement. Therefore
one can draw the paradoxical conclusion that
automated systems still are man-machine systems, for which both technical and human
factors are important'. The present paper
suggests that the increased interest in human
factors among engineers reflects the irony
that the more advanced a control system is, so
the more crucial may be the contribution of the
human operator.

This paper is particularly concerned with control in process industries, although examples
will be drawn from flight-deck automation. In
process plant different modes of operation may
be automated to different extents, for example
normal operation and shut-down may be automatic
while start-up and abnormal conditions are manual. The problems of the use of automatic or
manual control are a function of the predict-
ability of process behaviour, whatever the mode
of operation. The first two sections of this
paper discuss automatic on-line control where a
human operator is expected to take over in abnormal conditions, the last section introduces
some aspects of human-computer collaboration in
on-line control.

IRONIES OF AUTOMATION

The important ironies of the classic approach
to automation lie in the expectations of the
system designers, and in the nature of the
tasks left for the human operators to carry
out.

The designer's view of the human operator may
be that he is unreliable and inefficient, so
should be eliminated from the system. There
are two ironies of this attitude. One is that
designer errors can be a major source of
operating problems. Unfortunately people who
have collected data on this are reluctant to
publish it, as the actual figures are difficult to interpret. (Some types of error may
be reported more readily than others, and
there may be disagreement about their origin.)
The second irony is that the person who tries
to eliminate the operator still leaves him
to do the tasks which he cannot think how to
automate. It is this approach which causes
the problems to be discussed here, as it means
that the operator can be left with an arbitrary collection of tasks and little thought
may have been given to providing support for
them.

Tasks After Automation

There are two general categories of task left
for an operator in an automated system. He
may be expected to monitor that the automatic
system is operating correctly, and if it is
not he may be expected to call a more experienced operator or to take-over himself. We
will discuss the ironies of manual take-over
first, as the points made also have implications for monitoring. To take over and
stabilise the process requires manual control
skills, to diagnose the fault as a basis for
shut down or recovery requires cognitive

skills.

Manual control skills. Several studies
(Edwards and Lees, 1974) have shown the difference between inexperienced and experienced
process operators making a step change. The
experienced operator makes the minimum number
of actions, and the process output moves
smoothly and quickly to the new level, while
with an inexperienced operator it oscillates
round the target value. Unfortunately, physical skills deteriorate when they are not used,
particularly the refinements of gain and timing.
This means that a formerly experienced operator
who has been monitoring an automated process
may now be an inexperienced one. If he takes
over he may set it into oscillation. He may
have to wait for feedback, rather than controlling by open-loop, and it will be difficult
for him to interpret whether the feedback shows
that there is something wrong with the system
or more simply that he has misjudged his input.
He will need to make actions to counteract his
ineffective control, which will add to his work
load. When manual take-over is needed there
is likely to be something wrong with the process, so that unusual actions will be needed
to control it, and one can argue that the
operator needs to be more rather than less
skilled and less rather than more loaded than
average.

Cognitive skills. Long term knowledge. The
operator who finds out how to control the
plant without training uses a set of propositions about possible process behaviour,
from which he generates strategies to try (e.g.
Bainbridge, 1981). Similarly an operator will
only be able to generate successful new
strategies for unusual situations if he has an
adequate knowledge of the process. There are
two problems with this for 'machine-minding'
operators. One is that efficient retrieval of
knowledge from long-term memory depends on frequency of use (consider any subject which you
passed an examination in at school and have not
thought about since). The other is that this
type of knowledge develops only through use and
feedback about its effectiveness. People given
this knowledge in theoretical classroom instruction without appropriate practical exercises
will probably not understand much of it, as it
will not be within a framework which makes it
meaningful, or remember much of it as it will
not be associated with retrieval strategies
which are integrated with the rest of the task.
There is some concern that the present generation of automated systems which are monitored
by former manual operators, are riding on their
skills, which later generations of operators
cannot be expected to have.

Working storage. The other important aspect of
cognitive skills in on-line decision making is
that decisions are made within the context of
the operator's knowledge of the current state
of the process. This is a more complex form of
running memory than the notion of a limited

capacity short-term store used for items such
as telephone numbers. The operator has in
his head (Bainbridge, 1975) not raw data about
the process state, but of thinking about it
which will be useful in future situations,
including his predictions about future events
and pre-planned future actions. This information takes time to build up. Manual
operators may come into the control room $\frac{1}{4}$ -
$\frac{1}{2}$ hr. before they are due to take over control, so they can get this feel for what the
process is doing. The implication of this
for manual take-over from automatically controlled plant is that the operator who has to
do something quickly can only do so on the
basis of minimum information, he will not be
able to make decisions based on wide knowledge
of the plant state until he has had time to
check and think about it.

Monitoring. It may seem that the operator
who is expected solely to monitor that the
automatics are acting correctly, and to call
the supervisor if they are not, has a relative simple task which does not raise the
above complexities. One complexity which it
does raise of course is that the supervisor
too will not be able to take-over if he has
not been reviewing his relevant knowledge, or
practising a crucial manual skill. Another
problem arises when one asks whether monitoring can be done by an unskilled operator.

We know from many 'vigilance' studies
(Mackworth, 1950) that is is impossible for
even a highly motivated human being to maintain effective visual attention towards a
source of information on which very little
happens, for more than about $\frac{1}{2}$ hr. This
means that it is humanly impossible to carry
out the basic function of monitoring for
unlikely abnormalities, which therefore has
to be done by an automatic alarm system connected to sound signals. (Manual operators
will notice abnormal behaviour of variables
which they look at as part of their control
task, but may be equally poor at noticing
changes on others.) This raises the question
of who notices when the alarm system is not
working properly. Again, the operator will
not monitor the automatics effectively if
they have been operating acceptably for a
long period. A classic method of enforcing
operator attention to a steady-state system
is to require him to make a log. Unfortunately people can write down numbers without
noticing what they are.

A more serious irony is that the automatic
control system has been put in because it
can do the job better than the operator, but
yet the operator is being asked to monitor
that it is working effectively. There are
two types of problem with this. In complex
modes of operation the monitor needs to know
what the correct behaviour of the process
should be, for example in batch processes
where the variables have to follow a particular trajectory in time. Such knowledge requires either special training or special

displays.

The second problem is that if the decisions can be fully specified then a computer can make them more quickly, taking into account more decisions and using more accurately specified criteria than a human operator can. There is therefore no way in which the human operator can check in real-time that the computer is following its rules correctly. One can therefore only expect the operator to monitor the computer's decisions at some meta-level, to decide whether the computer's decisions are 'acceptable'. If the computer is being used to make the decisions because human judgement and intuitive reasoning are not adequate in this context, then which of the decisions is to be accepted? The human monitor has been given an impossible task.

Operator attitudes. The writer knows of one automated plant where the management had to be present during the night shift, or the operators switched the process to 'manual'. This raises general issues about the importance of skill to the individual. One is that the operator knows that he can take-over adequately if required. Otherwise the job is one of the worst types, it is very boring but very responsible, yet there is no opportunity to aquire or maintain the qualities required to handle the responsibility. The level of skill that a worker has is also a major aspect of his status, both within and outside the working community. If the job is 'deskilled' by being reduced to monitoring, this is difficult for the individuals involved to come to terms with. It also leads to the ironies of incongruous pay differentials, when the deskilled workers insist on a high pay level as the remaining symbol of a status which is no longer justified by the job content.

Ekkers and colleagues (1979) have published a preliminary study of the correlations between control system characteristics and the operators' subjective health and feeling of achievement. To greatly simplify: high coherence of process information, high process complexity and high process controllability (whether manual or by adequate automatics) were all associated with low levels of stress and workload and good health, and the inverse, while fast process dynamics and a high frequency of actions which cannot be made directly on the interface were associated with high stress and workload and poor health. High process controllability, good interface ergonomics and a rich pattern of activities were all associated with high feelings of achievement. Many studies show that high levels of stress lead to errors, while poor health and low job satisfaction lead to high indirect costs of absenteeism, etc. (e.g. Mobley and co-workers, 1979).

APPROACHES TO SOLUTIONS

One might state these problems as a paradox, that by automating the process the human operator is given a task which is only possible for someone who is in on-line control. This section will discuss some of the solutions to the problems of maintaining the efficiency and skills of the operator if he is expected to monitor and take over control; the next section will introduce recent proposals for keeping the human operator on-line with computer support.

Solving these problems involves very multi-dimensional decision making, suggestions for discussion will be made here. The recommendations in any particular case will depend on such factors as process size and complexity, the rate of process change and failure, the variability of the product and the environment, the simplicity and cost of shut down, and the qualities of the operator.

Monitoring. In any situation where a low probability event must be noticed quickly then the operator must be given artificial assistance, if necessary even alarms on alarms. In a process with a large number of loops there is no way in which the human operator can get quickly to the correct part of the plant without alarms, preferably also some form of alarm analysis. Unfortunately a proliferation of flashing red lights will confuse rather than help. There are major problems and ironies in the design of large alarm systems for the human operator, see Rasmussen and Rouse (1981).

Displays can help the operator to monitor automatic control performance, by showing the target values. This is simple for single tolerance bands, but becomes more complex if tolerances change throughout batch processing. One possible solution is to show the currently appropriate tolerances on a VDU by software generation. This does not actually get round the problems, but only raises the same ones in a different form. The operator will not watch the VDU if there is a very low probability of the computer control failing. If the computer can generate the required values then it should also be able to do the monitoring and alarms. And how does the operator monitor that the computer is working correctly, or take over if it obviously is not. Major problems may be raised for an operator who is highly practised at using computer generated displays if these are no longer available in an emergency. One ironic but sensible suggestion is that direct wired displays should be used for the main process information, and software displays for quantitative detail (Jervis and Pope, 1977).

'Catastrophic' breaks to failure are relatively easy to identify. Unfortunately automatic control can 'camouflage' system failure by controlling against the variable changes, so that trends do not become apparent until they are beyond control. This implies that the automatics should also monitor unusual variable movement. 'Graceful degredation' of performance is quoted in 'Fitts Lists' of man-computer qualities as an advantage of man

over machine. This is not an aspect of human
performance to be aimed for in computers, as it
can raise problems with monitoring for failure
(e.g. Wiener and Curry, 1980); automatic systems should fail obviously.

If the human operator must monitor the details
of computer decision making then, ironically, it
is necessary for the computer to make these
decisions in a way, using criteria and at a
rate which the operator can follow, even when
this may not be the most efficient method technically. If this is not done then,when the
operator does not believe or agree with the
computer he will be unable to trace back
through the system's decision sequence to see
how far he does agree.

One method of overcoming vigilance problems
which is frequently suggested is to increase
the signal rate artificially. It would be a
mistake however to increase artificially the
rate of computer failure as the operator will
then not trust the system. Ephrath (1980) has
reported a study in which system performance
was worse with computer aiding, because the
operator made the decision anyway and checking
the computer added to his workload.

Working storage. If the human operator is not
involved in on-line control he will not have
detailed knowledge of the current state of the
system. One can ask what limitations this
places on the possibility for effective manual
take-over, whether for stabilisation or shut
down of the process, or for fault diagnosis.

The straightforward solution when shut-down is
simple and low-cost is to shut down automatically. The problems arise with processes which,
because of complexity, cost or other factors
(e.g. an aircraft in the air) must be stabilised
rather than shut-down. Should this be done
manually or automatically? Manual shut-down is
usable if the process dynamics can be left for
several minutes while the operator works out
what is happening. For very fast failures,
within a few seconds (e.g. PWR nuclear reactor
rather than an aircraft), when there is no
warning from prior changes so that on-line
working storage would also be useless, then
reliable automatic response is necessary, whatever the investment needed, and if this is not
possible then the process should not be built
if the costs of failure are unacceptable.

With less fast failures it may be possible to
'buy time' with overlearned manual responses,
as in the USAF 'bold face' procedures. This
requires frequent practice on a high fidelity
simulator, and a sufficient understanding of
system failures to be sure that all categories
of failure are covered. If response to failure
requires a larger number of separate actions
than can be made in the time available then
some must be made automatically and the remainder by highly practised operator.

Long term knowledge. Points in the previous

section make it clear that it can be important to maintain manual skills. One possibility is to allow the operator to use hands-
on control for a short period in each shift.
If this suggestion is laughable then simulator practice must be provided. A simulator
adequate to teach the basic behaviour of the
process can be very primitive. Accurate fast
reactions can only be learned on a high
fidelity simulator, so if they are necessary
then this is a necessary cost.

Similar points can be made about the cognitive
skills of scheduling and diagnosis. Simple
pictorial representations are adequate for
training some types of fault detection
(Duncan and Shepherd, 1975), but only if
faults can be identified from the steady-
state appearance of the control panel, and
waiting for the steady-state is acceptable.
If fault detection involves identifying
changes over time then dynamic simulators are
needed for training (Marshall and Shepherd,
1981). Simple recognition training is also
not sufficient to develop skills for dealing
within unknown faults or for choosing corrective actions (Duncan, 1981).

There are problems with the use of any simulator to train for extreme situations. Unknown faults cannot be simulated, and system
behaviour may not be known for faults which
can be predicted but have not been experienced. This means that training must be
concerned with general strategies rather than
specific responses, for example simulations
can be used to give experience with low probability events, which may be known to the
trainer but not to the trainee. No one can
be taught about unknown properties of the
system, but they can be taught to practise
solving problems within the known information.
It is inadequate to expect the operator to
react to unfamiliar events solely by consulting operating procedures. These cannot cover
all the possibilities, so the operator is
expected to monitor them and fill in the gaps.
However it is ironic to train operators in
following instructions and then put them in
the system to provide intelligence.

Of course, if there are frequent alarms
throughout the day then the operator will
have a large amount of experience of controlling and thinking about the process as part of
his normal work. Perhaps the final irony is
that it is the most successful automated systems, with rare need for manual intervention,
which may need the greatest investment in
human operator training.

HUMAN-COMPUTER COLLABORATION

By taking away the easy parts of his task,
automation can make the difficult parts of
the human operator's task more difficult.
Several writers (Wiener and Curry, 1980;
Rouse, 1981) point out that the 'Fitts list'
approach to automation, assigning to man and
machine the tasks they are best at, is no

longer sufficient. It does not consider the integration of man and computer, nor how to maintain the effectiveness of the human operator by supporting his skills and motivation. There will always be a substantial human involvement with automated systems, because criteria other than efficiency are involved, e.g. when the cost of automating some modes of operation is not justified by the value of the product, or because the public will not accept high-risk systems with no human component. This suggests that methods of human-computer collaboration need to be more fully developed. Dellner (1981) lists the possible levels of human intervention in automated decision making. This paper will discuss the possibilities for computer intervention in human decision making. These include instructing or advising the operator, mitigating his errors, providing sophisticated displays, and assisting him when task loads are high. Rouse (1981) calls these 'covert' human-computer interaction.

Instructions and advice. Using the computer to give instructions is inappropriate if the operator is simply acting as a transducer, as the computer could equally well activate a more reliable one. Thompson (1981) lists four types of advice, about: underlying causes, relative importance, alternative actions available, and how to implement actions. When following advice the operator's reactions will be slower, and less integrated than if he can generate the sequence of activity himself; and he is getting no practice in being 'intelligent'. There are also problems with the efficient display of procedural information.

Mitigating human error. Machine possibilities range from simple hardware interlocks to complex on-line computation. Except where mandatory procedures have to be followed it is more appropriate to place such 'controls' on the effects of actions, as this does not make assumptions about the strategy used to reach this effect. Under manual control human operators often obtain enough feedback about the results of their actions within a few seconds to correct their own errors (Ruffell-Smith, 1979), but Wiener and Curry (1980) give examples of humans making the same types of errors in setting up and monitoring automatic equipment, where they do not get adequate feedback. This should perhaps be designed in. Kreifeldt and McCarthy (1981) give advice about displays to help operators who have been interrupted in mid-sequence. Rouse (1981) suggests computer monitoring of human eye movements to check that instrument scanning is appropriate, for example to prevent tunnel vision.

Software-generated displays. The increasing availability of soft displays on VDUs raises fascinating possibilities for designing displays compatible with the specific knowledge and cognitive processes being used in a task. This has lead to such rich veins of creative speculation that it seems rather mean to point

out that there are difficulties in practice.

One possibility is to display only data relevant to a particular mode of operation, such as start-up, routine operations, or maintenance. Care is needed however, as it is possible for an interface which is ideal for normal conditions to camouflage the development of abnormal ones (Edwards, 1981).

Goodstein (1981) has discussed process displays which are compatible with different types of operator skill, using a three-level skill classification suggested by Rasmussen (1979). The use of different types of skill is partly a function of the operator's experience, though they probably do not fall on a simple continuum. Chafin (1981) has discussed how interface design recommendations depend on whether the operator is naive/novice/competent/expert. He however was concerned with human access to computer data bases when not under time pressure. Man-machine interaction under time pressure raises special problems. The change between knowledge-based thinking and 'reflex' reaction is not solely a function of practice, but also depends on the uncertainty of the environment, so that the same task elements may be done using different types of skill at different times. It could therefore confuse rather than help the operator to give him a display which is solely a function of his overall skill level. Non-time-stressed operators, if they find they have the wrong type of display, might themselves request a different level of information. This would add to the work load of someone making decisions which are paced by a dynamic system. Rouse (1981) has therefore suggested that the computer might identify which type of skill the operator is using, and change the displays (he does not say how this might be done). We do not know how confused operators would be by display changes which were not under their own control. Ephrath and Young (1981) have commented that it takes time for an operator to shift between activity modes, e.g. from monitoring to controlling, even when these are under his control, and one assumes that the same problems would arise with changes in display mode. Certainly a great deal of care would be needed to make sure that the different displays were compatible. Rasmussen and Lind's recent paper (1981) was about the different levels of abstraction at which the operator might be thinking about the process, which would define the knowledge base to be displayed. Again, although operators evidently do think at different levels of complexity and abstraction at different times, it is not clear that they would be able to use, or choose, many different displays under time stress.

Some points were made above about the problems of operators who have learned to work with computer generated displays, when these are no longer available in abnormal contions. Recent research on human memory (e.g. Craik, 1979) suggests that the more pro-

cessing for meaning that some data has received
the more effectively it is remembered. This
makes one wonder how much the operator will
learn about the structure of the process if
information about it is presented so successfully that he does not have to think about it
to take it in. It certainly would be ironic
if we find that the most compatible display is
not the best display to give to the operator
after all! (As usual with display choice
decisions this would depend on the task to be
done. A highly compatible display always supports rapid reactions. These points speculate
whether they also support aquisition of the
knowledge and thinking skills needed in abnormal conditions.)

A few practical points can be suggested. There
should be at least one source of information
permanently available for each type of information which cannot be mapped simply onto
others, e.g. about layout of plant in space as
opposed to its functional topology. Operators
should not have to page between displays to
obtain information about abnormal states in
parts of the process other than the one they are
currently thinking about, nor between displays
giving information needed within one decision
process. Research on sophisticated displays
should concentrate on the problems of ensuring
compatibility between them, rather than finding
which independent display is best for one particular function without considering its relation to information for other functions. To
end on a more optimistic note, software displays
offer some interesting possibilities for en-
riching the operator's task by allowing him to
design his own interface.

Relieving human work-load. A computer can be
used to reduce human work-load either by simplifying the operator's decisions, as above, or
by taking over some of the decision making.
The studies which have been done on this show
that it is a complex issue. Ephrath and Young
(1981) found that overall control performance
was better with manual control of a single loop,
but was also better with an autopilot in the
complex environment of a cockpit simulator.
This suggests that aiding is best used at
higher work loads. However the effect of the
type of aiding depends on the type of workload. Johannsen and Rouse (1981) found that
pilots reported less depth of planning under
autopilot in abnormal environmental conditions,
presumably because the autopilot was dealing
with them, but more planning under emergency
aircraft conditions, where they suggest that
the autopilot frees the pilot from on-line
control so he can think about other things.
Chu and Rouse (1979) studied a situation with
both computer aiding and autopilot. They
arranged for the computer to take over decision
making when the operator had a queue of one
other task item to be dealt with and he was
controlling manually, or after a queue of 3
items if the autopilot was controlling. The
study by Enstrom and Rouse (1977) makes it
clear why Rouse (1981) comments that more sophisticated on-line methods of adapting computer
aiding to human work-load will only be poss-

ible if the work-load computations can be
done in real time. (It would be rash to
claim it as an irony that the aim of aiding
human limited capacity has pushed computing
to the limit of its capacity, as technology
has a way of catching up with such remarks.)
Enstrom and Rouse also make the important
point that the human being must know which
tasks the computer is dealing with and how.
Otherwise the same problems arise as in
human teams in which there is no clear allocation of responsibility. Sinaiko (1972)
makes a comment which emphasises the importance of the human operator's perception of
the computer's abilities: 'when loads were
light, the man appeared willing to let the
computer carry most of the assignment responsibility; when loads were heavy, the men
much more often stepped in (and) over-rode
the computer'. Evidently, quite apart from
technical considerations, the design of computer aiding is a multi-dimensional problem.

Conclusion. The ingenious suggestions reviewed in the last section show that humans
working without time-pressure can be impressive problem solvers. The difficulty
remains that they are less effective when
under time pressure. Hopefully this paper
has made clear both the irony that one is not
by automating necessarily removing the
difficulties, and also the possibility that
resolving them will require even greater
technological ingenuity than does classic
automation.

REFERENCES

Bainbridge, L. (1975). The representation
    of working storage and its use in the
    organisation of behaviour. In W.T.
    Singleton and P. Spurgeon (Eds.), Mea-
    surement of Human Resources. Taylor and
    Francis, London. pp. 165-183.
Bainbridge, L. (1981). Mathematical equa-
    tions or processing routines? In J.
    Rasmussen and W.B. Rouse (Eds.), Human
    Detection and Diagnosis of System Fail-
    ures. Plenum Press, New York. pp.259-
    286.
Bibby, K.S., F. Margulies, J.E. Rijnsdorp,
    and R.M.J. Withers (1975). Man's role
    in control systems. Proc. 6th. IFAC
    Congress, Boston.
Chafin, R.L. (1981). A model for the control
    mode man-computer interface. Proc. 17th.
    Ann. Conf. on Manual Control, UCLA. JPL
    Publication 81-95. pp. 669-682.
Chu, Y., and W.B. Rouse (1979). Adaptive
    allocation of decision making responsi-
    bility between human and computer in
    multi-task situations. IEEE Trans.
    Syst., Man & Cybern., SMC-9, 769-778.
Craik, F.M. (1979). Human memory. Ann.
    Rev. Psychol., 30, 63-102.
Dellner, W.J. (1981). The user's role in
    automated fault detection and system
    recovery. In J. Rasmussen and W.B.
    Rouse (Eds.), Human Detection and
    Diagnosis of System Failures. Plenum

Press, New York. pp. 487-499.

Duncan, K.D. (1981). Training for fault dia-
gnosis in industrial process plant. In J.
Rasmussen and W.B. Rouse (Eds.), Human
Detection and Diagnosis of System Failures.
Plenum Press, New York. pp. 553-573.

Duncan, K.D. and A. Shepherd (1975). A simulator and training technique for diagnosing plant failures from control
panels. Ergonomics, 18, 627-641.

Edwards, E. (1981). Current research needs in
manual control. Proc. 1st. European Ann.
Conf. on Human Decision Making and Manual
Control, Delft Univ. pp. 228-232.

Edwards, E. and F.P. Lees (1974). (Eds.) The
Human Operator in Process Control. Taylor and Francis, London.

Ekkers, C.L., C.K. Pasmooij, A.A.F. Brouwers,
and A.J. Janusch (1979). Human control
tasks: A comparative study in different
man-machine systems. In J.E. Rijnsdorp
(Ed.), Case Studies in Automation Related
to Humanization of Work. Pergamon Press,
Oxford. pp. 23-29.

Enstrom, K.D., and W.B. Rouse (1977). real-
time determination of how a human has
allocated his attention between control
and monitoring tasks. IEEE Trans. Syst.,
Man & Cybern., SMC-7, 153-161.

Ephrath, A.R. (1980). Verbal presentation.
NATO Symposium on Human Detection and
Diagnosis of System Failures, Roskilde,
Denmark.

Ephrath, A.R. and L.R. Young (1981). Monitoring vs. man-in-the-loop detection of
aircraft control failures. In J. Rasmussen and W.B. Rouse (Eds.), Human Detection
and Diagnosis of System Failures. Plenum
Press, New York. pp. 143-154.

Goodstein, L.P. (1981). Discriminative dis-
play support for process operators. In J.
Rasmussen and W.B. Rouse (Eds.), Human
Detection and Diagnosis of System Failures.
Plenum Press, New York. pp. 433-449.

Jervis, M.W. and R.H. Pope (1977). Trends in
operator-process communication development.
Central Electricity Generating Board.
E/REP/054/77.

Johannsen, G. and W.B. Rouse (1981). Problem
solving behaviour of pilots in abnormal
and emergency situations. Proc. 1st.
European Ann. Conf. on Human Decision
Making and Manual Control, Delft Univ.
pp. 142-150.

Kreifeldt, J.G., and M.E. McCarthy (1981).
Interruption as a test of the user-com-
puter interface. Proc. 17th. Ann. Conf.
on Manual Control, UCLA. JPL Publication
81-95. pp. 655-667.

Mackworth, N.H. (1950). Researches on the
measurement of human performance. Re-
printed in H.W. Sinaiko (Ed.), Selected
Papers on Human Factors in the Design and
Use of Control Systems (1961). Dover
Publications, New York. pp. 174-331.

Marshall, E.C. and A. Shepherd (1981). A fault-
finding training programme for continuous
plant operators. In J. Rasmussen and W.B.
Rouse (Eds.), Human Detection and Diagnosis
of System Failures. Plenum Press, New York.
pp. 575-588.

Mobley, W.H., R.W. Griffeth, H.H. Hand, and
B.M. Meglino (1979). Review and conceptual analysis of the employee turn-
over process. Psychol. Bull., 86, 493-
522.

Rasmussen, J. (1979). On the structure of
knowledge - a morphology of mental
models in a man-machine system context.
Riso National Laboratory, Denmark.
RISO-M-2192.

Rasmussen, J. and M. Lind (1981). Coping
with complexity. Proc. 1st. European
Ann. Conf. on Human Decision Making and
Manual Control, Delft Univ. pp. 70-91.

Rasmussen, J. and W.B. Rouse (1981) (Eds.).
Human Detection and Diagnosis of System
Failures. Plenum Press, New York.

Rouse, W.B. (1981). Human-computer interaction in the control of dynamic systems.
Computing Surveys, 13, 71-99.

Ruffell-Smith, P. (1979). A simulator study
of the interaction of pilot workload
with errors, vigilance, and decisions.
NASA TM-78482.

Sinaiko, H.W. (1972). Human intervention
and full automation in control systems.
Appl. Ergonomics, 3, 3-7.

Thompson, D.A. (1981). Commercial air crew
detection of system failures: state of
the art and future trends. In J. Rasmussen and W.B. Rouse (Eds.), Human
Detection and Diagnosis of System Failures. Plenum Press, New York. pp.37-48.

Wiener, E.L., and R.E. Curry (1980). Flight-
deck automation: promises and problems.
Ergonomics, 23, 995-1011.

MAN-MACHINE PROBLEMS OF PASSENGERS IN NEW URBAN TRANSIT SYSTEM

H. Miki and T. Watanabe

*Department of Engineering, Rolling Stock Division, Kawasaki Heavy Industries Co. Ltd.,
Wadayama-dori, Hyogo-ku Kobe, Japan*

Abstract. Highly automated systems which used to be topics of modern
factories and limited industries, are now widespread and are commonly
brought into daily life of citizen. In contemporary highly automated
systems, not only dedicated operator, but also unspecified kinds of public
users are participating partly in its application. The purpose of this
paper is to investigate these problems on the basis of actual data with
emphasis on the response of passengers in abnormal conditions, in fully
automated vehicles experienced in new urban transit system. We give a
discription about necessary conditions for highly automated system to be
accepted well in the society. For the further improvement of the safety
and reliability, it is essential to study the psychological situations of
the unspecified public users in case of abnormal conditions, to know their
response and behavior characteristics, and to design highly automated
system adapted for human use in the wide views. The study of the roles of
public users in the highly automated system, as we disscuss here, will
propose new aspect of man-machine studies.

Keywords. Man-Machine System; Human Factor; Traffic control;
Transportation control; Computer control

INTRODUCTION

Highly automated systems are advancing their
levels of automation, intelligence and
reliability based on rapid development of
elemental technologies such as computing,
communication and control. Highly
automated systems which used to be topics
of modern factories and limited industries,
are now widespread and are commonly brought
into daily life of citizens. This new trend
in automation is raising new significance
of man-machine studies. The most man-
machine problems studied so far, have dealt
the problems between complex systems and
dedicated operators as shown in Fig. la.
In contemporary highly automated systems,
not only the dedicated operator but also
unspecified kinds of public users are
participating partly in its operation.
Thus the structure of the man-machine
system has changed from two pole to three
as shown in Fig. lb. We also have to make
careful studies on the possible cooperation
between the public users and dedicated
operators in such systems. These highly
automated systems are required to operate
efficiently and safely not only in normal
but also in abnormal situations. Thus the
system should be ready to carry out
necessary dispositions in both normal and
every possible abnormal situations. At the
mean time the system should be capable of
informing the operator those processes under

excution and be ready for him to check
every sequence in progress. Moreover it is
essential to have a rational organization
for the operation of the system, and to
exercise periodic training to accommodate
the emergency conditions. For the further
improvement of the safety and reliability,
it is essential to study the psychological
situations of the unspecified public users
in case of abnormal conditions to know
their response and behavior characteristics.

It is reported that, in case of accident of
the atomic reactor in Three Mile Island,
the operator's mishandling happened owing
to faults in design policy, which gave few
attention to the mutual interaction between
man and machine in such circumstances, that
the condition of accident were changing
every moment.

The interaction between machine and
dedicated operator have been studied in the
various field of industries, i.e. the atomic
power plant, industrial manufacture plant,
the control of airplane. The following
items have been discussed in these studies.
(1) Layout analysis
(2) Psychophysical work load analysis
(3) Operator control characteristics
(4) Limitation of operator control
    characteristics

The new urban transit system is one of the three pole man-machine systems, shown in Fig. 1b, comprising of highly automated train, dedicated operator and unspecified passengers. In this system, passengers are participating partly in keeping normal operation and safety, by sending correspondence to the operator or by using emergency button.

In this paper we are discussing passengers response and behavior characteristics in abnormal conditions on the basis of experience in Kobe New Transit System. It is understood that the response and behavior of the passengers are deduced from motives owing to their anxiety or mental state of frustration against the system. The study of the roles of public users in the highly automated system, as we disscuss here, will propose new aspect of man-machine studies.

OUTLINE OF THE SYSTEM

The purpose of new urban transit system consist in developing an attractive public transportation system, bringing deadlocks in urban transportation to an end, solving environment pollution and traffic accident problem, and further more offering the means of transportation which can fulfill higher service level, convenience, and comfort, thus it has been developed by adopting new and sophisticated techniques such as computer control etc.

The new urban transit system, we are taking up here as an example, is a medium-capacity guideway transit system, which links the center of Kobe city "Sannomiya", with a man-made island "Port Island" which was reclaimed in the port area of Kobe, therefore called "Port-Island Line" Kobe New Transit System. The scale of this system is shown on TABLE 1. The train under operation is shown on Photograph 1, and the route map is on Fig. 2. Fig. 3. shows schematic diagram of traffic control and train operation aiming at man-machine interfaces. The main features of this system are as follows:
(1) The operations of train are fully computerized so that no driver on board is necessary. This lead to low operation cost, and high adaptability to change in transportation demand.
(2) In view of system control, full automatization of train control make it possible to provide passengers with short interval operation, and high quality of service.
(3) Abnormality in equipment and in operation are detected and systematically classfied, and are treated automatically, if possible.
(4) As for security, Automatic Train Protection (ATP) device is adopted and centralized control method realize high reliability and rationalized supervision.

Further details for this system are to be refered to other publications.

Kobe New Transit System, the Port Island Line, was opened on Feb. 5th, 1981. It served as the main transportation means to Kobe Port Island Exposition "PORTOPIA '81", which lasted from March 20th to Sept. 15th, 1981. During this period, over 21 million passengers were transported efficiently and in safe on fully automated vehicles. While fatal system downs responsible to equipments were few in number, the system frequently suffered from miss-uses and disturbances of passengers.

Having evaluated the characteristic of this system we found new problems in the design policy adopted during the planning stage. The purpose of this paper is to investigate these problems on the basis of actual data with emphasis on the responces and behaviors of passengers in emergency.

PASSENGERS PARTICIPATION
IN FULLY AUTOMATED SYSTEM

Classifying the transition stages of automatization of public transportation facilities into four classes: (1) contemporary railway (2) latest subway (3) latest subway in the USA (4) new urban transit system. Tasks of crew and ground personels have been decreased in the course of automatization step by step. Kobe New Transit System is the first public transportation system that accomplished fully computerized unmanned operation.

To establish both fully automated train operation and unmanned station control, it is necessary to automate human tasks existing in conventional system, and such items difficult to automate, should be constructed so as to be managed by staffs in control center. TABLE 2. shows how the various tasks alloted to each crew and ground personel in the latest subway, are to be managed functionally by each staff in case of new urban transit system. A large amount of tasks alloted to driver and conductor in the latest subways, has been automated in new urban transit system, consequently monitoring and confirmation are allocated to the staffs in control center. The important problem of new urban transit system is that it requires active participation of passengers to detect partial abnormality. Thus fully automated train operation and unmanned station control are made practicable.

The most important problem for unmanned operation of train and stations is to ensure safety for passengers, and accommodate in emergency. For the fully automated train operation, it is not only necessary to excute start control, speed control, and brake control for appropriate stops at stations, but also to make a systematic

style which make it to omit, to automate, or to construct a system handled by staffs in control center for the following four tasks.
(1) Monitoring the route
(2) Securing the safety for train and passenger
(3) Detection of abnormality in train and its urgent disposition
(4) Guidance to passengers in an emergency

In this system above mentioned tasks are carried out as follows:
(1) As for monitoring the route, both route control and train control are cooperatively automated.
(2) To secure train's safety is carried out by controlling automatically under the ATP. To endure passengers' safety, the station platform is built up in closed structure having transparent glass wall to prevent passengers from falling down. In the wall gate doors are built in, to be opened after arrival of train by the command from traffic control computer. CCTV camera located at the platform, transfers images showing passenger's getting on and off, to the central monitoring facility, which enable staffs to monitor the platform, and let them remotely to control doors or to restrain train departure if necessary.
(3) In case of abnormal state in equipments on board, various kinds of sensors installed on the vehicle, will detect them. These abnormal signs are indicated on the monitoring device on board, and then classified in eight levels and finally they are sent to traffic control computer as an abnormal information. Traffic control computer, through operating console, indicates abnormal information and recommends corresponding dispositions. The operator after confirmation, excutes corresponding disposition by pressing button for approval input.
When abnormal state proved to be serious and it can not be dispositioned automatically, personels in charge shall be sent to the site.
(4) For passenger guidance in an emergency, urgent announcing shall be delivered automatically, by the command from computer, to train and station concerned. The operator at the control center may squeeze himself into above urgent announcing as well as regular ones, and send staffs to the site if necessary.
By means of such process, train crews and station staffs were saved. Abnormality detection is made by the following methods:
(1) Information detected by sensors installed in vehicles, as stated above.
(2) Checking the rationality of input informations by means of traffic control computer
(3) Supervising traffic display pannel, control console, and CCTV monitors etc. by operator in control center.

However there are certain items of information which may be far more helpful to be sent from passengers, from the point of view of technical possibility, reliability, economy, and occurrence of unexpected situations. In TABLE 3. some examples of such items are listed. In order to assure active participation of passengers in sending correspondance about these information, following equipments are installed.
(1) In vehicle, telephone set and emergency train stop button for passenger's use
(2) On platform, telephone set and emergency button for passenger's use

RESPONCE AND BEHAVIOR OF PASSENGER

Since this system is designed intending unmanned train operation and station control, construction for system control and information transmission are of extremely high grade. The way of use by passengers, is designed to essentially same to the conventional manned public transportations. Thus in normal operations, the passengers behaved smoothly as usual, in abnormal cases their responses were quite different from that in conventional transportations.

Abnormal state of affairs occurred under extremely overcrowded transportation during the Exposition. These abnormal states were the first experience for manufacturer, operator, passenger, and also for the whole Japanese society. Considering such special circumstances, the problem concerning the features of passenger's response and behavior is very precious experience. After closing of the Exposition, reduction in the amount of transportation into the range of designed value, as well as passenger's habituation, contributed to remarkable decrease of such abnormal cases. It is worth reporting the results of investigation based upon this first experience.

Abnormal cases observed in this new urban transit system, compared with conventional public transportations, happened owing to following four reasons:
(1) Passenger's fingers, for instance, were drawn into door pocket
(2) Passenger's overloading
(3) Each member of a family was unable to get on a same vehicle
(4) Mischiefs carried out inside the train or at the station platform

Although emergency button are scarecely pressed in conventional railways, even in above situation, the buttons inside the train or at the station platform, in this system were relatively frequently pressed by passengers. This resulting to temparary system lock, that is, emergency brake of the said train is actuated, and it is necessary for a staff to unlock with the key. It has been the general principle to unlock the key manually by a staff in order to recover from the emergency. This principle should be applied to the new transportation system.

By the same thought, when the emergency button at station is pressed, emergency stop command is automatically effected, and feeding for the corresponding sections are cut down. To recover from cut down, emergency button should be unlocked manually at the site, first. Secondly feeding should be recovered by electric commander in the control center. Finally train operation command is sent under supervision of operators.

The typical and most frequent reason for the passengers to push the emergency button, was that a child had his fingers drawn into door pocket. This happened when the train is in complete stop at a station, and automatic excution of the succeeding procedures is blocked. It was by no means helpful for the passengers to push the emergency button in such situation. However, from the passenger's point of view, there are some grounds to worry about such serious conditions that the train might start again while his fingers were drawn into door pocket, and that he might not be able to get off the train at that station. Thus the button pressing is considered quite reasonable. On the contrary, from designer's standpoint, telephones were intended to be used in such situations, but it has never been used actually.

Fig. 4. shows the cases these conditions occurred. Accidents due to doors such as fingers are drawn into door pocket are concentratedly occurred from July to September during Exposition period. (Occurrences of mischiefs are also shown in the same figure. These are the cases of pressing the emergency button, or breaking open the door without necessity.) TABLE 4. shows response of passengers when doors are out of order. Only 41 % were cleared off by operation from control center and the time delayed were from one to three minutes per one accident, but in case of passenger's pressing button in the vehicle (36 %), button at the platform (19 %), button both in the vehicle and platform (4 %), more additional delay found inevitable.

SUGGESTION FOR THE SYSTEM DESIGN

Uneasiness of passengers on board in an emergency in an automated train

In conventional railway service, operator, conductor and station staff are struggling with machine called train, for security and reliability aiming at safe transportation for passengers. Namely, station staff performs to watch passengers getting on and off at the platform, to help passengers, and to inform conductor when safety is confirmed. Then conductor operates doors carefully after confirming it by himself. Station staff and conductor confirm complete closure of doors and passenger safety, and send the departure sign to driver, and then driver starts the train. As stated above, by such excellent human sense which could never be replaced by machine equipment — in this case two persons i.e. station staff and conductor — safety in wide range is attained. Depending upon this positive achievement, during long period of time, passengers have been able to take advantages of transportation facilities with spiritual peace.

On the other hand automated new urban transit system, where driver, conductor and station staff are not in service. Passengers should take case of themselves, and would be afraid that nobody would help them. Although image showing passengers getting on and off, is transfered by CCTV camera equipped at station platform to control center and observed, these facilities are only of remote existance. Passengers utilizing automated new urban transit system, do not know the control mechanism for train and platform door. For passengers these are entirely blackboxes and what they know is how to utilize this facility in due order in normal state. It is considered that, owing to lack of knowledge, passengers might press the emergency button even in case of trivial abnormality such as his fingers drawn into door pocket, worring about mechanism and reaching to the limit of emotional break down as shown on Fig. 5. This would correspond to theory on "characteristics and condition for apprehension" by Spielberger. This state of things shows that the passengers would be overcome by stimulus beyond the limit and feel as if they were subject to an air raid. Moreover, passengers are neither able to understand the circumstances located therein nor the computer is unable to note the occurrence of unfavorable condition, and then serial procedures for it are carried out and finally they would fall into a crisis. Consequently, it is considered that the passengers' feeling is thrown into confusion and lost the direction of movement, and so finally they would perform abnormal response and behavior.

The strongs and the weaks in information ages

In the present information ages, it is considered that between the strongs who absorb positively information as their own, and the weaks who are negative, there are some differences. Although the strongs are able sufficiently to use new system of higher level automation and information handling, the weaks use it with fear. The general public and student belong to the strongs and they constantly absorb abundant knowledge about new system from newspaper, magazine and TV. And they carefully catch hold of explanations on utilizing system which are noticed in vehicles and stations, and if they had opportunities, they are

intending to operate input devices of this
system. Once an accident due to somebody's
fingers drawn into door pocket is occurred,
they would take advantage of an opportunity
to operate them. It is general public and
student that excute mischiefs such as break
open the doors reflecting the stress accumulated in monotone and sultry vehicles.

On the contrary, aged person, children,
physically handicapped people and
foreigners, belong to the weaks, they can
not utilize information medium sufficiently
and can get only poor knowledge as their own
and so it is difficult to adapt themselves
to new system in short period of time. They
are hardly considered to excute mischiefs
but they might be unable to take care of
themselves in an emergency. Foreigners who
have difficulties in language may operate
input devices by misunderstanding without
understanding guide indicator and
announcement.

Possible Solutions

System manufacturer has offered man-machine
hardwares which can correspond to every
possible emergency. The investigation was
not enough on utilizing the hardware on the
variety of situations. The passengers'
psychophysical characteristics were not
taken into account sufficiently. As for
these problems, meaningful solution can be
derived only from careful investigations,
from the point of view of three parties,
passenger, operator and manufacturer. It
was necessary to make a guide indicator in
concrete form, saying such as, "when fingers
are drawn into door pocket, please notify by
telephone" or "Please press emergency button
in case accident."

It is considered necessary to give the
strongs among passengers some means that
they can get consciousness of participation
so as they may not feel frustrations, and it
would be effective to install equipments
which may not effect directly to the system
operation, but could be operated by passengers to let them know the available service.
We are proposing, for example, the installation of equipments that would offer
services to passenger and interactives about
the mechanism of system control, train
operation, and platform door control,
passenger's procedures to be performed in
emergencies, and explanations about geology
and history for adjacent regions by audio
and visual devices. We consider that, it is
necessary to investigate to offer indications and announcement easily understood
by the weaks including foreigners.

Information exchange would be necessary at
some international conference or equal, as
for equipment to be furnished in unmanned
vehicles to help passengers, and dispositions to be excuted by operator, since
these matters are not yet world-wide experienced. Nowadays, social life becoming more
and more international, and transportation

system should accommodate visitors from
various countries. Thus the way of response
required to the passengers in an emergency
should be unified to common and well understandable ones in every country. At the same
time it is also necessary to seek harmony to
the traditions and customs of the district.

PROPOSALS TO BE WELL ACCEPTED BY THE PUBLIC

For the automated system which has to
accommodate unspecified persons, to exhibit
expected functions and to be accepted by the
public, it is necessary to design the system
so as to adapt itself to human characteristics, in normal and abnormal conditions,
as we have mentioned in the proceeding
chapter on one side, while on the other
approaching of the public who utilize the
system with warm heart human fellowship is
of value.

In case of on-line system of bank network,
which is operated by some limited number of
personels only, crimes such as illegally
seize upon money by some insolent fellows
break out. Although efforts to adopt high
grade check mechanism so as to shut out
squeezing have been carried out, this method
should impose large amount of load on the
system side and make manager's check
mechanism more and more complicated, which
would not only necessitate to bear excess
charge and consequently the merit of automated system would be somewhat reduced.

Fully automated new urban transit system
enable to offer high quality service to
passengers. It should not be permitted that
mischiefs by some indiscreet users, hinder
the operation of the system. To make
protection mechanism enormous on the system
side, is not preferable however, and such
trend would lead to crisis of elevation of
perpetual struggle between man and machine.

Automated system is not perfect as human
being but has weak points. One limitation
of automated system is that it is unable to
construct a closed system which automatically operate properly to every possible
abnormal state, from the point of view of
contemporary technical possibility, reliability, economy, and occurrence of unexpected conditions. Those parts of disposition including detection of a few part
of abnormality, which are entrusted to
unspecified users, are unable to be
contained in the object of fool proof, so
they have to be dispositioned as an absolute
character. Accordingly, automated system is
very difficult to be put to practical use
unless we do not rest on the basis of human
nature as fundamentary good. We believe
that the future course for us, consist in
the attitude toward coexistence of man and
machine by compensating each other and not
making bad use of the weak point in automated system. If we can bring up the
society, to such direction, this would

promise to construct an effective society
where everybody can sufficiently utilize the
merit of highly automated system and enjoy
its benefit. It is necessary that social
dispositions such as public relations and
moral act should be adopted to school
education and equals, for making out society
to understand highly automated system
sufficiently and utilizing it with warm
heart as above mentioned.

CONCLUSION

In automated system, accommodating service
to many and unspecified persons, it is
necessary to impose no additional load to
the public users so that they may adapt themselves smoothly to the new system. As for
new urban transit system, since the way of
use by passengers is designed to essentially
same to the conventional manned public
transportations, passengers felt that no
additional load was imposed and so the
system was accepted in a comparatively short
period of time.

Fully automated new urban transit system is
now being operated safely and trustworthy in
high reliability. But it is dangerous not
only for operator but also for user i.e.
passenger, to entrust themselves that there
would not occur any unusual. At the same
time it is necessary to verify socially that
excess apprehensions is of no use, even if
emergency occurred. Passenger may not
afraid of the system in emergency, if he
understand its features sufficiently.

Fully automated new urban system has
possibilities in certifying the quality of
service by automation technology. Namely,
automated system realized frequent service
of operation, safe and highly reliable
operation, and high adaptability to the
change in transportation demand. Service
for users would be more and more increased
by assigning manpower saved by automation,
to guide and assist the aged, children, and
handicapped persons. Automation make it
possible for mankind to work more in direct
service for users, and to release from such
jobs as simple operation of machines.
Namely, it is possible to change human labor
to the same in the field of humanity. As
stated above, system automation for many and
unspecified persons, can offer abundant
service to the public users.

For the further improvement of the safety
and reliability, it is essential to study
the psychological situations of the
unspecified public users in case of abnormal
conditions, to know their response and
behavior characteristics, and to design
highly automated system adapted for human
use in the wide views.

The author would like to acknowledge for
continuing advices and encouragement to
Mr. Shozo Shimoura, Senior Manager,
Department of Engineering, Rolling-Stock
Division, Kawasaki Heavy Industries, and to
Dr. Hiroshi Tamura, assistant professor,
Faculty of Engineering Science, Osaka
University who inspired us to prepare this
paper.

REFERENCES

Shimoura, S., Miki, H. (1981). Outline of
    KOBE NEW TRANSIT SYSTEM and its
    Automatic Train Control Method.
    System and control. 25, 175 - 184.
Report of the President's Comission On THE
    ACCIDENT THREE MILE ISLAND (1980)
Imada, H., (1975). Emotional psychology
    170 - 185

TABLE 1.   Scale of system:Kobe New Transit Port Island Line

item	description
1. Route length (main line)	Single lane 3.5 km Double lane 2.9 km

2. Designed Value of
     transportation
	approx. 10,000 men/hour
	             70,000 men/day

3. Number of trains (main line)	Maximum 11 trains
4. Number of vehicles in train	6 Vehicles (fixed)
5. Minimum headway	approx. 2.5 minutes
6. Maximum Speed	60 km/hour
7. Train operation mode	usually unmanned
8. Number of station	Single lane 5 stations Double lane 4 stations
9. Station operation mode	usually unmanned
10. Yard	Centralized control by computer

Photo. 1. The train under operation

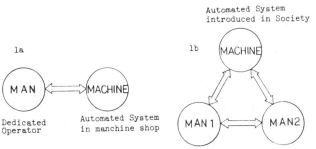

Fig. 1. Man-Machine System

Human factors of man-machine system is used to be dedicated operator only, but recently public user of many and unspecified persons become to be added.

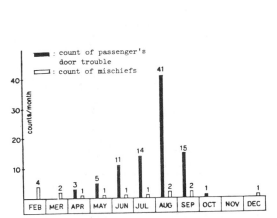

Fig. 4. Occurrence number of passenger's door trouble and mischiefs

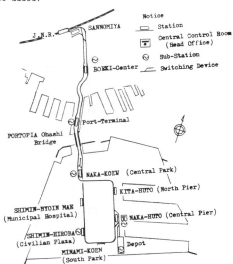

Fig. 2. Route Map : Kobe New Transit Port Island Line

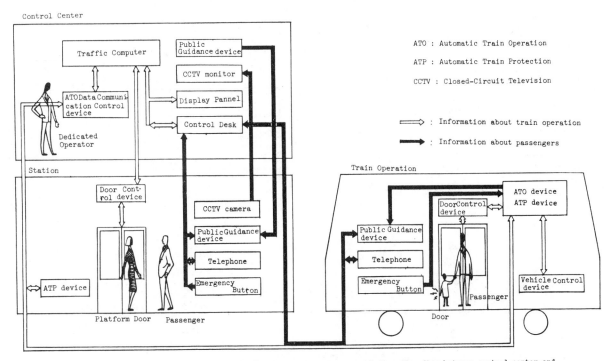

ATO : Automatic Train Operation
ATP : Automatic Train Protection
CCTV : Closed-Circuit Television

⟹ : Information about train operation
➡ : Information about passengers

Fig. 3. Schematic diagram of traffic control and train operation.

Information flow between control center and passengers in train and on station, are shown.

ADE-F

TABLE 2. Differences in tasks between the latest subway and new urban transit system.

In new urban transit system, most tasks is automated and operators in control center perform supervision. Passengers participate in detecting partial abnormal states.

item	contents	the latest subway Driver	Conductor	Staffs at station	Operator	Automatization	Passengers	new urban transit system Operator	Automatization	Passengers	description
normal condition — train supervision	a. Arrival supervision										(*1) installation of platform door; checking through CCTV monitor
	monitor route	O						(*1)	O		
	monitor passenger	O						O			
	monitor stop range	O	O						O		
	b. Stop supervision										(*2) checking through CCTV monitor
	monitor passenger on/off		O					(*2) O			
	monitor start time	O	O								
	c. Departure Supervision										
	monitor route	O						(*1)	O		
	monitor passengers		O					O			
normal condition — Train operation	a. Train operation										
	Start control	O							O		
	Speed control				O				O		
	Stop control				O				O		
	b. Door operation		O						O		
Service	Public access and guidance		O		O			(*3) O	O		(*3) automatic public address usually, manual guidance is possible
abnormal condition — Accommodation for abnormality	a. Detection of abnormal conditions and urgent disposition	O	O					O	O	O	
	b. Train protection	O	O					O	O		
	c. Communication and public address in train	O	O					O			(*4) in serious case, personel in charge shall be sent to the site, and move the said train off by manual operation
	d. Accommodation for abnormality	O	O		O			(*4) O	O		

TABLE 3. Items expected for passenger to be reported about detecting abnormality.

classification	abnormal item
Technical possibility	fingers are drawn into door pocket; crime , nuisance , mischiefs; emergency case
Reliability	fail in lightings; out of order in air conditioner
Economy	fail in broadcasting; fail in direction indicator; break down of window screen

TABLE 4. Responses of passenger in case of door trouble

classification of passenger's response	rate (%)	delayed time on traffic min/accident
cleared off by operation from control center	41	1 ~ 3
passenger pressed the button in vehicle	36	2 ~ 6
passenger pressed the button at platform	19	4 ~ 11
passenger pressed the buttons both in vehicle and platform	4	6 ~ 12

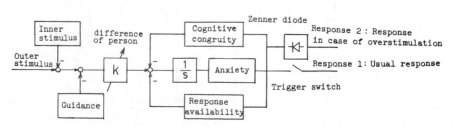

Fig. 5. Occurrence mechanism of passenger's anxiety

MAN-MACHINE SYNERGY IN HIGHLY AUTOMATED MANUFACTURING SYSTEMS

L. Nemes

Computer and Automation Institute, Hungarian Academy of Sciences, Budapest,
XI. Kende utca 13-17, H-1111 Hungary

Abstract. In computer controlled manufacturing systems men and
machines must act together to achieve an effect of which each is
individually incapable. In accordance with this aim they also
must *live* together so the machines must provide human-friendly
interfaces.
In this paper the human tasks in computer controlled workshops
are analyzed. The roles of operators, supervisors, service and
maintenance personnel are discussed and the technical facilities,
hardware and software support and social implications are also
discussed.

Keywords. Display systems; human factors; industrial control;
man-machine systems; manufacturing processes; numerical control;
machine tools.

INTRODUCTION

Users purchasing new types of goods are
always filled with mixed emotions. Everybody has an affinity for possessing new
products, but he also entertains fears
from the hazards of their misuse and
malfunctions and also from economical
detriments, like break-down time and
repairing costs. The engineering industry has progressed past this age of
infantile disorders: machines are
nowadays simple-to-use, highly reliable
articles, and their wide-spread use
can be observed in our everyday lives.

The demand for more complex and sophisticated machinery is growing, so
production must be *highly automated* to
ensure the necessary output. On the
other hand individual requirements vary,
therefore lot sizes must be small. The
scheme for a solution is the *flexible
manufacturing* system in order to satisfy both controversial requirements.

One thing must be realized: these are
not technical possibilities any more,
these are *necessary* characteristics of
the production processes in any industrial society. These are *basic economical constraints* as well. Nobody
can neglect them, otherwise one falls
short in the race for the market.

Consequently, the modern factory image
is defined. How can people be fitted
into this frame? According to the
hundred-year old pattern men and ma-

chines must *work* together. In the
later models, in the Automated Flexible Complexes, men and machines must
help each other to achieve an effect
of which each is separately incapable.
They must *act* together. They must
tolerate and *compensate* each other's
defeciences.

They must *live* together. These inter-
dependences can be expressed by the
concept of *men-machine synergy*, which
is becoming a leading factor of present day technology.

It is very difficult, however, to
formulate the elements or even their
numerical values, that circumscribe
this synergy. We shall approach this
concept from various directions. We
shall analyze functions to be carried
out, we shall investigate human
factors important for the evaluation
on the hardware and software components, and of workplace arrange-
ments for highly automated systems.

THE ANATOMY OF DEPENDENCE

For anybody who is working within a
frame of a large system, his actions
are deeply influenced by his relations
with the environment. He relies on his
own abilities, but also on the co-ope-
ration of his fellow-workers. He has
links with the technological processes
through various types of equipment.

In all his relations he is exposed to

constraints. Pressure accumulates on him, and his personality lies open to various influences.

On the other hand he also uses his power to rule the conditions encompassing him. To illustrate this problem, let me discuss the basic principles of this model of *dependence*.

Human Dependences

Nowadays only artists enjoy the privilege of being able to combine the pleasure of altering the shape of their material with the intellectual effort of their creative activity. They have the feeling of being masters of the entire "process". No one of the technical staff of even the most modern factory can boast of anything similar. The operator on the shop floor can serve a number of automatic machines and in most cases he will not even need the power of his judgement, let alone his creativity. The working conditions of supervisors are not much better. To obtain what is considered to be sufficient information about the flow of the process, they are usually seated in a control room packed with instruments, alarms and computer peripherals. Their task is to watch all the sources of information and to pick out the important ones from the entire mass.

The great variety of the products, multiplied by the number of parts incorporated in them, have innumerable physical characteristics, not to mention the great number of different properties of the material forming, production management, etc. processes. All these data can be arranged in tables or graphs and they are then displayed individually to the supervisor. It is certainly helpful if not all the information appears at the same time to perplex the operator, but in this case it is his task to call the subsequent frames. The methods which designers are today proud of, are based on the familiar interactive selection systems, where the computer offers possibilities and the person chooses the next picture from the suggested menu. In large, comprehensive systems, quite a few steps may be needed to get the report wanted, as the information is usually arranged in some kind of tree structure. The supervisor either has to watch separate display areas or to play on the keyboard for a while to change the display content, to collect data before exercising his perceptual, cognitive and decision--making capabilities.

The control room should instead be designed always to supply that information to the personnel, which they need at the moment of time, and in a form they can grasp without requireing too much unnatural human action.

When somebody is inquiring into a complex problem, he usually makes motions appropriate to the situation. If he wants to have a detailed view of a topic, he leans unintentionally closer, or on the contrary, if he prefers a more general oversight, he leans backwards in the chair. These movements of control personnel can be detected by sensors and transmitted to the computer, where th signals will be regarded as control commands for zooming [1].

Even in straight-forward graphical representation there are many details which are left out in the general view, but called in after having blown up the pictures. On a more symbolic plane, diagrams may include numerical values, where the numbers of digits are determined by "scale of magnification". With yet one further degree of abstraction, zooming from a block diagram could show, say, the logic diagram realizing it [2]. The sequences of the generated pictures and their respective content are compiled to satisfy the requirements of smooth continuity.

On the basis of the behaviour analysis of operators, new sets of input signals can be defined.

Sitting down in the controller's chair could effect an automatic login command to get detailed information, while standing up results in automatically switching to overall system messages.

All other movements of the supervisors, like turning right or left, can also be interpreted for changing the display content. Sensing the actions and force of fingers, writs, fore arms, are interpreted as reactions to certain stimuli [1].

An other way of communicating is the natural language. Many attempts [3,4] have been made with limited systems, mostly for data retrieval and management, and they seem by now to be sufficiently developed for manufacturing control application. In this case the operator need not type in preformulated texts, nor must he select actions by making successive selections from dialogue menues. Instead, he formulates questions of any kind, types them in and the context-sensitive data base answers on the display screen. (In actual fact, of course, the sentence is first analyzed and its meaning formulated, and the reply then compiled.)

Natural language communication does not necessarily mean voice utterance, but rather a free exchange of information in written form. It is not yet clear, whether in factory automation there will be enough computing power for simplified vocal communications. The few commercial systems [5, 6] available will have to be considerably extended if they are to be generally used.

In most cases the response of a control computer supplies precise information. Sometimes the overall supervisor does not want to get an abundance of figures about details of the technological process, which have to be mentally converted into judgements, but would prefer to obtain a "feeling" for the physical characteristics measured in the process. In this instance a model world could be built around the operator which simulates the effects in the real environment. The tolerances measured in the inspection area can, for instance, be illustrated in the control room on a colour display. If all the measured data are well in the middle of the tolerance field, then nothing can be seen on the screen, except the message "inspection on".

As a dimension creeps out of this ideal domain, the figure of the part will appear, but no numerical data will be added. Instead, the line representing the critical surface will be shown in colour. As the tone changes, this will mean that the value of a dimension is getting closer to the limit. From the speed of changing the tint the human operator can infer tool-wear.

The tooling for making a part must have been grossly erroneous for only one segment of the drawing to change, through all the colours of the rainbow. After a well optimized tool selection the whole contour will change colour segment by segment, but within a short period and into the same shade. It happens very seldom if at all, that the supervisor is really interested in the numerical value of the actual dimension. But he will call the attention of the tooling experts to check the effect in detail.

If all parts are well inside the tolerances but off the nominal value, then more than one part can be displayed in scaled down size without particular detail, just to indicate the type of parts and how many they are, and how serious the problems are. After having made a selection, the remaining parts will be deleted and the selected part will be presented enlarged and detailed.

The hardcopy, however, will always print numerical values for further reference on the line printer. The tightness of a fit can be made perceptible in a similar manner.

In weighing the human aspects, we must also consider how the structural changes of society might effect design criteria.

The countries undertaking advanced automation projects have widely different social problems in respect to employment, training, settlement,etc. They will accordingly build new manufacturing systems which will offer jobs either in the countryside for local residents or in towns for highly educated people, or both. However, the social question is not restricted to educational and training problems, although even here the experts are still not agreed on what qualification level should be expected from their employees. For instance, while they may well have studied computer programming, they do not need to know anything about how the control computer has been programmed. It is the manufacturing process itself which must be familiar to them, but it must not be forgotten that during normal operation they do not have enough occasion to practice their skills.In the automatic mode they are condemned mostly to be passive. Therefore special training modes must be developed which will be based on the simulation of the system and can be exercised from time to time. When designing interface facilities the tradeoffs between the off-line training system and software support for control must be carefully studied.

The members of the supervisory staff cannot be isolated from one another. They must work in a team, for which the system interface should offer adequate facilities. Information offered to supervisors should be overlapped, so everybody has the opportunity to consult about his reaction with somebody else.

There is also considerable dispute about how the control staff should be occupied. It is psychologically unsatisfactory to have people seated passively in front of instruments for long periods of time. Different kinds of activities should be combined and both hardware and software should be designed to be instrumental in mixing the active and passive roles.

Technical Dependences

The technical elements concern questions about *how* to control and *what* to control by humans during the manufacturing process. The most general of these is to decide on the operational modes of the system. Although the grouping of the functions is almost the same in all kinds of computer control - one might even say in all computer operations -, the implementation will differ widely according to the implications of the automated machine shop.

Set-up mode
When the machines, robots and other equipment are installed, modified, or reconfigurated, their new or remodelled characteristic data will have to be entered into the control scheme. In many cases this can be done by the designers only. In the case of flexibly reconfigurable systems, the operator should be provided with suitable interfaces which interpret the graphic modification of the layout, the description of new equipment, its technical data and new technological links. In the case of a manufacturing cell, for instance, the addition of a new machine tool is not a major modification. However, the exact place, orientation, and dimensions need to be known by the robot in order to avoid collisions and to be able to load parts into the chuck. All these factors need high-level actions and they modify the control software accordingly.

Start-up and shut-down procedures
These are repetitive actions and they are used when operations are periodic or when intermissions are necessary. In discrete part manufacturing systems these two control functions can run automatically, but mechanical components have to be checked to see whether their states have been changed by humans during the standstill. The more automated the system, the less importance these factors have as design criteria for man-machine communication. But for the time being, they are by no means negligible.

Operational mode
It is up to the designers to decide how many sub-modes they want to have, but two are essential:
automatic,   and
manual.

In the automatic mode, the question seems to be easy. All error messages are printed, but only those will be displayed, which cannot be tackled by the control system itself. The interface must be designed accordingly.

This could by analogy be called the "interface by exception" approach.

A careful analysis is necessary to establish how the information should be presented to the supervisor in order to render his inquiries easy, to help his interactions without requiring too much training. One of the fundamental features should be a clear distinction between "good" and "bad" trends and situations.

The manual mode is probably becoming less interesting. It is almost inconceivable for an entire discrete part manufacturing system to be operated without overall computer control. This is not a task that humans can accomplish. If somebody today argues for a manual mode, he will probably have functions in mind which should rather belong to the following mode.

Test-maintenance-repair mode
It is a matter of indifference to the determination of the design criteria whether these functions are collected in one mode or separated from one another. Probably it is their requirements which have the most severe impact on the new design methods. Depending on how much help and on what level of perceptual and cognitive decision-making capability is needed from humans, entirely different software backgrounds might be required, while variety in the hardware equipment, though not always essential, could also be important.

The first attempts at diagnosis will be made by people and computer support will only be used for documentation retrieval. Considering the usually huge pile of drawings and descriptions, manuals etc., even this is not to be depreciated. Very fast responses cannot easily be guaranteed.

EVALUATION METHODS

Anybody, who has tried to formulate postulates which should be satisfied at the end of a development project, will agree with us that the *methods* which will be used to accept and qualify the products must also be laid down as precisely as possible before beginning the designing phase. Human work-places, however, have many subjective elements, so not exact classification can usually be prepared. In our practice evaluation takes place in two separate phases of project development. Immediately after classifying all the conditions, a scenario is drawn up of how the operator can activate all the functions. There are versions for each purpose and only animation programs will at this stage be written to illustrate the effects. In this period not only can selection take place,but entirely new suggestions can still be made which alter the basic requirements. Even the hardware facilities are often simulated on a graphic display (Fig.1 is an example), so that many alternative constructs can be pondered. Beside the general, "higher level" proposals for system operation, a few key questions of "detail" can also be tested. Of these the most important are:

the intended solution has to be simplified because the mathematical background is inadequate;

- System tolerance towards human errors
would greatly improve working conditions, and though at present this can hardly be guaranteed, the risk of severe consequences can already be cut by carefully minimizing error possibilities;

Fig. 1

- Data representation,
mainly concerned with how the information should be displayed and which is the optimal proportion between graphic and textual information;

- Adequacy of communication and display facilities,
where the intended figures should be analyzed to establish whether the resolution, colour, speed etc. available, can satisfy the requirements of displaying the required drawings (alternatively, to see how illustrations can be simplified or stylized);

- The structure of the texts,
to find optimal lengths of questions and messages, or forms of questions and answers which do not force the operator to memorize large numbers of "sentences";

- The software developments need to meet the specifications,
i.e. establishing whether these are available, have to be created, or

- Teaching methods,
which are important to further development of the control software, can during the simulation stages be profitably compared and proposals tested;

- Error recovery
cannot be automated completely but different methods may be checked and requirements towards the control software stipulated;

- Operating technique,
both the operators and supervisors can operate on all the data - acquired from the shop floor or stored previously - with the help of various actions. The number of steps for manipulating with control software should be minimized. The proper proportions between tree structured interactivity, conversational methods and question-answer techniques must be determined;

Computer generated feedback,
to keep the human operators informed

the computer system must supply the
necessary feedback to them on how
the production system operates. This
can be done in the form of numerical
values (calculated from various inputs) or by generating just a feeling,
not only about the present status but
rather the swiftness of the changes
in the important factors;

- Daily activity,
  to keep the attention of supervisors
  awake, yet not to put too much burden
  on them, their daily activity must be
  carefully planned. The trade-offs
  of active and passibe involvement and
  off-line training are to be analyzed.

REFERENCES

1 Negroponte, N. On being creative
   with CAD.
   Information Processing 77, Toronto,
   Canada, 8-12. Aug. 1977. (Amsterdam, Netherlands: North-Holland
   1977) pp. 695-704.
2 Chang, S.K., Lin, B.S., Walser, R.
   Generalized zooming technique for
   pictorial data-base systems.
   Proc. of the National Computer
   Conference, 1979, pp. 147-156.
3 Harris, L.R. "ROBOT": A high
   performance natural language
   processor for data base query.
   Proceedings of the Fifth Inter-

Fig. 2

On the basis of the experience gained
with the simulated systems the real
software background can be set up.
However, the small scale testing of
each step can be successively checked
in the simulated environment. (Fig. 1)

The final test, after having completed
the whole control system, will be that
undertaken by a team of designers,
engineers, workers of different educational levels, who will operate the
interface and whose opinions will be
used to guide further development.
While this comparison is admittedly
subjective, no one has yet published
a better suggestion. (Fig. 2)

national Joint Conference on AI.
MIT Cambridge, Mass., 1977.
pp. 903-904.
4 Hendrix, G.G. "LIFER": A Natural
   Language Interface Facility.
   Techn. note 135, AI Center SRI
   International, Menlo Park, Calif.
   1977
5 Lowerre, B.T. The HARPY Speech
   Recognition System.
   Technical Report, Computer Sci.Dept.
   Carnegie-Mellon University,
   Pittsburgh, Penns. 1976.
6 Speech recognisers. Sales brochures
   of Nippon Electric, Threshold
   Technology, Inc.

HUMAN SOFTWARE REQUIREMENTS ENGINEERING FOR COMPUTER-CONTROLLED MANUFACTURING SYSTEMS

T. Martin

Manufacturing Technologies Program Management, Kernforschungszentrum Karlsruhe GmbH, P.O. Box 3640, D-7500 Karlsruhe, Federal Republic of Germany

Abstract. Computers as an integral system component determine the quality of work of operators in industrial production. This paper makes a contribution for improved design of man-machine systems by proposing a formal aid to human software requirements definition and design.

First, the need for considering human requirements is stressed. As technology is to be considered "non-deterministic", the designer is faced by a degree of freedom in his design decisions which can be and should be utilized for rising the quality of work of persons operating computer-controlled man-machine systems.

Then, a hierarchy of human quality criteria applicable for ergonomic judgement of work design measures is derived. Following the principle of prospective work design, these criteria must be considered just like technical and economical requirements as design goals. As a formal aid for making design decisions meeting these comprehensive requirements, the Requirements/Quality Criteria Matrix is proposed.

Finally, this method of considering human requirements is exemplified by designing some man-machine interface features for operators of a hypothetical flexible manufacturing system.

Keywords. Man-machine systems design; software requirements engineering; quality of working life; social effects of automation; computer aided manufacturing; flexible manufacturing system.

INTRODUCTION

In industrial production an ever increasing number of work places is influenced by computers partially or totally controlling manufacturing processes. The same is true in administrative and even private areas of work. The effects on human operator of such automation are determined by the programs controlling the built-in computers with which the operator works in a production unit. Computer software design, therefore, plays a key role in the design of man-machine systems with humane work places.

Are we able to specify human, i.e., person-oriented, software requirements for computer-controlled manufacturing systems which are operational for software engineers to work with? The author believes: hardly. He, therefore, wants to make a contribution towards solving this problem, by first indicating workable criteria for ergonomic judgement of work design measures, and then by applying a proved method of technical requirements specification engineering to these person-oriented criteria.

HUMAN CRITERIA FOR MAN-MACHINE SYSTEM DESIGN

Requirements specification in manufacturing systems design is usually limited to specifying WHAT the system shall do in detail (functions) under the constraints of the system's final purpose (output, quality and cost of parts to be manufactured by the system). The question is whether there is any need also to consider the future human operator of the system in such early stage.

Need for considering human requirements

The development of industrial production is characterized by growing mechanization, increasing automation and work becoming more and more fragmented and hierarchically organized. The fragments of work being left to one worker require a narrower range of skill. The content of work actually performed at site (where the workpiece is) is further reduced by taking the work planning and preparation away from it (separation of thinking from doing).

An example of this is the conventional NC machine tool operation being limited to machine loading and unloading and monitoring, leading to harmful qualitative underload of the work person (Noble, 1979).

To better understand such causes and effects, a number of studies were performed in Germany, a summary of which was published recently (Brödner and Martin, 1981).
Through these studies a number of social effects of automation in industrial production became visible which cannot be ignored any longer. In general, if counter-measures in the interest of the workers are not taken consciously, further technical changes may lead to de-skilling of most workers; physical and mental strains may rather grow and normally shift to the latter. The answer, therefore, is: Yes, there is an urgent need for considering human criteria in the early phase of requirements specification, indeed. Considering this, in the new German manufacturing technologies support program technical and social goals have been determined as being of equal rank (Martin, 1980).

The next question then is, whether there are any margins of design in developing computer-controlled manufacturing systems, which could be utilized for the sake of the work persons, possibly even without getting in contradiction with economical viewpoints.

Idea of technological determinism overcome?

People are easily led to believe that the historical path on which technology in its present shape has been developed was the only one that could have been pursued. This prejudice, commonly called technological determinism, has been refuted by one of the studies mentioned. For highly automated production plants a wide margin of designing man-machine interface and the role of the operator exists (Mickler, 1975).
Another well know example of alternative technology took place recently by re-enriching the content of work of NC machine tool operators. This is achieved by building data processing power into the NC control unit and, thus, becoming able to allow for part programming at the machine site by the operator himself (Martin, 1981). For small batch manufacutring this is being widely used with considerable economical advantage.

This shows that the new technology of microelectronics can be used to correct or even reverse the historical process of subordinating men and women to machines and of eliminating initiative and control in their work, and to develop a manufacturing technology which is subordinate to human skill and cooperates with it (CSS, 1981).

Yet, the prejudice still seems to prevail, and thinking in technical alternatives and making design choices according to comprehensive criteria, including person-related ones, is not widely spread amongst engineers.

Quality criteria for human work design

To be able to judge human work places ergonomically, one should resort to newest findings of work scientists, favorably of work psychologists. They found that a human being's work is of utmost importance for his self-experience, behavior and personality development in general. Quality criteria for human work must be derived from man himself as a thinking and acting being. The psychological action theory has developed a model conception of human action. One of the aspects derived therein is the notion of hierarchic-sequential organization (Großkurth and Volpert, 1975; Oesterreich 1981). This notion states that an action is regulated on one of several regulation levels depending on its complexity. Of these regulation levels, the highest is the intellectual level, the lowest is the level of psychomotor skill. The intellectual regulation level is employed for analysis and synthesis of complex situations representing new requirements to the individual. Thinking is exerted here as a simulated action on an imagined model, and by this, a planning strategy for future action can be developed. To be able to do this the individual needs a certain margin of action. If the worker is tied to repetitive work paced by the machine, he cannot (and need not) develop such action schemes, and, therefore, his action dexterity and motivation must vanish sooner or later. Thus, work psychologists have shown that personality promotion of the working person depends more on the work content itself than on work environment, social relations or wages.

Based on this, a hierarchical system of quality criteria for ergonomic judgement of work design measures has been developed, shown in Fig. 1 (Hacker, 1978).
In the left column there are four high level criteria with personality promotion on top of the hierarchy. The meaning of the hierarchy is that the minimum requirement of one level must be fulfilled before the next upper level can be reached.
Through consecutive improvements of the workplace one can proceed from level to level until one reaches the top. The four high level criteria are too high in abstraction to be operational for design engineers to work with. They must be broken down into operational criteria applicable for the individual manufacturing system. Some criteria are given in the right column of Fig. 1. More shall be given in the example later on.

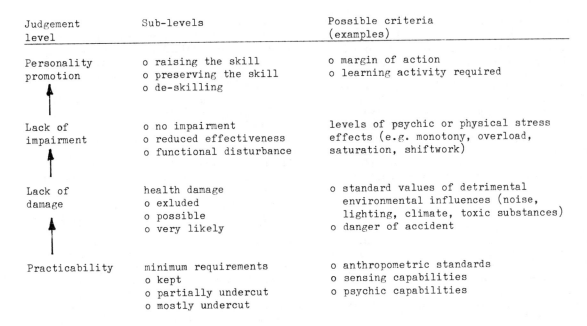

Judgement level	Sub-levels	Possible criteria (examples)
Personality promotion	o raising the skill o preserving the skill o de-skilling	o margin of action o learning activity required
Lack of impairment	o no impairment o reduced effectiveness o functional disturbance	levels of psychic or physical stress effects (e.g. monotony, overload, saturation, shiftwork)
Lack of damage	health damage o exluded o possible o very likely	o standard values of detrimental environmental influences (noise, lighting, climate, toxic substances) o danger of accident
Practicability	minimum requirements o kept o partially undercut o mostly undercut	o anthropometric standards o sensing capabilities o psychic capabilities

Fig. 1. Hierarchical system of quality criteria for ergonomic judgement of work design measures

HUMAN SOFTWARE REQUIREMENTS DEFINITION

Whenever a manufacturing technology is developed or a work organization is chosen, the content of work and, hence, the quality of work at the workplace is determined simultaneously. Unfortunately, work design very often takes place only after disadvantageous working conditions have become obvious (e.g., monotony is reduced by job enrichment).

Concept of prospective work design

Such corrective work design can be avoided by properly including human criteria into the requirements definition process in time. Such strategy has been called prospective work design (Ulich, 1980). This aims at anticipating possibilities for personality promotion and implementing them by creation of objective margins of action for the work persons involved.

In the process of work design there are four main occasions where the content of work is determined:
(1) Design of part or product to be manufactured
   A product should be designed so that it can be manufactured on a system which meets human quality.
(2) Division of functions between man and machine
   Computer technology - especially software - is in its nature flexible enough to be adjusted to meet human needs. Thus, it is possible to decide on what should and what should not be automated. Prac-

tically speaking, automation should be used to handle work that is routine and undemanding and cannot be enriched in any way, but no attempts should be made to let computers take complex decisions or operations; these should be given to the human worker (Mumford, 1979).
(3) Division of tasks between people
   During this phase fragmentation of work and separation of thinking from doing must be avoided.
(4) Providing tasks with margins of action
   Beside the advantages for personality enhancement mentioned above, one more advantage of this provision is that the individual qualification of the worker can be utilized.

Systematic definition of human requirements for software design

Trying to consider human aspects during design of computer-controlled manufacturing systems one can rely neither on any proved methodology nor on validated empirical data. So far only some priciples have been developed for the design of video display terminals. Most of these relate to anthropometric standards, and very few to mental requirements (e.g. dialogue flexibility).

As a contribution to overcome this methodological deficit, we propose to introduce the hierarchical system of human quality criteria given in Fig. 1 in a systematic way which fits into the general framework of modern software engineering principles.

Very often, the operation of a computer is
not consistent with what the user had asked
for. The reason for this is that the software
requirements were wrongly specified (incon-
sistent, incomplete). Considering this (as
being a main reason for the so-called "software crisis"), formal methods for software
requirements specification have been recently
developed in the technical field. One method,
the Requirements/Quality Criteria Matrix
(first introduced by BOEHM, 1974), is adopted
here for treating human criteria just like
functional and economical software quality
criteria and introducing them into the design
process.

The principle of this method is as follows.
On one hand, the functional requirements,
i.e., enumeration of WHAT the system shall
perform, is specified. On the other hand,
quality criteria to be fulfilled, such as
reliability, maintainability, resource constraints, are addressed. Formerly, no human
quality criteria were included in these, but
we demand that they are. From both of these
aspects - functional requirements and quality
criteria - design specifications answering
HOW the system accomplishes all of them must
be inferred.
In order to help check the consistency and
completeness of these requirements, and to
determine their design impications in a re-
constructible way, a proper formalism is
proposed. It is a matrix where functional
requirements are plotted versus quality criteria. For each matrix field a pointer points
to an entry in the catalogue of measures (design decisions).
This matrix requires that the designer thinks
about the design implications of each requirement with respect to each of the desired system quality properties, at least long enough
to make an appropriate entry in the table
(some criteria may be irrelevant to certain
requirements).

Example: Considering human requirements for
operators of a flexible manufacturing system

To demonstrate the method, it is applied to
the design of a hypothetical flexible manufacturing system (FMS) considering proper
criteria for the human operator. (A FMS is
a highly automated system of machine tools
linked together by material flow connections
with which a family of workpieces can be
manufactured completely.)

In Fig. 2 a partial matrix is shown containing only human criteria being of interest
here. Proper technical and economical criteria would have to be added in the horizontal axis. Likewise, all other functional requirements would follow in the vertical axis.
The high level human criteria (lack of damage
etc.) have been broken down into some applicable aspects. Some design decisions taken
shall be further commented.

M3: By early information of tool status the
worker is decoupled from the machine. Being
freed off the machine pace is one of the
fundamentals for improved work conditions.
If this information is available on request
also, a margin of disposition is granted to
the worker which promotes his personality.
M7: The operator, a machine tooling specialist, shall be responsible for maintaining
the hydraulic system himself, thus widening
his qualification.
M6: This measure is taken so that performance
control of the operator cannot be exerted by
anonymous data processing facilities.
M8: Freedback of the system after operational
action is necessary for the operator to learn
to do better next time (principle of learning
by doing).

These design decisions stand on a high level
of abstraction in the beginning. By refining
them in a stepwise fashion they become more
and more software specific, until the design
phase gradually turns into the implementation
phase.

CONCLUSIONS

The quality of work of persons working with
computers in a production unit can be designed
only, if human requirements are specified be-
forehand just as technical or ecnomical ones.
Such person-oriented software technological
approach has been hardly pursued before. The
proposed matrix and procedure should be tested
in close inter-disciplinary cooperation of
production and software engineers, work
scientists and the future operators. Presently, similiar endeavors are being pursued in
some projects within the German manufacturing
technologies program (Brödner, 1982).

FUNCTIONAL REQUIREMENTS / HUMAN CRITERIA	Lack of damage			Lack of impairment			Personality promotion			
	Danger of accident	Environmental influences	o o o	Decoupling from machine	Control of worker	o o o	Variety of skills	Possibility to learn	Margin of disposition	o o o
Change of tool magazine at manufacturing cell	M1	S		M3	O		O	M8	M3	
Automatic tool breakage recognition	S	O		M4	O		O	M4	O	
Monitoring of hydraulic system	S	M2		S	O		M7	M8	S	
o o o o										
Production planning and control	O	O		M5	M6		S	M8	M5	

Mi Catalogue of measures taken (design specification)

M1 Protection against manual operation in machining space.

M2 Limitation of noise.

M3 Automatic measurement of tool wear. Early output of message indicating the remaining tool life time (also on request).

M4 Automatic shut-down of machine at tool breakage with alarm signal.

M5 Manufacturing allotment data. Output on video display on request.

M6 Separation of progress of work data from person-related data (in the area of operating data acquisition).

M7 Hydraulic function indicator and diagnosis for the operator.

M8 Return message after each operator action allowing for checking of system reaction.

o   o

o   o

o   o

Fig. 2. Function/criteria matrix (example of considering human requirements of a hypothetical computer-controlled flexible manufacturing system)
Mi ... pointer from matrix field to catalogue of measures
S ... still to be specified
O ... irrelevant or self-explanatory

REFERENCES

Boehm, B.W. (1974). Some steps toward formal and automated aids to software requirement analysis and design. In Information Processing 74, North Holland, Amsterdam, 192-197.

Brödner, P., and T. Martin (1981). Introduction of new technologies into industrial production in F.R. Germany and its social effects - methods, results, lessons learned, and future plans. In Control Science and Technology for the Progress of Society, Pergamon Press, Oxford (Proceedings VIII. IFAC Congress Kyoto, to be published).

Brödner, P. (1982). Humane work design for man-machine systems - a challenge to engineers and labour scientists. (Refer to proceedings of this same conference.)

CSS (1981). New Technology: Society, Employment and Skill. The Council for Science and Society, London.

Groskurth, P., and W. Volpert (1975). Lohnarbeitspsychologie. Fischer Taschenbuch Verlag, Frankfurt.

Hacker, W. (1978). Allgemeine Arbeits- und Ingenieurpsychologie. Dt. Verlag der Wissenschaften, Berlin.

Martin, T. (1980). The new German Manufacturing Technologies Program. IFAC Committee on Social Effects of Automation Newsletter, No. 10, 10-13.

Martin, T. (1981). Anforderungen an die Software zur Gestaltung humaner Arbeitsplätze in computergesteuerten Fertigungsprozessen. In Informatik-Fachberichte Band 50, Springer-Verlag, Berlin.

Mickler, O. (1975). Arbeitsorganisatorische Gestaltungsmöglichkeiten bei automatisierter Produktion. Zeitschr. Arb.wiss., 29, 158-162.

Mumford, E. (1979). The design of work: new approaches and new needs. In Case studies in automation related to humanization of work. Pergamon Press, Oxford.

Noble, D.F. (1979). Maschinen gegen Menschen - Die Entwicklung numerisch gesteuerter Werkzeugmaschinen. Alektor-Verlag, Stuttgart.

Oesterreich, R. (1981). Handlungsregulation und Kontrolle. Urban und Schwarzenberg, München.

Ulich, E. (1980). Psychologische Aspekte der Arbeit mit elektronischen Datenverarbeitungssystemen. Schweiz. Techn. Zeitschr., 75, 66-68.

HUMANE WORK DESIGN FOR MAN-MACHINE SYSTEMS — A CHALLENGE TO ENGINEERS AND LABOUR SCIENTISTS

P. Brödner

Manufacturing Technologies Program Management, Kernforschungszentrum Karlsruhe GmbH, P.O. Box 3640, D-7500 Karlsruhe, Federal Republic of Germany

Abstract. In general, the relations between the technical components and the work structure of socio-technical systems are not completely determined. Therefore, regarding the socio-economic conditions, a man-machine system may be analysed and designed in such a way that the requirements of humane work design are achieved. The necessity to decide on different system alternatives during the design process constitutes the methodological problem to be able to predict and evaluate the transindividual behaviour of man working in the system and the impact of the working process on him.
Some theoretical approaches to these problems of prospective work design are shortly reported. The solution of these problems arises new interdisciplinary tasks for the engineering and labour sciences. Due to some changes in socio-economic conditions and social consciousness shortly represented, these tasks are getting rather more important. The paper then reports and discusses in some details the experience gained, the problems encountered, and the steps towards solutions on the basis of two projects supported by the German Federal Manufacturing Technologies Programm: a flexible manufacturing system and a production planning and scheduling system.

Keywords. Man-machine systems design; prospective work design; quality of working life; flexible manufacturing system; production planning and scheduling system.

INTRODUCTION

The German Federal Manufacturing Technologies Program aims at both improving technical and economical performance of the factories as well as humane design of the manufacturing technologies to be implemented. Therefore, the according man-machine systems to meet these goals have to be designed under two main aspects: the functional requirements of the technical system and the organization of work. As social analysis indicates (Altmann and others, 1978, Mickler, 1975, and Trist, 1981), the relation between a particular technology and the according work structure is not completely determined; there ordinarily exist different structures of man-machine systems, which satisfy the requirements and prove to be of equal rentability.

Consequently, the design of man-machine systems includes the task to work out alternative structures which differ in the division of functions and task allocation between man and machine, in the kind of man-machine interaction and in the shape of the interface between man and machine.

This task constitutes the following main problems that have to be solved under the perspective of the above mentioned goals and given economical conditions:

- Prospective work design - Since decisions about the suitable alternative have to be made already in the planning stage (i.e. ex ante), a theoretical concept of transindividual attributes of human being and behaviour must be developed. One promising access in finding adequate work structures (work content, margins of action) might be the action theory; other models are necessary to predetermine physical and psychic loads.

- Evaluation of alternatives - In order to find the most suitable alternative, all alternatives worked out have to be evaluated under criteria of technical and economical feasibility as well as under work aspects: which criteria are to be chosen and how should they be applied?

- Participation of workers and their representatives - In the past the management's interests often dominated the decision making process; in order to sufficiently consider the workers' interests in humane working conditions their participation in the decision making process is necessary: how can this be organized?

This paper reports and discusses in some detail the experience gained, the problems encountered, and the steps towards solutions

on the basis of two supported projects: a
flexible manufacturing system and a production planning and scheduling system.

THEORETICAL FRAMEWORK AND DEVELOP-
MENT CONDITIONS

In recent practice of analysis, design and
evaluation of man-machine systems, two opposite views of the underlying problems can be
stated: the technocentric and the anthropocentric view of man-machine systems (Fig. 1).
By the technocentric approach (which is supposed to be dominating) man-machine systems
are usually analysed and designed on the
basis of functional requirements to be realized with respect to the actual technological
state of the art, where man has to take over
those functions that are technically not yet
solved. The systems are evaluated only by
criteria of technical functionality, feasibility and economic rentability, while man
is regarded as being reduced to the functions
he can carry out better compared with the
machine. The anthropocentric approach, on the
other hand, regards man as acting consciously
and purposefully, being able to plan and
control his actions even under uncertainty.
This, in turn, requires the man-machine system to be analysed and designed in such a
way that man is enabled to interact with the
machine purposefully. Therefore, the division
of functions between man and machine, the
functional requirements of the technical
system and the man-machine interface have to
be derived from and adjusted at the specific
capabilities of man to produce certain results
and to his bounds to bear loads. For evaluating those systems, criteria of humane work
design (such as work safety, suitable work
loads, work contents, and margins of action)
besides those of technical feasibility and
rentability have to be considered (Fischbach
and others, 1980).

Fig. 1. Technocentric and anthropocentric
        view of man-machine systems
        (P. Hajnozcky).

In this paper the latter point of view is
taken. It arises some questions that seem to
be not fully answered until now. Since the
relations between the technical components
and the work structure of socio-technical
systems are not completely determined by itself, there exists a margin of decision in
designing those systems bounded by the economic and organizational conditions of their
environment. Therefore, the design process
includes the tasks to appoint

. the division of functions between man and
  machine
. the division of labour between the workers
. the man-machine interaction
. the man-machine inferface.

Thus, it becomes necessary to evaluate different system alternatives already in the
planning stage. This circumstance induces a
theoretical concept to be developed that can
explain the behaviour of man interacting with
the technical system and the impact of work
loads on him arising when working in the
socio-technical system. In explanation with
the rather simple behaviouristic stimulus-
response models, the psychologic action
theory gives a promising approach to answer
those questions, since it analyses human
actions as to be consciously and purposefully made and controlled within a hierarchy
of plans (Miller and others, 1960, Volpert,
1975, Hacker, 1978, Oesterreich, 1981).
According to this theory personality enhancement may be seriously restricted for instance,
if the worker is hindered to plan his activities, if heavy or frequent disturbances occur,
and if the information feedback on the consequences of his actions is inadequate. Also
other existing models of human working capacity and of the impact of different work
loads as well as empirical analysis of social
consequences of already realized man-machine
systems have to be further developed.

Far away from being already operational,
these approaches together with the mentioned
problems of a prospective work design they
claim to solve constitute the challenge to
the technical and labour sciences (compared
with conventional tasks such as allocation
of workers, work analysis etc.). The reasons
why these problems become more and more important have to be searched in some fundamental changes in social consciousness, which
may be characterised - as far as the Federal
Republic of Germany is concerned - as follows.

Although the improvement of working conditions has been a matter of concern ever since
the early days of the labour movement (for
instance the struggle for reduction of the
daily working time), workers and unions pre-
dominently aimed at securing an adequate
part of the economic surplus and extending
their participation. Especially during the
phases of reconstruction and expanding export
markets in the fifties and sixties the increase of personal income was the workers'
main concern, the firms commonly being able
to pay growing standard wages by using modern
machinery. On the other hand, these efforts
together with intensified competition have
been generating an extending pressure on the
management to force the division of labour,
to use even more efficient machinery and to
increase the working speed in order to raise
productivity.

These tendencies, however, led to individual
and collective reactions of the workers in
question, which both managers and government
found most apprehensive. To illustrate the

situation, the following spot lights which can be observed to a different extent and in various combinations may indicate working conditions getting worse:

Extending
- fluctuation   )
- absentism   )   increase the
- production sabotage)   firms' costs
- reduced product   )
  quality   )
- strike activities   ))   increase social
- working accidents   )   costs (Matt-
- premature retiring   )   höfer, 1978)

About 50 % of the employees are not able to work until normal retiring age; every 8 minutes a heavy, every 150 minutes a deadly working accident occurs (Hauff, 1980). The growing awareness of the interactions between industrial productivity and quality of working life caused the management, the unions and the government to take their part in improving the working situation. Consequently, at the decision making processes on investment, in new technologies and organizational changes more emphasis was put on the workers demand for humane working conditions (according to the field of action as shown in Fig. 2).

On the side of management several attempts for properly designed working processes have been made by applying such organizational principles as job rotation, job enlargement and enrichment, and working in groups, thus reversing some of Taylor's theorems. The unions' changed consciousness of the impact of social change and their attitudes towards the technical change can be characterized by the transition from agreements ("Rationalisierungsschutzabkommen"), which aimed at protecting the worker against reduction in wage as consequence of technical

changes, to a new kind of pay scale agreement, which also included minimum requirements for working conditions (minimum cycle time, working intermissions).
In the early seventies, the state on its turn took several legislative actions which profoundly changed the socio-economic environment (see Table 1).

TABLE 1   State Regulations on Human Work Design (Pöhler, 1979)

State Regulation	Year	Contents
Works Constitution Act	1972	Regulates cooperation between workers' council (i.e. shop comittee) and management; contains special rights of information and participation of workers concerning work design (workplaces, processes, technologies, environment)
Safety of Machines Act	1968	Obliges all users of machines and equipment to apply these only, if all safety instructions and technical rules are observed
Work Security Act	1973	Regulates the employment of security staff (medical and engineering) and the application of scientific findings in humanization of work
Workplace Decree	1975	Contains minimum requirements for the work environment (noise, lighting, climate etc.)
Decree on Toxic Substances	1975	Sets maximum workplace concentrations of toxic substances to be observed

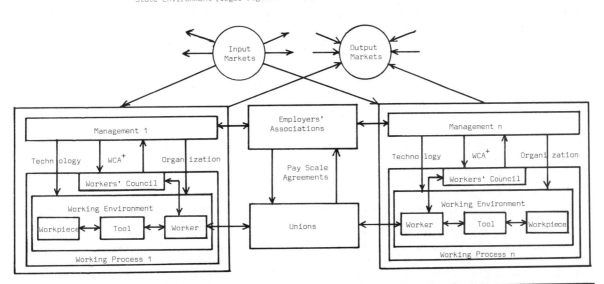

Fig. 2. Field of action for technological innovation ($^+$ Works Constitution Act)

TWO EXAMPLES

Flexible Manufacturing System

One of the major projects supported within
the Manufacturing Technologies Program is the
development of a flexible manufacturing system for rotatory parts in medium lot sizes.
According to the goals of this program, the
main objectives of the project are
o to raise productivity and reduce costs of
  the manufacturing process for medium lot
  sizes,
o to improve working conditions (i.e. security, loads, work content, margin of
  action) for the workers involved.

In order to meet these goals, the developing
team consists of members of different institutions, which cooperate as partners of the
project:
o the firm, which will test and use the pilot
  system and distribute such systems as general contractor, therefore cooperating with
  a number of subcontractors (namely machine
  tool builders),
o three technical research institutes,
  charged to design and to develop the
  materials flow system, the overall control
  system, and the control of the manufacturing cells (consisting of a NC machine
  tool with an integrated industrial robot
  and a positioning device for changing material magazines) as well as to work out
  suitable planning instruments for this,
o a social and a labour research institute
  charged to investigate the social system
  of the firm, where the flexible manufacturing system is going to be introduced,
  to study the social and work changes during
  the process of introduction, to design the
  new work organization and qualifying
  measures as well as to work out suitable
  instruments for this.

The project is being carried out since May
1977 and divided in three stages: a feasibility study, where the parts spectrum to be
manufactured and the structure of the technical system came out, a design and planning
stage to define the technical and social
components of the system (see Fig. 3) in all
details and an implementation and test stage;
actually the system is nearly implemented in
full and under test. The one year training
program for the workers is also nearly finished.
From the technical point of view, the flexible
manufacturing system is consequently designed
with a modular structure: each manufacturing
cell can be used autonomously, the random
access store can be arbitrarily enlarged
(with a possible restriction in the number
of handling devices, whose movements have
to be coordinated), and also the software is
accordingly structured. Each manufacturing
cell consists of a programmable automatic
machine tool with an integrated industrial
robot, a positioning device for material

magazines with a buffer for three magazines,
and a combined control unit. Due to the requirement of work rotation, the operating
panels of the control units of the cells
have been equally designed. For security
reasons the handling space is strictly separated from the operation area of the worker.
In part the machine tools are equiped with
automatic tool wear supervising and with
measuring devices. The magazines are able to
contain all sorts of parts of the whole part
family. The random access store for the
magazines is operated by a portal handling
equipment, which picks the magazines from and
brings them to the positioning stations of
the cells. The program for the overall control of the system (inventory, scheduling,
NC program management, diagnosis etc.) are
implemented on a PDP 11 process computer
(see Fig. 4).

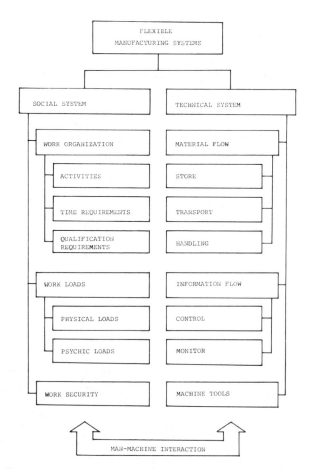

Fig. 3. Components of the flexible
manufacturing system.

By these basic features of the flexible
manufacturing system, a firm willing to use
it will be able to install the system step
by step with partially supplementary and
partially alternative manufacturing cells
according to the requirements of the part
family to be produced.

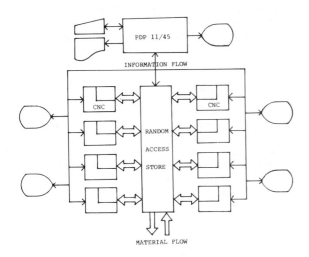

Fig. 4. Technical structure of the flexible
        manufacturing system.

From the social point of view, various alternative work structures for the system
have been worked out and evaluated, which
differ from each other with respect to
o the activities assigned to the workers and
  the demand for cooperation,
o the time requirements of the activities,
o the qualification requirements,
o the physical and psychic loads,
o the functional requirements for the man-
  machine interface.

One extreme work structure (which nevertheless is ordinarily found in most of the
existing flexible manufacturing systems,
Fraunhofer-Institut für Systemtechnik and
others, 1982) is strictly hierarchical in
three levels: On the top level the systems
foreman operates, who is responsible for the
overall performance of the system and capable
of all kinds of activities in the system,
which he carries out in cases of disturbance;
he owns the qualification of a technician.
The workers in the middle are specialized in
operating some specific manufacturing cells;
their qualification is below that of a skilled
machine tool worker. At the load and unload
station of the work pieces still a lower
qualification is required so that usually
unskilled are working at those places.
Another oppositional alternative work structure would be, that every worker can take the
place of everybody else in the system (with
the possible exception of the foreman's position); this means a high and equal qualification for all workers in the system, who
are to be highly cooperative and communi-
cative.

When analysing the different kinds of man-
machine interaction according to these various
work structures and when evaluating those
alternatives, some serious problems in the
labour sciences arise. In order to be able to
predict the effects of certain attributes of
for instance the man-machine interface on the
human being or the impact of psychic loads

of different work structures already in the
planning stage, it is necessary to develop a
theoretical concept or a model on these relations.

Another important question was to derive the
functional requirements that the technical
system had to fulfill for the humane design
of the work structure such as how to demonstrate the state of the system or cases of
disturbance, what kind of dialogue functions
should be implemented or what operations (for
instance tool wear supervising) should be
fully automated in order to decouple the
workers from the process.

Since those theoretically derived methods
and instruments of a prospective work design
are still lacking in general, during this
project some new approaches of this kind have
been developed mainly on the basis of the
psychologic action theory and some experience
from other similar projects. Especially some
design principles for the work structure have
been derived as well as an instrument for
grasping the psychic loads and their impact
on the workers has been worked out and tested
(the latter becoming a more and more important
problem, since with systems like this a strong
change of the work loads from reduced physical
towards new and partially unknown psychic
loads can be stated).

Based on this analysis of the social and
labour scientists, a bargaining process between the firm's management and the workers'
council on the evaluation of the different alternatives of the system has been established.
Finally the work structure that provided job
rotation between all work places turned out
to meet the different interests best: For the
workers it would enrich their work and widen
their margins of action on a relatively high
qualification level; the firm would take advantage in a high personal flexibility and
motivation and expect a high availability for
the whole system, which strongly affects its
rentability.

Once this decision has been made, a training
program to qualify the workers has been worked
out and is now practiced. It provides the same
qualifications - knowledge of the different
manufacturing and measurement technologies
used in the system and of the control devices - for all trainees independently of
their starting qualification.

Production Planning and Scheduling System

Since a couple of years there already exists
a variety of standard software packages in
the field of production planning and scheduling. In general, they comprise all or at
least the most important functions of a
firm's organization, but they differ from
each other in the way how these functions
can be coupled and what planning algorithms
are programmed. Their internal structure does
usually not allow major variations in the

programmed data processing runs without comprising and costly software adjustments.

The small and medium sized firms on the
other hand, which are typical for the machine
building industry, show great differences not
on the functional level of consideration, but
with respect to work organization, the internal data flow and forms, the key number
system etc. In order to keep the firm´s individual and organicly grown organizational
shape instead of adjusting it to the software system to be used, there would either
be arising a huge expense in adjusting the
available software packages in every single
case or a new conception of the according
software system would have to be considered.

Since the small and medium sized firms are
the main group of interest, where the supporting measures of the Manufacturing Technologies Program are aiming at, these are
the reasons why the development of suitable
structured production planning and scheduling
systems is one of the major items in this
program. According to the above situation
and to the goals of the program, these systems are to be designed in a modular structure (so that the necessary adjustment expenses are minimized) and they are to be
assisting a humane work design for the using
firms. Looking at the example of one of the
supported projects, the main features of the
system design to meet these objectives are
as follows.

From the technical point of view, the system
design provides an integrated approach that
comprises all organizational functions of a
firm. The software structure is strictly
modular with clearly defined interfaces between the modules operating on a common database. The software modules can be arbitrarily
combined in a wide range according to the
specific organizational needs of the using
firm. Thus, the system is open for being
embedded in different kinds of work organization (see Fig. 5). The design shows the
following further attributes:
o system access by terminals located at the
  work places,
o dialogue operation with user guidance,
o database management system,
o adjustable and extendable software by
  modules.

From the social point of view, the man-machine system design (hardware, software and
work organization) has to be oriented in two
directions. Looking at the system itself,
some requirements of humane work design have
been developed and applied for
o the ergonomic shape of the terminals,
o the structure of screen masks,
o the dialogue guidance.

Furthermore, much attention has been paid for
keeping the system open for being operated in
different work organizations. This has been
reached by a consequently modular structure

of the system, namely the software. In general, some principles for the division of
functions between man and machine have been
worked out on the basis of the psychologic
action theory. Thus, during the decision
making processes it is up to the computer
system to actualize the database, to give
full transparency into the situation, to
provide all relevant data, and to visualize
the consequences of decisions by simulation
means, while man makes the decision on this
improved information basis. Another principle
is to provide margins of action as wide as
possible at the positions where the decisions
in the manufacturing process are actually
made.

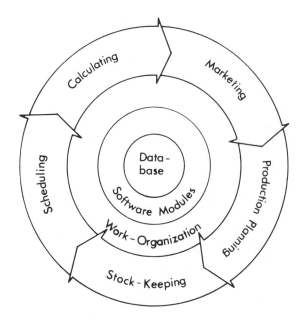

Fig. 5. Production planning and scheduling
        system.

Looking at a specific firm, organizational
analysis has been carried out to find a
humane work organization suitable to the firm
and to derive from this functional requirements for the man-machine interaction. In
this way the production planning and scheduling system can be adjusted to a firm´s
individual work organization.

According to the Works Constitution Act the
management of the firm that is going to use
a production planning and scheduling system
is under obligation to inform the workers´
council about this intention already before
final decicions on the introduction are made;
the workers´ council on his turn is legit-
imated to participate in the design and introduction process. By bargaining with the
management, he is thus enabled to take influence on the way the system is applied in
this particular firm with respect to work
organization, division of functions between
man and computer system, and the ergonomic

shape of the man-computer interface. Here it becomes obvious, that the potential influence of the workers´ representatives is the greater, the less predetermined the computer system is by its own structure.

In order to meet the objectives and to gather scientificly based experience (which actually exists only to a very little extend with respect to most of the above items) the developing firm of the system cooperates with a labour research institute and with the firms expected to be the pilot users. These partners congregate in a working group to discuss the results and to articulate their own requirements.

The main problems to be solved here again are the lack of theoretically founded trans-individual models of man interacting with computer systems (what relations betwen technical and social components of the system are personality enhancing, what attributes of the system are harmful, how to predict the impact of loads etc.) as well as the transformation of results of labour science into functional requirements of the technical system (namely the software, Martin, 1982). Another serious problem arises in case that the requisites of participation exceed the technical knowledge of the workers´ council. This constitutes the new task for technicians and labour scientists to advise the workers and their representatives on alternative system designs and their consequences, so that they become able to evaluate them properly.

PERSPECTIVES

As the discussed examples demonstrate, some steps have been made to use the degrees of freedom in developing man-machine systems with respect to humane work design. Although there is some theoretical and methodical progress in prospective work design, there still remains much work to do for engineers and labour scientists in developing more valid and broader useable instruments. The solutions found until now (at least in the examples discussed here) are rather empirically based than theoretically derived. Therefore, a demand exists for gaining operational design rules from theory. Probably the most knowledge has already been gathered in the field of ergonomic design of the man-machine interfaces, while only little experience and a lack of methods and instruments to define the man-machine interaction and the division of functions between man and machine (which strongly affect work content, margins of action and work loads) have to be stated.

On the other hand, only some relations between the system´s components and man co-operating and acting in a socio-technical system may be theoretically explainable, besides the variety of individual differences. For this reason, and since the evaluation of man-machine systems depends necessarily on the different interests of management and workers, participation of the workers and their representatives is required. Participation in evaluating such complex systems might exceed the technical knowledge of the workers; in this case the engineers and labour scientists have to be prepared to give them advice.

REFERENCES

Altmann, N., G. Bechtle, and B. Lutz (1978). Betrieb, Technik, Arbeit: Elemente einer soziologischen Analytik technisch-organisatorischer Veränderungen, Campus-Verlag, Frankfurt/Main, New York.

Fischbach, D., E. Nullmeier, and F. Reimann (1980). Planung und Gestaltung der Mensch-Maschine-Funktionsteilung. TU-Journal, Nov. 1980, 8-15.

Fraunhofer-Institut für Systemtechnik, Institut für Arbeitsmarkt- und Berufsforschung, and Institut für Werkzeugmaschinen und Fertigungstechnik (1982). Der Einsatz flexibler Fertigungssysteme. PFT-Bericht, Kernforschungszentrum Karlsruhe.

Hacker, W. (1978). Allgemeine Arbeits- und Ingenieurpsychologie, Hans Huber Verlag, Bern, Stuttgart, Wien.

Hauff, V. (1980). Humanisierung des Arbeitslebens, BMFT-Mitteilungen, 5.

Looman, J. (1978). Automatisierung der Fertigung durch flexibel verkettete Fertigungseinrichtungen, VDI-Z, 15/16, 700-704.

Martin, T. (1982). Human Software Requirements Engineering for Computer-Controlled Manufacturing Systems. (Refer to proceedings of this same conference)

Matthöfer, H. (1978). Humanisierung der Arbeit und Produktivität in der Industriegesellschaft, Europ. Verlagsanstalt, Köln, Frankfurt/Main.

Mickler, O. (1975). Arbeitsorganisatorische Gestaltungsmöglichkeiten bei automatisierter Produktion. Zeitschrift für Arb. Wiss., 3, 153-162.

Miller, G.A., E. Galanter, and K.H. Pribram (1960). Plans and the structure of behaviour, Holt, Rinehart & Winston, New York, Chicago, San Francisco, Toronto, London.

Oesterreich, R. (1981). Handlungsregulation und Kontrolle, Urban & Schwarzenberg, München, Wien, Baltimore.

Pöhler, W. (1979). In W. Pöhler (Ed.) (1979). ... damit die Arbeit menschlicher wird, Verlag Neue Gesellschaft, Bonn, Chap. 1, pp. 9-37.

Trist, E. (1981). The evolution of socio-technical systems. Ontario Quality of Working Life Centre, Occasional Paper No. 2.

Volpert, W. (1975). In P. Grosskurth und W. Volpert (Ed.) (1975). Lohnarbeitspsychologie, Fischer, Frankfurt/Main, Chap. 1, pp. 13-196.

DISCUSSION OF SESSION 3

3.2 T IRONIES OF AUTOMATION

Nzeako: Your treatment of cognitive skills,
working storage and monitoring tends to
negate the role of experience, long and continuous association (interaction) in control
and other human behavior. The level of automation not-with-standing, and provided that
the human operator is properly motivated, do
you not feel that the human operator can
always draw from his experience to meet any
unexpected behavior of the controlled system?

Bainbridge: I hope I have not given the im-
pression that the operator's experience is
not important. I consider that experience is
vital, and that retaining it is a major
problem in automation, because the operator
does not learn by watching automatics. She/he
can only learn by hands-on interaction with
the process, and in a successfully automated
system she/he has few opportunities for this.

3.4 I MAN-MACHINE SYNERGY IN HIGHLY AUTO-
MATED MANUFACTURING SYSTEMS

Chairman's Discussion Summary: The paper was
presented by dr. Joe Hatvany in the absence
of dr. Nemesz. There was a short discussion
of the operator. "How to keep the operator
active and alert when the machine can do
everything"? In some cases the company asks
the operators to fill in forms just to keep
them active, a thing which could be easily
done by the computer. Dr. Hatvany answered
that this is just the kind of thing that
makes the operator hate the new system. "Let
the machine do the stupid things and the man
do the clever things." Since a couple of
years dr. Nemesz' and dr. Hatvany's group
has designed systems where the machine asks
the operator: "Should I take this action
now?" so that it is up to the operator to
decide for the machine, not the other way
around.

Someone asked whether it was worthwhile to
design a complicated, interesting work place
in times of high unemployment. Won't people
take any job, good or bad?

Dr. Hatvany didn't believe in creating
meaningless and boring jobs independently of
the situation on the labour market. He said

that in the U.S. there is a shortage of
qualified machine operators despite an unemployment rate of 10%.

3.5 I HUMAN SOFTWARE REQUIREMENTS ENGIN-
EERING FOR COMPUTER-CONTROLLED
MANUFACTURING SYSTEMS

Discussion Summary prepared by Chair(wo)man
and Secretary: In relation to the presented
work organisation, there was a suggestion
that one should not overload the operators.
The author did not think this is a great
problem, as it is mostly the other way
round. It is just a matter of whether the
company wants to have qualified operators
or not. Next there was a comment that, when
the operators have learnt to operate the
system, there is no point still to have the
instructions within the system. Someone
said that with CAD system this has been
taken into consideration in the way that the
operator can let pass a lot of information
in the system.
Someone in the audience mentioned that the
social aspects of the work situation is
very important. Therefore these aspects will
have to be taken into account by the
designers.
There was a question of whether Martin's
hierarchical levels come from Maslow's
theory or not. The author answered that as
he is not a psychologist he had not gone
into these various theories.
Finally, the following question was put
forward by dr. Nzeako: How is the amount and
form of information (software information)
to be presented to the human operator related to his level of training and working
experience? In other words how do we avoid
the risk of giving irrelevant and/or redundant or too much information to the human
operator? The author replied: Your are completely right: the man-machine interface
must be adapted to the qualificational level
of the operator. So, at design time the
qualification of the future operator must
be known. Then, the engineer, a work scientist and, if possible, the future user can
work out the interface in an interdisciplinary approach. One further aspect to be
considered is that the operator becomes
smarter after having worked with the system
for a while. So one has to avoid to force

165

the operator to go through dull procedures
again and again which he needed only while
he learned to handle the system. The man-
machine system must be flexible in this
sense.

3.6 I   HUMAN WORK DESIGN FOR MAN-MACHINE
        SYSTEMS - A CHALLENGE TO ENGINEERS
        AND LABOUR SCIENTISTS

Discussion Summary prepared by Chair(wo)man
and Secretary: There were some questions on
the project of flexible manufacturing system.
One question was about the kind of information shown on the screens for the operators.
"Are the faults made by each operator shown
on the screen for all the others to see? If
so, this is a very negative feedback to
each individual operator." A following question was if the supervisor had information
on his screen that the operators did not
have; some kind of a control function? The
asnwer was that the supervisor has some
extra information. Presently the operators
are being trained, so that one does not yet
quite know how it will work out in reality.
Further it was the author's opinion that the
positive aspect of having the operators
rotate between the stations is sharing of
trouble shooting, as there is always a
machine to be repaired.
There was the question whether the operator
could also program his machines, as the more
one takes away from the shop floor the less
the work content will be. The author answered
that with the present set-up the programming
was prepared elsewhere, but that it could be
put on by the operator.
There was then a lengthy discussion on why
the operator could not do the foreman's job.
The author answered that this work place was
an island in the factory, hence the organization of this work group would have to be
fitted in with the rest of the factory,
which demands a foreman as being responsible
for the output and for the communication with
other parts of the production.
Someone suggested that if the operarors were
taught to do the foreman's job then it would
have been an autonomous group, while with
the present design the jobs have only been
enriched. However, it was also suggested
that to make the group fully autonomous, one
would have to put the economical responsibility on the group as well.
A further question was: "What are the proofs
that this is a good system?". The author
answered that all concerned agree about the
enrichment of the jobs as the operators have
been trained for 15 months to be able to
handle the different machines. There is also
a lower work load than with a conventional
system.

MODELS OF HUMAN PROBLEM SOLVING: DETECTION, DIAGNOSIS, AND COMPENSATION FOR SYSTEM FAILURES

W. B. Rouse

Center for Man-Machine Systems Research, Georgia Institute of Technology, Atlanta, Georgia 30332, USA

Abstract. The role of the human operator as a problem solver in man-machine systems such as vehicles, process plants, transportation networks, etc. is considered. Problem solving is discussed in terms of detection, diagnosis, and compensation. A wide variety of models of these phases of problem solving are reviewed and an overall model proposed.

Keywords. Man-machine interaction; human problem solving; detection; diagnosis; compensation; system failures.

INTRODUCTION

Man-machine interaction has been a topic of formal study for well over fifty years. The earliest investigations focused on the environment as it affected the human operator's safety and ability to perform his job. Later investigations began to consider also the design of equipment in terms of identifying possible limitations and devising potential enhancements of operator performance. Much progress has been made in these areas, although a great deal remains to be accomplished.

More recently, the impact of automation has come to be of increasing importance. In aircraft, ships, process plants, transportation networks, and other large-scale systems, more and more control loops that were once closed manually are now automatically controlled. As a result the human operator is becoming more of a monitor and supervisor of automation (Sheridan and Johannsen, 1976).

The possibility of failures is the primary reason for having humans monitor automatically controlled processes. If hardware and software failures could not occur and if the automation were capable of handling all contingencies, then human operators would be unnecessary. However, failures and design limitations are quite possible and therefore, a primary task of the human operator is to detect these events and deal with them appropriately. If the current trends continue, this task will come to dominate the human's responsibilities (Rasmussen and Rouse, 1981).

It seems reasonable to make the general claim that the manual activities of the human operator will increasingly be supplanted by problem solving activities. The purpose of this paper is to review models of human problem solving. While a brief review of the general area of human solving is presented, the models considered in most detail are those which are directly applicable to situations involving man-machine interaction in detecting, diagnosing, and compensating for system failures. The combined results of this brief general overview and the detailed review of the most relevant models are used as a basis for proposing a very general and widely applicable model of human problem solving.

GENERAL BACKGROUND

Much of the literature in the general area of human problem solving emphasizes the pattern recognition nature of human behavior in problem solving tasks. This position is argued from both a physiological basis (Albus, 1970) and using notions such as cognitive economy (Hormann, 1971). The pattern recognition need not be a concise one-to-one mapping. Familiar scripts (Schank and Abelson, 1977) or frames (Minsky, 1975) may evoke a sense of having seen a particular type of problem before. Of course, particular instances can also be recalled (Neimark and Santa, 1975).

The use of pattern recognition or visually-oriented approaches to problem solving has been advocated for a variety of domains including chess (Chase and Simon, 1973; Nievergelt, 1977), electronic troubleshooting (Dale, 1958; Burroughs, 1979), mental arithmetic (Hayes, 1973), and analogical problem solving (Sternberg, 1977). This view of analogies is rather interesting because it involves a double mapping; one to recognize the analogy and one to transform the solution.

Modes of problem solving

Not all problems can be solved by a direct mapping from observations to solution. Thus, modes of problem solving other than pattern recognition may be required. In an effort to describe possible multiple modes of problem solving various dichotomies have been suggested. Examples include pattern recognition vs. heuristics (Gerwin, 1974; Burroughs, 1979), intuition vs. analysis (Peters, et al., 1974; Simonton, 1975), remembering vs. solving (Rumelhart and Abrahamson, 1973; Jacoby, 1978), retrieval vs. search (Atwood, et al., 1978), imagistic vs. linguistic strategies (Wood, et al., 1974; Sternberg and Weil, 1980), and symptomatic vs. topographic strategies (Rasmussen and Jensen, 1974; Rasmussen, 1978, 1981).

The choice of mode of problem solving can be highly influenced by the way in which the problem is represented. Perceptual cues, even if they are irrelevant, can lead to a pattern recognition mode of behavior (Dale, 1958; Peters, et al., 1974). On the other hand, representations that preclude or inhibit pattern recognition may lead to a more analytical or heuristic approach (Peters, et al., 1974; Gerswin and Newstad, 1977) and, at least in the form of flow charts or functional diagrams, such representations have been shown to improve some aspects of

problem solving performance (Mayer, 1975; Brooke and Duncan, 1980). Some seemingly helpful forms of representation such as color coding (Neubauer and Rouse, 1979) and special formats (Brooke and Duncan, 1981) may, however, have surprisingly little effect, especially for highly practiced problem solvers. Another effect of experience is that humans tend to fixate on one form of representation even when multiple forms are available (Polich and Swartz, 1974)

Nature of expertise

Since multiple modes of problem solving are possible, it is quite natural to wonder if some modes are better than others. One way to approach this issue is to compare behaviors of novices and experts. Many authors have argued that expertise is synonymous with highly-developed pattern recognition abilities (Chase and Simon, 1973; Atwood, et al., 1978; Dreyfus and Dreyfus, 1979). Others have presented results that indicate expertise to be related to particular strategies (Goldbeck, et al., 1957; Wood, et al., 1974; Simon and Reed, 1976; Hayes-Roth and Hayes-Roth, 1979).

Depending on one's definition of pattern recognition, these two perspectives of expertise may or may not be conflicting. If one views expertise as being gained solely by the acquisition of a large repertoire of context-specific patterns, then strategy may be an irrelevant concept (Dreyfus and Dreyfus, 1979). On the other hand, if expertise in pattern recognition includes an ability to recognize useful context-free structural patterns in problems, then changes in strategy that reflect expertise can be described as changes in the perceived usefulness of structural patterns (Rouse, et al., 1980).

The distinction between patterns of context-specific observations and patterns of context free structures is central to the model proposed later in this paper. One school of thought emphasizes the dominance of context (Newell and Simon, 1972; Chase and Simon, 1973; Bree, 1975) while others give more credence to context-free aspects of problem solving (Kearsley, 1975; Mason, et al., 1975; Brooke, et al., 1980) and the importance of structure (Loftus and Suppes, 1972; Malin, 1979; Rouse, 1981b). This issue can be clarified and partially resolved by reviewing the results of a variety of transfer of training studies.

Transfer of training

If context dominates problem solving, then transfer of training should be negligible between problems that are structurally similar but contextually different. While there is some evidence of such a lack of transfer (Smith, 1973), most results indicate positive transfer of training (Shepard, et al., 1977; Siegler, 1977; Luger and Bauer, 1978; Rouse, 1981b). Thus, structure is clearly an important aspect of problem solving. However, context is also important and does probably dominate when the human is in a familiar problem solving environment.

Considering the general area of training, it is interesting to contrast the modes of problem solving that various training methods promote. There are methods that emphasize context-specific pattern recognition (Duncan and Shepard, 1975; Towne, 1981b; Johnson, 1981b), methods that stress inferential (i.e., searching) strategies with respect to particular systems (Glaser, et al., 1954; Landa, 1972; Brown, et al., 1975; Freedy and Lucaccini, 1981; Hunt and Rouse, 1981), and methods that promote context-free search strategies (Rouse, 1981b). Recently, it has been argued that a mixture of methods is probably the best overall approach to training (Johnson and Rouse, 1982a,b; Rouse, 1982).

Models

Many of the above results have motivated the development of models of human problem solving behavior. Some of the earlier efforts in this area compared human performance to optimal half-split strategies (Goldbeck, et al., 1957; Dale, 1958; Mills, 1971) or to time-optimal strategies (Stolurow, 1955). There were also efforts to model the problem solver as a Bayesian information processor (Bond and

Rigney, 1966; Kozielecki, 1972). These modeling endeavors have typically shown that the human is only optimal for simple problems unless extensive training is provided.

More recently, emphasis has come to be placed on the process (i.e., strategy) rather than the product (i.e., results) of problem solving (Gregg and Simon, 1967). A variety of methodologies useful for developing process models have

emerged (Newell and Simon, 1972; Waterman and Hayes-Roth, 1978; Rouse, 1980). As a result of this trend, considerable attention has been devoted to the concept of strategy (Wason and Johnson-Laird, 1972; Simon, 1975; Simon and Reed, 1976; Johnson, 1978). Some effort has been devoted to developing performance measures that are sensitive to differences in strategy (Duncan and Gray, 1975; Brooke, et al., 1980; Hunt and Rouse, 1981; Henneman and Rouse, 1982).

Process models have emerged for a wide range of tasks including chess (Newell and Simon, 1972), fitting of mathematical functions (Huesman and Cheng, 1973), water jug problems (Atwood and Polson, 1976), missionaries and cannibals (Jeffries, et al., 1977), errand planning (Hayes-Roth and Hayes-Roth, 1979), fault diagnosis (Rouse, et al., 1980; Rouse and Hunt, 1981; Hunt and Rouse, 1982), and many others. A common feature of these models is the extensive use of rule-based strategies (as opposed to algorithmic optimization) with, in a few cases, a mixture of probabilistic or fuzzy choices or transitions among modes. Another important aspect of these models is that they can actually perform the task of interest; this cannot be said of many of the earlier product-oriented models.

Summary

This brief review of the general area of human problem solving has served to point out several concepts and issues that are important to modeling human detection, diagnosis, and compensation for system failures. Perhaps the most important concept is the dichotomy between context-specific pattern recognition and structure-oriented searching, and the relationship of this dichotomy to the nature of expertise. It appears that training methods, forms of problem representation, and use of aids are important determinants of the mode of problem solving chosen by the human, at least initially.

While many of the models of human problem solving reviewed in this section are not directly relevant to the type of problem solving of interest in this paper, the notion of modeling the process rather than the product of problem solving is very important. Further, as later discussions will illustrate, much of the rule-based modeling methodology underlying these models has been quite useful for developing models that focus on detection, diagnosis, and/or compensation. Several of these models will now be reviewed.

MODELS OF DETECTION

Detection is defined as the process whereby the human operator decides that an event has occurred. There are four types of model of human performance in event detection. One type is based on signal detection theory; another type utilizes thresholds for error and error rate; another employs the residuals of a Kalman filter within a sequential decision theory algorithm; a final type is based on patttern recognition methods. These types of model are summarized in Table 1. The following discussion elaborates on the summaries in this table.

Model / Attributes	Basic Approach	Key Assumptions	Types of Failure	Experimental Results
1. Signal Detection Theory	detection threshold on likelihood ratio	separability of detectability and decision criterion	presence of a single known event obscured by noise; monitoring	typically used to describe hit vs. false alarm rates
2. Miller and Elkind (1967)	detection threshold on variance of changes in error rate	known set of possible failures	changes in gain and polarity for 1st order system; compensatory tracking	reasonable predictions of detection time and some false alarms; N=3
3. Phatak and Bekey (1969)	detection thresholds on error and error rate	known set of possible failures	changes between 2nd and 4th order aircraft dynamics; compensatory tracking	reasonable qualitative comparisons of detection time and thresholds; N=1
4. Niemela and Krendel (1975)	detection thresholds on error and error rate	known set of possible failures	changes in polarity for 2nd order system; compensatory tracking	analysis of detection threshold; N=5
5. Gai and Curry (1976)	detection threshold on cummulative filter residuals	known model of system and statistical properties of disturbances	step and ramp changes of mean disturbance of 2nd order system; monitoring	reasonable predictions of detection time; N=2
6. Greenstein and Rouse (1982)	detection threshold on likelihood ratio from discriminant function	features independent and linearly weighted	ramp changes of signal to noise ratio of 2nd order system; multiple process monitoring	reasonable predictions of detection time; N=8

Table 1.   Models of Human Failure Detection

Signal detection theory

Signal detection theory has been used extensively to describe the results of experimental studies of the human's abilities to detect infrequent signals in the presence of noise (Sheridan and Ferrell, 1974). The theory assumes that the human forms a likelihood ratio in terms of the conditional probability of observed data given there is a signal divided by the conditional probability of the observed data given there is only noise. This likelihood ratio is then compared to a threshold which is a function of a priori probabilities, values of correct responses and costs of incorrect responses. If the likelihood ratio exceeds this threshold, the theory predicts that the human will report the detection of a signal.

Results of signal detection studies are typically expressed in terms of "hits" and "false alarms." The probability of a hit is plotted versus the probability of false alarm. The resulting plot is called a relative operating characteristic or ROC curve. The shape of the curve can be expressed in terms of the human's sensitivity to the signal, while the human's operating point on the curve reflects the aforementioned response threshold.

Signal detection theory has been quite popular with experimental psychologists whose laboratory studies allow rather straightforward manipulations of probabilities, costs, etc. However, in more realistic settings it can be rather difficult to determine the values of these variables and therefore, the model is considerably less useful for realistic situations. Nevertheless, the ROC curve is still a useful way of summarizing human detection performance.

Error vs. error rate models

This type of model is attractively simple in that human failure detection decisions are assumed to be made solely on the basis of a two-dimensional threshold involving the displayed error and error rate in compensatory tracking tasks (Phatak and Bekey, 1969; Niemela and Krendel, 1975). A related model by Miller and Elkind (1967) assumes that detection is based on the variance of changes in error rate in compensatory tracking. While all three of these models were developed for manual control tasks, the notion of an error vs. error rate display is certainly very general and could be applied to other monitoring tasks.

The simplicity of this type of model is not without its disadvantages. In particular, the two-dimensional threshold on error and error rate is highly situation-dependent and the parameters of these models must be empirically adjusted for different types of dynamic processes and failures. Therefore, as with signal detection theory, this type of model is most useful for describing rather than predicting results.

Filter-based models

A Kalman filter is basically a method of resolving the conflict between prediction and subsequent observation (Rouse, 1980). While this conflict may be attributed to poor predictions or noisy observations, an alternative cause of conflict is system failures. Since a failure may change the input-output relationship of a system, predictions based on the pre-failure input-output relationship are likely to disagree with observations of variables produced by the post-failure input-output relationship. The extent of the disagreement can be used as a means of detecting failures.

Gai and Curry (1976) employed this concept to develop a model of human performance in failure detection. The sequence of differences between predicted and actual system outputs is obtained from a Kalman filter. This sequence of "residuals" is cumulated and compared with a threshold which is based on acceptable probabilities of missed events and false alarms. Beyond the step and ramp failures noted in Table 1, this model has also been applied to detecting changes in variance and bandwidth (Curry and Govindaraj, 1977).

The strength of this filter-based model is the invariance of its structure across a wide range of dynamic systems. Thus, the model need not be reformulated for each new task situation. There is a cost for this generality, however, in that the model must have explicit knowledge of an appropriate mathematical model of the dynamic process being monitored.

Pattern recognition models

In order to detect failures, humans may observe a wide variety of features beyond errors, error rates, and residuals. For example, sounds, vibrations, and smells may be part of the overall pattern of features relevant to failure detection. In general, the human's detection task is to recognize when the pattern of features is other than normal.

Greenstein and Rouse (1982) have developed a model of human performance in event detection based on discriminant analysis, one of the simplest pattern recognition methods. A discriminant function is used to linearly weight any number of task features deemed relevant. This function is used to generate a likelihood ratio which is then compared to a threshold similar to that used in signal detection theory.

The advantages of this model include its ability to overcome a key limitation of signal detection theory (i.e., the process whereby the likelihood ratio is determined), its generalization of the feature-based error vs. error rate models, and its ability to function without the mathematical model of the process required for filter-based models. The model's main disadvantages include the need to empirically determine discriminate function coefficients and, when a mathematical model is available, its inability to make direct use of this information.

Summary

Contrasting the four types of model reviewed here, it seems quite reasonable to conclude that the filter-based models are the best available if the requisite information to use them can be obtained. Otherwise, pattern recognition models are most appropriate. In either case, these models are certainly not available "off the shelf." For example, considerable thought would be needed before these models could be applied to resolving the issues associated with the human's relative abilities to detect failures in manually and automatically controlled processes (Ephrath and Young, 1981; Wickens and Kessel, 1981). Nevertheless, sufficient basic research has been performed to justify the effort necessary to apply these models to understanding and resolving these and other issues associated with failure detection.

MODELS OF DIAGNOSIS

Diagnosis refers to the process of identifying the cause of an event. Table 2 summarizes a variety of models of human performance in diagnostic tasks. These models roughly fall into two classes: perscriptive and descriptive. The following discussion elaborates upon this distinction and the summaries presented in Table 2.

Prescriptive models

Most of the earlier efforts to model human performance in fault diagnosis tasks involved comparing human performance to that of prescriptive models. One prescriptive method of diagnosis is the half-split or binary chop which attempts to choose tests that partition the feasible set into two halves in terms of uncertainty. The evidence is fairly conclusive that humans typically do not make optimal half-split tests (Goldbeck, et al., 1957; Dale, 1958; Mills, 1971). While training may help, its effect becomes limited as problem complexity increases. The primary difficulties appear to be humans' inabilities to identify the feasible set and tendancies to utilize irrelevant perceptual cues.

Another type of prescriptive approach to fault diagnosis utilizes probabilities of failure and average action times to find the minimum time solution (Stolurow, et al., 1955) or just probabilities to find the most likely fault (Bond and Rigney, 1966). Human's abilities to employ this type of strategy are highly dependent on their knowledge of a priori probabilities and average action times. This knowledge is often imperfect and therefore, humans are precluded from performing as well as the prescribed strategy.

From this set of comparisons of human performance with that of prescriptive models it can reasonably be concluded that humans are not optimal diagnosticians, at least not with respect to the criteria upon which these models are based. This suboptimality may be due to a lack of knowledge of the prescribed strategy or, due to a lack of knowledge of the requisite information for implementing the strategy or, due to an inability to process the information in the manner required by the strategy. Finally, of course, it could be that humans have performance criteria that include more than just number of actions or time. They may also be concerned with minimizing effort, risk, etc.

Attributes / Models	Basic Approach	Key Assumptions	Types of Failure	Experimental Results
1. Stolurow, et al. (1955)	consideration of minimum expected time strategy	repair actions based on probability and time, independent of structure	aircraft power-plants with unequal failure rates and repair times	significant disagreement among instructors about times and probalities; N=10
2. Goldbeck, et al. (1957)	comparison with half-split strategy	diagnosis based on decreasing size of feasible set	components in logic network	significant departure from optimality; difficulty identifying feasible set; N=130
3. Dale (1958)	comparison with half-split strategy	diagnosis based on decreasing size of feasible set	components in flow network	significant departure from optimality; irrelevant cues utilized; N=240
4. Bond and Rigney (1966)	comparison with Bayesian updating of probabilities	test choices based on probability, independent of structure	oscillator circuit	agreement on 50% of solutions; dependent on initial probabilities; N=39
5. Mills	comparison with half-split strategy	diagnosis based on decreasing size of feasible set	series electrical circuit with unequal failure probabilities	significant departure from optimality; N=6
6. Rasmussen and Jensen (1974)	description of strategies and selection among strategies	verbal protocols reflect strategy	electronic instruments	topographic and symptomatic strategies; strategy fixation; N=6
7. Rouse (1978, 1979)	formation of fuzzy feasible and infeasible sets	diagnosis based on decreasing size of fuzzy feasible set	components in two types of logic network	reasonable predictions of number of tests and effects of aiding; N=36
8. Rouse and Rouse (1979)	information theoretic measure of complexity	strategy of individual affects complexity	components in two types of logic network	reasonable predictions of solution time; N=88
9. Rouse et al. (1980)	rank-ordered set of situation-action rules	rank-ordering of rules fixed	components in two types of logic network	similar choices on 90% of actions; N=154
10. Wohl (1981)	description of possible causes of very skewed repair time distributions	exhaustive search in order of increasing time per action	several military electronics systems	high correlation with average field repair time; N=10 equipments
11. Hunt and Rouse (1982)	fuzzy rank-ordering of pattern recognition and network searching rules	rule choices governed by recall, applicability, usefulness, and simplicity	simulated powerplants and avionics systems	similar choices on 70% of actions; N=10

Table 2. Models of Human Failure Diagnosis

Descriptive models

Given that human performance departs
substantially from that of
prescriptive models, the next logical
consideration is describing how
humans actually do perform fault
diagnosis. While the studies of
prescriptive models, particularly the
studies of Dale (1958), did attempt
to describe deviations from
optimality, the studies were not
concerned with producing descriptive
models per se. Such models have only
more recently emerged.

The discussion of prescriptive models
indicated that humans have difficulty
identifying the feasible set. This
could be interpreted as meaning that
humans find it difficult to crisply
say "yes" or "no" about the
membership of each component in the
feasible set of failures. Rouse
(1978, 1979) has incorporated this
limitation into the half-split
strategy by using fuzzy set theory
and defining membership in terms of
the "psychological distance" between
components and symptoms. This model
was quite successful in predicting
the effects, in terms of number of
tests until solution, of providing
the human with aids for identifying
the feasible set.

Some fault diagnosis problems can be
solved quickly while others require
quite a long time. It seems
reasonable to argue that this
difference in problem solving time is
related to problem complexity. Rouse
and Rouse (1979) studied a variety of
measures of problem solving
complexity and found that an
information theoretic measure of the
uncertainty associated with all of
the connections among components in
the feasible set provided the highest
correlation between time and
complexity. Since this measure was
based on the feasible set as it
evolved through the course of each
individual's problem solving, this
measure can be said to incorporate
each individual's strategy. Thus,
the measure reflects the problem
solver as well as the problem.

Wohl (1981) has also studied problem
solving time as a function of
complexity. In contrast to the Rouse
measure which predicts the time for a
specific troubleshooter to solve a
particular problem, Wohl's model
predicts average repair time across
all failures and troubleshooters for
a particular piece of equipment. His
measure of complexity is based on the
number of connections among all
components in the piece of equipment.

The model also incorporates a
particular diagnostic strategy,
namely, exhaustive search in order of
increasing time per action. The
correlation between the average time
predictions of this model and data
from both field and laboratory
studies is quite impressive.

The number of tests and the time
until a fault is isolated reflect the
product of diagnosis. A much better
understanding of human problem
solving may be possible with models
that reflect the process of problem
solving. Rouse and his colleagues
(Rouse, et al., 1980) developed a
rule-based model that predicted the
sequence of actions chosen by the
troubleshooter. By appropriate
choices of rules and rank-orderings,
they were able to obtain a high level
of agreement between the behavior of
the model and that of humans. One
particularly interesting conclusion
was the fact that better
troubleshooters did not necessarily
have better rules than poorer
troubleshooters; they often simply
had a better rank-ordering of the
same rules.

Rasmussen's description of diagnostic
strategies (Rasmussen and Jensen,
1974; Rasmussen, 1978, 1981) has
shown that a variety of strategies
are adopted by troubleshooters. The
most important distinction found was
between strategies based on
context-specific pattern recognition
(i.e., symptomatic strategies) and
those based on relatively
context-free network searching rules
(i.e., topographic strategies).
While symptomatic strategies result
in a direct mapping from symptoms to
hypothesis, topographic strategies
involve searching through networks of
functional relationships. Rasmussen
has also studied the process whereby
humans choose and perhaps fixate on
strategies.

Hunt and Rouse have combined the
fuzzy set model, the rule-based
model, and the concepts of Rasmussen
to produce a fuzzy rule-based model
(Rouse and Hunt, 1981; Hunt and
Rouse, 1982). The model has S-rules
(symptomatic rules) which it prefers
to use if possible, and T-rules
(topographic rules) which it will
employ if necessary (i.e., if it
cannot find an appropriate S-rule).
Particular rules are chosen according
to membership in the fuzzy set of
choosable rules which is defined as
the intersection of the fuzzy sets of
recalled, applicable, useful, and
simple rules. This model was
reasonably successful in predicting
the sequences of actions chosen by

mechanics in troubleshooting simulated powerplant and avionics systems. Further, it was useful for illustrating the shift from S-rules to T-rules when unfamiliar problems were encountered.

Summary

Contrasting the variety of models of diagnostic behavior presented in this section, it is clear that the unconstrained prescriptive models do not provide good descriptions of human behavior. For the purpose of predicting repair time for a particular equipment system, averaged across types of failure and different troubleshooters, Wohl's model is probably the best choice. For more fine-grained predictions of human behavior, the model of Hunt and Rouse, as a derivative of earlier work by Rasmussen and Rouse, would seem to offer the most appropriate approach.

MODELS OF COMPENSATION

If a failure must be diagnosed during system operation, as opposed to during maintenance, then the human problem solver typically must be concerned with both keeping the system operating and diagnosing the source of the problem. The process of sustaining sytem operation in failure situations is termed compensation. In this section, two types of compensation will be considered: 1) compensating for symptoms and 2) compensating for failures.

Compensating for symptoms

There appear to be two general types of symptom. The first type includes abnormal and emergency events such as fires, leaks, system trips, etc. These events are often dealt with using standard procedures, which may or may not be formalized. In some situations, however, there are no written or unwritten procedures and humans must revert to problem solving.

The second type of symptom includes substantial deviations of state variables. Examples include pressures, levels, flows, and velocities that are too high or too low. This type of symptom can be dealt with in several ways. For loops that are normally automatically controlled, the operator may compensate by assuming manual control. Another approach to compensation is through reallocation of resources including switching to backup modes, rerouting of resources,

and reprioritizing or shedding of demands on the system. With this second approach, standard procedures are typically not available and problem solving is required.

Compensating for failures

Once the diagnostic process has proceeded to the point of having identified the failure, the operator must decide how to compensate for the failed component. The most obvious compensation is to repair or replace the failed component. However, this is often not immediately possible. As a result, the operator may have to plan for degraded mode operation.

Degraded mode operation involves continued operation with the loss or substantial impairment of one or more system functions. If the lost or impaired functions are critical (e.g., engines to an aircraft), then the only thing of interest may be that of bringing system operation to a halt in a safe and orderly manner (e.g., a safe landing). For critical situations such as these there are often standard procedures.

Many situations, however, do not call for or allow suspending of systems operations. In these cases, the operator may have to continue compensating for the symptoms and/or plan for operating with a long-term loss or impairment of functions. This may require problem solving.

Coordinating compensation and diagnosis

Compensation and diagnosis can be viewed as two separate task competing for the operator's attention (Rouse and Morris, 1981a, 1981b; van Eekhout and Rouse, 1981). Unfortunately, this situation can result in the operator focusing on one task to the exclusion of the other. If the effects of failures are cumulative or if symptoms tend to become aggravated, focusing on only compensation or diagnosis can have disasterous consequences. For example, at least one major airline crash has been attributed to the crew having focused on diagnosis to the exclusion of all other tasks.

The coordination of compensation and diagnosis is not simply the process of managing two tasks simultaneously. Since the two tasks are far from independent, they have the potential for being conflicting. For example, compensating for the symptoms may make diagnosis more difficult. On the other hand, the two tasks may be complementary in that information acquired for performing one task may provide information valuable to performing the other task. In either case (i.e., conflicting or complementary), this interdependence illustrates the potential complexity of dealing with problem solving at multiple levels.

Models

Unlike the discussion of detection and diagnosis, this section has no tabulation of models and their attributes. For those situations where compensation involves executing standard procedures or manual control, there are a variety of models available (Sheridan and Ferrell, 1974; Rouse, 1980; 1981a). However, these are not models of human problem solving. In fact, other than the aforementioned models by Rasmussen, Rouse, and Hunt, there are no directly applicable models of problem solving behavior in coordinating compensation and diagnosis, or for compensation itself. The next section will propose such a model.

AN OVERALL MODEL

From the foregoing discussion, it is obvious that considerable effort has been invested in the study of human problem solving in general, and human detection, diagnosis, and compensation for system failures in particular. However, most of the

models discussed thus far focus on a single aspect of problem solving. Only a few of the models (Newell and Simon, 1972; Minsky, 1975; Schank and Abelson, 1977; Rasmussen, 1979) consider the full breadth and robustness of human problem solving behavior. What is needed is a model that, at least conceptually, captures the whole of problem solving and, at the same time, can be operationalized within specific task domains. This section proposes such a model.

Pattern recognition orientation

A conclusion that surfaced repeatedly in the earlier discussions was that humans, if given a choice, would prefer to act as context-specific pattern recognizers rather than attempting to calculate or optimize. Obviously, life would be difficult indeed if one had to constantly recalculate various things in order to make choices. Thus, human preference for pattern recognition is justifiable both scientifically and practically.

However, if the human does not recognize a pattern, a mode of problem solving other than context-specific pattern recognition must be employed. The alternative modes may be called heuristic, analytical, topographic, etc. A common characteristic of these modes is that the human must go beyond the surface features of the problem. Since the focus of this paper is on system failures, this notion can be more precisely stated as the human must go beyond the system state and consider the system structure.

Figure 1 illustrates this fundamental concept. As can be seen, the human is assumed to have a clear preference for proceeding on the basis of state information. The use of structural information is definitely a less-preferred alternative. The mechanism depicted in Figure 1, which will be elaborated upon throughout the remainder of this section, is the central and only mechanism in the proposed model.

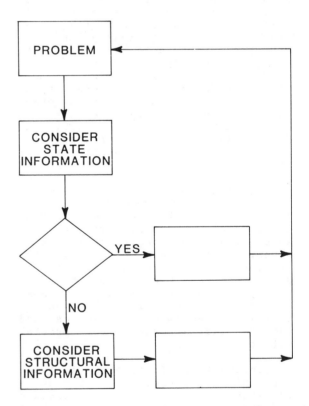

Figure 1. Basic Mechanism of Proposed Model

Levels of problem solving

Considering the literature reviewed earlier in this paper and a variety of studies of human problem solving in aircraft, ships, and process plants (Johannsen and Rouse, 1980, 1981; Rouse, et al., 1982; Johnson and Rouse, 1982 a,b; van Eekhout and Rouse, 1981; Rouse and Morris, 1981a,b), it seems reasonable to conclude that problem solving occurs on several levels. Perhaps the most obvious example of multi-level problem solving is the aforementioned coordination of compensation and diagnosis. The concept of multiple levels is, however, much more general than the idea of coordinating tasks.

It appears that three general levels of problem solving are needed to model human behavior: 1) recognition and classification, 2) planning, and 3) execution and monitoring. Recognition and classification involves detecting that a problem solving situation exists and assigning it to a category. Planning is the process whereby the approach to solving a problem is determined. Execution and monitoring is the actual process of solving the problem.

Table 3 summarizes how the basic mechanism in Figure 1 applies to the three levels of problem solving. At the highest level (i.e., recognition and classification), the human is assumed to identify the context and category of a problem. If the human finds the state information to match a familiar frame (Minsky, 1975), problem solving proceeds on that basis. If the frame is not familiar, structural information might provide clues to an analogy or be used to employ basic principles of, for example, the scientific method.

At the next level (i.e., planning), the human must decide how the problem will be attacked. Based on the state information, the human may conlude that the problem solving situation is familiar and the appropriate script (Schank and Abelson, 1977) or standard procedure can be employed. If no script is available, the human must use structural information to plan in terms of generating alternatives, imagining consequences, valuing consequences, and so on (Johannsen and Rouse, 1979).

Actual problem solving occurs at the lowest level (i.e., execution and monitoring) where scripts or plans are executed and monitored for success. Familiar patterns of state information may allow for the use of

Process / Level	Decision	State-Oriented Response	Structure-Oriented Response
1. Recognition and Classification	Frame Familiar?	Invoke Frame	Use Analogy and/or Basic Principles
2. Planning	Script Available?	Invoke Script	Formulate Plan
3. Execution and Monitoring	Pattern Familiar?	Apply Appropriate S-Rule	Apply Appropriate T-Rule

Table 3. Decisions and Responses for Three Levels of Problem Solving

context-specific symptomatic rules (S-rules) that map directly from observation to hypothesis or action (e.g., if the engine will not crank, check the battery). If the pattern is not familiar, structural information may allow the use of topographic rules (T-rules) for searching the structure of the problem (e.g., if a component's inputs are good and its outputs bad, the component has failed).

All of the responses noted in Table 3 (i.e., invoke frame, use analogy, etc.) invoke the same mechanism as shown in Figure 1. This mechanism is recursively invoked until actions are produced and the problem solved. Thus, in contrast to many of the models discussed earlier in this paper, the proposed model is very simple, involving a single mechanism that is recursively employed for all aspects of problem solving.

Hierarchical vs. heterarchical

If one views problem solving from an operations research or management science perspective, one should hierarchically consider goals, objectives, attributes, alternatives, etc. This hierarchical approach has often been adopted by computer scientists when designing knowledge-based expert problem solving systems (Sacerdoti, 1974). The model proposed in this paper certainly could perform hierarchically by first invoking a frame or an analogy, then invoking a script or planning, and finally acting via S-rules and T-rules.

However, it has been argued that human problem solving is heterarchical or opportunistic rather than hierarchical (Hayes-Roth and Hayes-Roth, 1979). In other words, the human does not solve problems in purely a top-down or bottom-up manner. Instead, it appears that the human operates on all levels almost simultaneously.

The proposed model can produce this type of behavior if one assumes that the three decisions (i.e., frame familiar, script available, and pattern familiar) are constantly, but not necessarily consciously, being re-evaluated. Thus, for example, the model might be using T-rules to plan on the basis of an analogy and suddenly realize the applicability of a script. This could result in the preemption of planning and the rapid application of a sequence of S-rules which provides new information and results in a familiar frame being recognized which leads to a new

script and so on.

If all three decisions are constantly being re-evaluated, it is possible that conflicts will arise in terms of which decision should take precedence. Such conflicts might be resolved by giving more credence to closer matches (e.g., a very familiar pattern is more captivating than a somewhat familiar script). Another method of resolving conflict is to place more weight on alternatives that maintain the current direction of the problem solving. In other words, the model should incorporate the assumption that the human would like to avoid changes in frame, script, etc.

Behavior of the model

Since the proposed model represents a synthesis of the many concepts and models reviewed in earlier sections of this paper, it should not be surprising that this model will produce the types of behavior noted in those discussions. The strength of the model is that a very simple mechanism and method of organization can produce such an impressive range of behaviors. Of course, this possibility, from a slightly different perspective, has been investigated by others (Newell and Simon, 1972).

A particularly interesting aspect of the model's behavior, as well as that of humans, is its potential for making errors. The model has two inherent possibilities for causing errors. The first possibility relates to the model's recursive use of the basic mechanism in Figure 1. As the model recursively invokes this mechanism, it needs a "stack" or some short-term memory for keeping track of where it is and how it got there. If short-term memory is limited, as it is in humans, the model may recurse its way into getting lost or, pursuing tangents from which it never returns. To constrain this phenomenon, it is probably reasonable to assume that lower level items in the stack are more likely to be lost first. For example, one is more likely to forget one's umbrella than to forget to go to work.

The second possibility for causing errors is the matching of irrelevant or inappropriate patterns. For example, the model, or a human, may be captured by an inappropriate but similar script or S-rule. As a result, the model may pursue an inappropriate path until it suddenly realizes, perhaps much too late to be able to recoup, that it has wandered far afield from where it thought it was headed.

The fact that the proposed model has inherent possibilities for making errors, particularly somewhat subtle errors, provides an interesting avenue for evaluating the model. Most models are evaluated in terms of their abilities to achieve the same levels of desired task performance as humans. A much stronger test would involve determining if the model deviates from desired performance in the same way and for the same reasons as humans. The proposed model can potentially be evaluated in this manner.

Summary

In this section, a model of human problem solving has been proposed. This model is based on a synthesis of a wide range of concepts and models as well as a variety of experimental results. The proposed model's main contribution is its potential ability to represent the full range of human problem solving behavior while also being readily implementable for evaluation.

DISCUSSION AND CONCLUSIONS

Considering the relationship of the proposed model to the other models discussed in this paper, this new model is, to a great extent, an outgrowth of earlier work by Rouse and Hunt, and also by Rasmussen. The frame, script, and S-rule aspects of the model are also fairly similar to concepts proposed by Newell and Simon (1972). The manner in which the model utilizes both state and structural information, and the recursive use of the same basic mechanism on all levels of problem solving are perhaps the model's most unique characteristics. In this way, the model is potentially capable of dealing with unfamiliar problem contexts via analogies, T-rules, etc.

Comparing the proposed model with the many models of detection and diagnosis reviewed in this paper, the pattern recognition oriented and rule-based models can be easily incorporated within the general framework outlined here. The models which assume the human to perform some type of calculation (e.g., Bayesian, filter-based, and fuzzy half-split models) best fit within the specific portions of this framework for planning and T-rules. It should be noted that the assumptions underlying these calculation oriented models differ substantially from those of the proposed model which assumes the human to avoid calculation if at all possible.

The model's ability to operate almost simultaneously on several levels provides the potential for representing the coordination of compensation and diagnosis. Because of short-term memory limitations, the model also allows for the possibility of errors in coordination (e.g., focusing on one task to the exclusion of the other). What is not clear at this point, and is the topic of several current investigations, is the nature of S-rules, T-rules, scripts, and plans relative to coordinating compensation and diagnosis. This important topic deserves considerable study.

The basic premise of this paper is that the responsibilities of the human operator will increasingly be dominated by problem solving. The design of systems to support the human operator in fulfilling these responsibilities should be based on knowledge of human abilities and limitations in problem solving. This paper has reviewed the state of the art in this area and proposed a new model which has the potential for being a vehicle for integrating and advancing understanding of human problem solving.

ACKNOWLEDGEMENT

This research was supported in part by the U.S. Army Research Institute for the Behavioral and Social Sciences under Contract No. MDA 903-79-C-0421 and in part by the National Aeronautics and Space Administration under Ames Grant No. NAG 2-123.

REFERENCES

Albus, J.S. (1970). Mechanisms of planning and problem solving in the brain. Math. Biosci., 45, 247-293.

Atwood, M.E., and P.G. Polson (1976). A process model for the water jug problem. Cog. Psych., 8, 191-216.

Atwood, M.E., P.G. Polson, R. Jefferies, and H.R. Ramsey (1978). Planning as a Process of Synthesis., Science Applications, Inc., Englewood, CO, Report SAI-78-144-DEN, December.

Bond, N.A., and J.W. Rigney (1966). Bayesian aspects of troubleshooting behavior. Hum. Fac., 8, 377-383.

Bree, D.S. (1975). Understanding of structured problem solutions. Instruc. Sci., 3, 327-350.

Brooke, J.B., and K.D. Duncan (1980). An experimental study of flowcharts as an aid to identification of procedural faults. Ergonomics., 23, 387-400.

Brooke, J.B., and K.D. Duncan (1981). Effects of system display format on performance in a fault location task. Ergonomics., 24, 175-189.

Brooke, J.B., K.D. Duncan, and C. Cooper (1980). Interactive instruction in solving fault finding problems - an experimental study. Intl. J. Man-Mach. Stud., 12, 217-227.

Brown, J S., R.R. Burton, and A.G. Bell (1975). SOPHIE: A step toward creating a reactive learning environment. Intl. J. Man-Mach. Stud., 7, 675-696.

Burroughs, M.M. (1979). An investigation of mental coding mechanisms and heuristics used in electronics troubleshooting., University of Oklahoma, Engr. D. Thesis.

Chase, W.G., and H.A. Simon (1973). The mind's eye in chess. In W.G. Chase (Ed.), Visual Information Processing., New York, Academic Press.

Curry, R.E., and T. Govindaraj (1977). The human as a detector of changes in variance and bandwidth. Proc. Thirteenth Ann. Conf. on Manual Control, Cambridge, MA, June, pp. 217-221.

Dale, H.C A. (1958). Fault finding in electronic equipment. Ergonomics., 1, 356-385.

Dreyfus, H.L., and S.E. Dreyfus (1979). The Psychic Boom: Flying Beyond the Thought Barrier., Operations Research Center, Berkeley, CA, Report ORC 79-3, March.

Duncan, K.D., and M.J. Gray (1975). An evaluation of a fault finding training course for refinery operators. J. of Occup. Psych., 48, 199-218.

Duncan, K.D., and A. Shepherd (1975). A simulator and training technique for diagnosing plant failures from control panels. Ergonomics., 18, 627-641.

Ephrath, A.R., and L.R. Young (1981). Monitoring vs. man-in-the-loop detection of aircraft control failures. In J. Rasmussen and W.B. Rouse (Eds.), Human Detection and Diagnosis of System Failures, New York, Plenum Press, pp. 143-154.

Freedy, A., and L.F. Lucaccini (1981). Adaptive computer training system (ACTS) for fault diagnosis in maintenance tasks. In J. Rasmussen and W.B. Rouse (Eds.), Human Detection and Diagnosis of System Failures., New York, Plenum Press, pp. 637-658.

Gai, E.G., and R.E. Curry (1976). A model of the human observer in failure detection tasks. IEEE Trans.Syst., Man, & Cybern., SMC-6, 85-94.

Gerwin, D. (1974). Information processing, data inferences, and scientific generalization. Behav. Sci., 19, 314-325.

Gerwin, D. and P. Newstad (1977). A comparison of some inductive inference models. Behav. Sci., 22, 1-11.

Glaser, R.D., E. Damrin, and F.M. Gardner (1954). The tab item: a technique for the measurement of proficiency in dianostic problem solving tasks. Educ. and Psych. Meas., 14, 283-293.

Goldbeck, R.A., B.B. Bernstein, W.A. Hillix, and M.H. Marx (1957). Application of the half-split technique to problem-solving tasks. J. Exp. Psych., 53, 330-338.

Greenstein, J.S., and W.B. Rouse (1982). A model of human decision making in multiple process monitoring situations. IEEE Trans. Syst., Man, & Cybern., SMC-12.

Gregg, L. W. and H.A. Simon (1967). Process models and stochastic models of simple concept formulation. J. of Exp. Psych., 4, 246-276.

Hayes, J.R. (1973). On the function of visual imagery in elementary mathematics. In W.G. Chase (Ed.), Visual Information Processing., New York, Academic Press.

Hayes-Roth, B., and F. Hayes-Roth (1979). A cognitive model of planning. Cog. Sci., 3, 275-310.

Henneman, R.L., and W.B. Rouse (1982). Measures of human performance in fault diagnosis tasks. Submitted for publication.

Hormann, A.M. (1971). A man-machine synergistic approach to planning and creative problem solving:I. Intl. J. Man-Mach. Stud., 3, 167-184.

Huesman, L.R., and C-M. Cheng (1973). A theory for induction of mathematical functions. Psych. Rev., 80, 126-138.

Hunt, R.M., and W.B. Rouse (1982). A fuzzy rule-based model of human problem solving. Proc. Am. Control Conf., Washington, June.

Jacoby, L.L. (1978). On interpreting the effects of repetition: solving a problem versus remembering a solution. J. of Verb. Learn. and Verb. Behav., 17, 649-667.

Jeffries, R., P.G. Polson, L. Razran, and M.E. Atwood (1977). A process model for missionaries-cannibals and other river-crossing problems. Cog. Psych., 9. 412-440.

Johannsen, G., and W. B. Rouse (1979). Mathematical concepts for modeling human behavior in complex and man-machine systems. Hum. Fac., 21, 733-747.

Johannsen, G. and W.B. Rouse (1980). A study of the planning process of aircraft pilots in emergency and abnormal situations. Proc. 1980 IEEE Intl. Conf. on Cybernetics and Society, Boston, October, pp.738-745.

Johannsen, G., and W.B. Rouse (1981). Problem solving behavior of pilots in abnormal and emergency situations. Proc. First Ann. European Conf. on Human Decision Making and Manual Control, Delft, The Netherlands, May, pp.142-150.

Johnson, E.S. (1978). Validation of concept learning strategies. J. of Exp. Psych., 107, 237-266.

Johnson S.L. (1981). Effect of training device of rentention and transfer of a procedural task. Hum. Fac., 23, 257-272.

Johnson, W.B., and W.B. Rouse (1982a). Analysis and classification of human errors in troubleshooting live aircraft powerplants. IEEE Trans. Syst., Man,& Cybern., SMC-12.

Johnson, W.B., and W.B. Rouse (1982b). Training maintenance technicians for troubleshooting: two experiments with computer simulations. Hum. Fac., 24.

Kearsley, G.P. (1975). Problem-solving set and functional fixedness: a conceptual dependency hypothesis. Canadian Psych. Rev., 16, 261-268.

Kozielecki, J. (1972). A model for dianostic problem solving. ACTA Psych., 36, 370-380.

Landa, S. (1972). CATTS: Computer-Aided Training in Toubleshooting , Rand Corporation, Santa Monica, CA, Report No. R-518-PR, May.

Loftus, E.F., and P. Suppes (1972). Structural variables that determine problem-solving difficulty in computer-assisted instruction. J. of Ed. Psych., 63, 531-542.

Luger, G.F., and M.A. Bauer (1978). Transfer effects in isomorphic problem situations. ACTA Psych. 42, 121-131.

Malin, J.T. (1979). Information processing load in problem solving by network search. J. of Exp. Psych., 5, 379-390.

Mason, E.J., W.J. Bramble, and T.A. Mast (1975). Familiarity with context and conditional reasoning. J. of Ed. Psych., 67, 238-242.

Mayer, R.E. (1975). Different problem-solving competencies established in learning computer programming with and without meaningful models. J. of Ed. Psych., 67, 725-734.

Miller, D.C., and J.I. Elkind (1967). The adaptive responses of the human controller to sudden changes in controlled process dynamics. IEEE Trans. Hum. Factors in Electron., HFE-8, 218-223.

Mills, R.G. (1971). Probability processing and diagnostic search: 20 alternatives, 500 trials. Psychonomic Sci., 24, 289-292.

Minsky, M. (1975). A framwork for representing knowldege. In P.H. Winston (Ed.), The Psychology of Computer Vision., New York, McGraw-Hill.

Neimark, E.D., and J.L. Santa (1975). Thinking and concept attainment. Ann. Rev. of Psych., 26, 173-205.

Nuebauer, H.N. and W.B. Rouse (1979). A study of the use of color graphics for representing schematic information. Proc. of the Fifteenth Ann. Conf. on Man. Control., Wright State University, March, pp.624-631.

Newell, A., and H.A. Simon (1972). Human Problem Solving., Englewood Cliffs, New Jersey, Prentice-Hall.

Niemela, R.J., and E.S. Krendel (1975). Detection of a change in plant dynamics in a man-machine system. IEEE Trans. Syst., Man & Cybern., SMC-5, 615-617.

Nievergelt, J. (1977). Information content of chess positions. SIGART Newsletter., 62, 13-15.

Peters, J.T., K.R. Hammond, and D.A. Summers (1974). A note on intuitive vs. analytical thinking. Organ. Behav. and Hum. Perform., 12, 125-131.

Phatak, A.V., and G.A.Bekey (1969). Decision processes in the adaptive behavior of human controllers. IEEE Trans. Syst. Sci. & Cybern., SSC-5, 333-351.

Polich, J.M., and S.H. Schwartz (1974). The effect of problem size on representation in deductive problem solving. Mem. and Cog., 2, 683-686.

Rasmussen, J. (1978). Notes on Diagnostic Strategies in Process Plant Environment. Riso National Laboratory, Roskilde, Denmark, Report M-1983, January.

Rasmussen, J. (1979). On the Structure of Knowledge - A Morphology of Mental Modes in a Man-Machine Context. Riso National Laboratory, Roskilde, Denmark, Report No. M-2192, November.

Rasmussen, J. (1981) Models of mental strategies in process plant diagnosis. In J. Rasmussen and W.B. Rouse (Eds.), Human Detection and Diagnosis of System Failures, New York, Plenum Press, pp.241-258.

Rasmussen, J., and A. Jensen (1974). Mental procedures in real life tasks: a case study of electronic troubleshooting. Ergonomics, 17, 293-307.

Rasmussen, J., and W.B. Rouse (Eds.), (1981). Human Detection and Diagnosis of System Failures, Plenum Press, New York.

Rouse, S.H., W.B. Rouse, and J.M. Hammer (1982). Design and evaluation of an onboard computer-based information system for aircraft. IEEE Trans. Syst., Man, & Cybern., SMC-12.

Rouse, W.B. (1978). A model of human decision making in a fault diagnosis task. IEEE Trans. Syst., Man & Cybern., SMC-8, 357-361.

Rouse, W.B. (1979). A model of human decision making in fault diagnosis tasks that include feedback and redundancy. IEEE Trans. Syst., Man & Cybern., SMC-9, 237-241.

Rouse, W.B. (1980). Systems Engineering Models of Human-Machine Interaction, North-Holland, New York.

Rouse, W.B. (1981a). Human-computer interaction in the control of dynamic systems. Computing Survey, 12, 71-99.

Rouse, W.B. (1981b). Experimental studies and mathematical models of human problem solving performance in fault diagnosis tasks. In J. Rasmussen and W.B. Rouse (Eds.), Human Detection and Diagnosis of System Failures, New York, Plenum Press, pp.199-216.

Rouse, W.B. (1982). A mixed-fidelity approach to technical training. J. of Ed. Tech. Sys., 11.

Rouse, W.B., and R.M. Hunt (1981). A fuzzy rule-based model of human problem solving in fault diagnosis tasks. Proc. Eighth World Congress of IFAC, Kyoto, August, pp. XV 95-101.

Rouse, W.B., and N.M. Morris (1981a). Human problem solving in a process control task. Proc. First Ann. European Conf. of Human Decision Making and Manual Control, Delft, The Netherlands, May, pp. 218-226.

Rouse, W.B., and N.M. Morris (1981b). Studies of human problem solving in a process control task. Proc. 1981 IEEE Intl. Conf. on Cybernetics and Society, Atlanta, October, pp. 479-487.

Rouse, W.B., and S.H. Rouse (1979). Measures of complexity of fault diagnosis tasks. IEEE Trans. Syst., Man & Cybern., SMC-9, 720-727.

Rouse, W.B., S.H. Rouse, and S.J. Pellegrino (1980). A rule-based model of human problem solving performance in fault diagnosis tasks. IEEE Trans. Syst., Man & Cybern., SMC-10, 366-376.

Rumelhart, D.E., and A.A. Abrahamson (1973). "A model of analogical reasoning. Cog. Psych., 5, 1-28.

Sacerdoti, E.D. (1974). Planning in a hierarchy of abstraction spaces. Art.Intell., 5, 115-135.

Schank, R.C., and R.P. Abelson (1977). Scripts, Plans, Goals, and Understanding., Hillsdale, New Jersey, Lawrence Erlbaum.

Shepard, A., E.C. Marshall, A. Turner, and K.D. Duncan (1977). Diagnosis of plant failures from a control panel: a comparison of three training methods. Ergonomics., 20, 347-361.

Sheridan, T.B., and W.R. Ferrell (1974). Man-Machine Systems: Information, Control, and Decision Models of Human Performance, MIT Press, Cambridge, MA.

Sheridan, T.B., and G. Johannsen (Eds.), (1976). Monitoring Behavior and Supervisory Control, Plenum Press, New York.

Siegler, R.S. (1977). The twenty questions game as a form of problem solving. Child Dev., 48, 395-403.

Simon, H.A. (1975). The functional equivalence of problem solving skills. Cog. Psych., 7, 268-288.

Simon H.A., and S.K. Reed (1976). Modeling strategy shifts in a problem solving task. Cog. Psych., 8, 86-97.

Simonton, D.K. (1975). Creativity, task complexity, and intuitive versus analytical problem solving. Psych. Rep., 37, 351-354.

Smith, J.P. (1973). The Effect of General versus Specific Heuristics in Mathematical Problem Solving, Columbia University, Ph.D. Thesis.

Sternberg, R.J. (1977). Component processes in analogical reasoning. Psych. Rev., 84, 353-378.

Sternberg, R.J., and E.M. Weil (1980). An aptitude-strategy interaction in linear syllogistic reasoning. J. of Ed. Psych., 72, 226-239.

Stolurow, L.M., B. Bergum, T. Hodgson, and J. Silva (1955). The efficient course of troubleshooting as a joint function of probability and cost. Educ. & Psych. Meas., 15, 463-477.

Towne, D. M. (1981). A general purpose system for simulating and training complex diagnosis and troubleshooting tasks. In J. Rasmussen and W. B. Rouse (Eds.), Human Detection and Diagnosis of System Failures., New York, Plenum Press, pp. 621-635.

van Eekhout, J.M., and W.B. Rouse, (1981). Human errors in detection, diagnosis, and compensation for failures in the engine control room of a supertanker. *IEEE Trans. Syst., Man, & Cybern.*, SMC-11, 813-816.

Wason, P.C., and P.N. Johnson-Laird (1972). *The Psychology of Reasoning.*, Cambridge, Massachusetts, Harvard University Press.

Waterman, D.A., and F. Hayes-Roth (1978). (Eds.), *Pattern-Directed Inference Systems.*, New York, Academic Press.

Wickens, C.D., and C.Kessel (1981). Failure detection in dynamic systems. In J. Rasmussen and W. B. Rouse (Eds.), *Human Detection and Diagnosis of System Failures*, New York, Plenum Press, pp. 155-169.

Wohl, J.G. (1981). System complexity, diagnostic behavior, and repair time: a predictive theory. In J. Rasmussen and W. B. Rouse (Eds.), *Human Detection and Diagnosis of System Failures*, New York, Plenum Press, pp. 217-230.

Wood, D., J. Shotter, and D. Godden (1974). An investigation of the relationships between problem solving strategies, representation, and memory. *Quarterly J. of Exp. Psych.*, 26, 252-257.

A METRIC FOR PROBLEM SOLVING IN MAN-MACHINE SYSTEMS

J. B. Brooke and K. D. Duncan

Department of Applied Psychology, University of Wales Institute of Science and Technology, Cardiff, UK

Abstract.   The function of man in man-machine systems is increasingly
that of problem-solver. The role of fault-finder is a case in point. The
extent to which training can provide problem-solving skill remains a comparatively neglected topic. Nor should one neglect the classical question
of how the problem is represented in the first place. The performance
differences attributable to different representations of a fault finding
problem are often subtle and require measures which go beyond time or
correctness of initial diagnosis. Examples of possible measures are
redundant checks or premature diagnoses. Some aspects of problem solving
quality, in these terms, are more likely to be improved by display formats
(e.g. reduced redundant testing) whereas others, in particular premature
diagnosis, are only substantially affected by training. Two sorts of
training interact quite powerfully. One is providing the problem solver
with information about his redundant tests and premature diagnosis during
problem solving; the other training intervention is to provide pretraining
in fault finding strategies. Both these training methods produce slight
improvements separately. In combination, they seem to be disproportionately effective. It is as if the subject can only make effective use of
rich feedback about his problem solving moves after first gaining insights
into the overall structure of the problem. These approaches to problem
solving are not new, but well specified techniques are lacking. There are
considerable technical problems in capturing the quality of problem solving
which is taking place, and to present it in such a way as to influence the
problem solver. The technical difficulty is two-fold: any metric must be
a) defensible in terms of information measurement, and b) intelligible to the
subject. Finally, measures may be context independent or specific to a
particular problem.

Keywords. Display systems; ergonomics; human factors; fault finding;
system failure; training.

INTRODUCTION

The function of man in a man-machine system is
increasingly that of problem-solver. The role
of fault-finder is a case in point. The extent to which training can provide problem-
solving skill remains a comparatively neglected topic. There is also the crucial
question of how the problem is represented
in the first place. Thus the development of
adequate measures of human problem solving
are necessary for two important purposes.

1. assessment of ergonomics, including job
    aids and procedural guides, i.e., the
    representation of the problem; and

2. the assessment of any differences
    between operators and of any improve-
    ments attributable to training.

Although our concern as a research group has
been primarily with the last of these, the
need for adequate measurement of problem
solving skill in other human factors applications is insufficiently recognised, and
research is lacking.

We are also concerned about the dearth of
empirical work (compared with the proliferation of models of human problem solving)
and that any empirical work should make use
of measures which are accurate and sensitive
to the problem solving process. Also, because we have reservations about the status
of most theoretical models of human performance, it seems to us crucial that any
measures of problem solving are "clean" in
the sense that they are independent of
theoretical context.

PRODUCT VERSUS PROCESS MEASURES

'Product' measures of diagnostic performance such as accuracy and diagnosis time are obviously important. However, they are relatively insensitive taken by themselves to differences in individual performance and could probably only serve as criteria when data for very large groups of people are available. In any case, they offer few insights into how correct or incorrect solutions are reached, and provide little basis for new display designs, training schemes or individual appraisal.

We first approached this problem of measuring the problem solving process by deriving indices based on the notion of "consistent fault set" (CFS). Three measures we have suggested in the past (Duncan and Gray, 1975) based on this notion are:-

- Premature Diagnosis or diagnoses attempted when the CFS contains more than one fault.

- Redundant tests or tests which do not reduce the size of the CFS when the size of the CFS is greater than one.

- Extra tests or tests made after the CFS has been reduced to one.

These measures are independent of psychological theories, being direct derivatives of the intrinsic logic of fault finding problems. We have applied them in various circumstances. For instance, the performance of novices differs from that of experienced refinery workers. Also, performance measured in these terms improves between tests before and tests following training courses (Duncan and Gray, 1975).

A more direct use of these measures in training is to feed them directly back to the trainee whilst he is attempting to solve the problem. This adds a further requirement for such measures, namely that they must be intelligible to the trainee. We have had some success in using graphic representations of these process measures in computer aided instruction; trainees learned more quickly when receiving computer aiding, an effect which persisted, and more quickly learned to solve problems in a different context (Brooke, Duncan and Cooper, 1980).

We are all too aware that more work to develop measures of the problem solving process is needed, especially when these measures are to be fed back to the subject during training. CFS based measures may have disadvantages to set against the advantage that they do not rely on any psychological hypothesis. We have been faced with an instance peculiar to a particular fault finding task where the CFS measure is unhelpful to the subject, if not actually misleading. The task in question is the identification of a fault in a network of interconnected logic units as used by Rouse and his associates and by ourselves (Rouse, 1981; Brooke and Duncan, 1981).

It is fairly common for subjects, at an early stage in these problems, to test a unit which is the 'centre of propagation'. See Figure 1. We have coined this expression to refer to the member of the CFS which is functionally nearest to the outputs of the network. Testing the signal at this unit does not reduce the CFS, and is therefore redundant. However, it is very sensible to make such a test if the trainee wishes to confirm that the nature of the problem has been correctly understood. Whether this is in fact what the subject is doing is nevertheless a matter for speculation, albeit reasonable. Unfortunately reasonable speculations are not always as defensible as measures which are intrinsic to the logic of the problem.

By definition, the CFS measures are not task-specific and we are only now beginning to explore the relative merits of task-specific process information. In the network problems just referred to, various assessments of subjects' testing behaviour suggest themselves, for example:-

- That unit must be OK, because it's connected to a working output.

- You know that unit is sending a bad signal because it's receiving a bad signal.

We will return to the training problem, and in particular, to the use of this more task-specific process information. But first there are general questions of problem representation to consider.

REPRESENTATION OF THE PROBLEM

The progress of problem solution and of the human's attempts to acquire problem solving technique through training or work experience is inevitably influenced by the way in which the problem is represented. Format and layout of job aids and diagrams may influence subsequent problem solving behaviour, as may the variety of instruction or explanation provided before the task is attempted.

Brooke (1982) has reviewed a number of studies of the effects of display format on fault finding performance. By and large, the only effects detectable in terms of product measures of performance are in time taken to solve problems and this can be seen to be attributable to changes in process measures of performance. This is easily understood since reduced testing of the faulty system decreases solution time. In general, improvement of display format leads both to reductions in the rate of redundant

testing and to improvements in the utilisation of positive information; the user is
apparently alerted to the relevance of particular parts of the system to the problem in
hand.   On the other hand, whilst improvements
in display format enhance perception of those
areas where tests are required, they do not
lead to better interpretation of information;
just as many diagnoses are made incorrectly
and with inadequate test information despite
improvements in display format.

In most situations, besides display format
there will usually be other influences on
how the problem is represented in the form
of instructions, perhaps by a superior, or
perhaps by an instructor as to what the task
is all about.   We suspect that such instructions, however casual or unsystematic, may
profoundly affect subsequent problem solving
performance.   This led us to consider
explicit carefully controlled pretraining in
the identification of the basic configurations
which go to make up network problems,
and appropriate strategies for dealing with
them.

PRETRAINING AND CONCURRENT PROCESS INFORMATION

One of our recent experiments has studied the
effect of combining process information
provided during problem solving with the type
of pretraining suggested above.

Rouse (1981) describes several studies of
computer aiding for fault finders using a
network task similar to that illustrated in
the Figure.   These studies achieved some
success in improving fault finding performance,
although this depended on the type of computer aiding used.   When Rouse used a
'bookkeeping' form of aiding, where the computer indicated which nodes of the network
were relevant to the problem, positive
transfer of training was found when subjects
later performed the task unaided.   On the
other hand, computer aiding relating to a
rule-based model of human diagnostic behaviour
led to negative transfer of training.   In this
form of training, a computer program examined
each test made, and identified which of a
number of rules in the model was probably
being used.   These rules had previously
been rank-ordered as 'excellent', 'good',
'fair' or 'poor' and this rating was fed back
to the subject.   This raises questions about
the value of the rules and their rank ordering,
but an equally important factor may have been
that the computer aiding did not tell the
subjects why a test was, for example,
"excellent" or "poor".

Brooke and Duncan (1982) have described
another form of computer aiding for this task.
In this program, each test is examined and
explicit information is given to the user
about how good a test it is in terms of
various process measures.   If a test is
redundant, an explanation is given which is

couched in terms of the logic units and the
connections between them, for instance:
"There is no point testing this unit, as you
have previously tested   unit X which sends
a signal to this one, and unit X was not
working either".   Other examples of this
task specific information have been given
above.

Besides information about the quality of
tests, the program also gives explicit
information when the user makes a premature
diagnosis.   He may be told, for instance,
that he ought not to have made a diagnosis
until he had checked all the inputs to a
particular unit.

Brooke, Cook and Duncan (1982) have investigated the effects of this aiding, both alone
and in conjunction with pretraining in fault
finding strategies.   The pretraining comprised instruction in the 'half-split' and
'bracketing' techniques (Myers, Carter and
Stover, 1964) and a discussion of the interpretation of output symptoms.   Four groups
of fault finders received the following
training combinations:

1. No pretraining/no computer aiding
2. No pretraining/computer aiding
3. Pretraining/no computer aiding
4. Pretraining/computer aiding

All groups solved a total of 32 problems;
the groups receiving computer aiding only
did so during the first 16 problems.

Most of the measures of performance used in
the study showed improvement with practice
for all four groups.   However, the group
receiving no training at all did not improve
in terms of accuracy of diagnosis.   Strategy
pretraining led to an immediate improvement
in accuracy;   computer aiding led to improved
accuracy after practice.

Strategy pretraining was most effective in
reducing the rate of errors where subjects
failed to take account of symptom information
(rather than information collected during
testing) and also reduced the rate of
certain types of premature diagnosis.

Computer aiding helped in reducing the rate
of redundant testing where subjects failed
to take account of information collected
during testing, as it might be expected to.
Furthermore, it was effective in reducing
the rate of premature diagnoses.

The combined effects of strategy pretraining
and computer aiding were significant in the
speedy reduction of redundant testing.
Either type of training on its own led to
reductions only after longer practice.

CONCLUSIONS

In general, improvement of display format leads both to reductions in the rate of redundant testing and to improvements in the utilisation of positive information. The user is apparently alerted to the relevance of particular parts of the system to the problem in hand.

Training is effective in reducing the performance deficits that exist even with improved display formats. However, a more important point is that the display format employed in the training study was the format previously found to be most effective in reducing process errors (Brooke and Duncan, 1981). Training therefore may improve performance beyond the levels achieved through better display ergonomics. On the other hand, we might speculate on whether the training would have been so effective had the trainees had to work with a bad display format; this is a question which deserves further investigation.

It remains for further research to determine the relative advantages of process information or measures which are specific to a problem compared with context independent measures such as those based on the CFS.

REFERENCES

Brooke, J.B. (1982). The Relative Contributions of Ergonomics and Training to Displays for Fault Finding. Proceedings of IEE Conference on "Man-Machine Systems".   UMIST, Manchester.

Brooke, J.B., and K.D. Duncan (1981). The Effects of System Display Format on Performance in a Fault Location Task. Ergonomics, 24, 175-189.

Brooke, J.B., and K.D. Duncan (1982). "Computer Aiding in Solving Fault Finding Problems". Department of Applied Psychology, UWIST, Cardiff.

Brooke, J.B., J.F. Cook, and K.D. Duncan (1982). "The Effects of Computer Aiding and Pretraining on a Fault Location Task". Department of Applied Psychology, UWIST, Cardiff.

Brooke, J.B., K.D. Duncan, and C. Cooper (1980). Interactive Instruction in Solving Fault Finding Problems - An Experimental Study. International Journal of Man-Machine Studies, 12 217-227.

Duncan, K.D., and M.J. Gray (1975). An Evaluation of a Fault Finding Training Course for Refinery Process Operators. Journal of Occupational Psychology, 48, 199-218.

Rouse, W.B. (1981). Experimental Studies and Mathematical Models of Human Problem Solving Performance in Fault Diagnosis Tasks. In: Rasmussen, J. and Rouse, W.B., (Eds) Human Detection and Diagnosis of System Failures, Plenum Press, New York.

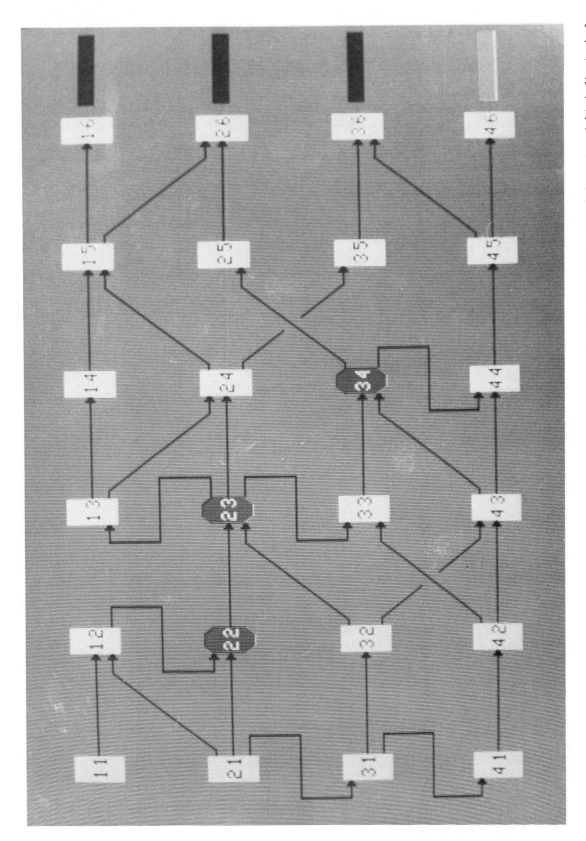

Figure 1   6 x 4 logic network.   Rectangular units are AND gates, octagonal units OR gates.   Black bars to right indicate bad final outputs of network;   the 'centre of propagation' given this symptom array is unit 24.

MODEL OF THE HUMAN OBSERVER AND DECISION MAKER — THEORY AND VALIDATION

P. H. Wewerinke

National Aerospace Laboratory NLR, Amsterdam, The Netherlands

Abstract. A model of the human decision maker observing a dynamic system is
presented. The decision process is described in terms of classical sequential
decision theory by considering the hypothesis that an abnormal condition has
occurred by means of a generalized likelihood ratio test. For this, a sufficient statistic is provided by the innovation sequence which is the result
of the perception and information processing submodel of the human observer.
On the basis of only two model parameters the model predicts the decision
speed/accuracy trade-off and various attentional characteristics.

A preliminary test of the model for single variable failure detection tasks
resulted in a very good fit of the experimental data.
In a formal validation program a variety of multivariable failure detection
tasks was investigated. A very good overall agreement between the model and
experimental results showed the predictive capability of the model.
In addition, the specific effect of almost all task variables (number, bandwidth and mutual correlation of display variables and various failure characteristics) was accurately predicted by the model.

Keywords. Human modelling; dynamic systems; estimation theory; decision
theory; human decision making; failure detection.

INTRODUCTION

Accident statistics and analyses clearly
indicate the necessity to consider more
explicitly the role of the human operator in
man-machine systems. Especially the increase
in technological capability is attended by
more complex and more automated systems. Yet
one rarely accounts (in the design stage)
for the inherent human capabilities and limitations (Moray, 1980).
Indeed, the last two decades considerable
research effort has been devoted to the study
of human control behavior. However, the
insight in higher mental functioning involved
in monitoring an automatic system, detecting
system failures, making decisions, etc., is
still rather incomplete. Fortunately, stochastic optimal estimation and control theory
has been shown to provide a general framework
adequately describing the continuous human
processing of information. Originally, most
of the research was related to manual control
(Kleinman and Baron, 1971; Baron and Levison,
1977; Wewerinke, 1977 and 1980), although
some attempts have been made to address
failure detection and simple decision making
behavior (Levison and Tanner, 1971;
Wewerinke, 1976 and Gai and Curry, 1976 and
1977).

In connection with this previous research a
theoretical and experimental study has been

performed of the human observer and decision
maker in multivariable failure detection
tasks. For an extensive presentation of the
research the reader is referred to Wewerinke
(1981) and v.d. Graaff (1981). This paper
summarizes the principal results.
In the next section a model of the human
decision maker is formulated in terms of
multivariable classical sequential decision
theory, accounting for the important effect
of correlated information. In the subsequent
section the model is tested against the
results of a single variable task experiment
reported by Gai and Curry (1976) and against
the results of a formal model validation
program.

MODEL OF THE HUMAN OBSERVER AND DECISION MAKER

Human Observer

It is assumed that the human observer (HO)
deals with a linear dynamic system which can
be represented by a Gauss-Markov random
sequence (Bryson and Ho, 1969)

$$x_k = \Phi_{k-1} x_{k-1} + \Gamma_{k-1} w_{k-1} \qquad (1)$$

where x_k is the vector of state variables at
time instant k, w_{k-1} is the vector of linear

independent, zero mean, gaussian, purely random sequences with covariances W_{k-1}, Φ_{k-1} is
the state transition matrix and Γ_{k-1} is the
noise distribution matrix.
The HO derives information y from this process, which is assumed to be linearly related to the state x

$$y_k = H_k x_k + v_k \qquad (2)$$

where H_k is the observation matrix, v_k represents the vector of linear independent,
gaussian purely random observation noise
sequences with covariances V_k; observational
delays are considered to be negligible.

Based on the known (learned) dynamics of the
system and the perceived information up to
time k the HO makes the best estimate, $\hat{x}_{k/k}$,
or simply $\hat{x}_k$, of the system state. This
central information processing stage is described in standard linear estimation theoretical terms (Kalman filter)

$$\hat{x}_k = \Phi_{k-1} \hat{x}_{k-1} + K_k n_k \qquad (3)$$

where K_k is the Kalman filter gain which is
essentially the ratio between the uncertainty
in the system state (in terms of the estimation error covariance P) and the reliability
of the incoming information, the so-called
innovation sequence n_k (in terms of its covariance N_k), where

$$n_k = y_k - H_k \Phi_{k-1} \hat{x}_{k-1} \qquad (4a)$$

and

$$N_k = H_k P_{k/k-1} H_k' + V_k. \qquad (4b)$$

Human Decision Maker

Now the statistics of this innovation sequence, which is in the normal mode (i.e.
when the internal model of the dynamic system
is correct) a zero mean, gaussian, purely
random sequence (Mehra and Peschon, 1971),
are utilized to decide about possible abnormal operation of the dynamic system. It is
assumed that abnormal system operation, as
caused by errors in instruments, malfunc-
tioning of the system and excessive system
disturbance levels (e.g., large windshears
in aircraft operation), can be represented by
a deterministic process. This process is not
explicitly known to the HO but detected on the
basis of a non-zero mean innovation sequence
whose statistic is sufficient to make decisions (test hypotheses) when the system is
completely observable.

This decision making process is mathematically
formulated in terms of classical sequential
decision theory (Wald, 1947; Sage and Melsa,
1971) by means of a generalized likelihood
ratio test. This test amounts to the comparison of the probability of a non-zero mean
with the probability of a zero mean innovation sequence. Herewith it is assumed that the

HO makes a short-term estimate of the mean of
the innovation sequence on the basis of the
sample mean of m past observations, $\tilde{n}_k$.
It can be derived (Wewerinke, 1981) that the
effect of each observation at stage k on the
(log of the) likelihood ratio is given by

$$\Delta L_k = \tfrac{1}{2} \tilde{n}_k' N_k^{-1} \tilde{n}_k \qquad (5)$$

under the assumption that the sample mean $\tilde{n}_k$
is constant during m observations. This short-
term operation can be related to the short-
term (working) memory functioning of the HO
for an appropriate value of m. The accumula-
ting effect of each observation on the total
(log of the) likelihood ratio can be related
to the long-term, permanent memory functioning
and is given by the recursive expression

$$L_k = L_{k-1} + \Delta L_k \qquad (6)$$

assuming that the innovation sequence is a
purely random sequence (independent samples)
which is exact in the normal mode of operation.
The number of observations, based on which
the decision is made, is chosen such that ΔL_k
is, on the average, not decreasing, to avoid
that there is, in the normal mode of operation, an accumulative effect of the positive
semi-definite elements on the likelihood
ratio. In other words, the HO uses only confirming evidence to make the decision.
When the likelihood ratio L_k (representing
the total evidence of abnormal system operation) is equal to, or larger than, (the log
of) a decision threshold T, the decision is
made that an abnormal condition has occurred.
This decision threshold can conveniently be
related to the accepted (or assumed) risk
according to (Sage and Melsa, 1971)
$T \doteq (1-P_M)/P_F$, with P_M the miss probability and
P_F the false alarm probability.

This decision accuracy is intimately related
to the time required to make the decision,
i.e. to detect the abnormal system operation.
The following expression is derived by
Wewerinke (1981) for the average number of
samples used to make the decision that the
system is operating abnormally, which is,
for a given sample rate, uniquely (linearly)
related to the average detection time

$$\bar{K} \doteq \frac{2 P_F T \ln T}{E\{\tilde{n}' N^{-1} \tilde{n}\}} \qquad (7)$$

where $(\bar{.})$ indicates the average over the
ensemble and $E\{(.)\}$ is the average over the
sequence.
Eq. (7) provides a relationship between the
average number of observations and the decision error probabilities, for a given innovation covariance N and the non-zero mean
failure sequence which is, however, a given
task variable. Thus the only human decision
model parameters are the short-term average
sample size m and the innovation covariance
which depends exclusively on the human observation noise covariance V_k (as indicated in
Eq. (4b)).

Optimal Allocation of Attention

Previous studies (Baron and Levison, 1977) support the hypothesis that the observation noise covariance scales with the mean-squared value of the corresponding signal. Thus for display variable j $V_j = P_o E\{y_j^2\}$. P_o represents a nominal, "full attention" noise ratio (typically 0.01π). In case a display variable is not looked at (foveally), it is assumed that the corresponding observation noise is infinite, thus neglecting peripheral viewing. Inserting this expression for V_j in Eq. (7) results, after some matrix manipulation (Wewerinke, 1981) in the following very simple expression for the average "detection time"

$$\overline{K} = \frac{2\,P_F\,T\,\ln T}{E_t\{\tilde{n}_e^2\}} \qquad (8)$$

where

$$\tilde{n}_{e_j}^2 = \tilde{n}_{r_j}^2 (1 - p_{r_j})$$

is the effective, relative, estimated sample mean-square; the subscript r indicates the normalization by the (constant) observation noise covariance V_j, p_j is the estimation error covariance of the display variable j and E_t indicates the average over the ensemble, and the sequence, and the display variables.

The expected value of $\tilde{n}_e^2$ over the display variables involves the probability distribution of the human attention to these variables. An optimal distribution of the human attention is assumed in the model, corresponding to a maximum average $\tilde{n}_e^2$ and minimal detection time. By Wewerinke (1981) it is shown that in the normal mode of operation $\tilde{n}_e^2$ is equal to the reduction in the normalized estimation error covariance of the display variables which is a measure for the total system uncertainty. So, the foregoing distribution of the human attention is also optimal in the more general sense of yielding the minimal total system uncertainty. This is in accordance with the afore-mentioned estimation process which is optimal in exactly the same sense (Baron and Levison, 1977).

The general model structure (the only assumptions are that the dynamic process is linear and that an abnormal condition can be represented by a deterministic process) accounts for the effect of a variety of task variables such as the number and bandwidth of display variables, the correlation among them and the failure characteristics.

On the basis of only two model parameters (short-term average sample size m and the overall level of attention P_o) the model predicts the effect of these interacting variables on the mean failure detection time (and corresponding decision error probabilities, P_F and P_M) and various attentional characteristics.

MODEL VALIDATION

The results of two experimental programs will be discussed and compared with the model predictions.

Preliminary Model Validation

The first experiment is reported in Gai and Curry (1976). The observers were required to detect the occurrence of step and ramp failures of various amplitudes which were superimposed upon a (second order) stochastic gaussian process. The experimental results in terms of the avarage failure detection times were considered for a preliminary validation of the human decision model and to "calibrate" the model with respect to the short-term average sample size m. Model results were obtained assuming the decision error probabilities of 0.05 which were also obtained in the experiment and on the basis of a typical overall level of attention P_o of 0.01π. The remaining model parameter m was selected so as to obtain the best overall match with the experimental results.

The resulting value of m of four seconds yields an excellent fit to the experimental data as shown in Fig. 1. For more details the reader is referred to Wewerinke, 1981.

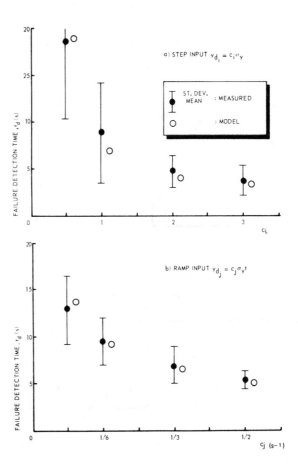

Fig. 1 Measured and model failure detection times for various step and ramp inputs

This value of m of four seconds yielding a
good fit to all experimental conditions seems
a human operator-related parameter. The same
value of four seconds, which ties in very
well with the short-term memory span typically
ranging up to five seconds for visual stimuli
(Sheridan and Ferrell, 1974), was assumed and
kept constant in the formal validation experiment.

Formal Model Validation

The formal validation program which is extensively described in Wewerinke (1981) and in
v.d. Graaff (1981) comprised a variety of
multivariable failure detection tasks. The
task variables were the number, the bandwidth
and the mutual correlation of display variables and failure characteristics (prior
knowledge, correlation and magnitude). The
results of the 16 experimental conditions
(in terms of failure detection times and
scanning measures) were compared with the
model predictions. The latter were obtained
on the basis of the two (constant) model
parameters: the overall level of attention
$P_O = 0.01\pi$ and the short-term average sample
size m = 4 seconds.

The failure detection times of two pilots
achieving an overall false alarm probability
P_F of 0.05 could be compared directly with
the model predictions. The result which is
shown in Fig. 2 reflects a very good overall predictive capability of the human
decision model (r = 0.86). A very plausible
explanation was found (Wewerinke, 1981) for
the clear discrepancy for one configuration.
The model result based on the alternative
assumption concerning the expected failure
type (refinement) is also indicated in
Fig. 2. In that case the linear correlation
coefficient is 0.95

The agreement between the model predictions
and experimental results with respect to the
effect of the specific task variables is
shown in Fig. 3.

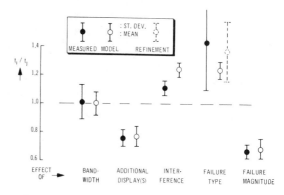

Fig. 3 Specific effect of task variables
         on failure detection times –
         comparison of model and experi-
         mental results

The effect of display bandwidth, of additional (correlated) displays and of the
failure characteristics (by pair-wise comparing the failure detection times t_i and
t_j) is excellently predicted by the model.
The predicted interference between uncorrelated displays is larger than obtained experimentally. However, these predictions are
based on a constant level of attention, while
heart rate measures which were obtained in
the experiment indicate an increase in attention with an increase in displays which can
explain the small difference in interference.
The experimental results of the third subject
reflect a distinctly different decision
strategy (Wewerinke, 1981). Yet his results
correlated rather well with the model predictions (r = 0.80).

The attention allocation model allowed various model predictions which could be compared with the experimental eye scanning data.
As shown in Wewerinke (1981), the predicted
optimal division of attention among displays
agreed very well with the experimental dwell
fractions. Also qualitative predictions concerning the optimal allocation of attention
in the time domain (scanning strategy)
agreed well with the experimental eye scanning data in terms of display link and scan
values.
This agreement illustrates the predictive
capability of the attention allocation model
which may be a powerful tool in the study of
human information processing tasks and display design problems.

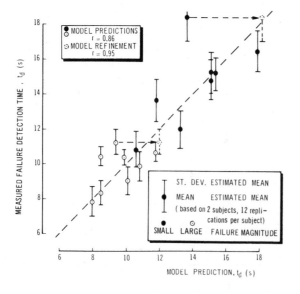

Fig. 2 Overall comparison of model pre-
         dictions and experimental
         detection times

CONCLUDING REMARKS

A model of the human observer and decision maker is formulated and experimentally validated for a variety of failure detection tasks. The model describes the human detection of abnormal operation of a multivariable system. On the basis of only two model parameters the model predicts mean detection times (for a given decision error probability) and various attentional characteristics.

The agreement between the results of a preliminary validation experiment and a formal experimental program and the model predictions shows the general structure and the predictive capability of the human observer and decision making model. It provides a meaningful description of the most important stages of perception, central information processing and decision making, at least for the type of monitoring tasks investigated so far.
Although additional model validation will be required for more complex monitoring tasks, the model will be useful to address many (e.g., display-related) questions concerning the design and evaluation of automated systems monitored by the human operator.

REFERENCES

Baron, S., and W.H. Levison (1977).
    Display analysis with the optimal control model of the human operator.
    Human Factors, 19(5).
Bryson, A.E., and Y.C. Ho (1969).
    Applied optimal control. Blaisdell Publishing Company.
Gai, E.G., and R.E. Curry (1976).
    A model of the human observer in failure detection tasks. IEEE trans. on systems, man and cybernetics. Vol. SMC-6 no. 2.
Gai, E.G., and R.E. Curry (1977).
    Failure detection by pilots during automatic landing: models and experiments.
    J. Aircraft, Vol. 14, no. 2.
Graaff, R.C. v.d. (1981).
    An experimental analysis of human monitoring behavior in multivariable failure detection tasks. NLR TR 81063 U.
Kleinman, D.L., and S. Baron (1971).
    Manned vehicle system analysis by means of modern control theory. NASA CR-1753.
Levison, W.H., and R.B. Tanner (1971).
    A control-theory model for human decision making. NASA CR-1953.
Mehra, R.K., and J. Peschon (1971).
    An innovation approach to fault detection and diagnosis in dynamic systems.
    Automatica, Vol. 7.
Moray, N. (1980).
    The role of attention in the detection of errors and the diagnosis of failures in man-machine systems. Nato Symposium on Human detection and diagnosis of system failures. Roskile, Denmark.
Sage, A.P., and J.L. Melsa (1971).
    Estimation theory with applications to communications and control.
    McGraw-Hill.
Sheridan, T.B., and W.R. Ferrell (1974).
    Man-machine systems: Information, control, and decision models of human performance. MIT Press, Cambridge.
Wald, A. (1947).
    Sequential analysis. J. Wiley, New York.
Wewerinke, P.H. (1976).
    Human monitoring and control behaviour
    - models and experiments. NLR TR 77010 U.
Wewerinke, P.H. (1977).
    Performance and workload analysis of in-flight helicopter missions.
    NLR MP 77013 U.
Wewerinke, P.H. (1980).
    The effect of visual information on the manual approach and landing.
    Paper presented at the 16th Annual Conference on Manual Control.
    MIT, Cambridge USA. (also NLR MP 80019 U).
Wewerinke, P.H. (1981).
    A model of the human decision maker observing a dynamic system.
    NLR TR 81062 U.

ON-BOARD FLIGHT PATH PLANNING AS A NEW JOB CONCEPT FOR PILOTS

P. Sundermeyer and L. Haack-Vörsmann

Institut für Flugführung, Technische Universität Braunschweig, Hans-Sommer-Straße 66, 3300 Braunschweig, Federal Republic of Germany

Abstract. This paper presents reasons and examples for increasing automation in flight guidance as well as resulting problem areas. Special attention is given to the change of the pilot's role. A novel concept of task division between pilot and air traffic control is introduced concerning the planning of the flight path in the terminal area. This concept, which is based on comparing the advantages and disadvantages of automation at different hierarchical levels in a flight guidance system, enables the pilot to plan the approach path himself with the help of a guidance computer.

A questionnaire was developed, which besides specific biographical data comprises the cognitive elements of the pilot's activity, which in particular are submitted to changes in the scope of automation. For the answering a "radar vectoring" approach with a certain degree of system automation is assumed on one hand, on the other hand the described possibility of flight path planning by the pilot is provided.

Statistical methods for evaluating the questionnaire are presented. The discussion of results assesses primarily the means of on-board flight path planning. Different dependencies between the judging of the pilot's activity according to the approach condition and biographical circumstances have been discovered.

Keywords. Air traffic control; automation; display systems; ergonomics; human factors; man-machine-systems; navigation.

INTRODUCTION

Modern microprocessors and recent progresses in display technology allow to automate functions in the aircraft cockpit which formerly were manually controlled. Not just simple control- and monitor-tasks are handled by the computer, but also complex flight management algorithms. In spite of all resulting and undisputed advantages, serious concern already exists about the implications of comprehensive automation on factors like pilot workload, performance, and air safety. This anxiety has been stirred up by reports about accidents and emergencies caused by automation and by simulator studies about "human factors", which deal with error recognition, response to manual takeover, work satisfaction, and self-concept of the pilot. In order to avoid negative consequences of automation with respect of the pilot's role, the aim should be to employ the pilot, where he can use his specific aptitudes like flexibility and experience and stays in the loop in spite of automation. Therefore it is necessary that the pilots will develop an awareness for pro-

blems which will emerge for the pilot's role from increasing automation.

AUTOMATION IN FLIGHT GUIDANCE

Reasons and Examples

The driving forces for an increasing automation are generally the two factors economy and safety. Particularly on short distance flights significant savings in fuel consumption can be achieved, if block times are shortened by automating air traffic and flight control and if climbs and descents are economically performed. Considering the explosively climbing fuel prices, the effect of fuel savings on the total air line profit is a well known fact. Salaries is another important portion of DOC (Direct Operating Costs). However, it is a controversial issue, whether the reduction of the flight crew due to automation is possible and justifiable without a curtailment in safety bandwith. In this context, it may be stated that most aircraft accidents are caused by human error (GABRIEL, 1979), where the development of an accident is usually a chain of causation. Outwardly recognizable causes often posses decisive psychological backgrounds, which are indispensable for the explanation and pro-phylaxis of an accident.

With the advances in microprocessor techno-logy and data processing, automation of flight-deck systems has become a real alter-native to manual control. Examples are Per-formance Data Computer and Flight Management Systems. Both systems supply data for an optimal performance- and fuel-management. The coupling of the Flight Management System with the autopilot permits to automatically follow an desired horizontal and/or vertical flight path.

In the next generation of aircraft conven-tional instruments will be substituted by CRT displays (BATEMAN, 1979). They permit to combine several instruments, which for example allows a condensed presentation of naviga-tional system diagrams, and various other types of aircraft data. The pilot can be directly informed, what kind of action is required in case of system malfunction.

Change of the Pilot's Role
Future tasks. The rising automation in flight guidance and the associated increase in complexity of display and control units more or less causes a fundamental change of the pilot's role or a change in task division between the participating groups, respec-tively. The activity of the pilot will be more and more limited to control and monitor displays and to manage information about the ongoing processes (CARAUX, WANNER, 1978). The pilot will only decide upon the basic para-meters and control modes, which afterwards are automatically executed. The direct contact with the aircraft will confined to irregular situations. The share of the human being is shifted from actively controlling the air-plane to perform tasks in the tactical and strategical domain.

Conceivable negative consequences. The pro-blems are rather complex, which are coming up for the pilot through the growing imple-mentation of automatic systems (WIENER, CURRY, 1980; BOEHM-DAVIS and others 1980). One of the most easily imagined consequences of automation is a loss of proficiency and expe-rience. Lack of familiarity with manual control of systems may lead to disastrous situations, when the pilot is forced to immediately react due to hardware or soft-ware failures in automatic processes. It is always stated that the pilot will be relieved through automation from fatiguing and exten-sive routine tasks in order to reduce his manual workload. Usually the high degree of complexity found in automation aircraft processes will reduce transparency and pre-dicticability of the systems behavior for the pilot. The resulting lack control possibility may increase the mental workload and stress

situations may even occur without specific incidents. So far an unsolved problem is how the decrease in physical and the increase in mental workload are related to each other and how the two different kinds of workload should be distributed. Again missing system transparency may produce negative conse-quences for a quick and precise manual take-over because the pilot is less and less inte-grated in the man-machine loop. The implemen-tation of further self-acting processes lowers work diversity at the place of work called cockpit. Along with it, the margin for autonomous planning, acting, and decision making is narrowed, where the computer will issue the operating instructions. In regard to this aspect, it has to be investigated to what extent the pilot is still able to use the aptitudes, which distinguish him from the computer. Furthermore the question is, whether the increasing limitation of the pilot's task to control and monitor displays will be sensed as a reduction in work quality and job satisfaction.

SHIFT OF THE PILOT'S ACTIVITY TO ON-BOARD FLIGHT PATH PLANNING

Introduction
Having the mentioned problems in mind, a deci-sion about technological innovations can only be made in a sensible and responsible manner, if the following questions are taken into consideration:
   In what area is the human being inferior or
   superior to the machine?
   To what extent and at which hierarchical
   level should the pilot be accordingly
   integrated in the man-machine system?
In brief it can be stated that man is superior, if decisions have to be made in unforeseeable and complex cases. He has also the capability to abstract, generalize, and to develop strategies. Further qualities of the human being are flexibility and creativity and the capacity to gain and apply experience.

Generally the machine is superior in the aquisition of information and the storage of large amounts of data, as long as the signals are clearly defined. It repeatedly executes routine tasks more reliable and precise. The reaction to control inputs is likely to be direct and without delay. Oversights and individual deviations in performance, which are typical of man for his physical and psychological work load limits, can be exclu-ded in most circumstances.

The decision to include the pilot in the pro-cess of flight path planning within the man-aircraft system is based on the assumption that the flight guidance system is structured according to the following hierarchical levels: state of the aircraft, flight path-, navigation-, and planning level. Each level can be related to an activity which perfor-mes the control and guidance of a machine, namely: manual control, monitoring, decisi-on making, and problem solving. That is to say the more one advances from the control loops on the level of aircraft state to the flight path-, navigation-, and planning level, the more tasks are placed in the foreground, which man due to his specific aptitudes can better fullfill than the computer. For example, in acute situations on-the-spot decisions have to be taken where adaptability and experience are required. These features can't be totally programmed into the machine.

Today the radar vectoring approach is the most frequently applied approach procedure. In this case the pilot primarily performs an executing function because the actual flight path planning has been realized on the ground. Air Traffic Control tells the pilot, which headings, altitudes, and speeds have to be established, but doesn't inform the pilot about the planned flight path and the manoeuvers which have to be expected. To avoid the mentioned negative consequences for the pilot being the "recipient of an order" (SUNDERMEYER, 1980) and to use the uncontested advantages of automation on the other side, thought must be given to a new division of tasks between pilots and ATC (Air Traffic Control), which will enable the pilot to perform a sensible task according to his aptitudes and to better understand automatic processes at the same time. This led to the development of a flight path guidance computer at the Institut of Flight Guidance and Control at the Technische Universität Braunschweig (SUNDERMEYER, 1980). In cooperation with ATC, this computer will provide the pilot with the capability of on-board flight path generation. The integration of the pilot in the level of flight path planning tries to be an alternative to his monitor function. It is the aim to relieve the pilot from all subordinate control tasks and to offer him the central and active role as a flight path manager which highly incorporates him in the loop.

Introduction to the Flight Path Guidance Computer

In this paper the technical details are presented in a simplified manner because the thematic emphasis is based on the philosophy of on-board flight path planning. The computer is designed in a way which to the pilots permits a flight path generation from numerous alternative functions. However, the developed hardware momentarily limited to horizontal profiles at a constant altitude. Usually it is a question of intervening in the spatial or timely formation of the flight path. The presented concept assumes that ATC is still coordinating all approaching aircraft by defining a coarse nominal flight path with waypoints and the time of arrival at the outer marker. Course and speed between fixed points or areas are up to pilot's discretion.

An input/output unit (see Fig. 1) and a chart display for the horizontal situation indication (see Fig. 2) have been developed. Numerical values and specific manipulations are entered on a keyboard while analog inputs are sensed by a mini-stick. The display further presents aircraft position, waypoints, and a control indication for monitoring the path progression with respect to time. This indicator consists of two speed columns; the left column displays the computed nominal True Airspeed and the right column displays the actual airspeed flown by the pilot. A difference in the two columns yields the time error which the aircraft would encounter at the outer marker. This time difference is displayed in another vertical column as well.

The basis for possible modifications of the flight path is the nominal track and speed which the guidance computer calculates from present position data and from ATC restraints. Nominal speed is determined from the total length of the flight path and the time of arrival at the outer marker. The pilot has the following options for modifying his approach path within the limits set by ATC:

1)   Inputs for speed control:
   a) speed at system entry
   b) final approach speed
   c) speed in the TMA
      (Terminal Manoeuvering Area)
   d) additional delay time

2)   Inputs for horizontal track situation:
   a) waypoint selection or deletion
   b) waypoint entry for holding procedure
   c) change of waypoint position within the area allocated by ATC, which is presented on the chart display.

The input of the desired speed values is fed into the computer either digitally on the keyboard or in analog form through the mini-stick. The input can be checked by observing the right speed column on the chart display. The effect of a speed change can be monitored on the time difference indicator. If both speed columns are at equal height, there is no time error at the outer maker.

The advantage of operating in the horizontal track mode is that the time error at the outer marker may be compensated by stretching or shortening the flight path without experiencing a speed change. In all of the three track modes (2a, 2b, 2c) the waypoint number displayed on the chart is entered on the keyboard while a change in waypoint position, mode 2c, is accomplished through the mini-stick and is directly plotted on the display. Modifying the track by any of the three choises (2a, 2b, 2c) will result in a change of flight path length which can be used to reduce the time error.

Normally the pilot will generate his desired flight path profile before leaving the top of descent and before entering the TMA. In this case the pilot will not change his preplanned profile unless unexpected incidents would occur.

It is the pilot's decision which choise of flight path modification will be taken. From any present position the guidance computer is offering to the pilot a physically flyable track which is a solution to reach the outer marker at the prescribed time. The chance to combine any of the modification possibilities provides the means to the pilot for a flexible flight path management in cooperation with ATC.

Besides the described possibility of manual flight path modification, the guidance computer can also furnish a fully automatic flight path planning and tracking. The pilot here only enters which aspect shall be optimized, e.g. fuel consumption, and otherwise monitors the chart display and control unit.

EMPERICAL STUDY

Issue and Methodics
In order to realize a satisfactory task division between pilots, ATC, and automation, as it is the aim in on-board flight path planning, the pilots have to build up a consciousness for problems arising from increasing automation. Only if the pilots realistically assess their position within the loop of the overall flight guidance system, an influence on the technical development and design can produce sensible results. The presented study tries to give an explanation, how pilots value their role during approach and how they rate the incorporation into the flight path management. Consequently two different conditions were taken as a basis for the inquiry. On one hand todays conventional radar vectoring approach was specified containing a certain degree of system automation: heading and altitude were to be flown by the autopilot according to ATC instructions. On the other hand the possibility of on-board flight path planning by the pilot was presented, as it was described in the preceding paragraph.

A questionnaire was developed which primarily was to acquire cognitive work elements that are subject to great change with advancing automation. In detail it concerned the following aspects:
decision level, relation of high intellectual performance to training level; flexibility and spontaneous decision making; effects of task division and the amount of information on factors like system transparency, takeover reaction in case of system failures, position of the pilot in the loop and the decision process; range of independent planning and management possibilities; extent of variation and structure of the working place cockpit.

The questions, which rate the pilot's activity during the radar vectoring approach and the approach with on-board flight path planning covered equal contents. In addition biographical data was inquired which might influence the evaluation of task division for the two approach conditions. Examples are: year of graduation from pilot school; total number of flight hours; type of aircraft formerly and currently flown; equipment of the aircraft with OMEGA radio navigation system, Inertial Navigation System (INS), and Performance Data Computer; private flying activity; experienced emergencies caused by human error.

The open answer kind of questions was chosen for the biographical background while the 6-graded Likert-attitude-scale was used for rating the pilot's activity. The questionnaire comprised 22 questions concerning biographical data, 37 questions relating to the pilot's role during the radar vectoring approach, and 46 questions dealing with on-board flight path management.

After the completion of a preliminary questionnaire, the final revision was distributed among 100 randomly selected pilots from four different German airlines. The rate of return turned out to be 61%.

Evaluation
Before analysing the answers, the raw data had been coded. For interpreting the returned questionnaire absolute, relative, and cumulative frequencies for all items were calculated. In addition median, average, and standard deviation were determined for the Likert-scaled items. The questions for the assessment of the pilot's role were submitted to a factor analysis. The analysis was seperately carried out for each of the two postulated approach conditions in order to find the number of dimensions to which the items may be reduced. In other words, the factor analysis was primarily used as a method for grouping a greater number of items. Each item was attached to the dimension, on which it received the highest factor loading or where the loading exceeded a specified value, respectively.

On the level of these factors, the relation of the assessment between the pilot's role and the professional situation of the pilot was examined. The Mann and Whitney U-Test was applied to biographical data, if the sample was classified into two groups, e.g. captain and first officer. A classification into more than two groups required the Kruskal-Wallis H-Test. Furthermore the Wilcoxon matched pairs signed rank test was used on the item level to value how the concept of on-board flight path management was viewed. Here the evaluation of the pilot's role during the radar vectoring approach served as a reference system.

Results
Structure of the sample. The paper will only present the more interesting results in more detail. The following frequencies occured: Captains made up for 72% of the pilots who had returned the questionnaire. 80% of the total sample has flown for at least 15 years. The querried pilots received their commercial license in 23% of the cases between 1961 and 1975, in 30% between 1966 and 1970, and about 33% between 1961 and 1965. The last quoted percentages closely correspond to the item which asked for the time period of flying transport category airplanes. Concerning the total hours of the first officer, there was a clear peak for the range of 2000 to 4000 hours, while the flying hours of the captain were evenly divided over all categories.

The mentioned biographical distribution reflects a degree of flying experience in a sense, where the pilots witnessed the technical progress and its consequences for ergonomical factors, as they were considered in the questionnaire. It can be supposed that this fact has a bearing on the evaluation of the pilot's role. A pilot who grew up with the changing task division in the cockpit due to automation, might be value his function and his possibility for on-board flight path planning differently from a pilot, who from the very beginning of his professional career was confronted with sophisticated technology.

A similar influence on the assesment was expected from the fact which aircraft was flown by the pilot at the time of the investigation or which types of aircraft had been flown before. The formerly piloted airplanes were grouped into the following categories: executive, military, and transport aircraft. The 16 pilots, who had flown military aircraft before, were of great interest because military airplanes often incorporate more complex and modern technology than commercial transports.

Concerning aircraft equipment, more than 95% of the pilots had experience with OMEGA and/or INS. In contrary only 30% used a Performance Data Computer on their aircraft. It may be assumed that pilots without experience of on-board computers and their associated control and display units distrust and negatively judge the changing pilot's role, as it is the case in on-board flight path management. Besides it is imagineable that pilots are not yet aware of the possible negative consequences in relation to automation where the creation of sensible tasks for pilots doesn't deem to be necessary.

Summarizing, one my state that the biographical data presents a sufficiently heterogenous group. Therefore it was possible to asses differences in the evaluation of the pilot's activity between various pilot groupings.

Assesment of on-board flight path planning.
The evaluation showed a general acceptance of the idea which stands behind the guidance computer. Especially, the task division between pilots, ATC, and automation was significantly more positively judged for on-board flight path planning than for a radar vectoring approach where the pilot is guided by ATC. The participating pilots were more content as a flight path manager, where they are able to make spontaneous decisions and to use their flying skills, experience, and human flexibility. The pilot's position within the decision process, the independant authority and high demands on decisions were rated higher. Furthermore more positively valued was the task division with respect to takeover behavior in case of system failures and with respect to system transparency of the approach situation, as well as the general information about the flight path. Therefore the intention was met to facilitate the transition from a passive monitoring to an active control task at any time by presenting the flight path to the pilot on the chart display.

The mentioned results were validated by the following pilot responses: 72% of the pilots stated that they would often choose a different flight path from the one prescribed by ATC, 92% would appreciate to influence the course of the flight path, and 87% of the pilots think it is necessary that they should be able to participate in the generation of a flight path.

Influence of the biographical data on the valuation of the pilot's task. The results of the U- and H-Test have shown that the following biographical facts affect the evaluation of the pilot's role under both approach conditions: system experience with OMEGA, INS and Performance Data Computer, aircraft type formerly and presently flown, and year of pilot training. In this context it was remarkable that most of the significant differences were found with regard to the opinion about on-board flight path planning, which obviously depends on the professional situation of the pilot.

Here are some examples: The number of hours flown as captain or first officer affected the view about further automation of approach procedures and the possibility of fully automated flight path planning. The valuation of the pilot's position within the decision process during the radar vectoring approach depended on the time of affiliation with the airline. However both results did not show a linear trend. The positive attitude to the pilot's competence in decision making in conjunction with on-board flight path planning declined the longer a pilot worked for an airline. The day-to-day routine - same type of aircraft, same destinations - is perhaps a reason why the corresponding pilots are more unable to imagine that there is enough room for independant decision making in on-board flight path planning.

Differences among the pilots with respect to the time of commercial pilot training affected all essential aspects of on-board flight path management. The following tendency can be outlined for factors, which in addition indirectly address the question, how some intentions of on-board flight path generation may be practically conceived: the opinion about flight path management negatively changed the earlier commercial pilot training took place. In other words, pilots are less sceptical about the realization of the presented novel approach of computer aided flight path planning, who received their pilot training during a time of already highly developed technology and automation.

The introduction of INS and OMEGA represents one of the most important revolutions which happened in the field of navigation during the last decade. The operation of the two systems went hand in hand with a change of the pilot's task. Therefore it was assumed that the degree of experience with the two systems has an influence on the evaluation of the pilot's role under the stated approach conditions. However, the results did not confirm the assumption that pilots without INS and OMEGA experience would rate the task of a flight path manager most negatively. In the contrary, these pilots were at least the most content persons with regard to the question of responsibility and flexibility in flight path planning. Pilots, who have worked with both systems, were the most discontent group during a radar vectoring approach concerning the degree of monitor activity, the range of routine activities, which are performed according to check lists, and the tendency to substitute check lists by computerized check routines. Also pilots, whose aircraft were equipped with a Performance Data Computer, were discontent with their functions on an approach prescribed by ATC. This fact obviously caused the significantly better view that these pilots had of the task division as provided by on-board flight path planning. Among other things this was expressed in a positive attitude towards on-board flight path generation with its flexibility and offer for autonomous decisions.

The fact, which type of airplane the pilot had flown before, influenced the valuation of on-board flight path planning in a significant way. As expected, the path management was viewed most positively by pilots, who had formerly flown only military aircraft or military and transport aircraft. The reason for this fact may be that many technical developments were originally conceived for military applications. It is imagineable that these pilots have had more experience with high technology aircraft and systems before, and therefore already gathered a broader consciousness for the consequences of increasing automation, which lead to a greater open-mindedness for technical and ergonomical novelties. Pilots, who had only flown general aviation aircraft before, were the most sceptical concerning the participation of the pilot in on-board flight path planning and the choices, which the pilot actually have.

For the radar vectoring approach pilots, who have experienced emergency situations, were most discontent with their position in the decision process and related aspects like degree of decision competence, exploitation of flying experience, and spontaneous decision capability. They also regarde the amount of information about the progress of the flight path as insufficient in order to continue the planned course in case of failures in automatic systems. Considering this fact, it can be explained that these pilots complained not to be able to autonomously determine the approach course. In an increasingly manner, they consider it to be necessary to actively participate in the flight path determination.

SUMMARY

The presented paper is partly a "pilot study". Therefore, as not otherwise expected, the results left many questions unanswered. For example, most biographical items did not show a linear trend concerning the evaluation of the pilot's role, even though significant differences existed. It is possible that some data contained so-called intervening variables, which were not picked up by the questionnaire. Therefore an uncontrollable influence on the pilot's opinion about their present role and their role with flight path management was responsible for the median sequence of several factors. Only for a few results justifiable assumptions could be stated for actual reasons why the biographical data affected several aspects of the pilot's role. Further investigations are required to clarify the relationship.

Summing up, it may be said that the questioned pilots apparently develop a consciousness for the problems arising form increasing automation. In spite of a critical attitude with regard to the realization of on board flight path planning within the limits of the overall flight guidance system, the pilots were open-minded for the novel concept of task division as in on-board flight path management. Further practical studies are necessary for assessing the pilot's role as a flight path manager in order to support statements, which can not be made from merely evaluating a questionnaire. It is of great interest, whether the acceptance of flight path planning by the pilot can be verified in practical trials. Possibly there is a greater variety in opinions, if the pilots have the chance to test the flight path guidance computer in a simulator or during flight.

REFERENCES

Bateman, L.F. (1979). The advanced flight deck. In Safety and Efficiency: the next 50 years, Symposium on Human Factors in Civil Aviation, Den Haag.

Boehm-Davis, D.A.; Curry, R.E.; Wiener, E.L.; Harrison, R.L. (1980). Human factors of flight-deck automation - NASA/Industry Workshop. NASA Technical Memorandum 81260.

Caraux, D.; Wanner, J.-C. (1978). The pilot's task in future years. In Navigation, Vol. 26.

Gabriel, R.F. (1979). Approaches to human performance improvement. In Safety and Efficiency; the next 50 years; Symposium on Human Factors in Civil Aviation, Den Haag.

Sundermeyer, P. (1980). Untersuchungen zur Verlagerung der Pilotentätigkeit auf eine höhere hierarchische Stufe der Flugführung. Dissertation, Technische Universität Braunschweig.

Wiener, E.L.; Curry, R.E. (1980). Flight deck automation: promises and problems. NASA Technical Memorandum 81206.

mini stick for   keyboard for digital input
analog input

Fig. 1. Control and Display Unit of the Flight Path Guidance Computer

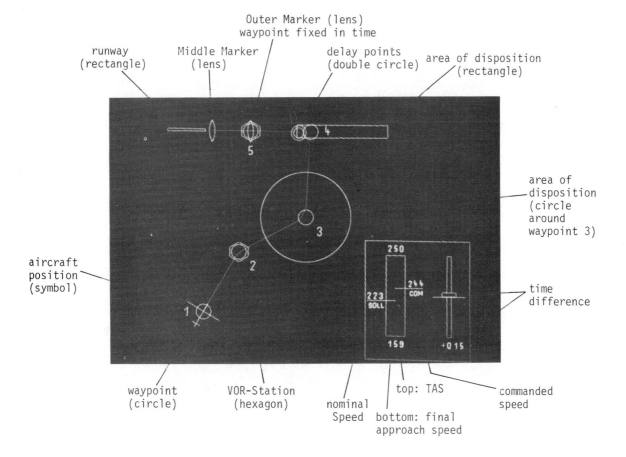

Fig. 2. Flight Path Representation on the Chart Display

OPERATION SIMULATION FOR THE EVALUATION AND IMPROVEMENT OF A MEDICAL INFORMATION SYSTEM

S. Trispel*, H. Klocke*, K. Günther*, G. Rau*, R. Schlimgen** and T. Redecker**

*Helmholtz-Institut für Biomedizinische Technik, Goethestrasse 27/29, D-5100 Aachen, Federal Republic of Germany
**Dept. Anästhesiologie der RWTH Aachen, Goethestrasse 27/29, D-5100 Aachen, Federal Republic of Germany

Abstract. Design and evaluation of a medical information system for the anaesthesiologist is the research application for which a two-operator design simulation configuration has been set up. A cognitive model is described to derive information structures and task levels for system development and operational simulation. The simulation configuration with the work station under test and the data generation station is described. Simulation objectives and procedures as well as experiments at relatively low task levels are discussed. We conclude that the two-operator simulation configuration with the expert operator in the modelling loop, though human-time-cost intensive helps considerably during the optimization of the human-system-interface and the information structure at different system development stages.

Keywords. Simulation; system design; man-machine-systems; human factors; medical information processing.

INTRODUCTION

The anaesthesiologist in the operating room is responsible for the patient's status before, during and after surgery. A multitude of patient and machine parameters have to be monitored and controlled, drugs and infusions have to be administered, and a protocol of all events, measurement data, and actions has to be established even during the course of operation. Monitoring of vital parameters (arterial and venous blood pressure, myocardial activity, body temperature, fluid balance, muscle relaxation, respiratory status) becomes increasingly supported by new devices and measurement techniques. Other data, derived from laboratory blood analysis (acid-base balance, gas partial pressures, blood components), are being gathered situation dependend on demand. Further information about preoperative care and patient conditions are contained in the (paper) files, which become supplemented by the manual protocol, as mentioned above.

The situation is that of a single operator (the anaesthesiologist) controlling and maintaining specific patient states. While patient state control, i.e. process control comprises diverse manual interactions with the patient and with machines, information acquisition and storage can become integral parts of an intraoperative information system. Thus our research and development goal consists in the analysis, design and evaluation of a task optimized information system with a human engineered interface between operator and system.

In this paper we outline the basic assumptions and ideas for our system design approach. The system is experimental in our laboratory. Construction and experimentation go hand in hand, and both will be described schematically here. The step towards operation simulation was dictated by the need of some kind of realistic environment for the evaluation of certain features of the information system. One or several models of patient parameter dynamics in the form of a combined process model would be most desirable to have for simulation. However, it shows that the process itself is so complex that only specialized physiological (e.g. the computer model of respiration by Dickinson, 1977) and diagnostic models (Rogers et. al., 1979, reviewed 58 medical diagnostic systems) could possibly be considered. Instead we chose to set up a two-operator simulation configuration, as in Fig. 7.

The following text is devided in three main parts: the investigation of decision making processes with two chapters on classification and cognitive modelling, the system design process with two chapters on information structuring and task categories, and the simulation situation with two chapters on the laboratory configuration and objectives and procedures. Since the system still is under construction, we assign (for the purpose of this paper) greater importance to task and operator analysis, followed by the implications for system and simulation design, than to the evaluation of the information system itself.

A DECISION MAKER CLASSIFICATION ATTEMPT

The decision maker in our case is the anaesthesiologist. It is important to judge his/her approach towards problem solving, which involves information acquisition, analysis, and interpretation, in order to provide a suitable information system and a relevant simulation environment. The following reflections use some of Sage's (1981) (and others') terminology and the references to the Janis and Mann decision process model in Sage's publication.

Due to the highly complex (and sometimes not fully understood) process and its time constraints we may assume that our decision maker processes information wholistically under considerable cognitive strain in some circumstances. From this reasoning we classify the thought process as concrete operational as opposed to formal operational and the information acquisition and analysis process as preceptive as opposed to receptive. In other words: the decision maker arrives at the appropriate actions through reasoning by analogy, utilizing standard operating procedures and referencing cause-effect relationships, often resulting in "symptomatic therapy". The decision rules at hand are generally heuristic and experience plays a dominant role. As a preceptive decision maker he/she uses concepts (precepts) to focus on patterns of information and to look for deviations from expected patterns. The choice of action is restricted by the set of available action alternatives.

The sources of information for our decision maker are direct observation of the patient and data from measurement (continous or occasional). The typical decision conflict arises whenever critical deviations of one or several parameters under control

occur and there is not enough or not the relevant information available to come to a satisfactory solution and subsequent action. The conflict must be solved, however, and the decision maker resorts to the least objectionable action in a "symptomatic" reaction.

In order to adapt the information system's structure to the needs of the so described decision maker it will be necessary to take a closer look at the decision mechanisms, as in the following chapter. From this exercise we may also and hopefully derive a framework for operation simulation.

A HYPOTHETIC THREE-LEVEL COGNITIVE MODEL

This model serves to illustrate some decision mechanisms which we derived in discussions with our medical co-workers (expert opinions), by observation (task analysis and standard procedures formulation) and from similar approaches in cognitive psychology. The three levels represent 1) the global course of action - 2) the act of supervising and comparing - 3) the verification of a hypothesis and action choice as the central decision making paradigm.

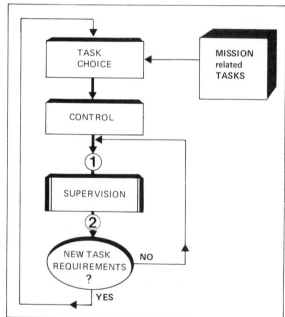

Fig. 1.   Level One: global course of action.

On the first level (Fig. 1) the anaesthesiologist performs tasks on the basis of a predeclared mission by implementing task elements as specific process control procedures and supervises process reactions by monitoring the appropriate parameters.

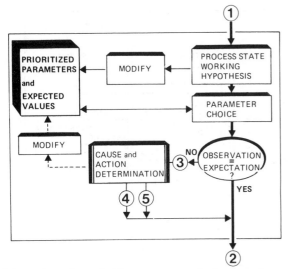

Fig. 2.   Level Two: supervision and
          comparison.

On the second level (Fig. 2) a conclusion is reached as to whether a
particular observed parameter conforms
with its expected value and behaviour
(trend). This level is active successively as long as all parameters under
observation behave as expected and no
new task requirements enforce controlling action.

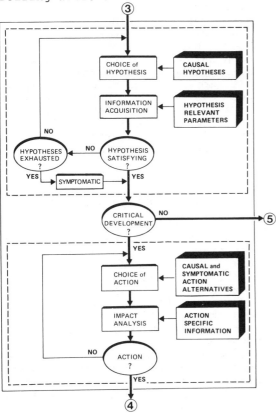

Fig. 3.   Level Three: hypothesis veri-
          fication and action choice.

On the third level (Fig. 3) a cause
for the observed parameter deviation
is being sought and the appropriate
action being selected. Note that the

decision to react to the parameter
deviation by going into the search
for an action becomes relevant after
one or more causes have been identified or, if this has not been possible,
after stating "symptomatic" (which
will initiate a "symptomatic action").
The second stage at this level, from
the verification of a critical development down to action commitment,
is similar to the basic line of the
Janis and Mann model (cited above).
We did separate, however, the ascertainment of an observation/expectation
misfit (on second level) from the
ascertainment of a critical development (on third level) because of the
intermediate steps of hypothesis verification, which we need to obtain a
normative measure for information
acquisition. This aspect will be dealt
with in the next chapter.

INFORMATION STRUCTURING

The large amount of process information as well as interactively generated data is typically segmented into
pages, as in our information system,
and the operator (henceforth, in this
context, we call the anaesthesiologist
"operator" or "work station operator")
accesses information segments by
selecting a dedicated page, which is
brought onto the monitor screen.

Fig. 4.   Planar page architecture
          with distributor page.

The page access scheme is illustrated
in Fig. 4. The question of what to
display on the individual pages can
be resolved by considering qualitative
data that has been established in the
knowledge units of the cognitive model.
Four such units have been filled on
the third level (Fig. 3) by the aforementioned analysis methods (medical
expertise, observation and task
analysis). Fig. 5 exemplifies the contents of these four units. The example
relates to the observation of blood
pressure deviation (drop) from the
expected value. Seven causal hypotheses
in the first unit are available. Each
can be accepted or rejected by looking
at the associated parameters in the

OBSERVATION: RR↓ (DROP of BLOOD PRESSURE)

CAUSAL HYPOTHESES			HYPOTHESES REL. PARAM.		ACTIONS		SPEC. INFORMATION
TYPE	REFER. to PARAM. NO.	REFER. to ACTION. NO.	NO.	PARAMETER	NO.	ACTION	CONTRA INDICATIONS
VOLUME DEFICIENCY	1,2,3,4,7	1	1	CENTRAL VENOUS PRESSURE	1	VOLUME SUBSTITUTION	RR↓ and HEART FAILURE
BLOOD LOSS	1,2,3,4,5,7	1	2	PULS	2	DRUGS	DRUG INTERACTION ALLERG. HYPERTON…
RESISTANCE LOSS	1,3,6,7	1,2	3	BLOOD PRESSURE	3	HEART–LUNG– MACHINE, PUMP	ECC > 3 HOURS ? AORTA PUMP ?
HEART FAILURE	1,2,3,7,8,10	2,3,4	4	CLINICAL INSPECT.	4	PACEMAKER DEFIBR.	
HEART–LUNG– MACHINE	1,6,9	3	5	HEMATOCRIT HEMOGLOBIN	5	VENTILATION	
VENTILATION	10	5	6	DRUGS			
NARCOTICS ADJUV.	4,6,12	2	7	L. A. PRESSURE P. C. PRESSURE			
			8	ECG			
			9	EXPERT OPINION			
			10	BLOOD GAS ANALYSIS			
			11	ELECTROLYTES			
			12	EEG			

Fig. 5. Knowledge unit contents exemplified.

second unit. On acceptance of a hypothesis, the associated action(s) in
the third unit can be considered, however, action specific information
(contra-indications) in the fourth
unit must also be considered before
the action is implemented.

Now it is clear that the information
structure should support this type of
single parameter problem solving
sequence. By working out this scheme
for all important parameters, one can
establish parameter groups and interrelationships. This is the basis for
information allocation to the pages
and for the page access structure.
The significance of operation simulation for test and evaluation of these
information allocation decisions will
become clearer with the next chapter
on task levels and task elements.

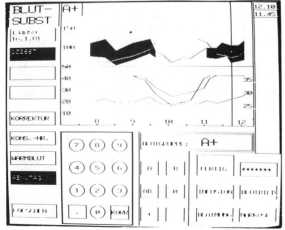

Fig. 6. Information page sample.

Fig. 6 gives an example of one information page.
A vital parameter window shows a number
of parameter values over time. The purpose of this page is to input blood infusion data to the system by using virtual input keys. Since this is a rare
event, more picture space was given to
input keys than on other pages. There we
have an additional parameter window for
respiration monitoring and, for example,
combinations of drug recording input
keys on special drug pages.

TASK LEVELS AND TASK ELEMENTS

When we talk about test and evaluation
of the information system or certain
features of the system, we must think
of the task complexity to be encountered in the real world and of how to
break up this complexitiy into quasi-
realistic task components. Here we consider four task levels for the simulation of operation phases. Table 1
shows these levels plus level Ø which
represents the task of data input and
system control common to all other
four levels. Parameter choice is the
observable search for a hypothesis
relevant parameter, whereas action
choice is the observable implementation
of an action.

The underlying notion of this classification is that we raise task complexity
from a predetermined single parameter
to be observed and the subsequent predetermined action to be implemented to
the full scale multiple parameter

observation with the associated multiple action choice. Low task complexity helps us to evaluate features
of the system's interface to the
operator, whereas high task complexity
lets us evaluate the system's information structure. Also we may test the
cognitive model - the question of
which type of human operator is needed
in each of these cases is touched upon
in the chapter on simulation objectives.

Table 1 Task Levels for Simulation

Task Level	Parameter Choice	Action Choice
1	Single	Single
2	Single	Multiple
3	Multiple	Single
4	Multiple	Multiple
∅	Data Input (Protocol) and Information System Control	

Global task elements typically are:
- search for an information item
- information item input to the
  system
- system control

Execution of these task elements can
be observed through the operator's
contact with the system's interface.
Other task elements like the indenti-
fication of new task requirements in
favor of a critical development are
of a cognitive quality and can only
be assessed by the observable (overt)
actions. It is by virtue of simulation, however, possible to disclose
much of the decision processes, since
all actions that would in reality be
implemented by various hardware and
patient contacts are now mediated to
the information system directly. In
other words: the operator knows and
contacts the world through the interface.

THE TWO-OPERATOR
SIMULATION CONFIGURATION

One of the best known and used simulators is the flight simulator,
classifying as analog simulator and
serving as a training device for
pilots. In industry, process simulators are becoming appreciated where
costly decisions are to be made.
Ritter et. al. (1981) describe a
training simulator for the operator
of an electric power dispatch control center. The energy net exists
as a digital model and certain net
failures and disturbances can be introduced by a human trainer, where-
upon the trainee feeds the model with
counteractions in a real-looking

environment. This is the basic two-
operator plus process model configuration, which we would like to have as
a training installation for the
medical operator. For two reasons we
fall short of such a training installation: because of the lack of a comprehensive patient model and because
of the need to build and expand our
information system. Looking at this
situation with Adams (1979), who evaluated training (flight) simulators,
we consider our configuration to be
a design simulator in the first place.
Fig. 7 shows the two-operator set-up.

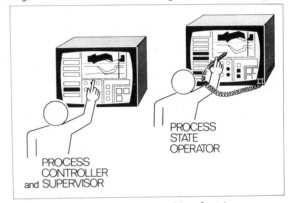

Fig. 7. Two-Operator Simulation
         Configuration.

For our purpose we couple a "work
station" for the process controller
and supervisor with a "data generation station" for the process state
operator. The work station is the
information system under test, whereas the data generation station serves
- just as the name indicates - as the
process data input and manipulation
system. The work station displays
information pages with process data
items and virtual (touch) input keys
and control elements. The data generation station displays much more and
condensed process information with a
light pen facility to enter process
data. Fig. 8. further clarifies the
issue.

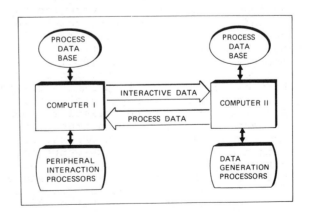

Fig. 8. Systems link and data
         transfer.

The operator at the work station generates interactive data (by protocolling his/her actions), which are deposited in the process data base and sent to the data generation station for display and likewise insertion in its data base. The operator at the data generation station generates process data (patient parameter values and operational events), which are deposited in the station's process data base and sent to the work station for display and likewise insertion in its data base. This tandem system supports the construction, evaluation and improvement of the information system by providing a quasi-realistic environment of specified complexity to the work station operator. It is obvious that the data generation station must feature many more data manipulation attributes than the work station to realistically cover the process complexity. On the other hand, the data generation station's interface need not necessarily be human engineered in such a painstaking way as the work station interface.

SIMULATION OBJECTIVES, PROCEDURES AND EXPERIMENTATION

The prime objective for our kind of design simulation is to identify design flaws at early stages of the system's construction and expansion. From the history of our development work our main concern is the optimization of the human-system-interface of the work station. For this we developed and realized specific user guidance strategies (Trispel, 1982). Experiments with untrained (novice) users and medical experts (also novices in respect to the information system, of course) were carried out at task level ∅. Process independent data entry tasks were performed. If you look at Fig. 6 you will find blood bag identification data in the top left-hand corner of that page. With the help of virtual keys and data input and correction sequences subjects had to generate these data. Keys activated and time needed were recorded and compared with the corresponding optimal procedure. Results were typically evaluated using an error graph. The graph's nodes are the input and control keys and the transition edges between nodes mark the optimal sequence for a specific data input procedure. Any deviation from the optimal sequence appears as specially coded transitions and may be interpreted as errors. Errors within a graph subset which represents a keypad, for example, suggest a suboptimal arrangement of the keys, whereas error transitions between graph subsets suggest a misinterpretation of a page function (misguidance). This technique leads to redesign and improvement of the (visual) interface.

At the time of writing this paper we had begun with task level 1 experiments with drug and infusion pages at the work station. Process data was generated by a somewhat simpler keyboard terminal data generation station. A typical task at this level is the surveying of available vital parameters and predetermined action implementation upon specific observations. Observing blood pressure, for example, the subject should apply a drug (Fentanyl) if the pressure rises significantly and apply a blood infusion if the pressure falls significantly. Since this happens at task level 1 and not 4, the lack of realism is still quite obvious. However, operator actions are now triggered by some features of the dynamic process and no larger by abstract instructions.

The type of operator we need in design simulation at both stations is without doubt the medical expert. We can, however, use non-medical operators (or medical students) in cases where very short operation phases are to be simulated, at least, this is true for task levels 1 and 2. Motivation of the medical expert operator reaches a sufficient height only at levels 3 and 4. Then the information system can be evaluated rather than tested.

At higher task levels we may also hope to validate the structural and material contents of our cognitive model. With sufficient simulation realism it should be possible to check the significance of the contents of the four knowledge units. Finally, we may consider the transfer from a design simulator to a training simulator if medical experts conclude that the operation simulation presents sufficient detail and realism to permit the training of medical students towards intraoperative decision making.

CONCLUSIONS

Operation simulation during the design process of a medical information system supports testing and evaluation at various development stages. By generating an operational environment with different levels of realism one can define meaningful tasks accordingly. This is valuable for the optimization of the human-system-interface and the system's information structure. The two-operator principle with the experienced operator "in the loop" provides insights into decision processes, which again adds to the task analysis that precedes system design. In this way system synthesis is an

iterative process with highly interactive components. The expert-time-
costs are high, of course, but due to
the research situation we are fortunately obtaining valuable assistance
from our medical co-workers. Prospec-
tives for training simulation are
being developed as the simulation
procedures become more sophisticated
at rising task levels.

REFERENCES

Adams, J.A. (1979). On the Evaluation of Training Devices. Human Factors,
21, 6, 711-720.

Dickinson, C.J. (1977). A Computer
Model of Human Respiration. MTP Press
Ltd., Lancaster, England.

Ritter, W., W. Neusel, and H. Carls (1981).
Rechnergestütztes Training am Leitstand
mit Prozeßsimulation, einsatzbereit für
eine Warte für elektrische Versorgungs-
netze (Computer-Aided Training für Operat-
ting Personnel in Process Control, Realized
for Dispatcher Training). FhG Berichte
1/2-81, 48-54.

Rogers, W., B. Ryack, and G. Moeller
(1979). Computer-Aided Medical
Diagnosis: Literature Review. Int.
J. Bio-Med. Comp., 10, 267-289.

Sage, A.P. (1981). Behavioral and
Organizational Considerations in the
Design of Information Systems and Pro-
cesses for Planning and Decision Support.
IEEE Trans. Syst., Man & Cybern., SMC-11,
9, 640-678.

Trispel, S., and G. Rau (1982). User
Guidance Strategies for the Visual Inter-
face with Virtual Control Elements.
Proc. 1982 Int. Zürich Seminar on Digital
Comm., Man-Machine Interaction, IEEE
Cat. No. 82 CH 1735-0.

ON MODELING TEAMS OF INTERACTING DECISIONMAKERS WITH BOUNDED RATIONALITY

A. H. Levis and K. L. Boettcher

Laboratory for Information and Decision Systems, Massachusetts Institute of Technology, Cambridge, Massachusetts, USA

Abstract. An analytical model of a team of well-trained human decisionmakers executing well-defined decisionmaking tasks is presented. Each team member is described by a two-stage model in which received information is first assessed and then responses are selected. An information theoretic framework is used in which bounded rationality is modeled as a constraint on the total rate of internal processing by each decisionmaker. Optimizing and satisficing strategies are derived and their properties analyzed in terms of organizational performance and individual workload. The relevance of this approach to the design and evaluation of command control and communications (C^3) systems is discussed.

Keywords. Decisionmaking, information theory, man-machine systems, organization theory, optimization.

INTRODUCTION

A command control and communications (C^3) system is defined as the collection of equipment and procedures used by commanders and their staff to process information, arrive at decisions, and communicate these decisions to the appropriate units in the organization in a timely manner. Implicit in this definition is the notion that the role of the human decisionmaker is central to the design of organizations and of the C^3 systems that support them. Therefore, in order to study the properties of alternative designs, it is necessary to develop a basic model of an interacting decisionmaker. Such a model, appropriate for a narrow but important class of problems was introduced by Boettcher and Levis (1982). In this paper, the work is extended to consider organizations consisting of several decisionmakers that form a team.

The basic assumption is that a given task, or set of tasks, cannot be carried out by a single decisionmaker because of the large amount of information processing required and because of the fast tempo of operations in a tactical situation. In designing an organizational structure for a team of decisionmakers, two issues need to be resolved: who receives what information and who is assigned to carry out which decisions. The resolution of these issues depends on the limited information processing rate of individual decisionmakers and the tempo of operations. The latter reflects the rate at which tasks are assigned to the organization for execution.

An information theoretic framework is used for both the modeling of the individual decision-maker and of the organization. Information theoretic approaches to modeling human decisionmakers have a long history (Sheridan and Ferrell, 1974). The basic departure from previous models is in the modeling of the internal processing of the inputs to produce outputs. This processing includes not only transmission (or throughput) but also internal coordination, blockage, and internally generated information. Consequently, the limitations of humans as processors of information and problem solvers, are modeled as a constraint on the total processing activity. This constraint represents one interpretation of the hypothesis that decisionmakers exhibit bounded rationality (March, 1978).

The task of the organization is to receive signals from one or many sources, process them, and produce outputs. The outputs could be signals or actions. Implicit in this model of the organization's function is the hypothesis that decisionmaking is a two-stage process. The first is the assessment of the situation (SA) of the environment, while the second is the selection of a response (RS) appropriate to the situation.

The input signals that describe the environment may come from different sources and, in general, portions of the signals may be received by different members of the organization. It has been shown by Stabile, Levis and Hall (1982) that the general case can be modeled by a single vector source and a set of partitioning matrices that distribute components of the vector signal to the appropriate decisionmakers within the organization. This model is shown in Fig. 1, where the input vector is denoted by X and takes values from a finite alphabet $\mathscr{X}$. The partitions x^i may be disjoint, overlapping or, on occasion, identical.

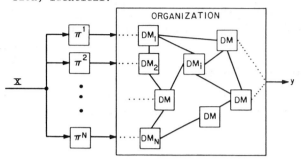

Fig. 1 The problem of information structures for organizations.

Many classes of organizational structures can be represented by Fig. 1. Consideration in this paper will be restricted to structures that result when a specific set of interactions are allowed between team members, as shown in Fig. 2. In this case, each team member is assigned a specific task, whether it consists of processing inputs received from the external environment or from other team members, for which he is well trained and which he performs again and again for successively arriving inputs. In general, a member of the organization can be represented by a two-stage model as shown in Fig. 2. First, he may receive signals from the environment that he processes in the situation assessment (SA) stage to determine or select a particular value of the variable z that denotes the situation. He may communicate his assessment of the situation to other members and he may receive their assessments in return. This supplementary information may be used to modify his assessment, i.e., it may lead to a different value of z. Possible alternatives of action are evaluated in the response selection (RS) stage. The outcome of this process in the selection of a local action or decision response y that may be communicated to other team members or may form all or part of the organization's response. A command input from other decisionmakers may affect the selection process.

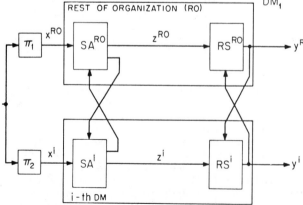

Fig. 2 Allowable team interactions

In the model of the organization developed in the following sections, internal decision strategies for each decisionmaker are introduced that determine the overall mapping between the stimulus (input) to the organization and its response (output). The total activity of each DM as well as the performance measure for the organization as a whole are expressed then in terms of the internal decision strategies. The locus of admissible strategies is shown in the performance-workload space. It is then possible to analyze the effects of the bounded rationality constraints on the organization's performance when either optimizing or satisficing behavior is assumed. The results indicate that the proposed model exhibits useful properties from the point of view of studying the information structure of decisionmaking organizations.

The paper is organized as follows. In the next section, the model of the interacting organization member is developed. In the third section an organization consisting of a team of two decisionmakers is described analytically. In the fourth section, the optimal and the satisficing decision strategies are obtained and analyzed.

MODEL OF THE INTERACTING ORGANIZATION MEMBER

The complete realization of the model for a single decisionmaker (DM) who is interacting with other organization members and with the environment is shown in Fig. 3. The detailed description and analysis of this model, as well as its relationship to previous work, notably that of Drenick (1976) and Froyd and Bailey (1980), has been presented in Boettcher and Levis (1982). Therefore, only the concepts and results needed to formulate the model of the organization will be described in this section.

Let the environment generate a vector symbol x'. The DM receives x which is a noisy measurement of x'. The vector x takes values from a known finite alphabet $\mathscr{X}$ according to the probability distribution p(x). The quantity

$$H(x) = - \sum_{x} p(x) \log_2 p(x) \qquad (1)$$

is defined to be the entropy of the input (Shannon and Weaver, 1949) measured in bits per symbol generated. The quantity H(x) can also be interpreted as the uncertainty regarding which value the random variable x will take. If input symbols are generated every τ seconds on the average, then τ, the mean symbol interarrival time, is a description of the tempo of operations (Lawson, 1981)

The situation assessment stage consists of a finite number of algorithms that the DM can choose from to process the measurement x and obtain the assessed situation z. The internal decisionmaking in this stage is the

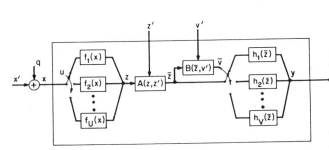

Fig. 3 Single interacting decisionmaker
         Model

choice of algorithm f_i to process x. Therefore, each algorithm is considered to be active or inactive, depending on the internal decision u. In this paper, it is assumed that
the algorithms f_i are deterministic. This implies that once the input is known and the algorithm choice is made, all other variables in
the first part of the SA stage are known.
Furthermore, because no learning takes place
during the performance of a sequence of tasks,
the successive values taken by the variables
of the model are uncorrelated, i.e., the model
is memoryless. Hence, all information theoretic expressions appearing in this paper are on
a per symbol basis.

The variable z', the supplementary situation
assessment received from other members of the
organization, combines with the elements of z
to produce $\bar{z}$. The variables z and $\bar{z}$ are of the
same dimension and take values from the same
alphabet. The integration of the situation assessments is accomplished by the subsystem
S^A which contains the deterministic algorithm A.

If there is no command input v' from other organization members, then the response selection
strategy $p(v|\bar{z})$ specifies the selection of one
of the algorithms h_j that map $\bar{z}$ into the output y. The existence of command input v' modifies the decisionmaker's choice v. A final
choice $\bar{v}$ is obtained from the function b(v,v').
The latter defines a protocol according to
which the command is used, i.e., the values of
$\bar{v}$ determined by b(v,v') reflect the degree of
option restriction effected by the command. The
overall process of mapping the assessed situation $\bar{z}$ and the command input v' into the final
choice $\bar{v}$ is depicted by subsystem S^B in Fig. 3.
The result of this process is a response selec-
tion strategy $p(\bar{v}|\bar{z}v')$ in place of $p(v|\bar{z})$.

The model of the decisionmaking process shown
in Fig. 3 may be viewed as a system S consisting
of four subsystems: S', the first part of the
SA stage; S^A; S^B; and S", the second part of
the RS stage. The inputs to this system S
are x, z', and v' and the output is y. Furthermore, let each algorithm f_i contain α_i variables denoted by

$$W^i = \left\{ w_1^i, w_2^i, \ldots, w_{\alpha_i}^i \right\} \qquad i = 1,2,\ldots,U \quad (2)$$

and let each algorithm h_j contain α_j' variables
denoted by

$$W^{U+j} = \left\{ w_1^{U+j}, \ldots, w_{\alpha_j'}^{U+j} \right\} \qquad j = 1,2,\ldots,V \quad (3)$$

It is assumed that the algorithms have no
variables in common:

$$W^i \cap W^j = \emptyset \quad \text{for } i \neq j$$

$$\forall i,j \ \varepsilon \ \{1,2,\ldots,U\} \text{ or } \{1,2,\ldots,V\} \quad (4)$$

The subsystem S' is described by a set of
variables

$$S' = \{u, W^1, \ldots, W^U, z\};$$

subsystem S^A by

$$S^A = \{W^A, \bar{z}\};$$

subsystem S^B by

$$S^B = \{W^B, \bar{v}\};$$

subsystem S" by

$$S'' = \{W^{U+1}, \ldots, W^{U+V}, y\}.$$

The mutual information or transmission or
throughput (Shannon and Weaver, 1979) between
the inputs x, z', and v' and the output y,
denoted by T(x,z',v':y) is a description
of the input-output relationship of the DM
model and expresses the amount by which the
output y is related to the inputs x,z',and
v':

$$\begin{aligned}
G_t &= T(x,z',v':y) \\
&= H(x,z',v') + H(y) - H(x,z',v',y) \\
&= H(y) - H_{x,z',v'}(y) \quad (5)
\end{aligned}$$

A quantity complementary to the throughput
G_t is that part of the input information
which is not transmitted by the system S.
It is called blockage and is defined as

$$G_b = H(x,z',v') - G_t \quad (6)$$

In this case, inputs not received or rejected by the system are not taken into account.
Blockage can also be expressed as the mutual
information between the inputs and all the
internal variables of S conditioned on the
output y, i.e.,

$$G_b = T_y(x,z',v':u,W^1,\ldots,W^{U+V},W^A,W^B,z,\bar{z}) \quad (7)$$

In contrast to blockage is a quantity that
describes the uncertainty in the output when
the input is known. It may represent noise
in the output generated within S or it may
represent information in the output produced
by the system. It is defined as the entropy
of the system variables conditioned[1] on x,

---

[1] The conditional entropy is defined as

$$H_x(z) = -\sum_x p(x) \sum_z p(z|x) \log_2 p(z|x)$$

that is,

$$G_n = H_x(u, w^1, \ldots, w^{U+V}, w^A, w^B, z, \bar{z}, \bar{v}, y) \quad (8)$$

The final quantity to be considered is the mutual information of all the internal and output variables of the system S. It reflects all system variable interactions and can be interpreted as the coordination required among the system variables to accomplish the processing of the inputs to obtain the output y. It is defined by

$$G_c = T(u{:}w_1^1{:}\ldots{:}w_{\alpha_V'}^{U+V}{:}w_1^A{:}\ldots{:}w_{\alpha_B}^B{:}z{:}\bar{z}{:}\bar{v}{:}y) \quad (9)$$

The Partition Law of Information (Conant, 1976) states that the sum of the four quantities G_t, G_b, G_n, and G_c is equal to the sum on the marginal entropies of all the system variables (internal and output variables):

$$G = G_t + G_b + G_n + G_c \quad (10)$$

where

$$G = \sum_{i,j} H(w_i^j) + H(u) + H(z) + H(\bar{z})$$
$$+ H(\bar{v}) + H(y) \quad (11)$$

When the definitions for internally generated information G_n and coordination G_c are applied to the specific model of the decisionmaking process shown in Fig. 3 they become

$$G_n = H(u) + H_{\bar{z}}(v) \quad (12)$$

and

$$G_c = G_c' + G_c^A + G_c^B + G_c'' + T(S{:}S^A{:}S^B{:}S'') \quad (13)$$

where

$$G_c' = \sum_{i=1}^{U} [p_i g_c^i(p(x)) + \alpha_i \mathbf{H}(p_i)] + H(z) \quad (14)$$

$$G_c'' = \sum_{j=1}^{V} [p_j g_c^{U+j}(p(\bar{z}|\bar{v}=j))$$
$$+ \alpha_j \mathbf{H}(p_j)] + H(y) \quad (15)$$

$$G_c^A = g_c^A(p(z)) \quad (16)$$

$$G_c^B = g_c^B(p(\bar{z})) \quad (17)$$

$$T(S'{:}S^A{:}S^B{:}S'') = H(z) + H(\bar{z}) + H(\bar{v},z)$$
$$+ T_z(x'{:}z') + T_{\bar{z}}(x',z'{:}v') \quad (18)$$

The expression for G_n shows that it depends on the two internal strategies p(u) and p(v|$\bar{z}$) even though a command input may exist. This implies that the command input v' modifies the DM's internal decision after p(v|$\bar{z}$) has been determined.

In the expressions defining the system coordination p_i is the probability that algorithm f_i

has been selected for processing the input x and p_j is the probability that algorithm h_j has been selected, i.e.. $p_i = p(u=i)$ and $p_j = p(\bar{v}=j)$. The quantities g_c^k represent the internal coordination of the corresponding algorithms and depend on the distribution of their respective inputs. The quantity $\mathbf{H}$ is the entropy of a random variable that can take one of two values with probability p:

$$\mathbf{H}(p) = - p \log p - (1-p) \log(1-p) \quad (19)$$

Relation (13) states then that the total coordination in system S can be decomposed into the sum of the internal coordination within each subsystem and the coordination due to the interaction among the subsystems. The subsystem coordinations are given by Eqs. (14) to (17) while the coordination among them is given by Eq. (18). The coordination terms for subsystems S' and S'' reflect the presence of switching due to the internal decision strategies p(u) and p($\bar{v}|zv'$). If there is no switching, i.e., if for example p(u=i)=1 for some i, then $\mathbf{H}$ will be identically zero to all p_i and Eq. (14) will reduce to:

$$G_c' = g_c^i(p(x)) + H(z)$$

and, similarly, Eq. (15) will reduce to:

$$G_c'' = g_c^{U+j}(p(\bar{z}|\bar{v}=j)) + H(y).$$

Finally, the quantity G may be interpreted as the total information processing activity of system S and therefore, it can serve as a measure of the workload of the organization member in carrying out his decisionmaling task.

A TEAM OF TWO DECISIONMAKERS

In the previous section, the information theoretic model of a decisionmaker interacting with other members of his organization was presented. In order to define an organizational structure, it is necessary to specify exactly the interactions of each DM with every other DM (if any) and the interactions with the environment. Then the Partition Law of Information can be applied to each DM. The expressions for total processing activity G and for its components can be derived then either from basic principles, or by specializing the expressions developed in the previous section. To demonstrate the procedure and, at the same time, keep the exposition brief, an organization consisting of two interacting decisionmakers will be analyzed.

The specific organizational structure is shown in Fig. 4. Both decisionmakers #1 and #2 receive synchronized signals from the environment -- they receive different partitions of the input X to the organization. Each member processes the external input through one of his algorithms f_i to obtain

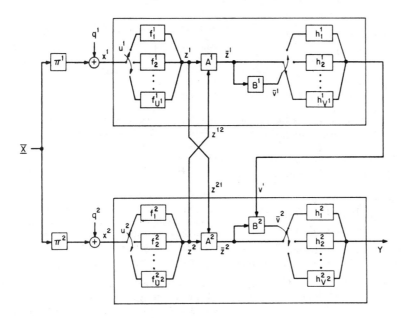

Fig. 4 A two decisionmaker team

his partial assessment z of the external situation. The partial assessments are then communicated to each other (variables z^{12} and z^{21} in Fig. 4). The first DM obtains his modified assessment $\bar{z}^1$ and then selects a response which is, in this case, a command input to the second DM. The latter receives the command input v' and, on the basis of that and his modified situation assessment $\bar{z}^2$, selects a response. The result is the output Y of the organization.

This particular configuration can be interpreted as follows: The second DM receives detailed observations about a small portion of the environment on which he has to act. He sends his estimate of the situation to the first DM who has a broad view of the situation. (DM_1 may be receiving situation assessments from other DMs that interact with him in the same way as DM_2).

Then DM_1 selects an overall strategy and communicates that to DM_2 (and to all others). This signal, v', restricts the option selection of DM_2 to be consistent with the overall strategy determined by DM_1. Finally, DM_2 generates a response to his (local) situation input that has been improved by the information he has received from DM_1.

The five quantities that characterize the information processing and decisionmaking activity of each decisionmaker are obtained directly by specializing Eqs. (5), (6), (12)-(18). The basic assumption that allows the derivations of the various expressions is that the graph showing the interactions between the DMs is acyclical.

Decisionmaker #1

$$G_t^1 = T(x^1, z^{12} : z^{21}, v') \qquad (20)$$

$$G_b^1 = H(x^1, z^{12}) - G_t^1 \qquad (21)$$

$$G_n^1 = H(u^1) + H_{\bar{z}^1}(v^1) \qquad (22)$$

$$G_c^1 = \sum_{i=1}^{U^1} [p_i g_c^i(p(x^1)) + \alpha_i \mathbf{H}(p_i)]$$
$$+ H(z^1, z^{21}) + g_c^A(p(z^1), p(z^{12}))$$
$$+ \sum_{j=1}^{V^1} [p_j g_c^j(p(\bar{z}^1 | \bar{v}^1)) + \alpha_j \mathbf{H}(p_j)]$$
$$+ H(v') + (H(z^1) + H(\bar{z}^1)$$
$$+ T_{z^1}(x^1 : z^{12}) \qquad (23)$$

Decisionmaker # 2

$$G_t^2 = T(x^2, z^{21}, v' : z^{12}, y) \qquad (24)$$

$$G_b^2 = H(x^2, z^{21}, v') - G_t^2 \qquad (25)$$

$$G_n^2 = H(u^2) + H_{\bar{z}^2}(v^2) \qquad (26)$$

$$G_c^2 = \sum_{i=1}^{U^2} [p_i g_c^i(p(x^2)) + \alpha_i \mathbf{H}(p_i)]$$
$$+ H(z^2, z^{12}) + g_c^A(p(z^2), p(z^{21}))$$
$$+ g_c^B(p(\bar{z}^2), p(v'))$$
$$+ \sum_{j=1}^{V^2} [p_j g_c^j(p(\bar{z}^2 | \bar{v}^2)) + \alpha_j \mathbf{H}(p_j)]$$

$$+ H(Y) + H(z^2) + H(\bar{z}^2) + H(z^{-2}, \bar{v}^2)$$

$$+ T_{z^2}(x^2 : z^{21}) + T_{\bar{z}^2}(x^2, z^{21} : v') \qquad (27)$$

It follows from expressions (20) to (27) that the interactions affect the total activity G of each DM. At the same time, these interactions model the control that is exerted by the DMs on each other. These controls are exerted either directly through the command inputs v' or indirectly through z^{12} and z^{21}.

Both decisionmakers in Fig. 4 are subject to indirect control. The supplementary situation assessment z^{12} modifies the assessment z^1 to produce the final assessment $\bar{z}^1$. Since $\bar{z}^1$ affects the choice of output, it follows that DM_2 has influenced the response of DM_1. Similarly, DM_1, influences through z^{21} the response of DM_2.

Direct control is exerted through the command input from DM_1 to DM_2. The variable v' modifies the response selection strategy $p(v^2 | \bar{z}^2)$ directly. Both direct and indirect control may improve the performance of a DM; they can also degrade it.

The values of the total processing activities G^1 and G^2 depend on the choice of internal decision strategies adopted by DM_1 and DM_2. Define a pure internal decision strategy to be one for which both the situation assessment strategy $p(u)$ and the response selection strategy $p(v | \bar{z})$ are pure, i.e., an algorithm f_i is selected with probability one, and an algorithm h_j is selected is selected with probability one when the situation is assessed as being some $\bar{z}$:

$$D_i^1 = \{p(u^1 = i') = 1 \ ; \ p(v^1 = j' | \bar{z}^1 = \bar{z}_m^1)\} \qquad (28)$$

for some i', for some j' and for each $\bar{z}_m^1$. Similarly,

$$D_j^2 = \{p(u^2 = i') = 1 \ ; \ p(v^2 = j' | \bar{z}^2 = \bar{z}_m^2)\} \qquad (29)$$

There are $n_1 = U^1 \cdot V^1 \cdot M^1$ possible pure internal decision strategies for DM_1 and $n_2 = U^2 \cdot V^2 \cdot M^2$ for DM_2. The quantity M is the size of the alphabet of $\bar{z}$.

All other internal decision strategies are mixed (Owen, 1968) and are obtained as a convex combination of pure strategies:

$$D^1(p_k) = \sum_{k=1}^{n_1} p_k \, D_k^2 \qquad (30)$$

$$D^2(p_\ell) = \sum_{\ell=1}^{n_2} p_\ell \, D_\ell^2 \qquad (31)$$

where p_k and p_ℓ are probabilities.

A pair of pure strategies, one for DM_1 and one for DM_2, defines a pure strategy for the organization:

$$\Delta_{ij} = \{D_i^1, \, D_j^2\} \qquad (32)$$

Independent internal decision strategies for each DM, whether pure or mixed, induce a behavioral strategy (Owen, 1968) for the organization

$$\Delta = \{D^1(p_k), \, D^2(p_\ell)\} \qquad (33)$$

Given such a behavioral strategy, it is then possible to compute the total processing activity G for each DM:

$$G^1 = G^1(\Delta) \ ; \ G^2 = G^2(\Delta) \qquad (34)$$

Alternatively, the distributions on u and v can be specified directly for each decisionmaker. This results in a set of behavioral strategies for the organization.

$$\Delta_b = \{p(u^1), p(v^1 | \bar{z}^1) \ ; \ p(u^2), p(v^2 | \bar{z}^2)\} \qquad (35)$$

that includes the set specified by Eq.(33) as well as strategies that are not induced by mixed internal decision strategies for each DM. Then, the total activity G can be computed from

$$G^1 = G^1(\Delta_b) \ ; \ G^2 = G^2(\Delta_b). \qquad (36)$$

These interpretations of the expressions for the total activity are particularly useful in modeling the bounded rationality constraint for each decisionmaker and in analyzing the organization's performance in the performance-workload space.

BOUNDED RATIONALITY AND PERFORMANCE EVALUATION

The qualitative notion that the rationality of a human decisionmaker is not perfect, but is bounded, has been modeled as a constraint on the total activity G:

$$G^i = G_t^i + G_b^i + G_n^i + G_c^i \leq F^i \tau \qquad (37)$$

where τ is the mean symbol interarrival time and F the maximum rate of information processing that characterizes decisionmaker i. This constraint implies that the decisionmaker must process his inputs at a rate that it as least equal to the rate with which they arrive. For a detailed discussion of this particular model of bounded rationality see Boettcher and Levis (1982).

As stated earlier, the task of the organization has been modeled as receiving inputs X' and producing outputs Y. Now, let Y' be the desired response to the input X' and let L(X') be a function or a table that associates a Y' with each member of the input X'.

The organization's actual response Y can be compared to the desired response Y' using a function d(Y,Y') which assigns a cost to each possible pair (Y,Y'). The expected value of the cost can be obtained by averaging over all possible inputs. This value, computed as a function of the organization's decision strategy , can serve as a performance index J. För example, if the function d(Y,Y') takes the value of zero when the actual response matches the desired response and the value of unity otherwise, then

$$J(\Delta) = E\{d(Y,Y')\} = p(Y \neq Y') \qquad (38)$$

which represents the probability of the organization making the wrong decision in response to inputs X.

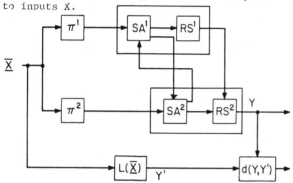

Fig. 5 Performance evaluation of organization

The information obtained from evaluating the performance of a specific structure and the associated decision strategies can be used by the organization designer in defining and allocating tasks (selecting the partitioning matrices π^i) and in changing the number and contents of the situation assessment and response selection algorithms.

The complete model of the team of two decisionmakers with bounded rationality is shown in Fig. 5. Two problems can be defined:

(a) Determine the strategies that minimize J;

(b) Determine the set of strategies for which $J \leq \bar{J}$.

The first is an optimization problem while the latter is formulated so as to obtain satisficing strategies with respect to a performance threshold $\bar{J}$. Since the bounded rationality constraint for both DMs depends on τ, the internal decision strategies of each DM will also depend on the tempo of operations. The unconstrained case can be thought of as the limiting case when $\tau \to \infty$.

A useful way of describing the properties of the solutions to the two problems is by introducing the performance-workload space (J,G^1,G^2). The locus of the admissible triples (J,G^1,G^2) is determine by analyzing the functional dependence of J, G^1, and G^2 on the organization strategy Δ, Eq. (33).

The total activity G^i of decisionmaker i is a convex function of the Δ in the sense that

$$G^i(\Delta) \geq \sum_{k,\ell} G^i(\Delta_{k\ell}) p_k p_\ell \qquad (39)$$

where $\Delta_{k\ell}$ is defined in Eq. (32). An equivalent representation of Δ is obtained from Eqs. (32) and (33):

$$\Delta = \sum_{k,\ell} \Delta_{k\ell} \, p_k p_\ell \qquad (40)$$

which describes the relative occurence of each pure organization strategy $\Delta_{k\ell}$.

The result in Eq. (39) follows from the definition of G^i as the sum of the marginal entropies of each system variable, Eq. (11), and the fact that the possible distributions p(w), where w is any system variable, are elements of a convex distribution space determined by the organization decision strategies, i.e.,

$$p(w) \; \varepsilon \; \{p(w) | p(w) = \sum_{k,\ell} p(w|\Delta_{k\ell}) p_k p_\ell \qquad (41)$$

The performance index of the organization can also be obtained as a function of Δ. Corresponding to each $\Delta_{k\ell}$ is a value $J_{k\ell}$ of the performance index. Since any organization strategy being considered is a weighted sum of pure strategies, Eq. (40), the organization's performance can be expressed as

$$J(\Delta) = \sum_{k,\ell} J_{k\ell} \, p_k \, p_\ell. \qquad (42)$$

Equations (39) and (42) are parametric in the probabilities p_k and p_ℓ. The locus of all admissible (J,G^1,G^2) triples can be obtained by constructing first all binary variations between pure strategies; each binary variation defines a line in the three dimensional space (J,G^1,G^2). These binary variations for a specific realization of the model in Fig.4 are drawn in Fig. 6. Each decisionmaker has only two pure strategies. Therefore, there are four triples that correspond to these pure strategies and four lines that join them under the assumption that one decisionmaker's pure strategy remains fixed while the other one considers variations between his two pure strategies.

For this particular example, the second decisionmaker's strategy does not affect the workload or total processing of the first DM; however, the first DM affects both performance and the total activity of the second decisionmaker through his command or direct control input. These properties are seen more closely if the locus of admissible triples is projected on the (J,G^1) and the (J,G^2) planes. The results for the second decisionmaker are shown in Fig.7; similar results are obtained for DM_1. As expected, these figures are similar to the ones obtained from the analysis of a

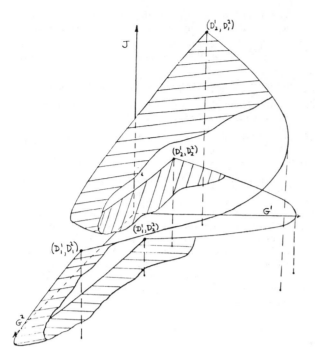

Fig. 6   The locus of binary variations
of pure strategies for a team
of two decisionmakers with two
pure strategies each

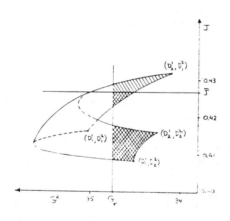

Fig. 7   Region of admissible (J, G^2)
pairs

the constraint for DM_1 is a plane parallel to
the G^2 axis and intersecting the G^1 axis at

$$G_r^1 = F^1 \tau \text{ with } G^1 \leq G_2^1$$

Similarly, the constraint for DM_2 is a plane
that intersects the G^2 axis at $G_r^2 = F^2 \tau$. For
fixed values of F^i, the bounded rationality
constraint is proportional to the tempo of
operations. As the tempo of operations increases the G_r^i become smaller and fewer of
the potential strategies are feasible.

The solutions of the satisficing problem can
be characterized as that subset of feasible
solutions for which

$$J(\Delta) = \sum_{k,\ell} J_{k\ell} \, p_k p_\ell \leq \bar{J} \tag{43}$$

The condition (43) defines a plane in the
(J, G^1, G^2) space that is parallel to the
(G^1, G^2) plane and intersects the J axis at $\bar{J}$.
All points on the locus below this plane,
which also satisfy the bounded rationality
constraints, are satisficing strategies.
While an infinite number of strategies can be
satisficing, the difference in total activity
between them can be quite large. This is
shown in the shaded region in Fig.7. The method of analysis presented in this paper is
readily extendable to teams of N decisionmakers whose interconnections can be represented by an acyclical graph.

ACKNOWLEDGEMENT

This work was supported by the Office of
Naval Research under contract N00014-81-k-0495.

REFERENCES

Boettcher, K.L., and A.H. Levis (1982). Modeling the interacting decisionmaker with bounded
rationality. IEEE Trans. Sys., Man & Cybern.
(USA), SMC-12, May/June 1982.

Conant, R.C. (1976). Laws of information which
govern systems. IEEE Trans. Sys., Man & Cybern.
(USA), SMC-16, 240-255

Drenick,R.F. (1976). Organization and Control.
In Y.C. Ho and S. K. Mitter (Eds.) Directions
in Large Scale Systems, Plenum Press, N.Y.

Froyd, J., and F.N. Bailey (1980).Performance
of capacity constrained decisionmakers. Proc.
19th IEEE Conf. Dec.&Control, Albuquerque,N.M.

Lawson, Jr., J.S. (1981). The role of time in
a command control system. Proc. 4th MIT/ONR
Workshop on C^3 Systems, LIDS-R-1159, LIDS,
M.I.T. Cambridge, MA.

March, J.G. (1978). Bounded rationality ambi-
guity, and the engineering of choice. Bell
J. Ecomc.(USA), 9, 587-608.

Owen,G. (1968). Game Theory. W.B.Saunders Co.
Philadelphia, PA.

Shannon,C.E. and W.Weaver (1949).The Mathematical Theory of Communication. The Univ.
of Illinois Press, Urbana, IL.

Sheridan, T.B., and W.R. Ferrell (1974). Man-
Machine Systems. The MIT Press, Cambridge,MA.

single decisionmaker (Boettcher and Levis, 1982).
The complete locus is obtained by considering
all combinations of mixed strategy pairs (Eq.
(33)); the surface generated in this example is
indicated in Fig.6. It is clear from the construction that the minimum value of the performance index is obtained for a pure organizational strategy as are the ones that minimize the
workloads of either decisionmaker or of both.
Therefore, the minimum error strategy is a pure
strategy when there are no bounded rationality
constraints.

The bounded rationality constraints can be realized in the form of planes of constant G^1 in
the three dimensional space (J, G^1, G^2). Thus

MODEL-BASED PREDICTION OF HUMAN PERFORMANCE WITH RESPECT TO MONITORING AND FAILURE DETECTION

W. Stein* and P. H. Wewerinke**

*Research Institute for Human Engineering (FAT), Wachtberg-Werthhoven,
Federal Republic of Germany
**National Aerospace Laboratory (NLR), Amsterdam, The Netherlands

Abstract. Human operator models for monitoring and failure detection are outlined. Corresponding experimental paradigms and extensive validation studies including eye movement results are explained. The state-space oriented models proceed with the information structure of the optimal control model (OCM) and consider single-observation as well as sequential decisions. The broad coverage of the models opens analysis and design applications in the area of supervisory control. The paradigm refers to multiple-process situations with optional dynamics, couplings and event characteristics and, thus, increases the practical utility of models. An outlook is given on design theory and respective methodologies for man-machine systems.

Keywords. Man-machine systems; mathematical analysis; human operator models; supervisory control; signal detection; decision theory; model-based design methodology.

INTRODUCTION

This paper addresses possibilities of a design or synthesis methodology for certain areas of man-machine systems. Emerging design procedures based on human performance models are emphasized that have the predictive potential of evaluating systems on a preliminary basis and to extrapolate into future concepts (Curry, Kleinman, Hoffman, 1977; Pew and others, 1977; Rouse, 1981). Certainly, analysis, design, and evaluation will be improved by satisfactory model-based methodologies in like manner, but the availability of analyses is a prerequisite to any systematic design. Analysis procedures might be more elaborated than design within the field of man-machine systems, too.

The design-methodological aspects of man-machine systems (Meister, 1971; Mc Cormick, 1976; Rouse, 1981; Topmiller, 1981) might be related to an arising and yet uncoherent design theory (Zwicky, 1967; Spillers, 1974, 1977; Director, 1981). Therein, so-called conventional approaches are regarded as fields of heuristics, i.e., decomposing a given problem to workable pieces (Himmelblau, 1973) and composing certain elements to a particular design alternative originate in a primarily heuristic taxonomy. Analytical as contrasted with conventional approaches imply steps to a more systematic design by using data bases, optimization procedures and specific metrics. Model-based approaches go beyond these by coherently incorporating analytical elements in human performance models. A prerequisite is a performance/workload metric. Experimental paradigms and appropriate validations may lead to predictive design tools and, furthermore, to a model-based task taxonomy.

The model-based approaches originate in display design for manual control situations (Clement, Jex, Graham, 1968; Clement, McRuer, Klein, 1972) and utilize frequency-domain models of the human operator combined with a rationale of visual scanning and an empirical metric of workload. Therein proceeding, the optimal control model (OCM) of human response has provided design approaches with a more elaborated formal structure (Curry, Kleinman, Hoffman, 1977; Baron, Levison, 1977; Hess, 1977; Schmidt, 1979; Hess, 1981). Its predictive applicability is based on (1) the framework composed of modules for separate human functions (e.g., perception, central processing, decision making, motor response), (2) the flexible information structure suited for multivariable, multiple process and/or multitask situations, (3) the comparably high level of validation, and (4) the underlying, normative modeling perspective. Hence, our approach employs the information structure of the OCM.

Our paper is purposed as a first step to extending model-based design procedures into the area of supervisory control. Correspondingly we examine tasks and models of both tolerance-band monitoring (TBM) and failure detection (FD) and aspire at a taxonomy originating in the modular framework of OCM-extensions. The experimental paradigm marks the area of validation and holds keyfactors of complex human operator tasks, that might be useful for analysis, design, and evaluation. The discussed results satisfy a high level of data/model correspondence. Early OCM-extensions to TBM- and FD-tasks refer to Levison (1971), Levison, Tanner (1971), and Gai,

Curry (1976), whereas the presented models have been developed by Wewerinke (1976, 1977a, 1981a, 1981b, 1982).

TASKS AND MODELS

Tolerance-band monitoring (TBM) involves observing a stochastic process $y_i(t)$ (see Fig. 1) in respect to exceeding the explicitly indicated tolerance band $[b_{li}, b_{ui}]$ that are represented by the indicator variable

$$h_i(t) = \begin{cases} H_i^o \text{ if } b_{li} \leq y_i(t) \leq b_{ui}, \\ H_i^1 \text{ else.} \end{cases} \tag{1}$$

The human has to duplicate the binary process $h_i(t) \in \{H_i^o, H_i^1\}$, where the intervals of exceeding, H_i^1, and the intervals related to the depressed keyboard, D^1, have to be synchronized. In case of m variables, the response $u(t) = D^1$ relates to the potentially up to m simultaneous exceedings of $y_i(t), i=1,...,m$. The TBM-performance metric involves the decision error $P_e = P_{fa} + P_{ms}$ and the error ratio $R_e = P_{fa}/P_{ms}$ where the time fractions P_{fa} and P_{ms}, caused by the incorrect responses $(H_i^o D^1)$ and $(H_i^1 D^o)$, denote the possibility of false alarm and missed exceeding, respectively. The binary variables $h_i(t)$ are generated by the Gauss-Markov processes $y_i(t)$. Thus, $h_i(t)$ and $u(t)$ are alternating renewal processes constituting a class of the point processes (Cox, Lewis, 1966). The TBM-task requires single-observation decisions testing one pair of hypotheses (Sage, Melsa, 1971).

Failure detection (FD) involves observing a stochastic process $y_i(t)$ with respect to the potential occurrence of abnormal events, where an event is defined as a change in the statistics of the displayed process that may be composed of changes in mean, standard deviation, and dynamic properties. The FD-performance metric contains speed and accuracy data with related trade-offs. The detection

time $T_d = (t_d - t_f)$ denotes the interval between failure detection t_d and failure occurrence t_f, whereas the false-alarm probability P_F and miss probability P_M describe the detection accuracy.

We restrict the class of events to deterministic time functions $z_{xi}(t)$ that are superimposed on stationary stochastic processes $x_i(t)$ (see Fig. 1). The display reference indications have a somewhat weak meaning in FD-tasks, since the failure-related information, e.g., the ramp function, is corrupted by a stochastic process. Hence, failure detection requires sequential decisions, where the number of observations used as input to a decision is not fixed, but greater than one (Sage, Melsa, 1971). Elementary failure detection is a binary task that consists in testing a pair of hypotheses. Situations with m displayed processes $y_i(t), i=1,...,m$ simultaneously admit m independent events or respectively composite events having redundant information.

We assume that the human observes an automatically controlled dynamic system, which is driven by white Gaussian processes $\underline{w}(t)$ with covariance $\underline{W}$,

$$\underline{\dot{x}}(t) = \underline{A}\ \underline{x}(t) + \underline{E}\ \underline{w}(t). \tag{2}$$

The system is assumed linear and time-invariant, with noise shaping states and automatic control system dynamics included in the system state description. The displayed variables $\underline{y}(t)$ involve combinations of the system states $\underline{x}(t)$ and, in case of failure detection, a superimposed deterministic time function,

$$\underline{y}(t) = \underline{C}\ \underline{x}(t) + \underline{z}_y(t), \tag{3}$$

$$\underline{z}_y(t) = \underline{C}\ \underline{z}_x(t), \tag{4}$$

where $\underline{z}_y(t)$ represents a specifiable abnormal event or a system failure to be detected. We assume that if a quantity $y_i(t)$ is displayed

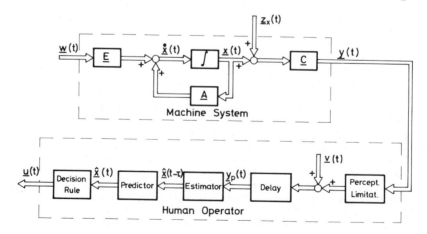

Fig. 1. Model of tolerance-band monitoring and failure detection

explicitly to the human, he can extract its rate of change, $\dot{y}_i(t)$. Thus, $\underline{y}(t)$ contains both position information $y_i(t), i=1,\ldots,m$ and rate information

$$\dot{y}_i(t) = y_{m+i}(t), i=1,\ldots,m, \qquad (5)$$

where $y_i(t)$ and $\dot{y}_i(t)$ are uncorrelated. The perceptual submodel reflects inherent limitations of time delay τ and observation noise $v_i(t)$,

$$y_{pi}(t) = y_i(t-\tau) + v_i(t-\tau), i=1,\ldots,2m, \qquad (6)$$

where $y_{pi}(t)$ is the perceived information upon which further processing is based. Frequently, the relatively invariant time delay $\tau \cong 0.2$ s is negligible in observation tasks, so that the predictor part of the model may be dropped. The covariance V_i of white Gaussian noise $v_i(t)$ scales with variance of σ_{yi}^2 of display variable $y_i(t)$ and is given by

$$V_i = \pi \, \sigma_{yi}^2 \, P_{yi}^o / f_{yi}, \qquad (7)$$

where P_{yi}^o is the reference value of the observation noise/signal ratio that relates to the level of full attention, $f_{yi}=1$. Hence,

$$P_{yi} = P_{yi}^o / f_{yi} = V_i / (\pi \, \sigma_{yi}^2), \qquad (8)$$

is the effective noise/signal ratio associated with variable $y_i(t)$. The attention sharing hypothesis of the OCM (Baron, Levison, 1977) assumes that the fractions $0 \leqq f_{yi} \leqq 1$, $f_{\dot{y}i} = f_{yi}$, applied to the variables $y_i(t)$, $i=1,\ldots,m$ obey the constraint

$$\sum_{i=1}^{m} f_{yi} = f_{yo} \leqq 1, \qquad (9)$$

where f_{yo} denotes the effective level of attention directed to the entire observation task. Losses due to visual scanning, for instance, cause $f_{yo} < 1$. In a great deal of situations the reference ratio P_{yi}^o equals $P_{yo} \cong 0.01 = -20$dB found in baseline studies (Kleinman, Baron, Levison, 1971), if specific physical conditions (i.e., zero-mean of $y_i(t)$, high resolution displays with zero-reference indication) and idealized viewing conditions (i.e., foveal viewing, full attention directed to $y_i(t)$, negligible threshold and saturation effects) are given. Some different P_{yi}^o have been found so far (Baron, Levison, 1980). Unlike the assumptions of Eq. (9), extensive studies indicate trade-offs within each pair of f_{yi}, $f_{\dot{y}i}$. Hence, the attention sharing hypothesis of Wewerinke (1977b,1981a) assumes, regarding the above baseline studies,

$$\sum_{i=1}^{m} (f_{yi} + f_{\dot{y}i}) \leqq 1, \; P_{yo} = -23\text{dB} \qquad (10)$$

that includes $f_{\dot{y}i} = f_{yi}$ as a special case, where f_{yi} and P_{yo} of Eqs. (7) and (10) differ by a factor of 1/2 and 2, respectively.

Accounting for thresholds associated with the perception of $y_i(t)$ (Baron, Levison, 1977) the describing function gain $N(.)$ of a threshold element is inserted in Eq. (7), so that the reference ratio P_{yi}^o is made a function of the appropriate standard deviation σ_{yi} and the threshold value a_{yi},

$$P_{yi}^o = P_{yo}/N^2(\sigma_{yi}, a_{yi}), \qquad (11)$$

$$N(\sigma_{yi}, a_{yi}) = \text{erfc} \, (a_{yi}/(\sigma_{yi} \, \sqrt{2})). \qquad (12)$$

Perceptual thresholds are frequently not negligible for $\sigma_{yi} < 3 \, a_{yi}$, i.e., in case of a low σ_{yi} and, since $\sigma_{\dot{y}i} \sim \omega_{oi} \, \sigma_{yi}$, a low bandwidth ω_{oi}. Typical threshold values are (Wewerinke, 1981a)

$$a_{yi} = 0.1 \text{ deg. visual arc}, \qquad (13)$$

$$a_{\dot{y}i} = 0.2 \text{ deg. visual arc/second}. \qquad (14)$$

The above observation noise model fits to FD- and not to TBM-situations, since it assumes continuously estimating an analogous quantity $y_i(t)$ with respect to a zero-reference indication. Performing a TBM-task turns out to be rather reading an uncorrupted binary-valued quantity $h_i(t)$ than estimating $y_i(t)$. The observation noise $v_i(t)$ might primarily relate to the intermittent indifference associated with the moments, when $y_i(t)$ crosses the thresholds b_{li} and b_{ui}. Studies including eye-movement recordings (Stein, 1981) let assume that the reference ratio P_{yi}^o is affected by $P(H_i^1)$, i.e., the probability of $y_i(t)$ exceeding the respective tolerance band $[b_{li}, b_{ui}]$. Thus,

$$P_{yi}^o = P_{yo} \, K(\sigma_{yi}, b_{li}, b_{ui}) \qquad (15)$$

is a reasonable expression, where $K(.)$ is empirically approximated by $(1-P(H_i^1))$, so that $\sigma_{yi} << |b_{ui} - b_{li}|$ yields $K(.) \cong 1$. Hence, $P_{yi}^o = -23$dB is a typical value of TBM-tasks having $P(H_i^1) = 0.5$.

Following the assumptions of the OCM (Kleinman, Baron, Levison, 1971), the perceived information $\underline{y}_p(t)$ is processed by an optimal observer (i.e., a Kalman-filter cascaded with an optimal predictor) that generates a best estimate $\hat{\underline{x}}(t)$ related to $\underline{x}(t)$ of the observed system; i.e., if the error covariance matrix

$$\underline{\Sigma} = E \, \{[\underline{x}(t)-\hat{\underline{x}}(t)][\underline{x}(t)-\hat{\underline{x}}(t)]^T\}, \qquad (16)$$

then $\hat{\underline{x}}(t)$ minimizes $\mathrm{Tr}\{\underline{\Sigma}\}$, where the trace $\mathrm{Tr}\{.\}$ indicates the sum of the diagonal elements of $\underline{\Sigma}$. The optimal observer yields $\hat{\underline{x}}(t)$ as well as $\underline{\Sigma}$, so that estimate $\hat{\underline{y}}(t)$ and other variables can be derived, since the matrices $\underline{A}$, $\underline{E}$, and $\underline{C}$ are defined by the experimental situation considered.

The TBM-model assumes that the human generates the response process $u(t)$ by maximizing the specifiable expected utility. Thus, applying an optimal Bayesian decision rule (Sage, Melsa, 1971), $u(t)$ is defined as a process of required single-observation decisions,

$$u(t) = \begin{cases} D^1 & \text{if } \dfrac{P(H^1|\underline{\alpha}_x)}{P(H^0|\underline{\alpha}_x)} \geq K_u, \\[2mm] D^0 & \text{else,} \end{cases} \qquad (17)$$

where $P(H^i|\underline{\alpha}_x)$ is the probability that hypothesis H^i is true, given information $\underline{\alpha}_x = (\hat{\underline{x}}(t), \underline{\Sigma})$ by the observer. K_u denotes the decision threshold

$$K_u = U_0/U_1 = (U_{00}-U_{10})/(U_{11}-U_{01}), \qquad (18)$$

where U_{ij} is the utility of responding $u(t) = D^i$, if $h(t) = H^j$ is given. The reference ratio P^0_{yi}, the attention indices f_{yi} and the decision threshold K_u are the adjustable model parameters. The decision error $P_e = P_{fa} + P_{ms}$ primarily depends on the parameters of the experimental situations as well as P^0_{yi} and f_{yi} (i.e., the given information $\underline{\alpha}_x$), whereas K_u affects the error ratio $R_e = P_{fa}/P_{ms}$ first of all.

The FD-model assumes that the human's decisions are based on the innovations process $\underline{n}(t)$ of the Kalman filter given by

$$\underline{n}(t) = \underline{C}\,\underline{x}(t) + \underline{z}_y(t) + \underline{v}(t) - \hat{\underline{y}}(t), \qquad (19)$$

whereby time delay $\tau \cong 0.2$ s is negligible in respect of detection time T_d. If no system failure has occurred (i.e., $\underline{z}_y(t)=0$), then $\underline{n}(t)$ is a zero-mean white Gaussian process having covariance $\underline{N}$. If a failure occurs (e.g., a ramp function $\underline{z}_y(t)$), then $\underline{n}(t)$ has a non-zero mean, but covariance $\underline{N}$ as before. Optimally detecting the system failure requires a number of sequential observations or, correspondingly, an observation interval $T_d = t_d - t_f$ that depend on the desired detection accuracy (e.g., $P_F = P_M = 0.05$). Regarding failure situations having $P_M = 0$ (e.g., ramp failures $\underline{z}_y(t)$), the expected detection time is given by

$$E\{T_d\} = \frac{-2 \ln (P_F)}{E_t\{\underline{n}^T(t)\,\underline{N}^{-1}\,\tilde{\underline{n}}(t)\}} \qquad (20)$$

where $E_t\{.\}$ indicates the expected value in respect of the ensemble and the time. The moving average

$$\tilde{\underline{n}}(t) = 1/T_{sm} \int_{t-T_{sm}}^{t} \underline{n}(t)\, dt \qquad (21)$$

has regard to the assumed span $T_{sm}=4s$ of short-term memory (Sheridan, Ferrell, 1974) and is affected by the failure $\underline{z}_y(t)$. The FD-metric includes the trainable speed/accuracy tradeoff $T_d(P_F,P_M)$. Detecting failures of multiple-process situations in a minimal time T_d^* presumes optimally allocated fractions of attention f^*_{yi}.

EXPERIMENTAL PARADIGMS

Two coherent experimental paradigms are formulated below covering six assumed keyfactors of tolerance-band monitoring and failure detection that are listed in Table 1. The paradigms can be related to various multiple-process situations (e.g., control of fast and slowly responding systems, vehicles, and industrial plants), since each factor is varied within a broad range.

TABLE 1 Experimental Factors

(indices i,j refer to the processes)

(1) N_a number of displayed processes
(2) ω_{oi} bandwidth
(3) $P(H^1_i)$ event probability
(4) G_a field of view
(5) ρ_{pij} level of process couplings
(6) ρ_{fij} level of failure couplings

Thus, questions concerning the analysis, design, and evaluation of man-machine systems can be considered on different levels of specification.

1. Information level.
2. Display level
3. Environmental and physical level.

The information level relates to all above factors except G_a, for the matrices $\underline{A}$, $\underline{E}$, $\underline{C}$, and $\underline{W}$ include intrinsic properties of the dynamic system under consideration. The display level primarily refers to the observation matrix $\underline{C}$ and the task-specific factors $P(H^1_i)$ and ρ_{fij}. Furthermore, the perceptual thresholds a_{yi} and the observation noise $v_i(t)$ are affected by the attributes characterizing the actual display devices (e.g., electromechanical versus electronic, level of scaling versus level of integration, monochromatic versus color). The physical level is connected with the workspace area and the control panel and, thus, relates to the field of view G_a that affects human attention sharing. We considered G_a running up to 34 by 34 degrees of visual arc. Taking into account these factors and levels is based on the performance/workload metric of the above human operator models.

Applying the paradigm to a given dynamic system requires approximating Eqs. (2) and (3) by a set of coupled second-order systems,

$$\ddot{y}_i(t) + 2\zeta_i\omega_{oi}\dot{y}_i(t) + \omega_{oi}^2 y_i(t) =$$

$$= w_i(t)\sqrt{4\zeta_i\omega_{oi}^3} , \qquad (22)$$

where $\sqrt{4\zeta_i\omega_{oi}^3}$ lets equivalent variances of $w_i(t)$ and $y_i(t)$. Assuming a steady state, the covariances $\underline{W}$, $\underline{X}$, and $\underline{Y}$ of $\underline{w}(t)$, $\underline{x}(t)$, and $\underline{y}(t)$ obey

$$\underline{A}\ \underline{X} + \underline{X}\ \underline{A}^T + \underline{E}\ \underline{W}\ \underline{E}^T = 0, \qquad (23)$$

$$\underline{Y} = \underline{C}\ \underline{X}\ \underline{C}^T, \qquad (24)$$

where the diagonal elements of $\underline{Y}$ equal σ_{yi}^2. Thus, the displayed processes $y_i(t), i=1,\dots,N_a$ are approximately characterized by bandwidths ω_{oi}, fixed damping ratios $\zeta_i=\zeta=1/\sqrt{2}$, variances σ_{yi}^2, and process couplings ρ_{pij} between $y_i(t)$ and $y_j(t)$,

$$\rho_{pij} = Y_{ij}/\sqrt{Y_{ii}\ Y_{jj}} , \qquad (25)$$

where $Y_{ij}, i,j=1,\dots,N_a$ denote the elements of covariance matrix $\underline{Y}$. Approximating the bandwidth of $y_i(t)$ is based on equivalent rectangulars of power spectral density functions and yields

$$B_i = \omega_{oi}\ \pi\sqrt{2}/4 \qquad (26)$$

for the above second-order system (Bendat, Piersol, 1971). The event probability $P(H_i^1)=1-P(H_i^0)$ of process $y_i(t)$ is given by the respective probability density function $p(y_i)$ and the tolerance band $[b_{1i}, b_{ui}]$. The failure couplings ρ_{fij} relate to the system failures $z_{yi}(t), z_{yj}(t)$ of Eq. (3) that are generated by ramp functions in this paradigm,

$$r_i(t-t_f)= \begin{cases} 0 & \text{if } t<t_f, \\ (t-t_f)c_{ri} & \text{if } t\geq t_f. \end{cases} \qquad (27)$$

The FD-performances depend on the slope ratio c_{ri}/σ_{yi}, so that $c_{ro}=0.1\ \sigma_{yi}$ per second is used to define the standard failure $r_o(t-t_f)$. Thus, the failure couplings are given by

$$\rho_{fij} = z_{yi}(t)/z_{yj}(t) = c_{ri}/c_{rj}, \qquad (28)$$

where $c_{ri}\leq c_{rj}$ and $c_{rj}\neq 0$; otherwise ρ_{fji} is used instead of ρ_{fij}. Similar coupling measures can be formulated for other failure types, too. Failure couplings are caused by

system failures affecting more than a single displayed variable, so that redundant information is yielded.

RESULTS AND INTERPRETATION

Human performance in situations according to the experimental paradigms is illustrated below, whereby the high level of data/model correspondence (Stein, 1981; Wewerinke, 1981a) justifies focussing on model results. Figure 2 summarizes TBM-results regarding the factors N_a, ω_{oi}, and G_a. The decision error $P_e=P_{fa}+P_{ms}$ represents a single-process situation and depends on the bandwidth ω_{oi} and the associated observation noise ratio $P_{yi}=P_{yi}^o/f_{yi}$, where the reference ratio $P_{yi}^o=-23dB$ is typical of a TBM-task having event probability $P(H_i^1)=0.5$. Human time delay τ is fixed at 0.2 s and perceptual thresholds are neglected.

Furthermore, Fig. 2 refers to monitoring N_a uncorrelated, homogeneous variables $y_i(t)$, $i=1,\dots,N_a$ having $\omega_{o1}=\omega_{o2}=\dots$ and $P(H_1^1)=P(H_2^1)=\dots$, so that $P(H^1)=0.5$ is the combined event probability of N_a processes. According to $f_{yi}=f_{yo}/N_a, f_{yo}=1$, and $P_{yi}^o=-23dB$, the full level of attention without losses is uniformly allocated to N_a processes. The P_e-curves arise at zero, increase monotonously with ω_{oi} and P_{yi}, and tend towards the asymptotic maximum $P(H_i^1)$ running up to 0.5 in the given example. P_e as a function of ω_{oi}, being approximately linear at $P_{yi}=-23dB$, curves with increasing noise/signal ratio P_{yi}.

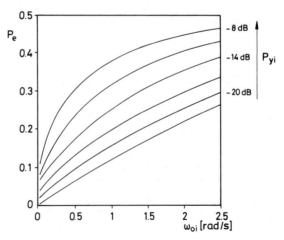

Fig. 2. Decision error P_e as a function of bandwidth ω_{oi}

The event probability $P(H_i^1)$ is an important factor of TBM-tasks reflecting aspects of rare or frequent events in man-machine situations. The here not figured decision error P_e versus $P(H_i^1)$ arises at zero, increases within $0 \leq P(H_i^1) \leq 0.5$, has the maximum at $P(H_i^1)=0.5$, and decreases within $0.5 \leq P(H_i^1) \leq 1.0$. A major outcome of TBM-investigations pertains to the effect of process couplings ρ_{pij} that is paralleled by the failure-detection results (see Fig. 5). Hence, the effect of redundancy due to ρ_{pij} seems to be negligible within the investigated range $0 \leq \rho_{pij} \leq 1/\sqrt{2}$, whereas $\rho_{pij} > 1/\sqrt{2}$ will hardly exist in actual systems.

Attentional losses, although neglected in Fig. 2, have to be considered, if the factors N_a, ω_{oi}, G_a, and, consequently, the frequency of visual scanning exceed a certain level. Given a TBM-task with $N_a=2$ homogeneous displayed processes, a field of view G_a having 34 degrees of visual arc, bandwidths $\omega_{oi} = 1$ rad/s, and a combined event probability $P(H_i^1)=0.5$, then the effective level of attention according to Eq. (9) is reduced to $f_{y1} + f_{y2} = f_{y0} = 0.5$, which has been supported by performance as well as eye-movement data.

Based on the assumption of ideal observation and signal reconstruction, monitoring a process $y_i(t)$ requires a fraction of attention, f_{yi}, scaling with its bandwidth ω_{oi} (Kleinman, Curry, 1977). Thus, monitoring $y_i(t)$ with a fixed f_{yi} may result in an error scaling with ω_{oi}. Respecting Fig. 2, signal reconstruction might be assumed, as far as P_e is approximately proportional to ω_{oi}.

Regarding the optimal allocation of attention, tolerance-band monitoring might parallel manual control (Kleinman, 1976; Wewerinke, 1977b) and failure detection tasks (Wewerinke, 1981a) in some respect. Details are given by monitoring multiple processes $y_i(t), i=1,\ldots,N_a$ that are inhomogeneous in bandwidth ω_{oi} and event probability $P(H_i^1)$. The process-related decision error P_{ei} is highly correlated with the product of associated parameters, $\omega_{oi}P(H_i^1)$, that scales with the dwell fraction f_{di} of foveally observing $y_i(t)$, as eye-movement data have shown. Hence, it is assumed that the optimal fractions of attention, f_{yi}^*, $i=1,\ldots,N_a$, allocated to the displayed processes $y_i(t)$, $i=1,\ldots, N_a$ minimize the cost functional

$$J_1 = 1/N_a \sum_{i=1}^{N_a} f(P_{ei}/P(H_i^1)). \qquad (29)$$

Results from failure detection relating to the factors ω_{oi}, N_a, ρ_{pij}, and ρ_{fij} are summa-

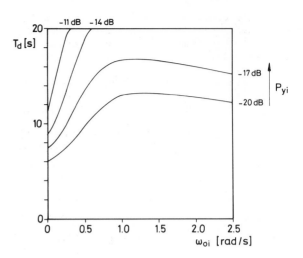

Fig. 3. Detection time T_d as a function of bandwidth ω_{oi}

rized in Figs. 3, 4, and 5. The considered situations involve idealized conditions having the perceptual thresholds $a_{yi}=a_{\dot{y}i}=0$ and the baseline ratio $P_{yo}=-20$dB. Hence, the effective noise/signal ratio P_{yi} is given by $P_{yi}=P_{\dot{y}i}=P_{yo}/f_{yi}$, where f_{yi} is the fraction of attention associated with the displayed variable $y_i(t)$. Attentional losses due to scanning are neglected. Superimposed failures $z_{yi}(t)$ are generated by the standard ramp $r_o(t-t_f)$ with a slope of 0.1 σ_{yi} per second. Furthermore, a fixed FD-accuracy is assumed that includes the probabilities $P_F=0.05$. (i.e., false alarm) and $P_M=0.0$ (i.e., missed failure).

Figure 3 illustrates detection time T_d versus bandwidth ω_{oi} of observing a single process $y_i(t)$. The noise/signal ratio $P_{yi}=-20$dB refers to the level of full attention, $f_{yi}=f_{yo}=1$. T_d versus ω_{oi}, generally being non-monotonous, is approximately constant in the upper region of bandwidth (i.e., beyond $\omega_{oi}\tilde{=}1$ rad/s). The level of T_d primarily depends on the effective noise/signal ratio P_{yi} that is a function of the fraction of attention, f_{yi}, and the perceptual threshold $a_{\dot{y}i}$. In any case, T_d is increased by $a_{\dot{y}i}$ (i.e., the threshold of rate information $\dot{y}_i(t)$) in the lower region of bandwidth ω_{oi}, since $\sigma_{\dot{y}i}<3a_{\dot{y}i}$ is given there due to $\sigma_{\dot{y}i} \sim \omega_{oi} \sigma_{yi}$ (see Eqs. (11) to (14)). It is to be added that T_d is sensitive to rate information rather in the lower than in the upper region of bandwidth. Consequently, T_d of actual situations is approximately constant. Due to $a_{yi}=a_{\dot{y}i}=0$, T_d reported in Fig. 3 is smaller than in actual situations, especially within the lower region of ω_{oi}. In sum, T_d is only weakly dependent on ω_{oi}.

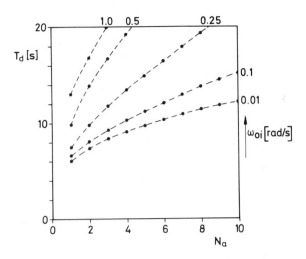

Fig. 4. Detection time T_d versus number
of processes N_a

Figure 4 represents monitoring N_a uncorrelated homogeneous processes $y_i(t)$,$i=1,\ldots,N_a$
(i.e., process couplings $\rho_{pij}=0$), whereby a
ramp failure $r_o(t-t_f)$ is given on a single
process. The level of full attention, $f_{yo}=1$,
is uniformly allocated resulting in $f_{yi}=1/N_a$
and $P_{yi}=P_{yo}/f_{yi}$, whereby $a_{yi}=\dot{a}_{yi}=0$. Certainly,
N_a increases the detection time T_d monotonous-
ly, but duplicating N_a effects a less than
duplicated T_d. Unlike Fig. 4, $a_{yi}\neq0$ of actual
situations causes T_d less depending on ω_{oi}.

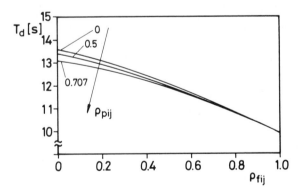

Fig. 5. Detection time T_d as a function
of couplings ρ_{fij} and ρ_{pij}

Figure 5 describes detection time T_d as a
function of failure couplings ρ_{fij} and process couplings ρ_{pij}. The situation involves
the displayed processes $y_1(t)$, $y_2(t)$ corresponding with $\omega_{o1}=\omega_{o2}=0.25$ rad/s and $P_{y1}=P_{y2}=$
$=-14$dB, whereby the two ramp failures $r_o(t-t_f)$
and $\rho_{12}r_o(t-t_f)$ are superimposed. Thus, redundancy generated by ρ_{fij} reduces T_d considerably. The effects of ρ_{pij}, paralleled by the

TBM-investigations, are weak and might be
negligible in actual situations. These findings are consistently supported by both data
and model, whereas a potential effect of
correlated processes is emphasized in the literature (Kleinman, Curry, 1977; Pew and others,
1977). Further findings relate to eye-movement data, the optimal allocation of attention and the speed-accuracy trade-off at failure detection.

Thus, tolerance-band monitoring and failure
detection are complementary tasks in respect
of the information and decision structure. A
TBM-decision presumes an appropriate observation only, whilst making a FD-decision with
a specified level of accuracy is based on an
amount of information that requires multiple
observations. An optimal FD-decision presumes
a sample of n sequential observations, whereby n depends on the information in the sample
and the specified FD-accuracy (Sage, Melsa,
1971). An indication to ideal observation and
signal reconstruction might be given by a task
error scaling with bandwidth ω_{oi}. Consequently,
signal reconstruction is likely in case of
TBM, but unlikely in case of FD.

CONCLUSIONS

Partly new concepts have been presented to
consider questions concerning analysis, design, and evaluation of man-machine systems.
The concepts are rooted in the elaborated
performance/workload metric of human operator
models that are based on the information
structure of the optimal control model (OCM).
The models pertain to fundamental multiple-
process tasks in the area of supervisory control that are given by fast and slowly responding systems. Thus, estimation, control,
and decision theory in connection with state-
space techniques provide the basis for a
multivariable system-theoretic approach to
man-machine systems. The presented paradigms
turn out to be appropriate tools for reducing
the complexity of man-machine problems in
theoretical and experimental respect. Developing design methodologies is considered a
problem of overwhelming size that presumes
long-termed scientific programs. The model-
based concepts discussed herein are viewed
rather complementing than substituting the
conventional design methodology.

228   W. Stein and P. H. Wewerinke

REFERENCES

Baron, S., and W.H. Levison (1977). Display analysis with the optimal control model of the human operator. Human Factors, 19, 437-457.

Baron, S., and W.H. Levison (1980). The optimal control model. IEEE Conf. Cybern. & Soc., pp. 90-100.

Bendat, J.S., and A.G. Piersol (1971). Random Data. Wiley, New York.

Clement, W.F., H.R. Jex, and D. Graham (1968). A manual control-display theory applied to instrument landings of a jet transport. IEEE MMS, 9, 93-110.

Clement, W.F., D.T. McRuer, and R.H. Klein (1972). Systematic manual control display design. In AGARD-CP96, Guidance and Control Displays.

Cox, D.R., and P. Lewis (1966). The Statistical Analysis of Series of Events. Methuen/ Wiley, London.

Curry, R.E., D.L. Kleinman, and W.C. Hoffman (1977). A design procedure for control/ display systems. Human Factors, 19, 421-436.

Director, S.W. (Ed.) (1981). Computer-aided design. Proc. IEEE, 69, 1185-1376.

Gai, E.G., and R.E. Curry (1976). A model of the human observer in failure detection tasks. IEEE SMC, 6, 85-94.

Hess, R.A. (1977). Analytical display design for flight tasks conducted under instrument meteorological conditions. IEEE SMC, 7, 453-461.

Hess, R.A. (1981). Aircraft control-display analysis and design using the optimal control model of the human pilot. IEEE SMC, 11, 465-480.

Himmelblau, D.M. (1973). Decomposition of Large Scale Problems. North-Holland, Amsterdam.

Kleinman, D.L., S. Baron, and W.H. Levison (1971). A control theoretic approach to manned-vehicle systems analysis. IEEE AC, 16, 824-832.

Kleinman, D.L. (1976). Solving the optimal attention allocation problem in manual control. IEEE AC, 21, 813-821.

Kleinman, D.L., and R.E. Curry (1977). Some new control theoretic models for human operator display monitoring. IEEE SMC, 7, 778-784.

Levison, W.H. (1971). A control theory model for human decision making. Ann. Conf. Manual Control, NASA SP 281.

Levison, W.R., and R.B. Tanner (1971). A Control-Theory Model for Human Decision Making. NASA CR-1953.

McCormick, E.J. (1976). Human Factors in Engineering and Design. McGraw-Hill, New York.

Meister, D. (1971). Human Factors: Theory and Practice. Wiley, New York.

Pew, R.W., and others (1977). Critical Review and Analysis of Performance Models Applicable to Man-Machine Systems Evaluation. Report No. 3446, Bolt Beranek and Newman, Cambridge, U.S.A.

Rouse, W.B. (1981). Human-computer interaction in the control of dynamic systems. Computing Surveys, 13, 72-99.

Sage, A.P., and J.L. Melsa (1971). Estimation Theory. McGraw-Hill, New York.

Schmidt, D.K. (1979). Optimal flight control synthesis via pilot modelling. AIAA J. Guid. & Cont., 2, 308-312.

Sheridan, T.B., and W.R. Ferrell (1974). Man-Machine Systems. The MIT Press, Cambridge, U.S.A.

Spillers, W.R. (1974). Basic Questions of Design Theory. North-Holland, Amsterdam.

Spillers, W.R. (1977). Design theory. IEEE SMC, 7, 201-204.

Stein, W. (1981). A monitoring and decision making paradigm. Eur. Conf. Hum. Dec. Mak. & Man. Cont., Delft, The Netherlands.

Topmiller, D.A. (1981). Methods. In J. Moraal, K.F. Kraiss (Eds.), Manned Systems Design. Plenum Press, New York.

Wewerinke, P.H. (1976). Human control and monitoring - models and experiments. Ann. Conf. Manual Control, NASA TM X-73, 170.

Wewerinke, P.H. (1977a). Human Monitoring and Control Behavior. TR 77010 U, Nat. Aerosp. Lab. NLR, Amsterdam, The Netherlands.

Wewerinke, P.H. (1977b). Performance and Workload Analysis of In-Flight Helicopter Missions. MP 77013 U, Nat. Aerosp. Lab. NLR, Amsterdam, The Netherlands.

Wewerinke, P.H. (1981a). A Model of the Human Observer and Decision Maker. TR 81062 L, Nat. Aerosp. Lab. NLR, Amsterdam, The Netherlands.

Wewerinke, P.H. (1981a). A Model of the Human Decision Maker Observing a Dynamic System. TR 81062 L, Nat. Aerosp. Lab. NLR, Amsterdam, The Netherlands.

Wewerinke, P.H. (1981b). A model of the human observer and decision maker. Ann. Conf. Manual Control, JPL 81-95, Calif. Inst. Techn., Pasadena.

Wewerinke, P.H. (1982). A model of the human observer and decision maker - Theory and validation. This Conference.

Zwicky, F., and A.G. Wilson (Eds.) (1967). New Methods of Thought and Procedure. Springer-Verlag, New York.

DISCUSSION OF SESSION 4

4.1 S MODELS OF HUMAN PROBLEM SOLVING:
DETECTION, DIAGNOSIS AND COMPENSATION
FOR SYSTEM FAILURES

Nzeako: In your specification of modes (or
what have you) of problem solving you tend
not to include "complexity" or "problem complexity" or to mention that whether hierarchical or heterarchical mode (or approach)
is adopted in problem solving depends to
some or greater extent on the complexity of
the problem, although there is no clear cut
line between which of these approaches is
applied to a particular problem. Can you,
please, comment on this.

Rouse: Certainly the particular rules and
sequences of behavior will depend on problem
complexity and thus, instances of transition
from state to structure-oriented problem
solving will be problem-dependent (as well
as individual dependent). However, my con-
jecture is that the overall formulation of
the model does not depend on complexity per
se.

4.4 T ON-BOARD FLIGHT PATH PLANNING AS A
NEW JOB CONCEPT FOR PILOTS

Venemans: You have spoken of three parts
working together: pilots-ATC-automation,
and your pilot study reported here has been
dedicated to the pilot's role and acceptance
of airborne flight path guidance automation.
1. How do you see the role for the ATC-controller with such a system? Increase or
   decrease of ATC workload?
2. How is the ATC-controller informed about
   the plans, chosen by the pilot?
3. Are you planning for asking ATC-controllers about this?

Haack-Vörsmann: 1. A ground-based computer
will be necessary for aiding the controller
on deciding the time of arrival at the Outer
Marker and transmitting the areas of the
disposition of way points, which guarantee a
collision-free flight path. So, it can not
be stated without an ergonomical investigation whether this task will decrease or increase ATC-controllers workload. The controller is still needed for coordinating all
arising flights.
2. Data link is one prerequisite for the
proposed system. ATC could be informed per
data link at a final stage of the system.
However, at a first stage ATC-controllers
definitely know that the pilot can choice
his approach course only within the limits,
for example area of dispositions, which had
been prescribed by ATC itself before.
3. No further investigation is planned at
the moment. However, ATC-controller should

definitely be asked and incorporated in
designing a new system allowing a new task
division between participating groups of the
overall air traffic systems.

4.5 T OPERATION SIMULATION FOR THE EVALUA-
TION AND IMPROVEMENT OF A MEDICAL
INFORMATION SYSTEM

Tainsh: I would like to know how you decided
on the requirements for a single screen work
station and the use of colour facilities.

Trispel: The requirement for a single screen
system was derived from task analysis and
subsequent paper- and pencil simulations,
while colour facilities were used to code
various types of switches to show their mode
during operations, which is one of the user
guidance features of the system.

4.6 I ON MODELING TEAMS OF INTERACTING
DECISION-MAKERS WITH BOUNDED RATIO-
NALITY

Chairman's Discussion Summary: The discussion
centered on limits of the method. One question was whether feedback was allowed. The
answer was no, it is an open-loop, straight-
through information processing model. Another
question concerned "coordination". The answer
was that coordination is a result, not given
as one of the situation assessment alternatives or a constraint. A third concerned
whether a sophisticated assessment element
need be watched by a sophisticated response
element. The speaker said no, the aim is
only to optimize performance under constraint. There was further discussion about
the notion of bounded rationality. In response to a question in how the calculations
were made the speaker answered "constructed,
not simulation in the Monte Carlo sense".

4.8 I MODEL-BASED PREDICTION OF HUMAN
PERFORMANCE WITH RESPECT TO MONI-
TORING AND FAILURE DETECTION

Chairman's Discussion Summary: Initial questions were to clarify "failure" and "failure
coupling". There were questions about applications of the research. The speaker answered that applications were to design of
displays and control systems for aircraft.
Further questions concerned the validity of
the authors'generalizations in view of the
specialized type of failure they used in
their experiments. The consensus was that
this was a useful generalizable experimental
result.

MAN-MACHINE INTERACTION IN COMPUTER-AIDED DESIGN SYSTEMS

J. Hatvany* and R. A. Guedj**

*Computer and Automation Institute, Hungarian Academy of Sciences, Budapest, Hungary
**Thomson-CSF, Central Research Laboratory, Corbeville, Orsay, France

Abstract. Many of the rapidly increasing numbers of computer users in the design environment are dissatisfied. The cardinal phases of design impose varying requirements, but recent research indicates that *skill, style* and *comfort* are the main categories for satisfaction. Advanced, long-term thinking is directed towards intelligent systems. However, a clearer perception of the essential features of interactive systems can help to achieve significant improvements even in the short term. Research on the symmetry (or its lack) of the human-machine partnership and on interfaces adaptable for and by the user, actions on collecting rules of thumb, on typography, on data-presentation, on experiments in complex environments and on establishing suitable levels of abstraction, can provide efficient means for rapid progress.

Keywords. Computer-aided design; man-machine systems; computer graphics; artificial intelligence; ergonomics; human factors; pattern recognition; speech recognition.

INTRODUCTION

Computer graphics and computer-aided design (CAD) are not equivalent terms, since the latter does not necessarily require the former. Nevertheless, since drawings are the designer's natural language, there is a close link between the two, and the 1200 million dollars worth of computer graphics equipment sold in 1981 (with a forecast of a fivefold increase by 1985), mostly served CAD purposes. Yet, while technical and economic reasons have led to a very rapid proliferation of CAD, the users of these systems are rarely satisfied.

In 1976 we said (Hatvany, 1977) that "most graphic display manufacturers today are not striving to meet the real needs of the design engineer engaged in practical CAD activity". Newman and Sproull (1979) remark that the designers of CAD systems do not seem to have learned much from experience, as they keep repeating the same mistakes. Smith and Green (1980) a year later state: "Communicating with a computer is typically a very unsatisfying experience. Most of the communication is constrained to rather clumsy devices like alphanumeric or graphic display terminals and usually consists of exchanges involving short, serially ordered strings of commands or

messages..." And finally, Courtieux (1981), in a most recent volume, writes: "We have little hope that these problems which limit computer aided design will soon be solved."

We propose to examine in somewhat more detail what is special about the man-machine systems involved in computer-aided design, what are the main causes of present user dissatisfaction, and how we can proceed significantly to improve human-machine interaction in this field of activity.

DESIGN AND THE COMPUTER

The Main Phases

Design is the generic name for a long chain of activities, leading from the establishment of a *product requirement* to the *generation of all the information* necessary for making the product (Hatvany, Newman and Sabin, 1977).

It might perhaps have been easier to write: "design is what takes place in a design office". This, however, would not have been accurate, for at present most design offices generate only a part of the information needed to manufacture and test a product.

Their primary concern has traditionally been with *shape* or *configuration*, whether the product be a machine, a building, or a chemical plant. The processes of *achieving* the shape, the tooling, the fixtures, the shaping (e.g. cutting, welding, casting) operations have, however, also to be designed, as have the intermediate and final quality checks and functional tests of the product. These design phases have in the past frequently been performed at the shop-floor level, often implicitly, as part of the skill of the operator.

The designers of a printed-circuit board, a machine-tool, a die, a building or a ship are obviously performing tasks of a widely differing kind. Yet in each of these areas, there are certain features that are common. The initial phase in each is the establishment of a *product requirement*, i.e. the specification of the design problem which the subsequent activities are expected to solve.

A product requirement may be motivated by a number of factors, such as

- the authoritative statement of a societal need (e.g. by a planning authority for a bridge)

- the close observation of market trends

- the recognition of an innovation opportunity (e.g. upon the emergence of a scientific research result).

The type of activity leading to its formulation will also be correspondingly different, though this phase will always have a number of common features as well. It will always involve intensive interaction with top management, with the industry's technical and business information background, with the non-technical worlds of costs, prices, market forecasts, development schedules. The technical content of the design is at this stage necessarily undeveloped, yet it has to be sufficient to allow management decisions on product development, production resource allocation, marketing strategies and market asessment to be made. The people concerned with formulating product requirements are often themselves non-technical, though such teams must also include a small sprinkling of very experienced, yet very imaginative engineers, who are able to outline heuristically what the product will look like, how it will be made and how much it will cost, before ever it has been designed, let alone made, One recent and rapidly growing factor in product requirement formulation is the impact of environmental and resource-conservation considerations, which are increasingly entering the design process right at this initial point.

The next step is that of *detailed design and analysis*, i.e. the solution of the design problem. This involves first the determination of the principles on which the solution is to be based (often requiring the investigation of several variants), then the outlining of conceptual designs, both for the product function and its outward appearance. These are jobs for engineers, scientists and industrial artists, working interactively and converging towards the outline of a technical and aesthetic plan of the overall product. This must next be elaborated by engineers and technicians into a technical design which embodies sufficient information for an engineering analysis of its quantitative parameters (strength, resonances, weight, stability, reliability, etc.). It is also at this stage that the production facilities (tools, jigs, fixtures) and the intermediate and final testing procedures (and, if necessary, equipment) must be designed.

Finally, there is the documentation phase, when the results of all the previous activities must be put into the form of drawings, instructions, tables, permitting the product to be manufactured, tested, sold, used and maintained. It is in this - extremely labour-intensive - phase that the majority of the personnel of a design office (draftsmen, typists, etc.) are employed. While it is today possible to output much of the design information in machine-readable form (e.g. paper tape for NC machine tools), the overwhelming mass of documentation is still in human-oriented drawings and descriptions which are - if there is a computer-related production facility, e.g. NC, around - carefully entered into it by human operators.

For each of these fields of activity the design office has evolved its own, traditional, non-computer tools. The reference library and the sketching board for the conceptual phase, the slide-rule and its successor, the calculator, to facilitate the dimensioning and analysis calculations, for which a host of set (by government, company or professional-concensus) procedures are often available. For drawing, there are computer-aided drafting systems, plotters, pastable symbol-sets, copying devices and - of course - the drawing board. And for marshalling and managing the results, there are microfilm techniques, microfiche and - most recently - the programmable video-disc.

The concern of contemporary design
system development is not merely to
improve upon one or other of these
techniques (note, that we have indeed
listed computer-aided drafting as one
of them), but to *integrate* them into
one continuous, organized, efficient
flow of information, from concept, to
final output. The integrating factor
is the *database*, which is the core of
every design *system*. It renders it
possible to utilize the results of
each previous phase in each subsequent
one, thus achieving the continuous
flow pattern which is its aim.

From this brief overview of the main
phases of design, it should be apparent
that each involves

- people of widely divergent skills,
  backgrounds and aims;

- activities of a fundamentally
  differing character.

The difficulty of creating a good CAD
system, lies precisely in the fact that
it must strive to *satisfy all these
people and assist in all these
activities* as part of a single,
homogeneous, integrated whole.

Recent Research

The Proceedings of the first conference
held specifically on the problems of
man-machine communication in computer-
-aided design and manufacturing (Sata
and Warman, 1981) highlighted the
current areas of prime concern in this
field.

Sambura (1981), Yoshikawa (1981) and
others discussed the *theory of design*
as a key to a more rational approach
to the construction of CAD systems.
Lewis (1981) and David (1981) put
forward new ideas on the utilization of
artificial intelligence techniques to
simplify and humanize the designer's
tasks - a subject which had previously
been raised at a special conference
devoted to the ties between artificial
intelligence and CAD research (Latombe,
1978). Hosaka and Kimura (1981), Bø
(1981) and others dealt with the *novel
input techniques*, enabling the designer,
for instance, to use free-hand
sketching and handwriting for communicating complex shapes to the computer.
Papers on the application of
recognition techniques (both of speech
and of objects) were presented by Kato
and Tsuruta (1981) and Shirai (1981),
respectively. The greatest number of
papers (Guedj, 1981, Courtieux, 1981,
Nemes and Hatvany, 1981 and others)
was devoted to improving the *techniques
of interaction*.

These recent research trends - on
theory, intelligence, input,
recognition and interaction -
adequately mirror the striving to
find better *ideas* and *means* for
satisfying the requirements of a
variety of, individuals, tasks and
environments. Their aims may be
summarized under three headings:

- the better matching of *skills*;
- an improvement in the *style* of CAD
  work;
- the provision of greater *comfort*
  to the user.

Skill

The designer is a person with skills
of his own. He is a trained architect,
an experienced electronic engineer, a
fine draftsman, a metal-cutting
expert, or whatever. The CAD system
should respect his skills. It should
not require him to become deeply
versed in an *alien skill* that of the
computer scientist, in order to be
able to use a computer in his job. He
should not be constrained to adopt
procedures and learn languages
extraneous to his profession, to
accept strange and novel description
modes of objects which are the
essence of his own experience. And he
should not be barred from applying
his own personal skills in his own,
freely chosen, personal and even
confidential way, unless this
restriction is really essential to
the integrated operation as a whole.

Style

The users of almost all CAD systems
require a style of work that conforms far better to their requirements.
Systems should be fast to act, but
patient to wait. They should afford
easy, unbureaucratic access to stored
data (e.g. without having to travel
down a long tree of decisions each
time). A good system should not
insist on the performance of all
design actions in a rigidly predetermined sequence, but allow for
individual decisions, for the interruption of one sequence and the
processing of another, for an easy
return to any activity that was left
half-done, permitting the designer to
proceed with it whenever he wishes.

Comfort

Ergonomic considerations in the
design of designers' work-stations
are familiar to the profession, but
often disregarded.(A quotation from·a
brand new scientific paper about a
future CAD system (Howard and von
Verschner, 1982) reads: "A4 upright
screens and reverse background will

have to prove acceptable to the user.")
Beside the well-known problems of glare,
flicker, poor contrast, inadequate
space, noise, etc., there are also the
phychological ones of adequate system
accessibility and of the consistent
maintenance of the correct master-slave
relationship between human and machine
in their dialogues.

INTELLIGENT COMPUTERS

As a long-term solution to the problems
outlined, researchers have proposed the
development of a fundamentally new
concept in CAD - that of the intelligent, problem-solving computer system.
which will store all the relevant
knowledge in a knowledge-base, all the
relevant rules for solving the problem
as "meta-knowledge" and require only
the input of the design *problem* to
produce a design *solution*.

One pioneering venture in this direction was that of Negroponte (1977),
who decided to exploit the possibilities of such a system to enhance
creativity. Using a vast background of
elaborate computer technology, he
finally strove to present to the user
an environment, where he could exercise
his creativity to a high degree of
freedom. A complete layman could set
about designing his own house by
inputing a series of unprofessionally
formulated *ideas*. Each of these was
regarded as a design problem formulation, solved on the basis of the
professional architectural knowledge
stored in the system, then returned to
the user as a technically feasible,
costed design.

Other workers in this field have
suggested either putting together
extant results in artificial intelligence, conversation theory, very high-
-level languages and the theory of
games to obtain systems that would show
a "human face" to the user (Hatvany,
1981), or the construction of an
entirely new systems architecture
(Moto-oka and others, 1981) to
create an entirely new concept in man-
-machine relations for all aspects of
computing, including CAD. While one of
these approaches may optimistically be
called "medium-range", the other is
assuredly in the long-range category.

BETTER INTERACTION NOW

In order to achieve a significant
improvement in interaction soon, there
are a number of difficulties to be
surmounted.

The lack of a clear and acceptable definition of human-machine interaction

Several signs indicate that we

experience great difficulties to
produce a clear and acceptable definition of human-machine interaction.
For instance, in the sub-field of
computer graphics, the tenacious
efforts of the scientific community
interested in a much needed standard
for computer graphics have achieved
in a short time an impressive result
as a draft proposal for ISO (GKS,
1982). However, from the inception of
that work at the IFIP Seillac-I
Workshop on Methodology of Computer
Graphics, as reported by Guedj and
Tucker (1976), the milestones that
followed, (GSPC, 1977, GKS, 1978,
GSPC, 1979), all reveal the
difficulty to give a clear and
acceptable definition of interaction.

It was not too difficult to get a
consensus on the primitives for
output, and to accept classes of
logical input devices following a
scheme initially suggested by Foley
and Wallace in (1974). However it is
still not quite clear how to
describe dialogues, and thus relate
input to output.

The IFIP Seillac-II Workshop on
Human-Machine Interaction in 1979,
whose basic goal was to increase our
understanding of what is interaction,
has brought several interesting
results, on interactive systems
(Guedj and others, 1980a). However,
the ultimate definition that was felt
acceptable is not a very formal one,
and does not lead us very far. It
says "Interaction is a style of
control, and interactive systems
exhibit that style".

Attempts at formal descriptions of a
dialogue are also given in papers
submitted to that workshop; (see
Crestin, Hopgood and Duce, Shaw, ten
Hagen, van den Bos' contributions in
the same volume). It is premature to
say whether those ways will be
fruitful.

Attempts at informal descriptions of
interaction sometimes shed more light
on this elusive and complex notion.
Such is the case with the interesting
study by Dzida and others (1978) on
user-perceived qualities of
interactive systems.

Other attempts from members of
disciplines seemingly far away may
provide us with the insight needed
for a necessary breakthrough in
understanding. Such seems the case
with the socio-philosopher J.Habermas
(1967), interaction is defined as a
"communicational activity" dealing
with symbolic exchanges. Instrumental
activity deals with the application

of technical means and technical rules in sequences (without being able to question the "why" of the activity).

For the french philosopher P.Ricoeur (1977), there is a difference between actions implied in the inter-action itself and the behaviour of the actor as described and explained by an outside observer. This leads to a careful look at the basic concepts of action, such as intention, motivation, cause, agent, etc.

Qualities of interactive systems involve several dimensions

Users, and unfortunately too often designers of interactive systems are not in general aware that several dimensions are involved in the evaluation of the quality of the system. Carol, Moran and Newell have been working on the question of what does 'a good interactive system' mean, There is no obvious metric for 'good', but systems can be measured against a set of criteria, such as the following (Moran, 1979):

Time:
How long does it take a user to accomplish a given set of tasks using the system?

Errors:
How many errors does a user make and how serious are they?

Learning:
How long does it take a novice user to learn how to use the system to do a given set of tasks?

Functionality:
What range of tasks can a user do in practice with the system?

Recall:
How easy is it for a user to recall how to use the system on a task that he has not done for some time?

Concentration:
How many things does a user have to keep in mind while using the system?

Fatigue:
How tired do users get when they use the system for extended periods?

Acceptability:
How do users subjectively evaluate the system?

Not all the dimensions play the same role for every user. It depends on the context, the history of the user with this system and other systems and the environment. However from a statistical study on user-perceived qualities of interactive systems by Dzida (1978) we know that two qualities are perceived as playing a major role for all users. Those are "the adequation to the task at hand" and "the self-teaching capability of the system".

Despite the study on user-perceived qualities, one is not sure of the real dimensions along which to assess an interactive system. How to caracterize them precisely? To what extent one accepts some trade-offs between two or several related dimensions, such as the case of learning traded to the difficulty of recall and the difficulty to learn traded to the ease to recall and to infer from what has been learned, etc. Obviously, this subject needs more investigation.

From what has been said of the difficulties, a much better and deeper understanding of what is involved in human-machine interaction is needed. We will suggest here several actions to improve our understanding and the quality of interactive systems. Of course, some actions can be taken in parallel.

Action on Research

More research should be done in such areas as cognitive psychology, user psychology, evaluation techniques of task performance in complex environments, design of human-machine interfaces, and other related areas. In a lucid and informative survey, Moran (1981), attributes the lack of progress in the field of user psychology to sloppy methodology. Of more concern is the persistence of "negative" results in this field. User psychology as this author mentions, "needs to move on all fronts, investing in long-term theoretical development, as well as cashing short--term gains by empirical methods". One of the difficulties to be overcome is to get the sort of researchers that are needed, people with a variety of backgrounds and allegiances.

We have suggested (Guedj, 1982) in particular two directions of research which seem to be a trail which may lead to some insight. Those are:

- Investigate to what extent human--machine partnership is symmetric in some ways and non-symmetric in other ways.
- Investigate to what extent user interfaces should be made adaptable

for the user and by the user.

Action on collecting Rules of thumbs

Collect known "rules of thumb" that
good interactive systems seem to
follow. This suggestion was made during
the Seillac-II Workshop, but to our
knowledge, has not been applied yet.

An attempt has been made (Guedj, 1980b)
for one specific aspect of interactive
systems, namely the style of echoes and
prompts.

Action on typography

Typography for grasping text, appears
very important, This factor should not
be   dismissed simply as a luxury,
especially with the coming of low-cost
bit-map display systems. Using adequate
typography may improve the overall
interactive process. Graphic designers
should be of great help. Cooperation
with computer graphics people and
interactive designers should be
encouraged.

Action on disseminating techniques for presentation of multi-dimensional data

Techniques for analysis of multi-
-dimensional data have provided a wide
body of knowledge for the presentation
of results of the analysis of data.
They have appeared in many fields, such
as economy, management, sociological and
technological applications. Several
theories underly the analyses, such as
factorial analysis, cluster analysis and
so on. There are not so many techniques
for the presentation of the results. One
body of techniques has been pionneered
by Bertin (1967). Techniques for data
presentation should be investigated,
spread, simplified and compared.

Action on disseminating simple and well-known laws of visual perception

(No comment)

Action on experiments in complex environments

Human interaction is sometimes better
at ease when several media are
simultaneously or alternatively
involved, auditory - speech and sounds -
visual - text, graphics, images,
sequences of images, etc. - Much
experience will be gained when
experimenting with mixing media in
human-machine interaction, such as the
work by Maxemchuck (1980).

Action on a search for the right abstractions for the interactive system being designed

Abstractions should be obtained once

one is familiar with many concrete
representations and not the other way
around. Symbols or expressions to
represent or to name an abstraction,
are secondary to whether or not it is
the right abstraction to deal with.
Let us not confuse the essential
problem of getting at the right
abstractions (or sometimes primitives)
with the problem of representing them.

CONCLUSIONS

Human-Machine interaction in Computer-
-Aided Design presents hard and
stimulating problems to computer
scientists as well as to all users.
Some actions have been suggested to
improve our understanding and
gradually build better interactive
systems now. We are all concerned by
this challenge. If we produce better
systems in the short range, we shall
have a sound base for producing
superlative ones in the future. If we
neglect the present, the future too,
will be in doubt.

REFERENCES

Bertin,J. (1967). *Semiologie Graphique: les Diagrammes, les Réseaux, les Cartes.* Mouton, Gauthier-Villars, E.P.H.E.

Bø, K. (1981). Compound logical input devices. In T.Sata and E.A.Warman (Eds.), *Man-Machine Communication in CAD/CAM,* North-Holland, Amsterdam. pp. 119-130.

Courtieux, G. (1981). Computer aided architectural design. In T.Sata and E.A. Warman (Eds.), *Man-Machine Communication in CAD/CAM,* North-Holland, Amsterdam. pp. 231-246.

Dzida,W., S.Herda and W.D.Itzfield (1978). User perceived quality of interactive systems. *IEEE Trans. Software Eng., SE-4.*

David, B. (1981). Man-machine communication in the specification, implementation and use of an integrated CAD/CAM system. In T.Sata and E.A.Warman (Eds.), *Man-Machine Communication in CAD/CAM,* North-Holland, Amsterdam. pp. 177-197.

Foley, J.D. and V.L.Wallace (1974). The art of natural graphic man--machine conversation. *Proc.IEEE,* 62, 462-471.

GKS (1978). *Functional Description on the Graphic Kernel System.* International Standards Organisation (ISO TC97/SC5/WG2), Bologna.

GKS (1982). *Graphic Kernel System,* Version 7. International Standards Organisation (ISO

TC97/SC5/WG2), Abingdon.

GSPC (1977). Status report of the Graphics Standards Planning Committee of ACM/SIGGRAPH. *Computer Graphics*, 11.

GSPC (1979). Status report of the Graphics Standards Planning Committee of ACM/SIGGRAPH *Computer Graphics*, 13

Guedj, R.A. and H.A.Tucker (Eds.) (1976). *Methodology in Computer Graphics*. North-Holland, Amsterdam.

Guedj, R.A., P.J.W. ten Hagen, F.R.A. Hopgood, H.A.Tucker and D.A.Duce (1980a). *Methodology of Interaction*. North-Holland, Amsterdam.

Guedj, R.A. (1980b). Guiding rules for dialogue construction in interactive systems. In *EEC Course on Interactive Graphical Man--Machine Communication*. CREST, Nantes

Guedj, R.A. (1981). Towards better interactive systems: methodology and problems in human-computer interaction. In T.Sata and E.A. Warman (Eds.), *Man-Machine Communication in CAD/CAM*. North-Holland, Amsterdam. pp. 89-93.

Guedj, R.A. (1982). Better human--machine interaction, a tough and unescapable challenge. *Computer Graphics*, 1, 20-25.

Habermas, J. (1968). *Technik und Wissenschaft als "Ideologie"*. Suhrkamp, Frankfurt-am-Main.

Hatvany, J. (1977). Interactive graphics hardware. In J.J.Allan (Ed.), *CAD Systems*, North-Holland, Amsterdam. pp. 409-415.

Hatvany, J., W.M.Newman and M.A.Sabin (1977). World survey of computer--aided design. *Computer Aided Design*, 9, 79-98.

Hatvany, J. (1981). L'intelligence artificielle, la théorie des conversations, les langages de très haut niveau et la théorie des jeux commes nouveaux outils de la conception des systèmes CAO. In *Troisièmes Journées Scientifiques et Techniques de la Production Automatisée*. ADEPA, Montrouge. pp.1-1.8.

Hosaka, M. and F.Kimura (1981). Interactive input methods for free-form shape design. In T.Sata and E.A.Warman (Eds.), *Man-Machine Communication in CAD/CAM*. North-Holland, Amsterdam. pp. 103-115.

Howard, R. and T. von Verschuer (1982). A performance specification for a European CAD workstation. In A.Pipes (Ed.), *CAD 82*. Butterworths, Guildford. pp. 221-227.

Kato, Y. and S.Tsuruta (1981). Voice input for CAD systems. In T.Sata and E.A.Warman (Eds.), *Man--Machine Communication in CAD/CAM*. North-Holland, Amsterdam. pp. 203-212.

Latombe, J.C.(Ed.) (1978). *Artificial Intelligence and Pattern Recognition in CAD*. North-Holland, Amsterdam.

Lewis, W.P. (1981). The role of intelligence in the design of mechanical components. In T.Sata and E.A.Warman (Eds.), *Man--Machine Communication in CAD/CAM*. North-Holland, Amsterdam. pp. 59-84.

Maxemchuck, N.F. (1980). An experimental speech storage and editing facility. *Bell Syst. Tech. I.*, 59, 1383-1395.

Moran, T.P. (1979). A framework for studying human-computer interaction. In R.A.Guedj, P.J.W. ten Hagen, F.R.A.Hopgood, H.A.Tucker and D.A.Duce (Eds.), *Methodology of Interaction*. North-Holland, Amsterdam. pp. 293-301.

Moran, T.P. (1981). An applied psychology of the user. *Comput. Surv.*, 14, 1-11.

Moto-oka, T. and 21 co-workers (1981). *Preliminary Report on Study and Research on Fifth-Generation Computers 1979-1980*. Japan Information Processing Cevelopment Center, Tokyo.

Negroponte, N. (1977). On being creative with CAD. In B.Gilchrist (Ed.), *Information Processing 77*. North-Holland, Amsterdam. pp. 695-704.

Nemes, L. and J.Hatvany (1981). Design criteria and evaluation methods for man-machine communication on the shop floor. In T.Sata and E.A.Warman (Eds.), *Man-Machine Communication in CAD/CAM*. North-Holland, Amsterdam. pp.217-225.

Newman, W.M. and R.F.Sproull (1979). *Principles of Interactive Computer Graphics*, second edition. McGraw Hill, New York.

Rioeur, P. (1977). *La Sémantique de l'Action*. CNRS, Paris.

Sambura, A.S. (1981). Conceptual framework for man-machine communication in CAD/CAM. In T.Sata and E.A.Warman (Eds.), *Man-machine Communications in CAD/CAM*. North-Holland, Amsterdam. pp. 21-29.

Sata, T. and E.A.Warman (Eds.) (1981). *Man-machine Communications in CAD/CAM*. North-Holland, Amsterdam.

Smith, H.T. and T.R.G.Green (Eds.) (1980). *Human Interaction with Computers*. Academic Press, New York.

Shirai, Y. (1981). Use of models in three-dimensional object recognition. In T.Sata and E.A.

Warman (Eds.), *Man-Machine Communication in CAD/CAM*. North-Holland, Amsterdam. pp. 137-156.

Yoshikawa, H. (1981). General design theory and a CAD system. In T.Sata and E.A.Warman (Eds.), *Man--Machine Communication in CAD/CAM*. North-Holland, Amsterdam. pp. 35-53.

HUMAN-COMPUTER DIALOGUE DESIGN CONSIDERATIONS

R. C. Williges and B. H. Williges

Department of Industrial Engineering and Operations Research, Virginia Polytechnic Institute and State University, Blacksburg, Virginia 24061, USA

Abstract. Various user considerations in the design of human-computer dialogues are reviewed. Topics dealing with existing dialogue design guidelines, behavioral research conducted to establish general dialogue principles, and computer-aided implementation of dialogue design are presented. Results of a compilation of over 500 user considerations are summarized as a means of describing the type of dialogue design information required. The need for empirically based dialogue guidelines relating to generic human-computer tasks is stressed, and results of two studies dealing with interactive text editing are presented as examples of research directed toward specifying dialogue design principles. One of these studies deals with a methodology for developing user models, and the second study deals with the design of HELP information. Computer-aiding in terms of rule-based systems for selecting appropriate dialogue considerations and software tools for authoring dialogues are discussed as means for implementing these dialogue design considerations. It was concluded that such human factor applications should improve the human-computer interface in terms of user acceptance, satisfaction, and productivity.

Keywords. Human factors; computer interfaces; man-machine systems; text editing; computer sofware; computer-aid design; dialogue design; human-computer interactions.

INTRODUCTION

The key to optimizing the human-computer interface is the appropriate design and management of dialogue. Traditionally, human factors considerations of the human-computer interface have been restricted primarily to hardware design and workplace layout considerations. Information related to user considerations in the design of computer hardware includes such topics as keyboard layout, system delays, and quality assessment of the visual display screens. Workplace design considerations, on the other hand, incorporate human factors information related to anthropometrics of the workplace and design of the working environment.

This research was supported by the Office of Naval Research and ONR Contract Number N00014-81-0143, and Work Unit Number NR SRO-101. The effort was supported by the Engineering Psychology Programs, Office of Naval Research, under the technical direction of Dr. John J. O'Hare.

Because computers are being used more frequently by the casual or computer-unsophisticated population, the information interface between the human and the computer is becoming increasingly more important. This information interface can be characterized as a communication or dialogue problem. The purpose of this paper is to discuss various human factors considerations in human-computer dialogue design. Specifically, topics dealing with existing dialogue design guidelines, behavioral research conducted to establish general dialogue principles, and computer-aiding for implementing dialogue guidelines are discussed.

COMPILATION OF EXISTING DIALOGUE CONSIDERATIONS

In the past several years a number of technical reports, journal articles, and books have offered guidelines in the form of user considerations for the design of computer-based dialogues. computer-based systems in the form of user considerations. Recently, Williges and Williges (1982a) compiled over 500 such considerations which deal directly with the human-computer dialogue primarily as it relates to software design. These considerations were compiled from sixteen source documents and were organized into six major sections including data organization, dialogue modes, user input devices, feedback and error management, security and disaster prevention, and multiple users as listed in Table 1.

TABLE 1 Classification Scheme for User Considerations Used by Williges and Williges, (1982)

1. DATA ORGANIZATION
1.1 Information Coding
    1.1.1 Color Codes
    1.1.2 Shape Codes
    1.1.3 Blinking Codes
    1.1.4 Brightness Codes
    1.1.5 Alphanumeric Codes
1.2 Information Density
1.3 Labeling
1.4 Format
    1.4.1 Prompts
    1.4.2 Tabular Data
    1.4.3 Graphics
    1.4.4 Textual Data
    1.4.5 Numeric Data
    1.4.6 Alphanumeric Data
1.5 Screen Layout

2. DIALOGUE MODES
2.0 Choice of Dialogue Mode
2.1 Form-Filling
    2.1.1 Default Values
    2.1.2 Feedback
    2.1.3 Screen Layout
    2.1.4 Data Entry Procedures
    2.1.5 Cursor Movement
2.2 Computer Inquiry
2.3 Menu Selection
    2.3.2 Selection Codes
        2.3.2.1 Letter Codes
        2.3.2.2 Number Codes
        2.3.2.3 Graphic Symbols
        2.3.2.4 Mnemonic Codes
    2.3.3 Menu Layout
    2.3.4 Menu Content
    2.3.5 Control Sequencing
2.4 Command Languages
    2.4.1 Command Organization
    2.4.2 Command Nomenclature
        2.4.2.1 Abbreviations
        2.4.2.2 Argument Formats
        2.4.2.3 Separators/
              Terminators
    2.4.3 Defaults
    2.4.4 Editor Orientation
    2.4.5 User Control
        2.4.5.1 Command Stacking
        2.4.5.2 Macros
        2.4.5.3 Immediate
              Commands
    2.4.6 Command Operation
    2.4.7 System Lockout
    2.4.8 Special Operations
2.5 Formal Query Languages
2.6 Restricted Natural Language

3. USER INPUT DEVICES
3.0 Data Entry Procedures
3.1 Selection of Input Device
3.2 Keyboards
    3.2.1 Special Function Keys
    3.2.2 Cursor Control
3.3 Direct Pointing Controls
3.4 Continuous Controls
3.5 Graphics Tablets
3.6 Voice Analyzers

4. FEEDBACK AND ERROR MANAGEMENT
4.1 Feedback
    4.1.1 Status Messages
    4.1.2 Error Messages
    4.1.3 Hard Copy Output
4.2 Error Recovery
    4.2.1 Immediate User Correction
    4.2.2 User Correction
          Procedures
    4.2.3 Metering and Automatic
          Error Checks
    4.2.4 Automatic Correction
    4.2.5 Stacked Commands
4.3 User Control
4.4 Help and Documentation
    4.4.1 Off-Line Documentation
    4.4.2 On-Line Documentation
4.5 Computer Aids
    4.5.1 Debugging Aids
    4.5.2 Decision Aids
    4.5.3 Graphical Input Aids

5. SECURITY AND DISASTER PREVENTION
5.1 Command Cancellation
5.2 Verification of Ambiguous
    or Destructive Actions
5.3 Sequence Control
5.4 System Failures

6. MULTIPLE USERS
6.1 Separating Messages/Inputs
6.2 Separating Work Areas
6.3 Communications Record

Data organization deals with aspects of structuring information on the visual display in an interactive environment. No consideration was given to auditory information displays. The major topics of consideration include methods of coding information displayed, the use of labeling to organize the information displayed, various techniques for formatting the display, and considerations for the overall layout of information fields on the display screen.

Dialogue considerations dealing with dialogue mode were organized into six general types of interaction including form-filling, computer inquiry, menu selection, command languages, formal query languages, and restricted natural languages. The first three represent primarily computer-initiated dialogues, whereas the last three are primarily user-initiated dialogues. As yet mixed-initiative dialogues have seen little application, and few data exist concerning their optimal design. Form-filling is a structured dialogue mode in which the user provides information in designated fields on the interactive display. Considerations in form-filling involve the choice of default values, feedback to the user, screen layout, data entry procedures, and cursor movement to designated fields on the form. Computer inquiry dialogue is a computer-initiated query mode. Menu selection is a type of structured dialogue in which the user must select among a variety of options. Dialogue design considerations include the order of options, the selection codes for the options, the display layout of the menu, the content of the menu, and the control sequence of the menu. Command language dialogues allow the user to communicate with the computer by providing specific commands which specify various functins to be performed. Topics discussed by Williges and Williges (1982a) include command organization and nomenclature, default values, editor orientation, user control, command operation, system lockout, and special commands. Formal query languages are a set of syntactical and lexical rules with which the user can question the computer in order to retrieve information from a data base. Restricted natural language is the most unstructured dialogue and is used as a flexible method to query a data base. Although sentence-like commands are used, vocabulary size and/or syntax may be restricted.

Various user considerations are necessary to select the device by which the user makes a dialogue entry to the computer. The guidelines compiled by Williges and Williges (1982a) are concerned with the selection of an input device and considerations related to the use of special function keys and cursor control, pointing controls such as light pens and touch panels, continuous controls such as trackballs and joysticks, the choice of graphic tablets for graphical data entry, and voice analyzers.

Dialogue considerations pertaining to feedback and error management deal primarily with communications from the computer to the user. These communications include feedback, error recovery, user control of the transaction sequence, help/documentation, and computer aids. Feedback information provided by the computer includes status messages, error messages, and hard copy output. Actions required to correct an error involve user correction procedures, computer metering of transactions, automatic correction, and stacking of multiple commands. Guidelines are considered for the level, amount, and type of user control of feedback and error messages. User consideration for on-line and off-line documentation as well as help information to enhance feedback and error management are presented. And, finally, guidelines for computerized aids for program debugging, decision making, and graphical inputs are listed.

Security and disaster prevention considerations deal with human-computer transactions that are directed toward the prevention of catastrophic situations, such as the inadvertent deletion of important files or the premature termination of a computer session. Topics included are the cancellation of data entry and command sequences, the requirement to confirm ambiguous or destructive actions, the control of destructive actions, and the handling of system crashes.

Although most human-computer dialogue considerations are concerned with a dialogue between a single user and the computer, a few design considerations have been offered for the multiple-user environment. These considerations are concerned primarily with separating messages and inputs of multiple users, the use of cursors in multiple-user displays, and computerized record keeping for inter-user messages.

Fundamentals of Human-Computer Dialogue Design

In general, the designer of any system sets out to minimize equipment as well as personnel costs. However, these goals are often not compatible. Until recently the high cost and relatively limited speed of computer systems have dictated a design tradeoff favoring the capabilities and limitations of the computer and not the human. With the rapid advances in computer technology and the subsequent cost reductions in computer hardware, designers can now effectively attempt to optimize the human aspect of the interface. In terms of human performance, the system designer must work toward an acceptably low error rate and an acceptable cost in personnel time. In addition, user acceptance and satisfaction with the computer system seem to be critical to effective use.

It is quite likely that many basic human factors considerations found to be fundamental to good system design in other applications will be equally important in the design of computer-based systems. Indeed, many of the user considerations proposed for computer-based systems appear to be nothing more than a restatement of basic human factors design guidelines as they specifically relate to systems involving computers. The general human factors principles that seem to be present in the specific human-computer dialogue design considerations reviewed by Williges and Williges (1982a) include compatibility, consistency, brevity, flexibility, immediate feedback, and operator workload.

Compatibility. The principle of compatibility predicts high information transfer when the amount of information recoding necessary is minimal. Translated to the human-computer system this would suggest that the input required of the user should be compatible with the output of the computer and vice versa. Compatibility implications for human-computer dialogue can take several forms. The data organization of system output should be compatible with the data to be entered. Both the input required of the user and the output of the system should be consistent across the display, module, program, and the information system. The choice of terminology, format, and system action should be consistent with user population stereotypes. The input required of the user should not be ambiguous, and the output of the computer should be clear and, therefore, useful. To minimize the information processing requirements of the user, information should be presented in a directly usable form. Movement of displayed objects should parallel the direction of the input movement of the user. In summary, the need to translate, transpose, interpret, or refer to documentation should be minimized.

Consistency. Problem solution occurs most readily in a consistent environment. This permits the user to develop a conceptual model of the operation of the system. To insure consistency it may be necessary at times to require operations which appear to decrease system throughput, such as requesting information from the user already known by the system in order to insure that similar procedures require identical user actions. The system should perform in a generally predictable manner, without exception.

Brevity. Theories of human memory suggest the existence of some upper limit of information that can be received in a given period of time. The limit of short-term memory is generally accepted to be seven or eight items. When longer input is required, chunking should be used such that meaningful units of information are grouped together. To increase the number of bits of information that can be included in one input sequence, larger chunks each containing more information should be built. In computer-based dialogues this would suggest that both the input required of the user and the output of the system should be brief to minimize both the short-term memory load on the user and the probability of input errors by the user. In addition, user input and computer output should be grouped into meaningful chunks, whenever possible.

Flexibility. Individual differences among users necessitate system flexibility to insure optimum performance of all users. In many systems a decision must be made as to whether the system should be designed to

accommodate the extreme individuals or the average individual. However, capitalizing upon the capabilities of the computer, one can provide a flexible or adaptive system that suits most potential users equally. In computer-based dialogues this would suggest that both the input required of the user and the output provided by the system should depend upon the user's expectations of the system, past experience, and capabilities.

Immediate feedback. A human-computer system should be closed-loop with information feedback to the human about the quality of user performance and the condition of the system. Without immediate feedback which is readily understandable, the user cannot make decisions regarding the necessity for corrective action and the form it should take. In computer-based systems, users should at all times be aware of where they are, what they have done, and whether or not it was successful. The user should be given every opportunity to correct errors. System feedback also provides the user assurance that the system is available and information concerning whether user queries are being procesed or delayed.

Operator workload. An assessment of potential operator workload should be one of the first tasks in the design of human-computer dialogues. Because the probability of human failure increases in overload situations, the overall goal should be to keep the workload of the user within acceptable limits. Consideration should be given to the limited channel capacity of the human to define the operator's task and interactive dialogue requirements. If the human operator acts as a single-channel capacity device, information from various sources arrives and is queued until processing can occur. Displayed output should be organized to minimize the scanning required of the user. Workload considerations in human-computer interactions have implications for determining the information density on display screens, providing redundant information in multiple information channels, determining the appropriate size for a command language, and allocating tasks between the human and the computer.

Evaluation of Existing Dialogue Guidelines

Although the Williges and Williges (1982) compilation postulate a large variety of dialogue design guidelines, they are restricted primarily to alphanumeric information. Some consideration is given to dialogues dealing with graphic information. Only a few user considerations deal with voice input/output configurations or with "intelligent" systems incorporating rule-based or other artificial intelligence techniques. A need clearly exists to extend and develop dialogue considerations dealing with adaptive-human computer systems, computer display of graphic information, and dialogues incorporating various aspects of speech technology.

In addition to the lack of postulated guidelines for several human-computer dialogue configurations, the compilation by Williges and Williges (1982a) also resulted in some direct conflicts. Seven conflicting considerations were offered. These areas of conflict deal with blinking codes, letter codes, number codes, command organization, separators/terminators, system lockout, and error messages. In regard to error messages, for example, one reference states that it is permissible to refer the user to off-line documentation to specify what remedial action to take. Another reference, however, indicates that the user should not have to translate error messages via an off-line reference system.

Both the neglected dialogue areas and the direct conflicts of existing considerations suggest needed areas of research. For example, Williges and Williges (1982b) discuss several of the research gaps in structuring human/computer dialogue using speech technology.

In cases where conflicting user considerations are offered, the designer must determine which to adopt until behavioral research can resolve these conflicts and/or establish the appropriate context for each. Even the non-conflicting considerations have only sparse research support as to their validity. Consequently, a great deal of human factors research needs to be directed toward developing and validating empirically based dialogue design guidelines for generic human-computer tasks.

RESEARCH ON HUMAN-COMPUTER DIALOGUE PRINCIPLES

To provide examples of empirical research directed toward the development of general human-computer dialogue principles, the results of two behavioral studies are summarized. Both of these studies evaluated dialogue in interactive text editing environments which represent fundamental task of locating, retrieving, and manipulating alphanumeric information. The first study deals with the development of a user's model for command language selection, and the second study concerns the evaluation of procedures for the retrieval of HELP information.

User Models of Command Language Selection

Most interactive text editors have been developed with little regard to the user's model of the editing task. Consequently, the novice user is often faced with the difficult task of determining the appropriate system model of a particular text editor. Recently, Folley and Williges (1982) developed a method for determining novice and expert user's models of interactive text editing by empirically analyzing patterns of command language usage using statistical clustering algorithms. These models, in turn, can form the basis for developing an interactive editor which is designed from the user's, rather than the system's, point-of-view.

Method. A set of 40 command names were used by 20 computer-naive subjects and 20 computer experts to perform 20 different edit changes representing various editing tasks at the character, word, line, and general level. All subjects were instructed on the 40 editing commands before beginning the text editing. Subjects were then given two copies of the text to be edited. A BEFORE text had 20 corrections marked with circles, arrows, or underlines. An AFTER text showed the manuscript in its corrected form. The subject's task consisted of listing the commands to be used in the order in which each would be used to make the given edit change. The commands chosen were marked on an optical scan form for subsequent computer analysis.

Results. A series of hierarchical cluster analyses was conducted to determine the expert and naive users' models. First, the 20 editing changes were clustered on the basis of the commands used to perform them. Both the naive users' and the expert users' data yielded the same four general clusters as shown in the top portion of Figure 1. These edit clusters included changing and inserting characters and words; deleting characters, words, and lines; moving and copying lines; and changing and inserting lines. After determining each of four common edit task clusters, each of these clusters was analyzed separately to determine the unique naive and expert user's clusters of commands. The commands which were appropriate for each of the four general edit tasks are listed separately for naive and expert users in the bottom portion of Figure 1. Clearly, the list of commands in the two user clusters differ quite markedly both across the four general edit tasks and in terms of the unique commands. Essentially, the expert user added more powerful commands to those included in the naive users' model, but neither groups' cluster incorporated the full set of available commands. Consequently, both naive and expert users agreed quite closely as to the four general edit tasks, but each group had a different user model for the command repertoire within each of the four general edit tasks.

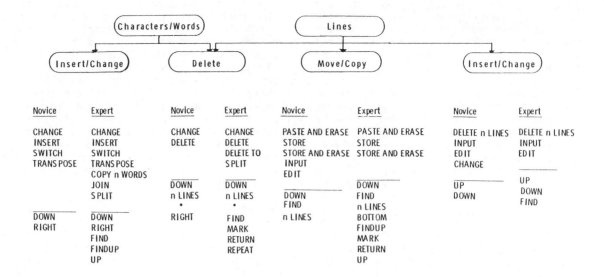

Fig. 1. Summary of naive and expert user commands within each of the four editing
          clusters (from Folley and Williges, 1982).

Implications. The results of this study suggest that cluster analysis may provide a useful method for inferring the user's model of text editing. Given this method can be validated and interactive editors built around user models are more effective than those currently built around system models, then an important design principle could be stated. Namely, such a principle would recommend the development of a users' model to form the basis of the command language dialogue design of interactive text editors. Additional research is needed to evaluate the validity of this guideline.

Retrieval of HELP Information

A second study which was designed to provide behavioral data for dialogue design guidelines investigated methods to enhance the retrivel of information from computerized HELP displays. Cohill and Williges (1982) evaluated various procedures for initiating, presenting, and selecting HELP on interactive computer systems. Their primary concern was to determine what roles the user and the computer should perform in the retrieval of HELP information.

Method. A control condition in which no HELP was available was compared to eight experimental conditions formed by the factorial combination of initiation (user vs. computer), presentation (hard copy manual vs. on-line), and selection (user vs. computer) of HELP information. Computer novice subjects learned a version of an experimental line editor SAM (Ehrich, 1980) and used SAM to perform both text and data file editing tasks either by using a constrained or unconstrained set of editing commands. Various measures of time, accurracy, and commands used to complete the various editing subtasks were recorded automatically by the metering system in SAM.

Results. Two general results are evident from the data analysis as summarized in Table 2. First, HELP information of any configuration manipulated in this study resulted in improvement in operator performance when compared to the no HELP configuration. As shown in the last row of Table 2, the time to complete a subtask, the errors per subtask, and the number of commands used were all significantly larger ($p<.05$) when no HELP was provided as compared to the various HELP configurations. Even more striking was the result in Table 3 which showed that subjects who did not receive any HELP could only complete an average of 3.4 of the 5 subtasks, whereas subjects who received any form of HELP could essentially complete all the editing subtasks.

The second general finding of this study was related to a relative comparison of the various HELP configurations. As summarized on the first row of Table 3, the most effective HELP configuration was user initiation and selection of HELP presented in hard copy manuals. This configuration resulted in significantly ($p<.05$) less time to complete an editing subtask, fewer errors per subtask, and fewer number of commands used per subtask.

Implications. The finding of this study aid in resolving conflicting dialogue guidelines and suggests that user initiated and selected hard-copy HELP yields the best performance. In this particular HELP configuration, the subjects spent most of

TABLE 2 Comparison of various HELP configurations evaluated by Cohill and Williges (1982)

	HELP Configurations					
Initition	Presentation	Selection	Time in Subtask	Errors per Subtask	Commands per Subtask	Subtasks Completed
User	Manual	User	293.1	0.4	8.4	5.0
User	Manual	System	442.2	2.0	17.7	4.9
User	Online	User	350.9	1.1	13.5	5.0
User	Online	System	382.2	1.8	17.6	4.9
System	Manual	User	367.9	1.3	13.1	4.8
System	Manual	System	399.1	0.9	13.3	4.9
System	Online	User	425.9	2.8	15.1	5.0
System	Online	System	351.5	1.2	13.7	4.9
Control:	No HELP Available		679.1	5.0	20.2	3.4

their time browsing the HELP information and looking at a variety of information contained in the HELP file. All of the computer initiated and selection configurations provided quite specific information and perusing of other information was not possible. Consequently, a dialogue design consideration based on the results of this study would recommend that HELP information be constructed such that the user can browse the various information files. Enhancements of these automatic browsing features still need to be developed.

IMPLEMENTATION OF HUMAN-COMPUTER DIALOGUE GUIDELINES

The two studies reviewed in this paper provide examples of the type of human-computer dialogue investigations that need to be conducted in order to generate empirically based dialogue design guidelines. Williges and Williges (1982b) discuss the need for using both theoretical and empirical model building procedures to build a comprehensive data base of design guidelines.

Even if a completely comprehensive and non-conflicting data base were available, it would be necessary to determine how this information can be best conveyed to the designer of dialogue software. The usual approach is merely to compile these considerations into a handbook with little retrieval assistance beyond a table of contents and/or index. Due to the complexity and overlapping nature of any comprehensive dialogue data base, the organization of these handbooks and subsequent search for relevant guidelines quickly becomes unmanageable. Consequently, various forms of computer aiding should be considered. For example, the computer aiding may be no more than tree searching procedures for data retrieval or may incorporate sophisticated rule-based procedures to aid in decision making.

Computer-aided implementation may include four basic stages. First, the complete set of empirically derived dialogue guidelines must be available in a computerized data base. Obviously, this data base will expand and require updating as additional research is completed. Second, an on-line acquisition procedure is needed to retrieve the dialogue principles relevant to the dialogue configuration required. Third, decision aiding may be helpful to select the appropriate set of dialogue guidelines for a particular application environment. Rule-based systems need to be developed to select the optimal dialogue configuration from the set of dialogue guidelines available. And, finally, the fourth stage of computer-aiding requires a set of software tools for dialogue implementation. One such approach to providing these implementation tools is the Dialogue Management System described by Ehrich (1982) which provides a flexible environment for the dialogue author.

Research is needed both to develop computer-aided dialogue design guideline implementation and to evaluate the efficacy of each stage of this computer aiding. The conduct of systematic research which is directed toward these issues might prove to be quite useful in improving the implementation of human factors dialogue design guidelines during the early stages of systems design of interactive computer systems as well as in the redesign of existing systems.

CONCLUSIONS

Major improvements can be made in the human-computer interface through human factors design of user-initiated, computer-initiated, and mixed-initiative dialogues. Design oriented guidelines and principles which are based on research support need to be developed and organized into a retrievable data base. Computer-aided procedures should be developed to facilitate the implementation of the dialogue guidelines into the design of interactive systems. Such an approach should improve the human-computer interface in terms of user acceptance, satisfaction, and productivity.

REFERENCES

Cohill, A. M. and Williges, R. C. (1982). Computer-augmented retrieval of HELP information for novice users. Submitted to Human Factors Society 1982 Annual Meeting, Seattle, Washington.

Ehrich, R. W. (1981). SAM--A configurable experimental text editor for investigating human factors issues in text processing and understanding. Blacksburg, Virginia: Virginia Polytechnic Institute and State University, Technical Report CSIE-81-4.

Ehrich, R. W. (1982). DMS--A system for defining and managing human-computer dialogues. Proceedings of IFAC/IFIP/IFORS/IEA Conference on Analysis, Design, and Evaluation of Man-Machine Systems, Baden-Baden, FRG.

Folley, L. J. and Williges, R. C. (1982). User models of text editing command languages. Proceedings of Human Factors in Computer Systems Conference, Gaithersburg, Maryland. pp. 326-331.

Williges, B. H. and Williges, R. C. (1982a). Dialogue design considerations for interactive computer systems. In B. Shackel (Ed.), Usability of Human-Computer Systems.

Williges, B. H. and Williges, R. C. (1982b). Structuring human/computer dialogue using speech technology. Proceedings of Workshop on Standardization for Speech I/O Technology, National Bureau of Standards, Gaithersburg, Maryland.

ONLINE INFORMATION RETRIEVAL THROUGH NATURAL LANGUAGE

G. Guida[1] and C. Tasso[2]

Istituto di Matematica, Informatica e Sistemistica, Università di Udine, Udine, Italy

Abstract. The task of constructing a natural language interface for the access to online information retrieval systems is faced. The peculiarities of this application, which requires both natural language understanding and reasoning capabilities, are first discussed. The general architecture and the fundamental design criteria of a system presently being developed at the University of Udine are then presented. The system, named IR-NLI, is aimed at allowing non-professional users, to directly access through natural language the services offered by online data bases. Attention is later focused on the basic functions of IR-NLI, namely, understanding, strategy generation, and reasoning. Knowledge representation methods and algorithms adopted are shortly discussed and illustrated. A brief example of interaction with IR-NLI is presented. Perspectives and directions for future research are also discussed.

Keywords. Artificial intelligence; man-machine systems, information retrieval; natural language dialogue; reasoning; expert systems.

INTRODUCTION

Natural language understanding and expert systems constitute two of the most challenging and promising topics within artificial intelligence. Their technologies are often combined together into high-level systems, where natural language is used to provide easy access to consultation services for casual users. Still the issue of conceiving in an unitary way understanding and reasoning capabilities is quite new and shows several interesting facets. From the point of view of the foundations of artificial intelligence, it would be very desirable to face and model in an unitary way these activities, which appear strongly connected in humans (Smith, 1980). In fact, understanding is often aimed at acquiring knowledge to be utilized in deductive and inductive activities; furthermore, reasoning on domain-specific and common-sense knowledge is required in several cases for triggering the comprehension process. From a more technical point of view, it is argued that a common knowledge representation could be appropriate for both understanding and reasoning activities in order to ensure a good level of integration and performance. Moreover, a unitary internal representation could suggest the adoption of homogeneus, mainly rule-based, algorithms for the basic processing steps of understanding and reasoning.

In this paper we face the above topics in the frame of a particular application, which fits quite well the combined issues of natural language comprehension and knowledge based consultation: the access to online data bases through natural language (Doszcos, 1979). In fact, it is well known that this task requires not only a basic natural language understanding capability, but also the ability of performing the intermediary's role: analyzing and deeping user's requests, reasoning on them, and generating an appropriate search strategy (Politt, 1981). Particular attention will be devoted in this paper to the problems related with the general architecture and organization of an experimental system, called IR-NLI (Information Retrieval - Natural Language Interfa-

---

(1) also with: Milan Polytechnic Artificial Intelligence Project, Milano, Italy.

(2) also with: CISM, International Center for Mechanical Studies, Udine, Italy.

ce), which is presently being developed at the University of Udine to sup
port the new issue of natural language reasoning.

IR-NLI takes strongly into account
previous experience of the authors in
the area of natural language processing (Guida and Tasso, 1982a,b,c) acquired in the development of the NLI
system for the natural language access to a relational data base. From
this work the original methodology of
goal oriented understanding is fully
inherited into IR-NLI, although the
technical details concerning its implementation are here revised according to the new knowledge representation adopted.

In the paper the motivations underlying the design of IR-NLI are first
analyzed and the general goals and fea
tures of the application are illustra
ted. The technical specifications of
IR-NLI are then defined and the overall architecture of the system is pre
sented. The mode of operation of the
basic modules of IR-NLI is shortly
described and an example is presented.
Perspectives and directions for future research are also discussed.

MOTIVATIONS

The development of effective and sophisticated   telecommunication techniques, which has taken place in the
last decade, has contributed to a wide
diffusion   of telematics. One of the
most utilized services available
through computer networks is online
information retrieval, that allows in
terested users to access very large
amounts of information stored in remo
te computer installations. The kinds
of accessible data bases range over
almost all fields of science, technology,economics, and humanities.

Information retrieval systems always
include a formal query language, which
is used for selecting from the data
base the desired information through
a sequence of commands (search,display,
boolean operations, etc.) combined to
gether to form a search strategy. Gen
erally   the end user is unwilling to
search personally, but resorts to the
assistance of a qualified information
professional, the intermediary, who
knows how to select the appropriate
data base, to use the formal query lan
guage, and to devise an appropriate
strategy. Usually the interaction between user and intermediary starts with
a dialogue aimed at precisely clarify-
ing the objectives of the request.On
the base of the information gathered,
the intermediary chooses the most suit
able data bases and, with the help
of searching referral aids such as the

sauri,directories, etc., he plots the
search strategy to be executed on the
system. The output of the search is
then evaluated by the user, who may
propose a refinement and iteration of
the search for a better matching with
his request.

Going deeper in analyzing the peculia
rities of online information retrieval, let us contrast this application
to the enquiry of a data base management system. Let us consider two
main points. First, the information
stored in a data base is precisely
described by a definite logical struc
ture and, therefore, the navigation
in the data base in order to get desired information items is quite easily and univocally identified.On the
other hand, in the information retrie
val case, the stored records are
identified through descriptive information (brief abstracts,summaries,key-
words, etc.), which only partially cap
tures the precise content of the document and, also, the logical organization of the files is very poor. This
implies that retrieval of requested
information is much more difficult,
since it relies on several loosely
defined factors,such as domain speci
fic knowledge, knowledge about inde-
xing criteria, availability of updated and complete searching referral
aids, working experience on the particular data base, etc. This first
feature is clearly captured by the
two traditional criteria of precision
and recall (Salton,1968) which are
used to evaluate the quality of a
search strategy. Second, while a user
of a data base generally looks for
precise information to be extracted
from the stored files,a user of an on-
line service desires to get information on a given topic which does not
immediately correspond to some stored object; what records to extract
in order to match the user's request
is usually a non-trivial problem,that
has to be solved by the intermediary.

The above illustrated peculiarities
motivate the design of a natural lan
guage expert system for interfacing
online data bases. In fact,the IR-NLI
project has among its long-term goals
the implementation of a system to be
interposed between the end-user and
the information retrieval system,
which should be capable of fully sub
stituting the intermediary's role
(Anderson, 1977; Waterman, 1978). IR-
NLI should behave as a skilful consul
tant in restricted, well-defined domains, and translate natural language
user's requests into appropriate
search strategies expressed in a
formal query language, in such a way
to make technical details of the information retrieval process fully

transparent to the user.

SYSTEM SPECIFICATIONS AND ARCHITECTURE

IR-NLI is conceived as an interactive interface to online information retrieval systems supporting English language interaction. Its specifications may be divided into two classes:

- at a basic level the system is only required to understand the user's requests (possibly expressed through a limited dialogue) and to translate them into a correct sequence of formal search statements;

- at an higher level of performance the system must also be able of reasoning on the user's requests in order to devise an appropriate search strategy, which could generally be not immediately derivable from the user's query.

Clearly, the first class of specifications is mainly intended for trained users, which are able of expressing appropriate requests that can be directly translated into a search program. For casual users, on the contrary, the reasoning capability is strictly needed, since their way of expressing queries is often very far from the actual search strategy that must be utilized to satisfactory answer them. Let us note that both classes of specifications may be suitable both for interactive users, who are willing to access the available data bases through an online worksession, and for professional users, who often prefer to prepare the most appropriate search strategy during an offline session with IR-NLI and to access the online data base only later in order to save access time.

Let us now illustrate in some detail the overall architecture of IR-NLI, which is shown in Fig. 1.

The kernel of the system is constituted by the UNDERSTANDING and REASONING modules which are devoted, respectively, to translate the natural language user's requests into a formal internal representation called basic internal representation (BIR) and to reason on this representation in order to gather the information needed for generating an adequate search strategy. The result of the reasoning activity on the basic internal representation is called extended internal representation (EIR). Let us note that both UNDERSTANDING and REASONING modules may engage a dialogue with the user, whenever necessary for a more effective interpretation of his requests and a more precise identification of his needs. UNDERSTANDING and REASONING

utilize two knowledge bases containing linguistic and domain specific knowledge, respectively. These bases are not fully separate, but generally share a common subset of information which may be used by both modules (e. g., thesauri, indexes, etc.). Both UNDERSTANDING and REASONING modules are organized as bags of special-purpose functions which are available for parsing the user's requests and developing reasoning activities on them (e.g., deduction, induction, generalization, particularization, analogizing, etc.).

The activity of UNDERSTANDING and REASONING modules is managed by the STRATEGY GENERATOR, which is devoted to devise the top-level choices concerning the overall operation of the system and to control their execution. It utilizes for its activity a base of expert knowledge on how to evaluate user's requests, to select the suitable approach for generation of the search strategy, and to schedule the activities of the lower modules. The operation of the STRATEGY GENERATOR is organized around a basic pattern-directed algorithm which, taking into account the appropriate expert rules that apply to the different situations, fires a suitable sequence of understanding and reasoning functions until the extended internal representation is completely expanded and validated.

After the STRATEGY GENERATOR has completed its activity, the FORMALIZER generates from the extended internal representation the output search strategy to be executed for accessing the online data base. The FORMALIZER utilizes for its operation knowledge about the formal language needed to interrogate the online data base and operates through a simple syntax-directed translation schema. It is conceived as a parametric translator capable of producing search strategies in several languages for accessing online services, such as Lockheed DIALOG, Euronet DIANE, SDC ORBIT, etc.

BASIC MODE OF OPERATION

We analyze in this section the basic features of STRATEGY GENERATOR, REASONING, and UNDERSTANDING modules, and provide a short description of their mode of operation.

The STRATEGY GENERATOR is devoted to two main tasks:

- to perform the activity proper of the intermediary's role;

- to manage and control the activity of REASONING and UNDERSTANDING modules.

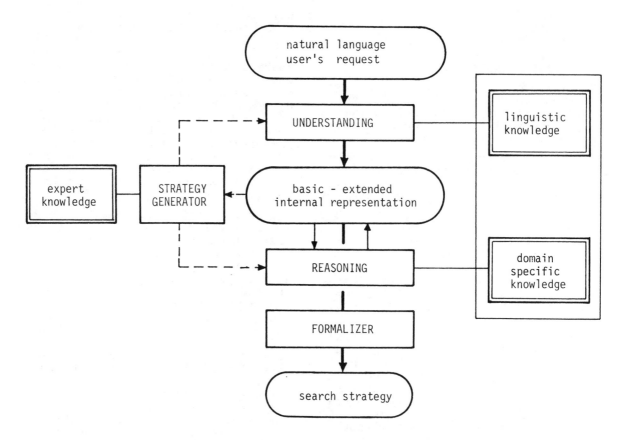

Fig. 1. General Architecture of IR-NLI.

From this point of view it must embed expert capabilities and behave as a consultation system for information retrieval (Politt,1981). The basic mode of operation of the STRATEGY GENERATOR is organized around the following main step:

1. activate the UNDERSTANDING module and construct the BIR of the user's request;

2. select a suitable approach for the construction of the search strategy;

3. carry out the chosen approach through activation of the REASONING module for execution of the involved tactics; incrementally expand the BIR into the EIR;

4. possibly engage a dialogue with the user for refining and deeping his request and validating the EIR currently being constructed (the UNDERSTANDING module may be newly activated in this step, if necessary for supporting the dialogue with the user);

5. when the EIR is fully expanded and refined, activate the FORMALIZER for generation of the search strategy.

To go further in our description, let us introduce precise definitions of three technical terms above used in an informal way:

search approach: the abstract way of facing a search problem, reasoning on it, analysing its facets, and devising a general mode of operation for having access to desired information stored in an online data base;

search tactic: a move, a single step or action, in the execution of a search approach;

search strategy: a program, written in an appropriate formal query language, for obtaining desired information from an online system; the result of the execution of a search approach through application of appropriate search tactics.

Within IR-NLI, a search approach is represented as a non-deterministic program that defines which tactics to utilize, among the available ones, and how to use them in the construction of a strategy. An approach is not, therefore, a fixed, well-defined procedure, since it does not specify at each step which particular tactic to execute, but only suggests a set of candidate tactics, whose execution may or may not be successful. Four approaches are considered among the most classical and commonly utilized ones: building blocks approach, citation pearl growing, successive fraction approach, and most specific (or lowest postings) facet first approach.

Tactics are represented at two different levels of abstraction:

- high-level representation < name, preconditions for applicability, goals-effects on EIR > is provided for use by the STRATEGY GENERATOR;

- a low-level representation < name, reasoning rules involved, action on EIR > is supplied for use by the REASONING module.

The connection between the two types of representations is through the 'name' descriptor. About 15-20 tactics are considered, taken from the very rich discussion by Bates (1979): SELECT, SPECIFY, EXHAUST, REDUCE, PARALLEL, PINPOINT, SUPER, SUB, RELATE, NEIGHBOR, VARY, FIX, REARRANGE, CONTRARY, RESPELL, RESPACE, etc.

The specific role of STRATEGY GENERATOR (steps 2 and 3 above defined) can now be stated in a more detailed way:

2'. evaluate the promise of each approach to be a successful one through weighted pattern-matching between the BIR and the high-level representation of the tactics involved in the approach itself; select that approach which best fits the search problem currently faced;

3'. carry out the chosen approach through execution of the non-deterministic program that represents it; activate the REASONING module for trying the execution of all the candidate tactics that are suggested at each step (choose the tactic to execute first according to a fixed priority order).

Note that the choice of the best fitting approach performed in step 2' is a merely static one, since it only involves the high-level representation of tactics, and does not try to guess which of them will actually be executed with success during carrying out of the approach.

The REASONING module has the main task of trying the execution of the tactics requested by STRATEGY GENERATOR. It utilizes both the high-level and low-level representations of a tactic and the current EIR to select and fire the appropriate reasoning rules which are necessary to execute the requested tactic. Among the basic capabilities of the REASONING module we consider analogizing from experience, generalization to broader terms, particularization to narrower terms, extension to related concepts, analysis of synonimi and omonimi, domain specific in-

ference,etc.

The mode of operation of the REASONING module may be viewed as structured around the following main steps:

1. accept from the STRATEGY GENERATOR a tactic to be executed;

2. verify its executability through pattern-matching between the 'preconditions for applicability' descriptor in its high-level representation and the current EIR; if not executable return a failure to the STRATEGY GENERATOR, otherwise continue;

3. try firing all the reasoning rules specified in the 'reasoning rules involved' descriptor of the low-level representation of the current tactic;

4. when all firable reasoning rules have been executed, refine the EIR according to the 'action on EIR' descriptor of the low-level representation of the current tactic.

Note that firing of a reasoning rule does not change the content of the EIR, but only provides new information that will be utilized later for the execution of the 'action on EIR' descriptor of the tactic to which the rule belongs. This prevents that execution of a reasoning rule may affect, by changing the EIR, firability of other rules belonging to the same tactic. In this way, the order in which firing of reasoning rules is tried is immaterial to the global effect of execution of a tactic.

The UNDERSTANDING module is devoted to provide the correct BIR, which can be constructed from the natural language user's request. Its conception strongly relys on the experience developed by the authors with the NLI system,and is organized around the concept of goal-oriented parsing (Guida and Tasso 1982a,1982c).

The operation of UNDERSTANDING module is mainly semantics-directed and is organized around a rule-based mechanism: a main parsing algorithm performs the most elementary steps of the analysis (search in the linguistic knowledge base, construction of a tentative internal representation,validation of the basic internal representation), and manages a pattern-directed invocation of heuristic rules for the resolution of critical events (e.g., ambiguities, ellipsis,anaphora, indirect speech, etc.). An important feature of the UNDERSTANDING module is the ability to solve critical situations by engaging the user in a clarification dialogue of bounded scope (activated by some of the above

mentioned rules) to gather additional
information which is necessary to cor
rectly understand the input natural
language queries.

AN EXAMPLE

In this section we present a short
example of the basic mode of operation
of IR-NLI. Fig. 2 shows a sample session in which, in addition to the user-
system dialogue, the basic and extended internal representations and the
search strategy generated (in Euronet
DIANE EUROLANGUAGE) are reported. The
example is quite selfexplanatory and
needs only some comments on knowledge
representation. Let us note that one

of the most distinctive features of
IR-NLI is the adoption of a unitary
internal representation for both understanding and reasoning activities.
The knowledge representation designed,
which is only partially shown in Fig.
2, is basically constituted by a frame, which is first instanciated by
the UNDERSTANDING module, and then
further refined and expanded by the
REASONING activity. This frame is
structured in such a way to contain,
classified under different semantic
types, any information that is relevant for searching an online data
base, and to allow an effective and
sophisticated pattern-matching with
tactics and reasoning rules.The flex-

```
> PLEASE SPECIFY YOUR REQUEST

* I would like to have information about intelligent terminals for online
  information retrieval

> BIR : CONCEPT-1 : INTELLIGENT TERMINALS
        CONCEPT-2 : ONLINE INFORMATION RETRIEVAL

> BUILDING BLOCKS APPROACH SELECTED

> TRYING TACTICS  : PARALLEL
                    RELATE
                    RESPELL
                    RESPACE

> EIR : CONCEPT-1 : INTELLIGENT.TERMINAL$
        CONCEPT-2 : ONLINE, ON.LINE, INTERACTIVE, REALTIME, REAL.TIME
        CONCEPT-3 : INFORMATION, REFERENCE, BIBLIOGRAPHIC$, RETRIEV$, SEARCH$

> DO YOU LIKE TO HAVE A GENERAL VIEW OR DETAILS ON SPECIFIC FACETS ?

* costs

> TRYING TACTICS  : PARALLEL
                    RESPELL

> EIR : CONCEPT-4 : COST$, PRICE$

> ANY MORE REQUEST ?

* no

> DO YOU LIKE ONLINE OR OFFLINE PRINTING ?

* offline

> PLEASE SPECIFY FORMAT : WITH OR WITHOUT ABSTRACT ?

* with abstract

> SEARCH STRATEGY GENERATED

> STRATEGY       : 1 FIND INTELLIGENT.TERMINAL$
                   2 FIND ONLINE OR  ON.LINE OR INTERACTIVE OR REALTIME
                       OR REAL.TIME
                   3 FIND INFORMATION OR REFERENCE OR BIBLIOGRAPHIC$
                       OR RETRIEV$ OR SEARCH$
                   4 FIND COST$ OR PRICE$
                   5 FIND S=1 AND S=2 AND S=3 AND S=4
                   6 PRINT  F1
```

Fig. 2. Sample dialogue with IR-NLI.

ibility allowed to the frame structure adopted is limited to the possibility of dealing with the four search approaches mentioned in the previous section, which require slightly different organizations of the internal knowledge representation.

PERSPECTIVES AND CONCLUSIONS

In the paper the main features of the IR-NLI system for natural language access to online data bases have been discussed. The project described is to date in an advanced design stage, and a bounded scope experimental activity aimed at concrete evaluation of the most critical parts of the system is planned for the next months. The authors also plan to implement in the next future a complete prototype version of the system to be connected to a real online service.

Several topics in the project reported in the paper deserve further research and will be faced in future works. Among them we mention:

- the design of dynamic techniques for evaluating and selecting the most suitable approach to a given search problem, through the implementation of guessing capabilities;

- the development of more flexible and robust dialog capabilities,including limited justification of the mode of operation of the system (Hayes and Reddy,1979; Webber,1982);

- the design of new tactics (e.g., PATTERN,RECORD, BIBBLE (Bates,1979)) and reasoning rules, that enable the system to keep track of previous search sessions and to analogize from experience in devising and executing a search approach;

- the refinement of the current representation of a search approach in order to enable the STRATEGY GENERATOR to go beyond the fixed set of possibilities presently offered and to change an approach during execution if its performance reveals unsuccessful.

REFERENCES

Anderson, R.H.,(1977). The use of production systems in RITA to construct personal computer "agents". ACM SIGART Newsletter 63, 23 -28.

Bates,M.J. (1979). Information search tactics. J. of the Am. Soc. for Information Science 30, pp. 205-214.

Doszcos, T.E. and B.A. Rapp (1979). Searching Medline in English: A prototype user interface with natural language query, ranked output,relevance feedback. In Proceedings A.S.I.S. Annual Meeting, vol.16, 131-137.

Guida,G. and C. Tasso (1982a). NLI: A robust interface for natural language person-machine communication. Int. J.Man-Mach.Stud., 17.

Guida,G. and C.Tasso (1982b). Natural language access to online data bases. In Proceedings 6th European Meeting on Cybernetics and System Research, Wien, Austria, April.

Guida,G. and C. Tasso (1982c) Dialogue with data bases: An effective natural language interface. In Proceedings Informatika 82, Ljubljana, Yugoslavia, May.

Hayes, P. and R. Reddy (1979). Graceful interaction in man-machine communication. In Proceedings 6th Int. Joint Conference on Artificial Intelligence, Tokio, Japan, August, pp. 372-374.

Politt, A.S. (1981). An expert system as an online search intermediary. In Proceedings 5th Int. Online Information Meeting, London, England, December, pp. 25-32.

Salton, G. (1968). Automatic Information Organization and Retrieval. McGraw-Hill, New York.

Smith, L.C. (1980). Implications of artificial intelligence for end user use of online systems. Online Review 4, 383-391.

Waterman, D.A. (1978). Exemplary programming in RITA. In D.A. Waterman and F.Hayes-Roth (Eds.), Pattern Directed Inference Systems, Academic Press; New York, pp. 261-279.

Webber, B,and A. Joshi (1982). Taking the initiative in natural language data base interactions: justifying why. In Proceedings COLING-82, Prague,July, North-Holland Linguistic Series 47, pp.413-418.

NATURAL-LANGUAGE COMMUNICATION WITH COMPUTERS: A COMPARISON OF VOICE AND KEYBOARD INPUTS

E. Zoltan*, G. D. Weeks** and W. R. Ford**

*Department of Psychology, The Johns Hopkins University, Baltimore, Maryland, USA
**Prism Associates, 7402 York Road, Baltimore, Maryland, USA

Abstract. This study investigated people's natural spoken and typed language as they communicated with a computer during a checking account management task. No restrictions were placed on the subjects' vocabulary, syntax, or manner of speaking. Results show that spoken entries, as compared to typed entries, are more verbose, contain more unique words, and take significantly less time to transmit. Regardless of the communication mode, the subjects' messages were clear, to the point, and consisted of simple grammatical structures. The results are compared with those obtained in studies of person-person communication.

Keywords. Computer interfaces; man-machine systems; artificial intelligence; telecommunication; language analysis; voice-input.

INTRODUCTION

For some time now there has been hope for widespread development of natural-language information systems for computers. The failure of such systems to materialize has brought a rash of scepticism over the possibility of developing natural-language systems. Dreyfus (1972) pointed out that computer techniques for natural-language understanding were inefficient, costly, and below performance requirements for practical use. Moreover, some have even questioned the suitability of natural language for person-computer communication. Fitter (1979) and others believe that natural language is too unstructured to ever serve as a medium for person-computer interaction. Recent advances, however, give cause for reconsideration. Ford (1981) has developed a promising algorithm for accommodating natural language. In a test of performance, the program responded correctly to 94% of the 1697 natural-language entries typed by 10 computer-inexperienced users. This algorithm and others like it may well lead to the widespread development and use of natural-language information systems.

Concurrent with these developments have been developments in voice input systems (Lea, 1978). With such advances, it is not hard to envision the marriage of these two technologies in natural-language voice systems. The obvious question becomes: How will people naturally communicate by voice with computers?

One approach to answering this question is to use person-to-person telecommunication as a model for person-computer interaction. We have found that keyboard inputs to computers are similar in many respects to keyboard communications between people. For example, Chapanis and others (1977) observed that grammatically correct messages were the exception rather than the rule in teletypewriter communication. Similarly, Ford (1981) reported that only 2.2% of subjects' typed computer entries were grammatically correct with respect to proper capitalization and punctuation. One characteristic of telecommunications keyboard input that does not parallel person-computer keyboard input is the frequent occurrence of misspelled words. In an unpublished analysis of the teletypewriter transcripts from the Chapanis and others study (1977), 4.5% of words transmitted by subjects were found to be misspelled. In Ford's analysis of person-computer interaction, however, only .4% of the words transmitted by subjects were misspelled. While the reason for this difference is not clear, it is clear that transference of keyboard communication descriptors between telecommunications and person-computer interaction is not direct in all cases.

Interpersonal communication as a model for person-computer communication may be even further limited in the voice mode. Telecommunications studies comparing voice and keyboard inputs reveal consistent differences between the two modes. For example, problems are solved twice as quickly when subjects communicate by voice as when they communicate by typing. Also, voice communication is more verbose and redundant than keyboard communication (Ford, Chapanis and Weeks, 1979; Weeks and Chapanis, 1976). However, it is not known whether these mode effects will also be found in person-computer communication.

The primary purpose of this study was to

obtain a precise description of the natural language used by speakers interacting with a "perfect" voice recognition system. The system was perfect in the sense that it imposed no restrictions on a person's vocabulary, syntax, or manner of speaking. A second purpose of the study was to compare the spoken inputs with the keyboard inputs of other subjects who were preforming the same task.

METHOD

Subjects

Eighteen subjects, four males and fourteen females, from the Baltimore area, were recruited for the study through informal advertisements. Persons who responded to the "participants needed" advertisement were screened for previous computer experience. Only those respondents without computer programming experience were selected. The subjects were paid for participating in the study.

Apparatus

The laboratory and the apparatus used in this study are shown in Figure 1. Subjects interacted with an IBM 370/168 computer through an IBM 3270 series communication system. The subject's room (6'5" X 8'9") contained a computer terminal (IBM 3277 model II), an adjustable terminal work station (NKR Environments, Ltd.), two chairs, and communications equipment necessary for both communication modes described below.

Keyboard mode. In the keyboard mode, subjects interacted with the applications program through an IBM keyboard and terminal display. That is, subjects typed their inputs into the computer and received the program's output over the CRT screen. While the majority of the alphanumeric and terminal function keys were left unchanged, two modifications were made to the IBM keyboard to enhance its usability. First, a label was placed on the space bar ("SPACE BAR") to remind subjects of the functional difference between spacing and using the cursor control key. Second, since the dual function of the ENTER key (i.e., entering information and paging), can be confusing to novice users, the paging function was assigned to a separate key. This was done by reprogramming the CLEAR key and relabelling the key cap as NEXT PAGE. When a screen was full, users were instructed to push the NEXT PAGE key.

Voice mode. In the voice mode, subjects spoke into a minature headset microphone (Electro-voice, model 651) and received the application program output over the CRT screen. The subjects' microphone was ostensibly connected to a Compu-voice II Voice Analyzer which in turn appeared to be connected to the 3277 display. In reality, the subject's microphone was connected to an audio tape recorder (Kenwood, model KX-1030) located in the observation room. One of the experimenters listened to subjects' messages through headphones connected to the recorder, keyed in simplified versions of the messages, and sent the application program's output to the subject's display. For increased reliability, a second experimenter also monitored the subject's inputs.

Subjects' Task

In both conditions, the subjects' task was to manage their personal checking accounts. Each subject used the CHECKBOOK program (see Ford, 1981) once a week for four consecutive weeks. The first session occurred shortly after the subject received his or her monthly bank statement. The last session occured after a subject received the next bank state-

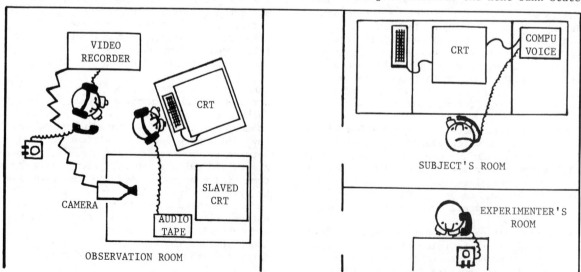

Fig. 1. The three experimental rooms and associated apparatus.

ment. Typical sessions involved recording transactions that had been made since the last visit (i.e., checks written, deposits made), asking the computer for the account balance, or requesting a display of some information such as all the deposits or just one check. During the fourth session subjects balanced their checkbook against their new statement.

Experimental Design

The experimental design included one between-subjects factor, communication mode, with two levels: keyboard and voice. Nine subjects were randomly assigned to each mode. It is possible to consider the four experimental sessions per subject as a within-subjects variable. Since subjects varied greatly from session to session in the number of their checking transactions, the four sessions for each subject were pooled and analyzed as only one task.

Procedure

The first sessions were scheduled to occur as closely as possible after the subjects received their bank statements. The subjects were told to reconcile their statements in the customary manner before the initial session.

When a subject arrived for the initial session, he or she was told that the purpose of the study was to test some newly developed computer equipment that allowed people to communicate with a checking account management program in ordinary, everyday English. The experimenter then read a description of the check management task to the subject and informed the subject that he or she would be paid $100.00 for participating in the study. A subject then read and signed an Informed Consent form.

After the subject had an opportunity to ask questions, the experimenter familiarized the subject with the computer equipment. For the keyboard condition, the experimenter described (1) the use of the alphanumeric and shift keys; (2) the difference between the space bar and the cursor control keys; (3) the NEXT PAGE key; and (4) the procedures for correcting errors. The subject then practiced using the keyboard.

For the voice condition, the experimenter described the operation of the microphone, the COMPU-VOICE II analyzer, and the CRT display station. The voice subject then "trained" the computer to recognize his or her voice by reading standardized phrases from the CRT screen (e.g., "Now is the time for all good men to come to the aid of their country").

After the experimenter was satisfied with the subject's proficiency with the computer equipment, he answered any remaining questions and then informed the subject that he would be in

the next room should any additional questions or problems arise during the session. The session began when the subject asked the computer (through typing or speaking) for the CHECK BOOK program and ended when the subject told the computer that he or she was done.

When the subject returned for his or her second, third, and fourth sessions, the experimenter reiterated the instructions to tell the computer about checking account transactions made during the previous week.

Data Collection and Analysis

A 'slaved' CRT in the observation room (see Fig. 1) displayed the material present on the subject's CRT. A video recording was made of the 'slaved' CRT to preserve the computer program's output in all sessions and the subject's inputs in the keyboard mode. Audio recordings were made of the subject's spoken inputs in the voice mode. All recordings were subsequently transcribed onto hardcopy and then entered into a computer for later data analyses.

Performance between the subjects in the two communication modes was assessed for five dependent variables: time to complete the task and four verbal measures (words, unique words, messages, and mean message length).

Time to complete the task. For each subject, the time for each session was measured in minutes from the subject's request for the CHECK BOOK program to the subject's statement that he or she was finished. Total time to complete the task was the sum of the subject's four session times.

Words. The total number of words used by each subject was determined on the basis of the following guidelines:
  1. Utterances in the voice mode such as "ah," "uh," "huh," and "mm-hm" were counted as words if they were perceived to be at the same audible level as other, more meaningful utterances;
  2. Mispronounced words in the voice mode and misspelled words in the keyboard mode were counted as words;
  3. Contractions, such as "I'd," were counted as single words;
  4. Colloquialisms and slang, such as "'em" and "yeah," were counted as words;
  5. Numbers such as 320 were counted as one word in the keyboard condition. For the voice condition, however, the word count varied as a function of the subject's speech. For example, 320 could be uttered as "three hundred twenty" (three words) or "three hundred and twenty" (four words).

Unique words. The total number of unique words used by each subject was recorded. Mispronounced and misspelled words were counted as different words from their correct prototypes.

Messages. For the keyboard condition, a message began when a subject started to type and ended when the subject pressed the ENTER key. For the voice condition, a message began when a subject started to talk and ended when the subject's intonation fell slightly or when the subject paused thereby breaking the communication flow.

Mean message length. Mean message length was the total number of words used by a subject divided by the total number of messages he or she transmitted.

RESULTS

The data were analyzed with a Hotelling's T^2 to control for experiment-wise alpha. The resulting T^2 value was highly significant (T^2 = 30.94, p < .001), indicating that at least one contrast exists between the two modes of communication. Individual t-tests were then conducted on each dependent measure to determine which effects differed significantly between the two groups. Four of the five dependent measures analyzed yielded significant results. Table 1 presents the means and the probability levels for all dependent measures.

keyboard and voice subjects are available from the authors.

Figure 2, shown on the next page, illustrates the functional relationship between unique and total word usage. Plotting the data in this cumulative manner, it can be seen that only 7, or about 2% of the unique words accounted for 50 percent of the total number of words used by keyboard subjects. Similarly, 29, or 4.5%, of the unique words accounted for one half of the total number of words used by voice subjects. As may be seen in Fig. 2, a little less than half of the vocabulary words accounts for 95 percent of all words used in each mode.

Revised Number of Words and Number of Unique Words

Using our method for counting words, a number such as "three hundred and twenty" when spoken was counted as four separate words, while the same number when typed (e.g., 320) counted as one word. Because this method could disproportionately increase the word and the unique word counts for voice sessions,

TABLE 1   Mean Values and Probability Levels (p) for the Dependent Measures

Dependent measure	Communication Mode		
	Voice	Keyboard	p
Time to complete task (min.)	39.8	85.4	.0096
Number of words	1146.6	452.1	.0141
Number of unique words	190.7	78.8	.0003
Number of messages	172.9	170.0	.8903
Mean message length	7.1	3.0	.0015
Revised number of words	817.7	288.1	.0228
Revised number of unique words	155.8	74.4	.0027

Time to Completion

On the average, the keyboard subjects spent more than twice as much time to complete the task as did the voice subjects.

Number of Words and Number of Unique Words

Subjects in the voice condition used, on the average, 2.5 times as many words and 2.4 times as many unique words as did subjects in the keyboard condition.

Summing over all subjects, a total of 840 unique words were used in the study. Of that total, 642 were used by subjects in the voice condition and 373 were used by subjects in the keyboard condition. There was, then, only a moderate overlap in the vocabularies used by both groups of subjects. Lists of the unique words used by keyboard, voice, and both

the protocols were reanalyzed with all numbers and all word representations of numbers removed. The new means and probability levels for these revised analyses are given below the dashed line in Table 1.

The difference between the two modes of communication still remains highly significant for these two measures, although the ratios between the modes changed slightly.

Number of Messages and Mean Message Length

The mean number of messages transmitted to the computer did not differ significantly between the two modes. The messages entered by the voice subjects were, however, significantly longer than those used by the keyboard subjects. Voice subjects used, on the average, 2.4 times as many words per message as did keyboard subjects.

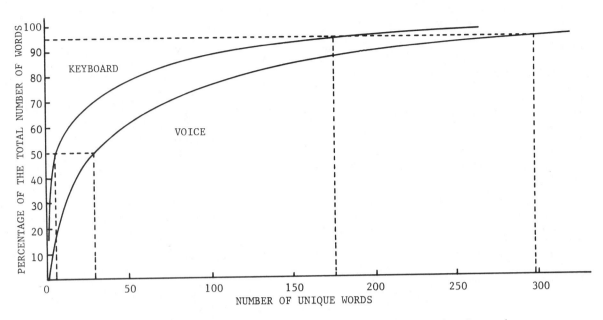

Fig. 2. Cumulative percentages of words as a function of communication mode.

DISCUSSION

At the beginning of this paper, we suggested that person-to-person telecommunication could be used as a model for person-computer interaction. From the information presented in this paper, this statement holds for several of the measures analyzed. For example, the finding that it takes people one-half as long to complete a task when speaking as opposed to typing has been reported in numerous telecommunications studies (Chapanis and others, 1972; Chapanis and Overbey, 1974; Ford, Chapanis, and Weeks, 1979; Ochsman and Chapanis, 1974; Weeks and Chapanis, 1976; Weeks, Kelly, and Chapanis, 1974). Similarly, the results of the present study parallel those of telecommunications studies in that voice subjects use more words and more unique words than do keyboard subjects.

There are some contrasts between the findings of this study and studies of telecommunication. The several studies cited above consistently report that voice subjects use a significantly greater number of messages than do keyboard subjects, but the messages are about the same length regardless of mode. Our two groups of subjects used about the same number of messages, but the voice subjects used a significantly greater number of words per message (see Table 1). There are at least three plausible hypotheses to account for this reversal in findings. The shift may be a function of: (1) the fact that one of the communicators was a computer; (2) the particular experimental task; or (3) the way in which we structured the computer's output. A definitive answer would require further study (e.g., replicating the experiment with a different task in order to test hypothesis 2).

At the present time, however, we tend to favor hypothesis 3. The computer program generates messages that are, on the average, several words long (e.g., "What is the date of the check?"). Voice subjects were inclined to respond in kind (e.g., "The date was August tenth."), while a typical keyboard subject's response would be "8/10." If our program's queries had been more succinct (e.g., "Date?"), we suspect that the voice subjects' responses would also have been terse.

Another difference observed was that subjects in our study made fewer typographical errors and spelling errors in the keyboard condition and fewer stutters and false starts in the voice condition then had previously been observed in telecommunications studies. We believe this phenomenon to be directly linked to communicating with computers. Users seem to demonstrate a need for more exact communication than is seen in person-to-person communication.

An additional observation was a subjective one on our part. That is, the style of communication for users in both groups was clear and to the point. Users were never chatty in their communications. Most messages in both modes were simple grammatical constructions or incomplete, simple grammatical constructions. There were, however, in the voice mode some cases of very complex grammatical constructions. For example, one subject said during her session: "Okay, so that's right, these are the ones that were not on the last time. I would like to see the check numbers that I just put into the computer that appear on my statement this month beginning with check 1428. Please go on."

Further analyses are required in order to more clearly determine the features of natural-language voice communication with computers. We plan to reinvestigate the language measures for this study after removing that information pertaining to checks, deposits, and cash withdrawals. It may be that without this information, the unique words used by keyboard subjects will be a subset of unique words used by voice subjects.

In addition, we plan to categorize and compare the messages in both conditions by function. This will be done by classifying messages by two schemes. The first scheme will be the interactive classification scheme used in past telecommunications research (Ford, Weeks, and Chapanis, 1980). The second scheme will be the one presented by Ford (1981). These two schemes differ in that the Ford scheme (1981) classifies messages by computer operations of data base retrieval, creation, and manipulation. The interactive classification scheme classifies messages as they contribute to the process of dialog exchange.

These proposed analyses along with those analyses presented here should lead to a thorough understanding of the differences and similarities between voice and keyboard natural-language communication with computers.

ACKNOWLEDGEMENT

This research was supported by a contract between GTE Laboratories, Inc. and The Johns Hopkins University. Preparation of this paper was supported by a contract between GTE Laboratories, Inc. and Prism Associates.

REFERENCES

Chapanis, A., R.B. Ochsman, R.N. Parrish, and G.D. Weeks (1972). Studies in interactive communication: I. The effects of four communication modes on the behavior of teams during cooperative problem solving. Hum. Factors, 14, 487-509.

Chapanis, A. and C.M. Overbey (1974). Studies in interactive communication: III. Effects of similar and dissimilar communication channels and two interchange options on team problem solving. Percept. Mot. Skills, 38, 343-374.

Chapanis, A., R.N. Parrish, R.B. Ochsman, and G.D. Weeks (1977). Studies in interactive communication: II. The effects of four communication modes on linguistic performance or teams during cooperative problem solving. Hum. Factors, 19, 101-126.

Dreyfus, H.L. (1972). What Computers Can't Do. Harper & Row, New York.

Fitter, M. (1979). Towards more "natural" interactive systems. Int. J. Man-Mach. Stud., 11, 339-350.

Ford, W.R. (1981). Natural-Language Processing by Computer: A New Approach. Doctoral dissertation, The Johns Hopkins University.

Ford, W.R., A. Chapanis, and G.D. Weeks (1979). Self-limited and unlimited word usage during problem solving in two telecommunication modes. J. Psycholing. Res., 8, 451-475.

Ford, W.R., G.D. Weeks, and A. Chapanis (1980). The effect of self-imposed brevity on the structure of dyadic communication. J. Psychol., 104, 87-103.

Lea, W.A. (1978). Trends in Speech Recognition. Prentice-Hall, Englewood Cliffs, N.J.

Ochsman, R.B. and A. Chapanis (1974). The effects of 10 communication modes on the behavior of teams during co-operative problem-solving. Int. J. Man-Mach. Stud., 6, 579-619.

Weeks, G.D. and A. Chapanis (1976). Cooperative versus conflictive problem solving in three telecommunication modes. Percept. Mot. Skills, 42, 879-917.

Weeks, G.D., M.J. Kelly, and A. Chapanis (1974). Studies in interactive communication: V. Cooperative problem solving by skilled and unskilled typists in a teletypewriter mode. J. Appl. Psychol., 59, 665-674.

ERGONOMIC ASPECTS FOR IMPROVING RECOGNITION PERFORMANCE OF VOICE INPUT SYSTEMS

H. Mutschler

*Fraunhofer Institute for Information and Data Processing, Karlsruhe,
Federal Republic of Germany*

Abstract. Present voice input systems have a recognition rate less than 100%.
Beside the quality of the voice recognition algorithms some ergonomic factors
relating to the speaker, the vocabulary and the environmental noise determine
the recognition rate. In a series of experiments 7 parameters (speaker
training, long-term speech consistency, system training, speaker sex, vocabulary phonetics, -size, background noise) were tested in a simple word input
task. The experiments were performed with different vocabularies (mostly the
ICAO alphabet) and with one or both of the voice input systems Threshold
T600 and SpeechLab 20A. The results are reported and discussed in the context
of the literature.

Keywords. Speech recognition; computer interfaces; human factors.

INTRODUCTION

Due to the technological development of
voice recognition systems voice input can be
embodied in man-machine systems as an oral
information input in the machine. Voice input gives the chance to relieve the overloaded visual/manual channel and to make
better use of the less stressed aural/oral
channel. Present voice input systems permit
the input of isolated or few coherently spoken and previously determined words (word
recognition) and the system has to be trained
by each speaker several times (speaker dependency). After being processed by the system
the recognized words have to be presented to
the speaker for verification.

Beside these characteristics the incomplete
recognition performance is a main weakness
of present voice recognition systems. Therefore, parallel to the further technological
development of voice recognition, some ergonomic aspects have to be taken into account
so as to get an improvement of recognition
performance. This imposes certain restrictions on the user. In this paper a series
of experiments with ergonomic parameters
concerning the speaker, the vocabulary, and
the environmental noise are reported which
promise to improve recognition performance.

EXPERIMENTS

Parameters

Voice recognition is based on matching the
spoken word to one of the reference patterns.
Therefore, the recognition rate much depends
on a consistent articulation of words between
system training and recognition mode. Conse-

quently, speaker- dependent factors prove
to be important parameters for improving
recognition performance.

Speaker training, the learning of the user
how to work with voice input (not to be con-
founded with system training), is studied
in Experiment 1. Apart from learning the
mere handling of the devices (microphone,
switch), speaker training comprises mainly
adherence to a natural fluent speech so as
to avoid any speech inconsistency between
system training and recognition mode.

Even if speaker training is completed and
the short-term speech consistency is maxi-
mized, we have to be aware of a bad long-
term speech consistency, which is the object
of Experiment 2. Spoken words would be more
and more mismatched with previously trained
reference patterns if voice features would
show a steady drift with time, e.g. within
weeks. Both short-term and long-term speech
inconsistency can be taken partly into
account by more word repetitions in the
system training mode, which is a parameter
in Experiment 3.

Another speaker-dependent factor is the sex
of the speaker. The higher pitch of a female
voice of 230 Hz on an average makes us expect another recognition rate than the lower
pitch of a male voice of 130 Hz on an
average (Peterson and Barney, 1952) relating
to the fixed algorithm of voice input
systems. The speaker sex is the second parameter in Experiment 3.

In voice recognition a choice is made between several vocabulary words of the reference patterns. From the ergonomic point of

view, this discrimination process involves vocabulary-dependent factors for improving the performance of voice input systems.

In a separate series of experiments, called Experiment 4, the vocabulary phonetics were investigated relating to articulation type, articulation location and voicedness. The aim was a set of rules to constitute a vocabulary out of the phoneme classes with a minimum of confusion in voice recognition.

With a growing size of any undesigned vocabulary more confusions between words are probable because of a higher probability of resembling features. The vocabulary size is object of Experiment 5.

A further ergonomic design possibility for improving recognition performance of voice input systems is the control of the background noise. Generally spoken, it is assumend that discrimination between words becomes more difficult with a lower signal-to-noise ratio. In Experiment 6, background noises of different types were varied in recognition mode with a constant low-noise level in the system training mode. The resulting recognition rates were measured.

Vocabulary

Different vocabularies were used in the experiments: ICAO alphabet (Experiments 1,2,4, 6), Digits (Experiment 3) and German Christian names (Experiment 5). The ICAO alphabet is the NATO spelling alphabet and comprises 26 words of mostly two syllables. The words were pronounced chiefly as in German (as in parenthesis): Alpha (alfa), Bravo (brawo), Charlie (dscharlie), Delta (delta), Echo (echo) ... For the Experiment 3 with a small vocabulary size the following 10 German digits of mostly one syllable were used: Null, Eins, Zwo, Drei, Vier, Fünf, Sechs, Sieben, Acht, Neun. For the Experiment 5 with different vocabulary sizes 10, 30 or 90 German Christian names were used including the German spelling alphabet, such as

Anton, Berta, Cäsar, Dora, Emil, ...

Apparatus

Essentially, the apparatus consists of the voice recognition systems Threshold T600 and Speechlab 20A.
Both systems recognize isolated words and are speaker-dependent. They cover a large price spectrum and with it - as hypothesized - a wide recognition performance spectrum.

Threshold T600 represents a current standard voice recognition system costing approx. DM 50.000 which made expect a good recognition performance. It consists of a headset microphone, a preamplifier, the processor, a tape-recorder, a keyboard, and a monitor, linked with a PDP 11/34. Its maximal vocabulary size is 250 words, with a speaking duration of .2-2 seconds. The silence interval between words must be .1 seconds at least.

According to the manufacturer, the response times of .1 - .5 seconds depend on the actual vocabulary size.

For voice input the microphone had to be switched on. First, volume adjustment was accomplished with the aid of the preamplifier to match the signal amplification to the loudness of the speaking voice. Then, the system had to be trained by 10 repetitions of each word. A green light indicated the readiness for each voice input, simultaneously indicating the end of the necessary silence interval. The recognized word was presented on the monitor. An audible "beep" announced a rejection.

The SpeechLab 20A is a board, here placed in an Apple II-computer, costing approx. DM 5.000 as a whole. Thus, a factor 10, a price lower than that of Threshold T600, a lower recognition performance had to be expected. The system consists of a fixed microphone. Apple II, including the SpeechLab 20A-board, a keyboard, and a monitor. Its maximal vocabulary size is 32 words, with speaking duration of .1 - 1.5 seconds. The silence interval between words must be .1 seconds at least. The manufacturer did not specify any response times.

For voice input with SpeechLab 20A, a push-button had to be operated. No volume adjustment could be accomplished. The system was trained by one or more repetitions of each word, depending on the actual vocabulary size (s. Experiment 5). The resulting reference patterns were stored on a floppy disk. A flashing cursor indicated the readiness for each voice input, simultaneously indicating the end of the necessary silence interval. The recognized word was presented on the monitor. No feedback on the screen announced a rejection or - because of no volume adjustment - too weak a speaking voice.

The experimental room was lined with sound absorbing material to control the background noise. With the exclusion of Experiment 3 with variations of background noise, the noise level in all other experiments never exceeded 50 dB(A) which had no influence on voice recognition performance. The main noise source was the ventilator of Threshold T600.

Subjects and Procedure

To examine the 7 parameters listed above a series of 6 experiments with 1 or 2 factors was performed. As subjects 6 male and 6 female persons participated, mostly schoolboys, schoolgirls and students (17 years old upwards) all untrained at the beginning of the experiments.

At the beginning of each session the subjects read the instructions on the monitor explaining the experiment. In particular, they were instructed to speak normally, though rather quickly and loudly when recognition performance was low. The subjects familiarized

themselves with the voice input, according
to microphone position, vocabulary, time
organization, and speaking manner. After
system training, the randomized words of the
vocabulary were presented several times, to
be simply repeated by the subjects after
operating the manual device starting voice
input. Verification of the recognized words
was effected by means of pushbuttons "Yes"
and "No" constituting the data samples for
calculating the recognition rate. In the
case of rejections, or when the speaking
level was below threshold, the utterance was
repeated. The experiments were controlled
and the recognition rates were calculated by
the respective computer, PDP 11/34 or
Apple II.

RESULTS

Experiment 1 (speaker training)

Experiment 1 for studying the training effect
of the speaker in voice input was a one-
factorial experiment with 3 identical sessions and a repeated measure design. One
session consisted of the system training and
the 2-fold recognition mode of the ICAO
alphabet (26 words), with SpeechLab 20A. The
3 parts were scattered among all other experiments: At the very beginning of the experimental series, after 4 weeks with 2
other sessions in between, and after 1 year
with another 7 sessions in between. The hypothesis was a growing learning effect from
session to session, according to the repeated instructions, and according to the instant feedback of recognition accuracy in
all experiments.

The results are shown in Figure 1. Starting
from a recognition rate of 72%, speaker
training produced an increase of 7% after
about 3 hours, and an additional increase of
7% after about 8 hours training of voice input. The differences proved to be statistically significant in a conservative analysis of variance (Greenhouse and Geisser,
1959), $F = 5.1$, F_{krit} $(1,11; p<.05) = 4.8$.
Thus, even after a few hours of voice input
training, the short-term speech inconsistency between system training and recognition
mode was markedly reduced. Especially too
marked a pronunciation in system training
mode on the analogy of a poor telephone line
was abolished. Stops within words (e.g. Vic-
tor, Foxtrot) are decreased by speaking
faster than usual. Moreover, subjects
learned to suppress the common breathing
noise immediately before and after speaking.

The early effect of speaker training agrees
with results of Clapper (1971). A fast adaption to the speaking requirements of voice
input systems increased the recognition rate
in his experiments from initially 88% by 7%
with 1/2 hour of experience. Doddington and
Schalk (1981) report that a new user tends
to speak too softly because of the microphone being close to the mouth. But low
speaking levels lead to inconsistency and

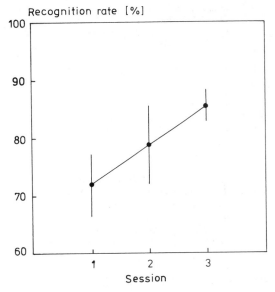

Fig. 1. Mean recognition rates with standard-deviations as a function of
speaker training within 3 sessions.
There were 3 hours training
between pass 1 and 2, and 8 hours
training between pass 2 and 3
(SpeechLab 20A).

reduced signal-to-noise ratio, resulting in
low recognition rates. Beside speech consistency, speaker training includes an improved time organization in a series of voice
inputs. The multiple-task operations involving voice input, an operator will achieve
maximal data entry rates after a 3-4 months
training (Martin and Collins, 1975). Then
he will know the proper speaking rate, when
to verify the data, and what words should be
used at any particular instant.

Experiment 2 (long-term speech consistency)

Experiment 2 for studying the long-term
speech consistency in voice input was a one-
factorial experiment with a unique system
training at the beginning and 3 recognition
passes of the 26 ICAO alphabet words. The
1st and 2nd passes were carried out in one
session immediately after, and after 1/2 hour
after the system training. The 3rd pass was
carried out in another session, 4 weeks later
(repeated measure design). The voice input
systems were Threshold T600 and SpeechLab
20A. Though not really a parameter, both
voice input systems were used by all subjects
in a counterbalanced order. The hypothesis
was a decrease of recognition rate between
the 2nd and the 3rd passes, according to a
long-term drift of voice features.

Figure 2 shows the medians and the 25% and
75% quartiles of the recognition rates. Regarding the interesting differences between
pass 2 and 3, the recognition rate remains
constant over 4 weeks, when using Threshold
T600 and decreases from 81% by 4%, when using
SpeechLab 20A, which does not prove to be
statistically significant with a Wilcoxon

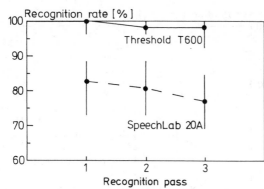

Fig. 2. Medians and 25%, 75% quartiles of the recognition rates within 3 recognition passes in a range of 4 weeks. The same reference patterns were used (long-term speech consistency).

test (T=41.5,μ_T=39,T_{krit} (12, p<.05) = 14).

This result is confirmed by Poock (1981) who demonstrated a relative stability of voice recognition rate over 20 weeks. He used a Threshold T600 with up to 240 words. Moreover, he mentioned his own two-year-old reference patterns any of which he hardly ever re-trained. Despite no detrimental effect of long-term speech inconsistency,one has to be aware of an extraordinary speech varying because of physiological (hoarseness) or psychological reasons (stress), as the idea of Peckham (1979) shows to assess the strain of pilots by tracking the fundamental frequency of speech.

Experiment 3 (system training, speaker sex)

Experiment 3 was a two-factorial experiment for studying the effect of a different number of word repetitions in the system training mode and the sex of the speaker on voice recognition rate. 6 male and 6 female subjects had participated in all 3 parts of the session with 1, 2 and 3-fold word repititions of the ten German digits (repeated measure design). The order of word repetition number was counterbalanced across the subjects. The voice input system was the SpeechLab 20A. Recognition rate was measured by the recognition mode, with 30 voice inputs per subject in each session part. Hypothesis 1 was an increasing recognition rate with a greater number of word repetitions because speech inconsistency would be included in the reference patterns. Hypothesis 2 was a different recognition rate between male and female speakers according to different pitches and formants.

The results are shown in Figure 3. Starting from a recognition rate of 79% with a 1-fold system training of each word (averaged across speaker sex), the doubling of the number of word repetitions revealed an increase of 8%, and another 7% by tripling. The differences proved to be statistically significant in a two-factorial conservative analysis of variance (Greenhouse and Geisser, 1959),

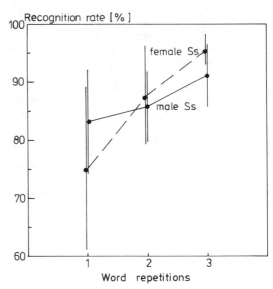

Fig. 3. Mean recognition rates and standard-deviations as a function of the number of word repetitions in system training mode (SpeechLab 20A).

$F_{word\ rep.}$ = 11.4, F_{krit} (1,10; p<.01) = 10.0. There was no difference of recognition rate relating to the speaker sex $F_{speaker\ sex}$ = .02).

As a subsequent experiment with 1 subject and a vocabulary of 6 words revealed, no further improvement of recognition rate could be obtained when words were repeated 4 and 5 times during system training. Using Threshold T600 with a vocabulary size of 50 words, Batchellor (1981) found an improvement from 97.3% to 99.0% when a 5-fold instead of a 3-fold word repetition was chosen. A 7-fold word repetition brought no further improvement which is not surprising in view of the high level of the recognition rate. Runge (1980) recommends a 7-fold-10-fold word repetition for the Interstate VRM. Summing up, it is suspected that the necessary number of word repetitions during system training mode between about 3 to 7 depends positively on vocabulary size.

Concerning the repetition order, Connolly (1977) has also indicated that there may be an advantage by randomizing the order of the repetitions to eliminate rhythmic patterning and to provide a variety of coarticulatory environments.

The results relating to the influence of speaker sex on recognition rate are not conform in literature. Whereas Batchellor (1981) with a Threshold T600 and Doddington and Schalk (1981) with a Nippon DP-100 did not find any difference between male and female speakers, as in Experiment 3, Doddington and Schalk (1981) with 6 other voice input systems, inclusive a Threshold T500, found a little higher recognition rate for men.

Experiment 4 (vocabulary phonetics)

Experiment 4 was a pilot experiment with one subject for exploring the effect of vocabulary phonetics onto the recognition rate of voice input. Articulation type, articulation location and voicedness were studied by varying systematically artificial words differing only in one phoneme at the beginning or in the middle of the word. The voice input system was Threshold T600. Because of the lack of a statistical representance only qualitative results are reported (Lieberknecht, 1980).

The results show that equally-positioned phonemes can be better distinguished among one another when different articulation types are used (e.g. vocal-consonant, plosive-fricative). Articulation location and voicedness are of secondary importance. Concerning consonants, misrecognitions were caused mostly due to erroneous interpretations of plosives (especially "b" instead of "d"), then fricatives (especially "s" instead of "c"). Different vocals tend to be good features of words to be recognized if they do not belong simultaneously to the high vowels "e" and "i" and to the deep vowels "o" and "u". Generally, the extended versions are better, excepting the "i" whose short version pronounces the characteristics of the coarticulated consonants. Comparing the ICAO alphabet and a spelling alphabet which was constructed by using these rules, a final experiment with 12 subjects yielded a recognition of 99% for both. Presumably, such phonetically balanced vocabularies are advantageous only when a normal vocabulary fails to be recognized accurately such as with a large vocabulary size.

Experiment 5 (vocabulary size)

Experiment 5 for investigating the voice recognition (= accuracy) rate as a function of vocabulary size included a comparison with a general purpose alphanumerical keyboard (= alpha keyboard) and a function keyboard. (The accuracy of the keyboards depended on the user, the accuracy of voice input depended mostly on the system.) The voice input system was Threshold T600. With a vocabulary size of 10, 30 and 90 words followed a 3x3 factorial design. 9 high trained subjects, 3 subjects per system, made 30 entries presented at random at each level of vocabulary size. The levels were counterbalanced across subjects. (The accuracy per alpha keyboard was assumed to be independent of vocabulary size so that these 3 subjects were tested only under the 30 words condition.) The vocabularies consisted of German Christian names with equally distributed capital letters.

The results are shown in Figure 4. The recognition rate of voice input decreases monotonously over vocabulary size from 100% to 92%, while accuracy with the function keyboard remains approximately at 100%. The

accuracy with the alpha keyboard runs up to 94%; this value was measured only at 30 words and was extrapolated to 10 and 90 words. Thus, the mainly system-related recognition rate of voice input competes with the user-related accuracy of a function keyboard at a small vocabulary size and drops to the lower level of the accuracy of an alpha keyboard at larger vocabulary size.

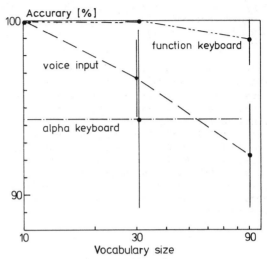

Fig. 4. Mean recognition rates with their average-deviations (Threshold T600) vs. accuracy with keyboards as a function of vocabulary size.

The finding, that voice recognition rate decline with increasing vocabulary size, is confirmed by Martin and Welch (1980). Consequently, the actual vocabulary should reduced in size by splitting the whole vocabulary according to the syntax of the data entry problem. Regarding the accuracy comparison between different data entry devices, McSorley (1981) also found more system-related errors with voice input than user-related errors with a keyboard at a high level of vocabulary size (170 words). He did not specify the keyboard he used, but it seemed to be a mixture between an alpha and a function keyboard.

Experiment 6 (background noise)

For investigating the influence of background noise on recognition rate with a practically noise-free system training the two-factorial Experiment 6 was performed. The first factor was the type of a noise with a white noise and a so-called speech noise only during recognition mode; the latter of which simulating the frequency spectrum of many persons speaking. There were no noise peaks. The second factor was the level of noise which was varied in four 1-2 dB(A) steps. Because of different speaking levels of the subjects the absolute noise level was individually chosen so as to get the critical fall-off of the recognition rate of around 50% with each subject. There was a completely repeated measure design concerning noise type and noise level steps (disregarding the different noise levels). Both noise type and noise level steps are counterbalanced across subjects.

The noise generator "Revox A78 MkII" with a connected amplifier produced the noise. The sound radiation was done by two loudspeakers standing 2 meters away, on the right of the subjects, at an angle of 60° between them. Noise measurement was performed by the sound level meter "Brüel & Kjaer, Typ 2209", with an omnidirectional microphone. Both systems - Threshold T600 and SpeechLab 20A - were used as voice input systems by all subjects.

The results are shown in Figure 5. The gradients at a recognition rate of 50% were averaged (solid lines). The dashed line represent the standard deviations of the gradients (oblique lines), and 50% noise levels (horizontal lines). There were 50% noise levels of 56-57 dB(A) with white noise (both systems), and 65.4 dB(A) (Threshold T600), and 57.5 dB(A) (SpeechLab 20A) with speech noise. Beyond the 50% point, all noise levels produced increasing false alarms and rejections.

port(1980) has done extensive measurements of voice recognition rates with noise levels during system training equal to or greater than during recognition mode. With an Interstate Electronics Voice Data Entry System (VDES) and white noise he found a 50% noise level of 82 dB(A). A further marked improvement of recognition performance was achieved with system training in higher noise level environments than in the ambient one in recognition mode. Davenport stated that recognition rate is not really a function of noise level, but of the difference between the operating noise level and the noise level at which the system was trained. Using different microphones he concluded that voice input works well even in factory areas of about 80 dB(A).

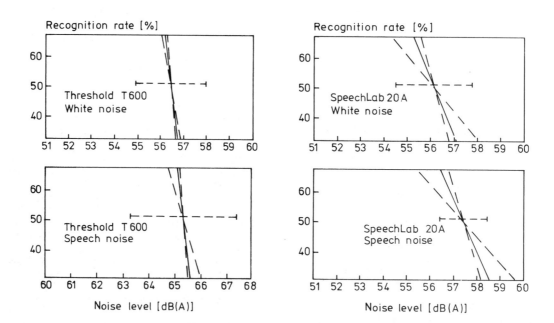

Fig. 5. Mean gradients of the recognition rates at the 50% level and standard deviations with different voice input systems and types of noise as a function of noise level.

According to Völker (1976), office machines (white noise) and male persons (speech noise) produce at a distance of 1 meter a noise level of 65 dB(A), of both. With this, no voice input would be possible - if system training was carried out in a rather silent environment as shown in Experiment 6. Daven-

DISCUSSION

These experiments show that ergonomic aspects have to be taken into account to reduce the main weakness of present voice input system: the possibility of misrecognitions. It is foreseeing that the further rapid development of algorithms will not only produce systems for recognition of connected-speech or speaker-independent systems but also speaker-dependent isolated-word voice input systems with recognition rates near 100% being more robust against variations of the user, the vocabulary and the environment. Until then following rules should be attended to:

1) A short speaker training of several hours is necessary.

2) The reference patterns once created in system training mode can be used for months at minimum.

3) If possible, a 5-fold word repetition in system training should be carried out, approximately.

4) Equally-positioned phonemes of the vocabulary words should be out of different articulation types, especcially high or deep vowels.

5) Vocabularies should be splitted down even in smaller subvocabularies than specified.

6) System training must be performed with the operational noise at minimum.

ACKNOWLEDGEMENTS

This research was supported by the German Federal Ministry of Defense.

REFERENCES

Batchellor, M.P. (1981). Investigations of parameters affecting voice recognition systems in C3 systems. Naval Postgraduate School, Monterey (CA), Master Thesis.

Clapper, G.L. (1971). Automatic word recognition. IEEE Spectrum, Aug., p.57.

Connolly, D.W. (1977). Voice data entry in air traffic control. In 1977 Proc. of Conference on Voice Technology for Interactive Real-Time Command/Control Systems Applications, NASA Ames Research Center, Moffett Field (CA).

Davenport, D. (1980). Voice data entry research/evaluation report: noise tests. Boeing Aerospace Co, Seattle (WA), Technology Evaluations, ASTAD, unpublished report.

Doddington, G.R. and Schalk, T.B. (1981). Speech recognition:turning theory to practice. IEEE Spectrum, Sep., 26-32.

Greenhouse, S.W. and Geisser, S. (1959). On methods in the analysis of profile data. Psychometrika, 24, 95-112.

Lieberknecht, H. (1980). Zusammenstellung eines maschinenangepaßten Buchstabieralphabets zur Spracheingabe. Fraunhofer-Institut für Informations- und Datenverarbeitung, Karlsruhe, Studienarbeit.

Martin, T.B. and Collins, J.C. (1975). Speech recognition applied to problems of quality control. In 1975 ASQC Tech. Conf. Trans. (San Diego, CA), 8-15.

Martin, T.B. and Welch, J.R. (1980). Practical speech recognizers and some performance effectiveness parameters. In W.A. Lea (Ed.), Trends in Speech Recognition, Prentice-Hall, INC., Englewood Cliffs, 24-38.

McSorley, W.J. (1981). Using voice recognition equipment to run the warfare environmental simulator (WES). Naval Postgraduate School, Monterey (CA), Master's Thesis.

Peckham, J.B. (1979). A device for tracking the fundamental frequency of speech and its application in the assessment of 'strain' in pilots and air traffic controllers. Royal Aircraft Establishment, Technical Report 79056.

Peterson, G.E. and Barney, H.L. (1952). Control methods used in a study of vowels. JASA 24, p. 175.

Poock, G.K. (1981). A longitudinal study of computer voice recognition performance and vocabulary size. Naval Postgraduate School, Monterey (CA), Report-No. NPS 55-81-013.

Runge, R. (1980). VRM-Product specification and reference manual. Interstate Electronics Corporation, Anaheim (CA).

Völker, E.-J. (1976). Zur Verdeckung von Sprach- und Störschall durch Rauschen am Büro-Arbeitsplatz. DAGA '76, VDI-Verlag, 243-246.

COMPUTERIZED TEXT PROCESSING: NEW DEMANDS AND STRAINS?

E. Haider

*Institut für Arbeitswissenschaft, Technische Hochschule Darmstadt, Petersenstr. 30,
6100 Darmstadt, Federal Republic of Germany*

Abstract. With new technologies in text processing new demands arise
for the employees. Technologies are classified and related demands
are investigated by an Ergonomic Job Description Questionnaire.
The shift of demands with new technologies is described. For typical work systems physiological strain and performance are measured over 8 hours. Strain reactions above physiological tolerance
limits result from
- text processing with preponderant input functions
- close linkage of man and computer
- high demands from handling work means
- unfavourable environment.

Recommendations for ergonomic design of text processing systems
are given.

Keywords. Text processing; technology; stress; strain; ergonomics.

NEW TECHNOLOGIES IN TEXT PROCESSING

The proportion of workers and employees
which changed from 9:1 in 1882 to 1:1 in
1980 illustrates the expansion of office work. The enormous increase of
costs for personnel is one of the main
reasons why automation has been introduced in today's office work.
New technologies are often combined with
a new work organisation. We should be
aware of the danger of organizing office
work in exactly the same way as industry
work. Division of control-, planning-
and manual tasks and partition of tasks
have already shown some negative effects
(e.g. high fluctuation rates or absenteeism).

The introduction of new technologies
(e.g. computer controlled text processing) and new work organisation may result in new tasks and changed task demands. Therefore two problems will be
dealt with in this paper:
- are demands in office work changing
  with new technologies?
- if so, what does this mean for the
  human operators in terms of stress
  and strain?

METHODS

Technologies are classified due to the
functions of input, processing, output
and control of information being taken
over by machines and computers.

Five steps of technologies emerge:
Technology A: manual information pro-
    cessing system (IPS). Example: elec-
    tric typewriter
Technology B: low automatized IPS with
    text memory. Example: electric type-
    writer with disks
Technology C: low automatized IPS with
    text memory and control system. Ex-
    ample: typewriter with row-display
Technology D: middle automatized, pro-
    grammable IPS. Example: IPS with dis-
    play and programmable software
Technology E: middle automatized, pro-
    grammable IPS with control system.
    Example: IPS with pre-programmed
    text and data elements.

Job Demands

Job demands were registered by an Ergonomic Job Description Questionnaire
(EJDQ) (Rohmert, Landau 1979) which
has been adapted to data- and text processing tasks (Haider, Rohmert, 1981).
This Questionnaire represents the job
demands in the areas of
- informatory work (perception, processing, output)
- energetic work (static, dynamic)
- work organisation (duration, rest
  periods, flow of information, workplace design, work means)
- environment (climate, noise, illumin-
  ation, social contacts).

269

Data are inquired by watching the task and filling out 365 standardized items in terms of
- duration
- importance
- frequency
of the demands involved.

Performance and Strain

Performance and strain measurements were done for work systems with typical demands. These "typical work systems" were selected by cluster analysis. Measures were:
- typing rate
- electrocardiogram
- electromyograms
- blinking rate.

RESULTS

Technologies versus Demands

Technologies were classified for 71 text processing systems in industries and offices. Demands were compared for each group of tasks within the five steps of technology (A-E). All demands were relativated to technology A (electric typewriters without data storage).

The following results emerge with increasing technology:

Informatory work (Fig. 1):
Importance of visual perception increases while frequency of work specific communication decreases.

Energetic work (Fig. 2):
Time and intensity of finger/head movements decrease.

Work organisation (Fig. 3):
The linkage between operators and machines becomes closer. Work speed and sequence of actions are determined more and more often by computers. Unpredictable interruptions due to computer process times or time sharing become more frequent.

Environment (Fig. 4):
Climatic and accustic stresses decrease.

Demands versus Strain

Evaluation of typing rate and electrophysiologic measurements in 10 work systems showed, that decrease of performance or muscular fatigue (increasing reactions of electromyograms) is not primarily associated with higher steps of technology. Intolerably high fatigue (Laurig, 1973) was found in traditional typewriting as well as in computerized text processing. Technology is only one of the factors which determine strain.

For those work systems for which strain reactions above tolerance limits were found, we superposed the related demands from EJDQ. The analysis of these demand profiles indicates that critical demands for the human operator are
- text processing with preponderant input functions
- close linkage of man and computer
- high demands from handling work means
- unfavourable environment.

WORK DESIGN

In order to avoid the critical demands mentioned above, recommendations for ergonomic system design are given on the basis of workplaces found in this research project:

Perception should be improved by ergonomic design of visual display units and other work means. Literature provides findings on design of contrasts and reflections. The signal to noise ration of dictating machines is often bad and tactile perception should be improved by giving feedback from keyboards.

Work means are often arranged in a way that leads to unnecessary static muscular work load and fixed working positions (see Fig. 5). In all text processing systems we type on keyboards which are known to be suboptimal (Kroemer, 1972 ; Rohmert, Haider, 1982). The research study shows that static work of the hands (abduction and pronation) leads to strain (see Fig. 6) which could easily be reduced by bending the keyboard.

Work process is strongly determined by computers in high steps of technology, software is mostly not adapted to human characteristics. User oriented software and implementation of job enlargement strategies should be able to avoid those unfavourable conditions.

DISCUSSION

Having in mind the shift of demands with new technologies environmental demands decrease while social isolation (man machine interaction) increases. Energetic demands, especially static work, which is physiologically disadvantageous, are not reduced.

Some effort should be taken to adapt work organisation, work means and software to human abilities in order to enable the operators to derive benefit from new technologies in text processing and to prevent them from just being a filler for not (yet) available reading machines.

REFERENCES

Haider, E., and W. Rohmert (1981). An-
    forderungsermittlung für Tätigkei-
    ten der Daten- und Textverarbeitung
    mit dem DTV-AET (AET Supplement
    für den Bereich der Daten- und
    Textverarbeitung). In: Landau, K.;
    Rohmert, W. (Hrsg.), Fallbeispiele
    zur Arbeitsanalyse. Hans Huber,
    Bern, pp 135-173

Kroemer, K.H.E. (1972). Human Engineer-
    ing the keyboard. Human factors,
    14, 51-63

Laurig, W. (1974). Beurteilung einseitig
    dynamischer Muskelarbeit. Beuth-Ver
    trieb, Berlin, Köln, Frankfurt (Main)

Rohmert, W., and K. Landau (1979). Das Ar-
    beitswissenschaftliche Erhebungsver-
    fahren zur Tätigkeitesanalyse (AET).
    Handbuch mit Merkmalheft. Hans Huber,
    Bern.

Scholz, L., M. Reinhard, J. Günther, W.
    Hetzler, V. Müller, and G. Schien-
    stock (1982). Arbeitswissenschaftli-
    che und soziale Auswirkungen neuer
    Technologien im Bereich der Textver-
    arbeitung - Wirtschafts- und sozial-
    wissenschaftliche Analyse. For-
    schungsprojekt A 135 im Auftrag des
    RKW. RKW Frankfurt.

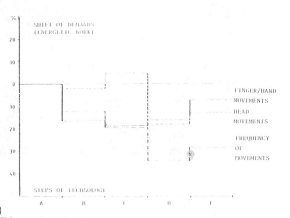

Fig. 2: Technologies and shift of ener-
    getic demands. The demands are
    related to those of technology
    A (= 0 %)

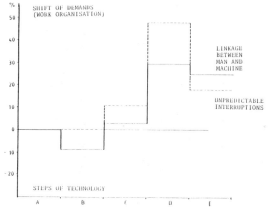

Fig. 3: Technologies and shift of organ-
    isatoric demands. The demands
    are related to those of tech-
    nology A (= 0 %)

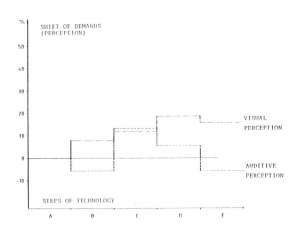

Fig. 1: Technologies and shift of per-
    cepted demands. The demands are
    related to those of technoly A
    (= 0 %)

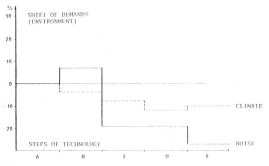

Fig. 4: Technologies and shift of en-
    vironmental demands. The demands
    are related to those of technol-
    ogy A (= 0 %)

Fig. 5: Static muscular work load of
        shoulder, arm and back with text
        processing

Electromyographic
reaction (c.V)

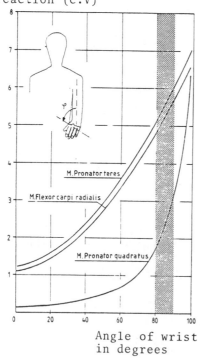

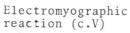

Angle of wrist
in degrees

Fig. 6: Electromyographic reactions of
        muscles used for pronation while
        working on a keyboard. Usual
        wrist position for text typing
        is marked.

INTERACTIVE IMAGE PROCESSING FOR OBSERVER PERFORMANCE ENHANCEMENT

G. Nirschl

*Fraunhofer Institute for Information and Data Processing, Karlsruhe,
Federal Republic of Germany*

Abstract. Image processing techniques may be applied to assist a human
observer in TV-reconnaissance tasks. Various enhancement methods have been
investigated experimentally with regard to their measurable influence on
observer performance. Global contrast enhancement by means of linear
histogram stretching proved to be most suitable in target detection,
in comparison with histogram equalization or histogram hyperbolization. In
case of target classification local enhancement techniques and zoom
operations resulted in an essential increase of observer performance. When
implementing interactive image processing techniques in observation tasks
the design of a convenient man-machine interface is a focal point. An
operator station is put forward, whereby an observer may specify procedures
and parameters by means of interactive tools (joystick, lightpen, function
keys) using a user-guiding menu technique.

Keywords. Picture processing; aerial reconnaissance; man-machine systems;
human factors; observers; interactive systems; television; remote sensing.

INTRODUCTION

The task of a human observer evaluating
aerial pictures essentially is to detect
certain objects (e.g. military vehicles)
and to classify them within different classes
of objects. The main difficulties of this
task result from the potentially poor contrast of the image, the naturally low conspi-
cuousness of targets, and the possible occurrence of jamming noise. As a result a human
observer may soon be overloaded. Therefore,
image preprocessing procedures may be applied
in order to enhance the 'intelligibility'
of images, e.g. to show off potential targets
by improving the target-to-background contrast, thus making the observer's decisions
faster and more reliable.

The various techniques available for image
enhancement are very much problem-oriented.
According to the aspired objective of enhancement the approaches may be divided
into three broad categories:

- contrast enhancement techniques

- contour sharpening techniques

- noise reduction techniques.

Contrast enhancement techniques are based
on modifications of the grayscale of an
image, for example, contour sharpening
techniques rest on highpass filter mechanisms and noise reduction is achieved by
methods based on lowpass filtering.

With respect to the hardware implementation
of enhancement procedures another classification of the approaches is essential:
frequency-domain methods and spatial-domain
methods. Processing techniques in the first
category are based on modifying the Fourier
transform of an image. The spatial-domain,
on the other hand, refers to the image itself, and approaches in this category are
based on direct manipulation of the pixels
of an image. Because of the costly transform algorithms preferably spatial-domain
methods come into question in case of real-
time or near real-time applications.

INTERACTIVE IMAGE ENHANCEMENT

With an interactive image processing system
the user may specifically convey pictures
from various input media (e.g. TV-camera)
to a computer, look at pictures via convenient output media (e.g. TV-monitor), gather
specific information from the image, and
modify further evaluations according to the
previously effected measures. Real-time processing is given, if an approach is fast
enough to process one image within the frame
time of television images, i.e. 40 ms.

The basic structure of an interactive computer display system, as implemented in our
laboratory (Krause, Nirschl, and others,
1982), is depicted in Fig. 1. The relevant
components are drawn out for one memory
plane and video output channel. In the case
of color representations the configuration
of the color channels may be controlled by

273

the computer in such a manner, that each
single color channel is attached to another
memory plane, for example.

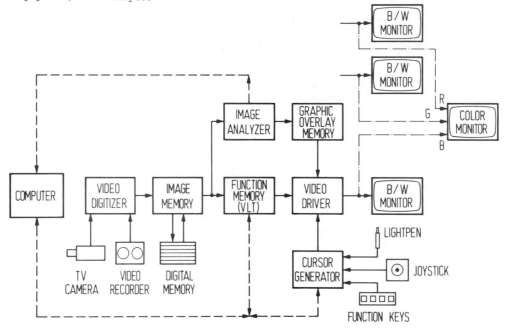

Fig. 1.   Basic structure of an interactive computer display system

Interactive Grayscale Manipulation

The implementation of an approach for interactive grayscale manipulation in real-time
demonstrates the interactions between the
computer, the display system, and the human
user exemplarily (see Fig. 2). By means of
the function memories (video look up tables
= VLTs) the gray values stored in the image
memory can be transformed into new gray
values to be displayed on the monitor.

The relation between input and output values
of the VLTs is represented by a transformation curve, which can be faded in from a
graphic overlay memory. In addition, the
actual frequency distribution (histogram) of
the gray levels in the image, which is calculated via an image analyzer, can be figured (Fig. 2.(a)). The shape of the transformation curve is given by the location of
three independent cursors, which can be controlled by means of interaction tools (joystick, function keys). The effect of the
manipulation in Fig. 2.(b) is a stretching
of the histogram in the middle range and a
compression in the outer ranges.

(a) Original

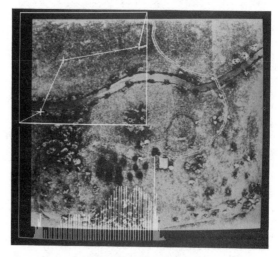

(b) Modified image

Fig. 2.   Interactive grayscale manipulation

Histogram Modification Techniques

A histogram of gray level content provides
a global description of the appearance of
an image. The methods discussed in these
sections achieve enhancement by modifying
the histogram of a given image in a specified manner (Gonzalez, Wintz, 1977). To this
either the transformation function is given,
and the conversion of the gray values is performed immediately by the computer, or the
shape of a destination histogram is given,
and the computer determines the shape of the
transformation function. Figure 3 shows
examples of image enhancement by histogram
modification. As a starting point, a picture
with evidently too narrow a dynamic range
is depicted in Fig. 3.(a) with its related
gray level histogram.

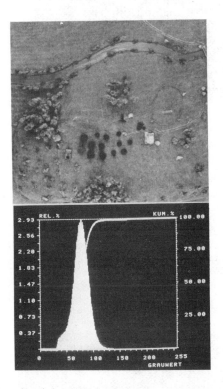

Fig. 3. (a) Original image

Linear histogram stretching. Figure 3 (b)
demonstrates the contrast enhancing effect
by linear histogram stretching. The narrow
input range of Fig. 3 (a) is spread out to
the complete available range of gray levels,
i.e. 255 in case of 8 bit gray level resolution. The linear transformation function
is given by

$$z' = az + b$$

where z' : transformed gray value
          z' : original   gray value

$$a = 255/(z_{max} - z_{min})$$
$$b = (255 * z_{min}) / (z_{min} - z_{max})$$

The calculation of the lower and upper input limits z_{min}, z_{max} was achieved automatically by the computer

to $$z_{min} = INT(z_s - 3*s)$$
       $$z_{max} = INT(z_s + 3*s)$$

where z_s : mean gray level

       s   : standard deviation

       INT : next lower integer gray level

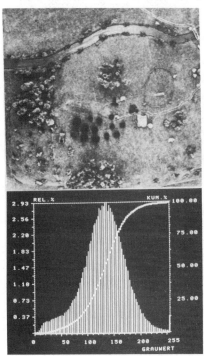

Fig. 3.(b)   Linear histogram stretching

Histogram equalization. An approach coming
from a given shape of a destination histogram is the so-called histogram equalization. The underlying assumption is that
a human observer can extract maximum information from an image, if the distribution of gray levels is uniform.
Figure 3 (c) shows an example of this
principle.

The computation of the transformation
function is achieved by the following
steps (according to Hoyer, Schlindwein,
1979).

$$h(z') = 1 / K$$

where K $= 1 / (z'_{max} - z'_{min})$

to

$$z'   = z'_{min} + n(z) / K$$

where $n(z)$ $= \sum_{z=0}^{z} h(\bar{z})$

where h(z) is the original distribution of gray levels and h(z') is the destination histogram. The limit values z'_{min}, z'_{max} in the destination histogram have to be chosen empirically.

In Fig. 3 (c) the uniform distribution is reflected in the shape of the cumulative histogram, whereas the relative frequencies are unequal because of the unequal intervals.

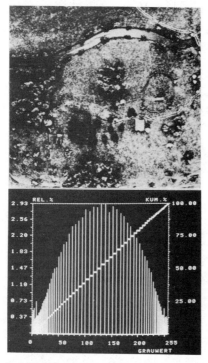

Fig. 3.(c) Histogram equalization

Histogram hyperbolization. An approach further refined coming from histogram equalization, the so-called histogram hyperbolization, additionally takes into account the sensibility function of the human eye (Frei, 1977). Assuming a logarithmic shape of this sensibility function the equalization of the gray level distribution takes place 'behind the eye', if the shape of the histogram of a given image is hyperbolic.

The transformation function can be calculated (according to Hoyer, Schlindwein, 1979) by

$$h(z') = 1/(z'+\alpha)*1/(\ln(z'_{max}+\alpha)-\ln(z'_{min}+\alpha))$$

to

$$z' = -\alpha + (z'_{max} + \alpha)^{n(z)} * (z'_{min} + \alpha)^{1-n(z)}$$

whereby α is a free parameter in order to improve the outcome, if need be.

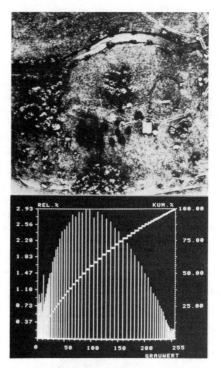

Fig. 3.(d) Histogram hyperbolization

Zonal Image Enhancement

The abovementioned approaches are based on a global modification of an image. If, however, an operation is restricted to a certain area of an image (e.g. target), the result of a processing procedure may be much better, than in the case where the data of the entire image enter into computation (Guildford, 1979).

Figure 4 shows an example of this principle, which is called 'zonal image enhancement'.

Fig. 4. Zonal image enhancement

In this example a contrast enhancing operation, i.e. linear histogram stretching, is restricted to a window, including a target. The position and size of the window can be controlled by means of interaction tools, e.g. joystick, lightpen, and function keys.

EXPERIMENTAL INVESTIGATION OF
OBSERVER PERFORMANCE

When evaluating reconnaissance pictures obtained by electro-optical sensor systems
the task of a human observer essentially is
to detect and to classify or identify definite targets as exactly as possible and
within a time as short as possible.
According to this, the performance of an observer may be described by the following
parameters:

- detection rate   : %-quota of correctly
                      detected targets

- search time   : time needed for target
                      detection

- classification   : %-quota of correct-
  rate   ly classified targets

- classification   : time needed for
  time   classification of
                      detected targets.

When searching in natural scenes the performance parameters are determined fundamentally by the scene complexity (context), the
contrast of the target, and the resolution
of the opto-electronic transmission chain
(Mutschler, 1979; Scanlan, 1977).

Detection experiments have been carried out
in order to investigate how far interactive
image enhancement techniques, such as histogram modification techniques, may involve a
measurable increase of human observer performance.

Classification experiments should prove, how
far an observer may be assisted in the classification phase by means of interactive
local processing procedures.

Detection Performance at Various Histogram
Modification Techniques

Method. Digitized images were displayed on
a TV monitor after they had been subjected
to different contrast enhancement procedures:
Linear histogram stretching, histogram equalization, histogram hyperbolization.
The pictures were presented on a standard
625-line TV monitor with a 30 cm screen diagonal. The subject's task, sitting in front
of the monitor at a distance of 0.5 m in
a darkened surrounding, was to mark potential targets by means of a lightpen. The
presentation duration was 4 s at the most.
Controlling of the trial course, recording
and evaluating of spatial and time coordinates of the subject's reactions where
achieved by a digital process computer
(VAX 11/780).

Results. The results of the measurement of
detection rates and search times at various
preprocessing procedures are depicted graphically in Fig. 5.

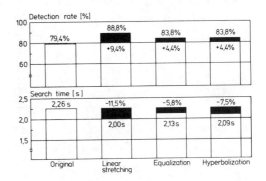

Fig. 5. Detection rate and search time at
         various histogram modification
         techniques

A significant increase of the detection rate
(9.4%) was achieved by linear histogram
stretching. The methodically more costly procedures of histogram equalization or hyperbolization led to an increase of the detection rate only half as big (4.4%) as in case
of linear stretching.

These results are accordingly reflected in
the search times: A signigicant shortening
of the search time (11.5%) could be measured
by linear histogram stretching; relatively
less decreases by equalization (5.8%) and
hyperbolization (7.5%).

The outcome indicates, that contrast enhancement procedures essentially have to take into
consideration the reproduction dynamics of
the TV monitor (i.e. $\approx$ 17 dB).
A linear transformation function, which
spreads out the gray levels within the linear
range of the monitor may be more suitable in
terms of an improvement for a human observer
than modification techniques based on model
conceptions like histogram equalization or
hyperbolization.

Classification Performance at Various
Preprocessing Procedures

Method. Subjects had to evaluate digitized
images with marked targets. The pictures had
been preprocessed by different combinations
of contrast enhancement and zoom operations:
A linear histogram stretching operation,
either global or local, eventually combined
with a zoom operation with an enlargement
factor of 2 or 4.

Figure 6, for example, shows a local histogram stretching operation in combination with
a zoom operation (factor 2), whereby a target
originally covering 30 TV lines was enlarged
from 1 degree of visual angle to 2 degrees
of visual angle.

Subjects had to classify the targets by
marking a corresponding field in an inserted
menu. Subject's reactions were registered and
evaluated by the computer again.

Fig. 6. Local contrast enhancement + zoom
(factor 2)

Results. The results of the classification
experiment are depicted graphically in Fig.7.

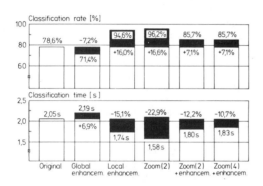

Fig. 7. Classification rate and time at
various preprocessing procedures

All preprocessing operations with the exception of the global histogram stretching led
to an enhancement with respect to classification rate and time. This may be explained
by the fact that a global contrast enhancement operation yields a masking effect for
details necessary for target classification.

A local contrast enhancement operation restricted to the surroundings of the target
involves a distinct improvement of performance
parameters: classification rate increases by
16%, classification time decreases by 15%.

It is a remarkable fact, that an enlargement
of the target section by the factor 2 led to
about the same quantitative improvement of
classification performance as a contrast enhancement operation without the implication
of target enlargement. This may be interpreted as an indication, that contrast and
resolution are exchangeable within a certain
range.

Combinations of contrast and resolution enhancement oprations resultet in a relatively
reduced performance enhancement, which may be
explained by the increasing visibility of
the pixel raster.

MAN-MACHINE INTERFACE FOR
INTERACTIVE IMAGE EVALUATION

The results of the experimental investigations show, that an observer, evaluating reconnaissance pictures, may be assisted effectively by interactive image enhancement
techniques.

Contrast enhancement is an efficient measure
to improve the detection process.
Local contrast or resolution enhancement is
helpful in the classification phase.
Here special interaction tools, like joystick, lightpen, or function keys are necessary in order to specify image sections for
local operations.

Furthermore, the user should be able to
select and control processing procedures and
parameters via convenient interaction
facilities.
We have realized an implementation, whereby
an observer is enabled to fade in a graphic
table on the TV monitor (Fig. 8.).

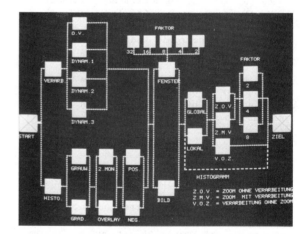

Fig. 8. Table for interactive selection
of procedures and parameters

The observer is guided through this table via
a blinking path and makes his decisions by
marking corresponding fields with the lightpen. On a parallel monitor he may pursuit the
effects of his manipulations (Fig. 9.).

Fig. 9. Observer station for interactive
image evaluation

For a more direct transposition of the observer's specifications the implementation
of a speech input facility is scheduled for
the nearer future.

CONCLUSIONS

A fully automatic evaluation in target
acquisition tasks is not in sight within the
near future because of the complexity of
this process.
Interactive image processing techniques may
be applied in order to combine the capabilities of man and computer aiming at an improvement of the observer's performance. As experimental investigations have shown, global
image enhancement operations may be useful
in the detection phase, local enhancement
operations may improve classification performance.

Convenient interaction tools have to be made
available to the human observer for specification of procedures and parameters. This
could be achieved by means of joystick,
lightpen or function keys and a graphic function selection technique, as shown above. A
further improvement of this man-machine
interface is foreseeable by the additional
implementation of a speech input facility.

ACKNOWLEDGEMENT

This research was supported by The German
Federal Ministry of Defense.

REFERENCES

Frei, W., (1977). Image enhancement by
    histogram hyperbolization. Computer
    Graph. Image Process., 6, 286-294.
Gonzalez, P.W., Wintz, P. (1977).
    Digital Image Processing. Addison-Wesley,
    Reading, pp. 115-136.
Guildford, L.H. (1979). Das "Dot-Scan-CCTV"-
    System, eine vielseitige Apparatur für
    Echtzeit-Bildverarbeitungsexperimente,
    Phillips techn. Rdsch., 38, 324-339.
Hoyer, A., and Schlindwein, M. (1979).
    Bildverbesserung durch digitale Nachver-
    arbeitung. Phillips techn. Rdsch., 38,
    311-323.
Krause, P., Nirschl, G., Herzog, H., Freytag,
    R. (1982). Interaktive Bildverarbeitung
    mit dem V.I.P.-Bildlabor. FhG-IITB-Be-
    richte 1/2-82, im Druck.
Mutschler, H. (1979). TV operator perfor-
    mance in real time air-to-ground missions
    under task-loading conditions. AGARD
    Conf. Proc., 267, 17.1-17.11.
Scanlan, L.A. (1977). Target acquisition in
    realistic terrain. Proc. Human Factors
    Society, 21st Annual Meeting, San
    Francisco, 249-253.

CONTROL OF INPUT VARIABLES BY HEAD MOVEMENTS OF HANDICAPPED PERSONS

A. Korn

*Fraunhofer Institute for Information and Data Processing, Karlsruhe,
Federal Republic of Germany*

Abstract. For people with only limited muscle functions the delegation of
alpha-numeric or graphic information is sometimes very difficult. Considering seriously handicapped persons who cannot use their arms or legs, head
or eye movements can often be used for written communication. For such persons we have developed a "Visually Coupled System (VCS)" which consists of a
suitable frontlet and luminous sources (light dots) fastened to the head. A
light dot is superposed on the visual field and serves as a mark clearly
determining the line of sight. Another source of light is pictured on the
target of a TV camera in order to measure the head position without making
any contact. The task of the handicapped consists in shifting the optical
mark until it coincides with a preset environmental target. Because of the
great accuracy of the eye-head coordination the measured precision for the
fixation of a small target or for tracing a simple optical pattern is about
20 minutes of arc. When using the VCS as an aid for writing letters the
performance for selecting a character out of a 5x6 matrix was 0.3-0.4 s per
letter.

Keywords. Rehabilitation; communication; visually coupled system; eye-head
coordination; head dynamics; writing aid.

INTRODUCTION

Thanks to the rapid progress made in the
development of electronic measuring systems,
seriously handicapped persons are offered
technical aids providing more advantages
than the mechanical aids so far used. Such
conventional aids such as suction and blow-
ing devices or mouth guided pens, very often
call for an unnatural posture or head movements and prevent any verbal communication.
Assuming that the handicapped cannot write
but he is able to speak then the present
possibilities of automatic speech recognition
may be considered for the translation of the
spoken words into writing. It is however
difficult in this case to create graphic
information such as drawings. In this case
and for paralyzed patients who have lost
their voice and writing/typing ability, a
communication device with transducer facilities to detect the line of sight has been
developed. This device is,essentially, a
so-called "Visually Coupled System (VCS)"
which is a control system activated by head
and/or eye movements, with a visual feedback (Russo, 1978; Korn, 1980). The interest
in VCS is mainly given by its many possible
applications in the field of flight control.
The VCS, developed in our institute, will
subsequently be presented as an aid to the
handicapped.

A fast and accurate coordination of eye and
head movements is, similar to the eye-hand
coordination, a natural physiological
activity of man, closely linked to his perception of and reaction to environmental
stimuli. In visual search tasks eye and
head movements generally will occur. Here
the process of visual acquisition can be
divided into three motor outputs: saccadic
eye movements, very fast head movements and
compensatory eye movements (Zangemeister
and others, 1981). Head shifting has,
normally, a share of 80 to 9o% in total
line-of-sight shifting.

Provided it will be possible to carry out
coordinated head movements, a task of the
handicapped, e.g. of those suffering from
paraplegia,may consist in shifting an optical
mark until it coincides with a preset environmental target. After this, the measured
coordinates of the head position can be
assigned to the line of sight.

In the process, four different problems
have to be solved:

- How to measure the head position without
  incommoding the handicapped by the
  measuring process.

- The visual feedback of the head position.

- The verification of the called-up
  information.

-   The starting of the process by the
    handicapped.

These problems are individually solved as
follows

METHODS

Measurement of the line of sight

The scheme of the experimental arrangement
is shown in Fig. 1. The head position is
measured without making any contact. A
light-weight frontlet and a luminous source
L_1 are fastened to the head. This source
of light is pictured on the target of a
TV camera. Any shift of the light dot caused
by head movements is electronically measured
by means of a Video-Analyzer which signals
the x,y-coordinates of the light dot. The
measurement of two degrees of freedom,
namely shaking and nodding the head, is
sufficient in most cases (Korn, 1981).

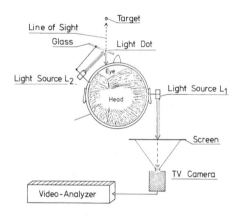

Fig. 1. Schematic view of the apparatus
         to determine the line of sight
         by measuring head movements.

The visual feedback of the head position is
realized by superposing on the visual field
a light dot which is the reflected beam of
the light source L_2. This light source is
coupled with the head and its optics can be
adapted in order to get a sharp image in
the plane of the target. Our investigations
have been performed with the prototype shown
in Fig. 2. Here the head had to be carry a
load of 160 grammes.

The two frames of reference which must be
considered for the interpretation of the
measured data are shown in Fig. 3. On the
left hand side the positions of the point of
head rotation D, the eye A and the target F
are described in the laboratory system or
working space. On the right hand side all
these points, together with the light dot,
are related to a reference system which is
firmly coupled with the head. The output of
the Video-Analyzer in Fig. 1 is a measure
for the head rotation around two axes in
this reference system.

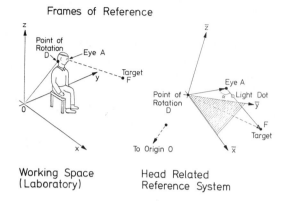

Frames of Reference

Working Space
(Laboratory)

Head Related
Reference System

Fig. 2. The two frames of reference which
         must be considered for the deter-
         mination of the line of sight. In
         the working space a target F is
         watched by the subject whose head
         has the point of rotation D. This
         point is the origin of the second
         reference system which is the head
         related reference system. Only
         rotations of the head around the
         two axes $\bar{z}$ and $\bar{y}$ are measured.

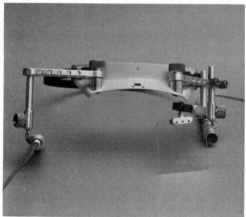

Fig. 3. A simple Visually Coupled System
         which is composed of two sources
         of light and a simple optics. The
         visual feedback is reached by a
         light dot superposed on the visual
         field by means of the glass and the
         focusing adjustment pictured on the
         right hand side. The head has to
         carry a load of about 160 grammes.

Control of a typewriter by head movements

In many cases the visual feedback of the
head position can be performed by observing
a cursor on a monitor which may be a home
TV monitor. The cursor is shifted by head
movements and must be made to coincide with
a desired character on a foil which is fixed
on the monitor screen. This procedure has
the advantage of immediately marking the
relevant target. If its position relative
to the foil is stored in a computer, the
measured data can be used in order to

control technical devices such as a typewriter or a speech synthesizer as indicated
in Fig. 4. Here the measured analog signals
are fed inot a computer where this input is
transformed into suitable control signals.
We have investigated the control of a typewriter. For safety reasons, an acknowledgement of the called-up information will
frequently be necessary.This may be effected
by optical or acoustic means. We have
integrated a voice output unit (voice
synthesizer) producing an acoustic acknowledgement by an appropriate access to the
word-organized storage.

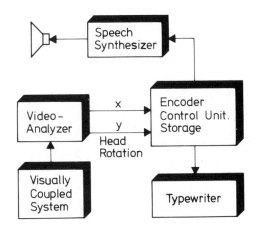

Fig. 4.   Electronic devices for analyzing
          a video signal and the generation
          of acoustic and optical pattern.

Considering that, for the release of a
desired function, hand or foot movements are
not feasible, we have coupled the starting
pulse with the dwell time of the cursor in
the different fields of our 5x6 input matrix.
With a simulated typewriter a dwell time
between 0.3 and 0.4 s has proved to be
suitable.

RESULTS

When using the VCS, the fixation precision
was measured with the aid of a grid which
was watched at a visual angle of 8° in a
horizontal, and of 6° in a vertical direction. The angular changes of the head-
related reference system as compared with
the laboratory system were measured with
the aid of the experimental device shown in
Fig. 1. Two analog voltages corresponded to
the rotations of the head. A registration
of these voltages with the aid of an x-y
plotter is shown in Fig. 5. Some elements of
the pattern were fixed for approx. 1 s. The
registration shows that the amplitudes of
the fixation movements in both x and y
directions are, i.a., recorded at a visual
angle of 10', this corresponding to space
of less than 3 mm between elements at a
distance of 1 m.

Fixation of a Grid

Fig. 5.   Fixation accuracy when the subject
          has to shift an optical mark by
          head movements until it coincides
          with a point of the grid.

Whereas with the registrations shown in
Fig. 5 the writing process after each
fixation is interrupted by lifting the
writing aid at the plotter, the curves shown
in Fig. 6 - 8 show a continuous registration
of the head movements following the lines
of the meandering pattern at bottom left in
Fig. 6, the individual characters at the
top of Fig. 7, and the individual circuits
of a printed board in Fig. 8. Also with
these continuous registrations the deviations
from the should-be-value do not exceed a
visual angle of 10'.

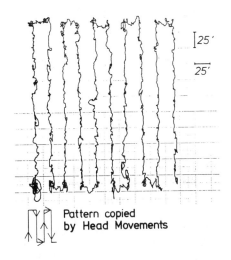

Fig. 6.   Continuous registration of the head
          movements following the lines of
          the meandering pattern at bottom
          left.

are successively fixed by means of saccadic eye movements.

The release of the starting pulse for printing a character via the dwell time when fixing the corresponding matrix element was found to be agreeable. The subjects so far used for experiments were members of the Institute.

Fig. 7. Continuous registration of head movements following the lines of the individual characters at the top.

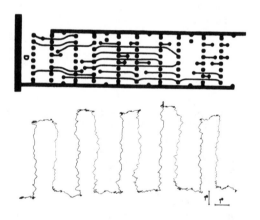

Fig. 8. Continuous registration of head movements following the individual circuits of a printed board at the top.

When controlling a typewriter by head movements, the characters of the simulated keyboard must be fixed successively with the aid of the VCS. For the words "Guten Morgen" (good morning) the registration of the corresponding head movements is presented in Fig. 9. For selecting 11 characters, a time of approx. 3.5 s was required, equal to 0.3 s per character. A negligibly shorter time is needed when the single characters

Fig. 9. Controlling a typewriter by head movements. The letters of the words "GUTEN MORGEN" (good morning) are fixed successively by head movements with the aid of the Visually Coupled System. A time of approx. 3.5 s was required for selecting 11 characters.

CONCLUSION

Our investigations with a prototype of a simple Visually Coupled System show that a fast selection of alpha-numeric characters and a faithful reproduction of graphic pattern can be performed by head movements. This prototype is still uncomfortable in order to be used by seriously handicapped persons in daily life. Nevertheless we are sure that the optics, the light sources, and the device for the measurement of the two degrees of head rotation can be reduced to an acceptable size. If the visual feedback is achieved by a cursor, that is a light spot for example on a home TV monitor, than nothing but the rotation of the head around two axes must be measured which is also possible by simple non-optical methods like potentiometers or inertia sensors. Further improvements can only be reached by a close cooperation with handicapped persons.

ACKNOWLEDGEMENTS

This report was supported by The German Federal Ministry of Defense

REFERENCES

Korn, A. (1980). Verfahren zum indirekten
    Messen der Blickrichtung. Patent No.
    26 29073 of the German Patent Office.
Korn, A. (1981). Visual Search: Relation
    between detection performance and
    visual field size. Proc. of the
    1st European Annual Conference on
    Human Decision and Manual Control,
    Delft, May 1981, 27-34.
Russo, L. (1978). Helmet mounted visually
    coupled systems. Proc. of the S.I.D.
    Vol.19/4, 181-185.
Zangemeister, W.H., Lehmann, S., Stark, L.
    (1981). Sensitivity analysis and opti-
    mization for a head movement model.
    Biol. Cybern. Vol.41, 33-45.

EXPERIMENTAL STUDIES OF MAN-COMPUTER INTERACTION IN FINANCIAL ACCOUNTING SYSTEMS

B. Lüke

Lehrstuhl für Betriebsinformatik, Universität Dortmund, D-4600 Dortmund 50, Federal Republic of Germany

Abstract. Business application systems with online operations obviously need good man-computer interfaces. Experimental studies seem to be a useful approach to face the problem of designing user-adequate interfaces. With special regard to the design of dialogues for online accounting, a subset of an application system was redesigned to give several different features of dialogue design, while all other parameters like hardware and application-bound functions remained identical. The differences in the constructed dialogues can be divided into three categories, which are (A) the presentation of information by the system, (B) the handling of input and user errors and (C) the means of dialogue operations including control functions and online documentation. 48 professional accountants carried out a series of tasks, involving the handling of accounts receivable entries, with each of the three systems. All interactions were recorded. Results are given for the improvements achieved by the modifications in dialogue design. The first part concerns the behaviour of the subjects with respect to their different online experiences. In the second part the general effects of the special design features are shown. Finally the user's own judgements based on the evaluation of a post-test questionnaire are discussed.

Keywords. online operation; man-machine systems; interactive systems; online accounting; software ergonomics; design of dialogues; human engineering; experimental evaluations.

SOFTWARE INTERFACE FOR BUSINESS APPLICATIONS

Nowadays more and more people in the business environment are directly faced with computers, working with a display terminal for data-entry, inquiries and/or other kinds of tasks. Video-display-units are typical equipment for commercial applications such as order entry or accounting. They are utilised in conjunction with suitable software packages. The amount of work as a whole is divided up into user's tasks and computer's tasks with a high rate of interaction between user and computer. In this process of dialogue operation the need for a good man-computer interface is obvious. The state of the art in computer technology allows the machines to be adapted to the user in an ergonomically acceptable manner. In the design of workstations, and especially of terminal hardware, much progress has been made during the last years, where definitive guidelines and checklists have been developed (Cakir, Hart and Stewart, 1979). Growing attention is now given to software aspects of the interface. Shackel (1980) uses the term "software ergonomics" for this type of research, while the more psychologically oriented approach is referred to as "cognitive ergonomics".

DESIGN OF DIALOGUES

Various recommendations have been published by several authors for the design of dialogues in commercial applications, e.g. Martin (1973), Engel and Granda (1975), Stewart (1976), Hebditch (1979), Gaines (1981). Their recommendations or guidelines for implementation are mainly based upon the designer's experiences and on the more informal feedback from the users. Empirical or experimental validations are given only in some cases (e.g. Hirsch, 1981).

It has been shown that standard online accounting systems for smaller business computers have quite similar structures and power of functions (Griese, 1980). However, in spite of this similarity, it can be recognized that, with respect to the design of dialogues, there is a wide variety of different products. This may be due to a lack of the designer's general knowledge and of thorough and covering guidelines (besides the constraints caused by different hardware and operating systems). It seems to be useful to give some contributions to the augmentation of knowledge about user behaviour in such kinds of interactive systems. Thus, in this study, emphasis is placed on the evaluation of special features in the design of dialogues.

RESEARCH METHODS

Classical forms of ergonomic research are empirical and experimental studies. For man-computer interaction in commercial applications the empirical results of Eason, Damodaran and Stewart (1974) show the variety and problems which arise from the differences in user job types. In other fields of interactive computer usage valuable experimental results have been achieved, e.g. by Ledgard, Singer and Whiteside (1981) who studied the syntax of text editors. They started with the hypothesis that a command language should be as natural and familiar to the user as possible. For the experimental evaluation two editors were used: one already existing editor and another one with a structure and syntax of commands based on legitimate English phrases avoiding unfamiliar words and the use of special delimiting characters. The results, based upon 24 sessions with student subjects, show that the performance using the second editor was much superior on the basis of three measures (percentage of task completed, erroneous commands and editing efficiency). Many other experiments with various kinds of user tasks have been carried out and are assessed in the current literature, e.g. by Embley and Nagy (1981), Shneiderman (1980) and Reisner (1981).

THE EXPERIMENT

The general idea of our research project was to let users carry out comparable tasks in different online accounting systems, where the hardware and the specific application-bound functions remained identical and only the design of dialogues was changed between the systems. In this situation a laboratory experiment seemed to be the most promising approach. Three accounting systems on a small business computer with disk storage and two VDU-workstations were investigated: a supplier-made open item method accounting package and two other systems constructed under the direction of the author. For the sake of a manageable experimental environment the development efforts were restricted to tasks connected with accounts receivable entries. Among all frequent operations in accounting these seem to be best suited for our purposes, because they provide reasonable handling problems for the user, and are neither too complex nor too specialized. It was expected that every accountant should be able to understand the contents and procedures of such operations.

Design Goals

Tasks like those operations in accounting can be regarded as rather 'closed tasks' following the classification of Eason (1980). This means, that input and required output have a rather fixed structure and vary only within predictable limits. In a simplifying example with accounts receipts we always have an entry of the amount of money for two accounts,

one of these being a cash or bank account and the other being a customer's or eventually a supplier's account, where the existing open items may fit to the payed amount or not. The knowledge about the structure and contents of the task makes it possible to create dialogues with prescribed performance procedures, at least as far as the limited variations of tasks do not require free choices or selection of alternatives by the user himself. These restricted assumptions are helpful by making the area of design problems smaller, but not even the general goals for user-adequate design are completely determined by them alone. For our design purpose we used a set of design goals with some similarity to those developed by Dzida, Herda and Itzfeld (1978) and Dehning, Essig and Maass (1978). Having in mind that the dialogue should fit to the frequent and accustomed user as well as to the less frequent and untrained user we stated three categories of design goals: the dialogue should be (A) adequate for required functions, (B) capable of explanation and (C) secure in utilization.

Adequate for required functions means that the form of the dialogues should be suited, as far as possible, to the contents and structure of the tasks.

Capable of explanations. Any instruction and information within the dialogue should be understandable and relate to the user's task. The user should see what the system is doing as a result of his operations, it should be clear what the system is expecting from him, and, finally, the user should be able to recognize the whole set of his possible alternatives.

Secure in utilization. During operation - mainly on data input - the user should have the lowest possible feeling of uncertainty about the correctness of his actions. This means especially the user-perceived security against unexpected and unwanted reactions by the system.

Features of the Three Systems

Oriented to those design goals the two new accounting systems were developed. A selection was taken from among those existing techniques and additional features, which in our belief would lead to some improvements. The implementation was carried out with different sets of those selected techniques and features in the new systems. Some constraints on the designer's freedom were set by the known limits in hardware and operating system of the machine to be used. However the major common factor of the three dialogues results from the decision to use the screen in a form-filling mode with both the new versions too. This decision allowed for the fact that all comparable online accounting systems we know use this technique as interaction mode. Taking into account this global common feature, the differences in the constructed dialogues can be classified in three catego-

ries. These being (A) the presentation of in
formation by the system, (B) the handling of
inputs and user errors (C) the means of dialogue operation.

Presentation of information. This comprises
everything which is displayed on the screen
to the user. It includes the type of input-
form used, the spacing and grouping of the
input fields, and the feedback that is given
to the user as information about the current
state of the dialogue. In addition, it includes the frequency in changing of complete
screen-images and the location of explanatory
text. The kind of presentation determines
essentially the visible surface of the dialogue.

Handling of inputs and user errors. Within
each system there is a unique manner of designing all the fields that are used for data
input by the user. The differences between
the systems are the kinds of preformatting or
indicating the length of the input fields,
and the availability and visibility of default values. Handling of user errors is done
by error messages in a fixed partition at the
bottom of the screen, where the reason of the
error is indicated and a repetition of the
input is required.

Means of dialogue operation. All features of
the control functions available to the user
are categorized here, i.e. backtracking and
reset facilities as well as the type and
completeness of online assistance on the syntactical and semantical level. Syntactical
means for control actions are implemented as
commands, function keys or menus.

Each of the systems can be described as consisting of a join of elements from all the
three categories. Figures 1 to 3 show the
treatment of an identical set of data and
dialogue state by each of the three systems,
which are referred to as S1, S2 and S3. A
short verbal characterization is given in the
following.

Features of system S1 (Fig. 1). The dialogue
enables the user to carry out the whole task
without further written material such as application manuals and tables for coding of
input. Special aids for input operations such
as visible defaults or abbreviations are not
provided. Consecutive input fields are organized column by column instead of line by
line. Compared to the other systems we have
here the highest flexibility because operations can be controlled by a small set of
commands. The dialogue is split into several
frames of up to 5 different screen-images for
one task. The available special options
(e.g. scrolling or other control functions)
are always visible at the top of the screen.
All further material is put together in a
permanently available subsystem for on-line
assistance. This help-system can be reached
by issuing one of the offered specific commands. A syntactical and semantical explanation of the current input field is then dis-

played by default. Guided by a set of menus
within the help-system other facilities can
be activated, i.e. further explanations,
tables of coded items (e.g. account numbers)
and finally the end of the subsystem to continue in the interrupted main dialogue.

Features of system S2 (Fig. 2). The main features in dialogue design of this system are
the operator instruction facilities and the
assistance for input operations. The operator
instructions are located at the bottom of the
screen and show the possible user actions by
explaining the meaning of the function-keys.
According to the current input field, up to 7
different actions are offered by short keywords. Visible default values in the input
fields allow the re-use of data items for the
current or following tasks. An additional
feature is the use of line numbers instead of
the complete coded values in those cases
where codes are not known and have to be
looked up in internal dictionaries (e.g. the
code of a customer of whom only the name is
known from the source document). The line
number with its one digit causes the entire
code to be moved into the input field. The
control within the dialogue is generally
handled by the use of function-keys; accordingly the meaning of single keys may change
from one input field to another. However
fixed key numbers are used for backtracking
and reset functions, which are available with
every single step.

Features of system S3 (Fig. 3). In this system the dialogue provided by the supplier of
the standard package has not been changed.
Within the dialogue the explanatory texts on
the screen are essentially restricted to the
descriptions of the input fields. Sometimes a
single description is followed by more than
one input field. Because all the input fields
are packed rather closely together and the
explanatory text is quite short, most of the
tasks can be carried out without a complete
change of the screen-image. For the necessary
control functions some of the function-keys
are used, but the facilities differ from
field to field and are not available every-
where. In some situations a short text is is
sued as a hint for the possible control actions. Two of the dictionaries to be used for
finding codes of input items consist of
printed lists and are not accessible within
the dialogue. In addition to the operation
manual a printed list of those fields is
provided, where backtracking is possible.

Subjects

For experimental testing 48 professional accountants were available for participation.
The majority of them was delegated by their
department managers. Only rough information
about the desired procedure was given prior
to the experimental session itself. The subjects were classified according to the kind
of computer-assistance for the accounting
system of their own affiliation.

```
-----------------------------------------------------------------------------
I      B U C H E N   Z A H L U N G S E I N G Ä N G E   O P - A U S G L E I C H      I
I---------------------------------------------------------------------------I
I FELD ZURÜCK : F1  ENDE : F2  INFORMATION : *H  VORBLÄTTERN : +  ZURÜCKBL. : -  I
I---------------------------------------------------------------------------I
I  16 NEU-OP/SKONTO (N/S)?  _                      2736,50 ZAHLBETRAG          I
I  17 REFERENZ-NUMMER       : _____               54,82 SKONTOBETRAG        I
I  18 AUSGLEICHSBETRAG      : _____           2018,29 RESTBETRAG          I
I  19 ÄNDERUNG OK   (J/N)?  _                                                  I
I---------------------------------------------------------------------------I
I      REF-NR     RE-DATUM      FÄLLIG       ZU ZAHLEN      SKONTO   NOCH OFFEN  I
I                                                                             I
I       17420    78.07.05     78.10.05        732,46        12,98  AUSGEGLICHEN I
I       17438    78.07.07     78.10.07       2777,77       138,89               I
I       17512    78.07.09     78.10.09          9,19         0,46               I
I       17580    78.07.11     78.10.11        299,66         0,00               I
I       17700    78.07.13     78.08.03       1402,11         0,00               I
I       17782    78.07.15     78.10.15          0,00         0,00               I
I       17783    78.07.17     78.10.17       2061,43       103,07               I
I       17922    78.07.19     78.10.19        902,66        45,13               I
I                                                                             I
I                                                                             I
-----------------------------------------------------------------------------
```

Fig. 1   Treatment of open items in system S1

```
-----------------------------------------------------------------------------
I        BUCHUNG VON ZAHLUNGSEINGAENGEN FUER    BLECHALBRECHT GMBH            I
I===========================================================================I
I UNTERNEHMEN/WERK  101        SACHKONTONUMMER 1130000  BELEGNUMMER   19438   I
I DEBITORENNUMMER  1432600  BERHERMANN GMBH            ZAHLBETRAG    2736,50  I
I---------------------------------------------------------------------------I
I                    LISTE DER OFFENEN POSTEN                                 I
I  INDEX    REFNR    RE-DATUM      OP-BETRAG       SKTOBETRAG    ZAHLBETRAG    I
I    1      17420    78.07.05        732,46          12,98                     I
I    2      17438    78.07.07       2777,77         138,89                     I
I    3      17512    78.07.09          9,19           0,46                     I
I    4      17580    78.07.11        299,66           0,00                     I
I    5      17700    78.07.13       1402,11           0,00                     I
I    6      17782    78.07.15        100,00           2,00                     I
I    7      18113    78.08.12       2131,43         103,07                     I
I    8      18422    78.08.14        902,66          45,13                     I
I---------------------------------------------------------------------------I
I INDEX / OP-NR      +1         +                  RESTBETRAG     2736,50      I
I===========================================================================I
I F3 = VORBLAETTERN             F4 = RUECKBLAETTERN         F5 = OP-NEUANLAGE  I
I FA = ABB BUCHUNG    FB = ABB OP   F2 = ZURUECK BUSCHLUE  ETB = INDEXAUSWAHL  I
I===========================================================================I
-----------------------------------------------------------------------------
```

Fig. 2   Treatment of open items in system S2

```
-----------------------------------------------------------------------------
IUNT/WERK 101  BUCHERZEICHEN BI        BUCHUNGSMONAT 7803                      I
I                                                                             I
IBELEGDATUM       78.08.30   BELEGNR           35267   BELEGSUMME    4189,20   I
ISTORNO-KZ             0      SACHKONTONR.    1000000                          I
IRESTSUMME       4189,20      KONTOBER/-NR. 1 1432600  SCHLUSSEL          31 2 I
IZAHLBETRAG      2736,50      SKTO/STEUER        0,00  GEGENWERT               I
ISTEUER 1                     STEUER2                  TEXT    ZE 1            I
IWECHSELDATUM                                                                  I
IREST-AUFTBET   REF-NR.  RE-DATUM    OP-SALDO     SKONTO   ZAHLBETR. REST-OP   I
I               17420    78.07.05      732,46     12,98 (3,00%)                I
I               17438    78.07.07     2777,77     83,33 (3,00%)                I
I               17512    78.07.09        9,19      0,28 (3,00%)                I
I               17580    78.07.11      299,66      0,00                        I
I               17700    78.07.13     1402,11      0,00                        I
I               17782    78.07.15      100,00      2,95 (3,00%)                I
I               18113    78.08.02     2061,43     61,84 (3,00%)                I
I               18422    78.08.04      902,66     27,08 (3,00%)                I
I               18631    78.08.09     1087,40     32,62 (3,00%)                I
I                                                                             I
I   2736,50     17420    78.07.05      732,46     12,98   719,48       0,00    I
I                                                                             I
-----------------------------------------------------------------------------
```

Fig. 3   Treatment of open items in system S3

Group 1, termed the 'manual'-group, consisted of 6 subjects, who do all accounting without computer-assistance at all.

Group 2, termed the 'batch'-group, consisted of 18 subjects, who use batch systems by filling coding-sheets as input and receiving printed lists as output.

Group 3, termed the 'online'-group, consisted of 18 subjects, who use online accounting systems by themselves (generally several hours per day). These subjects had no prior knowledge about any of the three systems used in the experiment.

Group 4, termed the 'S3'-group, consisted of 6 subjects, who use the standard package referred here as S3 for daily work.

Procedure

Each session started with an interactive tutorial at the workstation. This tutorial consisted of instructions and practices about the use of the keyboard, especially the locations and meanings of the function-keys. It included at its end a multiple-choice questionnaire about the characteristics of the subjects' clerical work.

After that, a description of the first system to be used was given to the suject. It was about 7 pages long and contained a summarized presentation of the structure of the implemented dialogue. After the reading of the material was completed the system was started and the set of prepared tasks was given to the subject. This set was composed of 26 tasks with a fixed order. Each task consisted of the "original" source documents of a single accounts receivable entry as received from the bank or the cashier's office. Additional written remarks were given if necessary for the understanding of the documents. A member of the experimental staff was present during the whole session, giving advice for the first two tasks but restricting his further communication with the subject to exceptional cases. When the set of tasks was finished it was offered to the subject to have a break without prescribed time limits.

The second and the third system were treated in the same manner. The order of sequence between the three systems was randomly determined so that the six possible sequences were balanced within in each group of subjects.

After a subject had completed all tasks in all systems, individual judgements about the different features in the dialogues were collected via a post-test questionnaire. The end of the session was normally reached in the early afternoon with a time taken for the whole session ranging from 4 up to 9 hours (breaks included).

Measurement of Interactions

All user input was recorded by a monitor program especially implemented for that purpose. For every interaction (i.e. one dialogue step represented by the input of the user into one field) a record was created containing the following data: field identification, the input itself, the end-key for completion of data entry, the user time and a code for indication of errors or branches in connection with that step. Accumulation to the overall and the task level resulted in a set of three variables for each of the systems. These being (A) the performance time, (B) the number of errors, (C) the overhead on steps.

Performance time. This is the time the system has to wait for the end of the user's input action. It covers the time from the moment the cursor is moved to the input field by the system until the user presses an end-key for transmission of the field. Measurement is in units of 0,15 seconds. The performance time for a single step may be very short and even zero, because the workstation allows some characters to be typed in advance. Otherwise performance time can exceed one minute or more, if the user has to look at manuals or other additional material (or is just thinking about what to do next).

Number of errors. By the monitor program only those errors were logged, which could be found by a syntactical and semantical check of the single input items. Other errors (e.g. the use of a wrong but valid account number) were evaluated by manual procedures. If a user finds and corrects an error by himself, this action counts to the overhead of steps.

Overhead on steps. For each task the minimum number of necessary steps was calculated as the best possible solution. The overhead on steps is then defined as the percentage of steps additionally performed by the user within one task or at the overall level.

RESULTS FROM MEASUREMENT OF INTERACTIONS

Performance times

The total performance times for the three systems are shown in table 1. It should be mentioned again, that these times do not cover the processing time taken by the systems themselves, so that performance time is always lower than the whole session time. The processing time does not differ much from system to system and is ordinarily so short, that it does not cause the user to wait for completion of processing. Therefore processing time is neglected in further evaluation. The average performance times required for the use of system S1 or S2 are considerably shorter than those needed for the use of S3. Subdivision into the predefined user groups shows that

TABLE 1 Average of Performance Times at the Overall Level

system	all subjects		'manual' group		'batch' group		'online' group		'S3' group		all subjects exc. 'S3'-group	
S1	97,8	88%	94,9	90%	115,2	81%	90,4	89%	70,2	129%	101,7	85%
S2	90,7	81%	90,0	86%	108,3	76%	84,1	83%	58,6	108%	95,3	80%
S3	111,5	100%	105,2	100%	142,9	100%	101,3	100%	54,4	100%	119,7	100%

The first number of each column is the performance time scaled with the value of 100 for the overall average of performance times. The second number is the percentage of performance time of system S3.

this is not true for the 'S3'-group, where the lowest performance time was recorded with the known system (i.e. S3), while the new systems required more time. However all times for this user group remain far lower than those for the other groups. If the subjects of the 'S3'-group are excluded, the average of improvement of total performance time by systems S1 and S2 over system S3 is about 15 % and 20 % respectively.

Further comparisons of the values between the groups show unexpectedly low times for the 'manual'-group. These times are not much higher than those for the 'online'-group whose subjects could be expected to have some advantage due to their previous experiences with that kind of online environment.

Analysis of errors

The average for the number of errors made by the subjects has the best value with system S2 (see table 2), the averages for the other two systems are up to 50% higher. With respect to these number of errors it has to be taken in consideration, that the possibilities of making errors were somewhat higher in systems S1 and S2, where the additional features have replaced the use of printed material, partially or completely, by equivalent online operation. This is especially true in S1.

TABLE 2 Averages for Number of Errors and Overhead on Steps

system	number of errors	overhead on steps
S1	20,4	29,5 %
S2	13,4	18,7 %
S3	18,6	13,6 %

A more sophisticated analysis of the total number of errors shows that in all three systems about 70 % of the errors are concentrated upon only 5 of the more than 20 input fields. The most frequent error in S1 and S3 results from the interconnection of two input fields: the user has first to select a code to distinguish between customers, suppliers and cash receipts, an then, in the next step, the proper code number from the chart of

accounts is required as input. The selection code is not always required but is copied automatically from one task to another. Moreover sometimes the same code applies for several tasks in sequence. Thus it happens quite frequently, that the correctness of the selection code for the actual task is not carefully enough controlled and causes errors in the interpretation of the code number in the next input field. Another frequent error in S1 occurs at the point where a return from the help-system to the main dialogue is intended but the input is not a valid selection from the menu table. The two most frequent errors in S2 are wrong inputs during the described one-digit selection from internal dictionaries.

It is of course true that a user learns to avoid errors, and that errors are an unavoidable consequence of trying to operate with an unknown system. Thus a system's designer has to worry more about errors that a user makes after some time of experience than about those errors made at the very beginning of the operation. For the analysis of persistence of errors in our study three sections of tasks near the beginning, the middle and the end of the operation with the systems were formed. Each section consists of an equal number of tasks. The first section contains comparably simple tasks, the second section contains rather complicated tasks which cannot be completely carried out with the experience gained in the first section, and the third section contains only tasks with no additional difficulties. The average for the number of errors per section is shown in table 3. Due to the additional online operation that is required in the use of S1 and S2 the corresponding values in these systems are at first higher than those for S3

TABLE 3 Average for Number of Errors in three Sections of a Session

system	first section simple tasks	second section complicated tasks	third section known tasks
S1	4,4	5,3	1,9
S2	3,2	2,6	1,0
S3	3,0	6,2	2,5

but the situation changes in the second and third section. Thus it seems especially with system S2 to be possible to overcome more complicated tasks with fewer problems.

Learning

It was intended to give special emphasis to the progress that users make by learning during operation. In order to measure learning progress, the performance of the subjects was tested when they carried out a series of four pairs of tasks each of which had identical structures of documents. These four pairs were fairly evenly distributed within the complete set of the 26 tasks and had the numbers 3/4, 11/12, 18/19 and 25/26. The results from evaluation of the performance times for those selected tasks are shown in Fig. 4. The thick lines indicate the average performance times measured for all subjects, while the thin lines above and below indicate the standard deviation between the subjects. It can be seen that the progress made by learning is similarly structured in all the three systems. The highest differences between the systems occur in the first period of measurement. These differences become smaller from period to period. In the second period the level of performance time is reduced to about one half of the value of the first period. The further reduction of time is much slower. That means that the subjects seem to have remembered quite well the structure of a task that they have performed once, since most of the progress for execution of the identically structured tasks was achieved already with the first repetition.

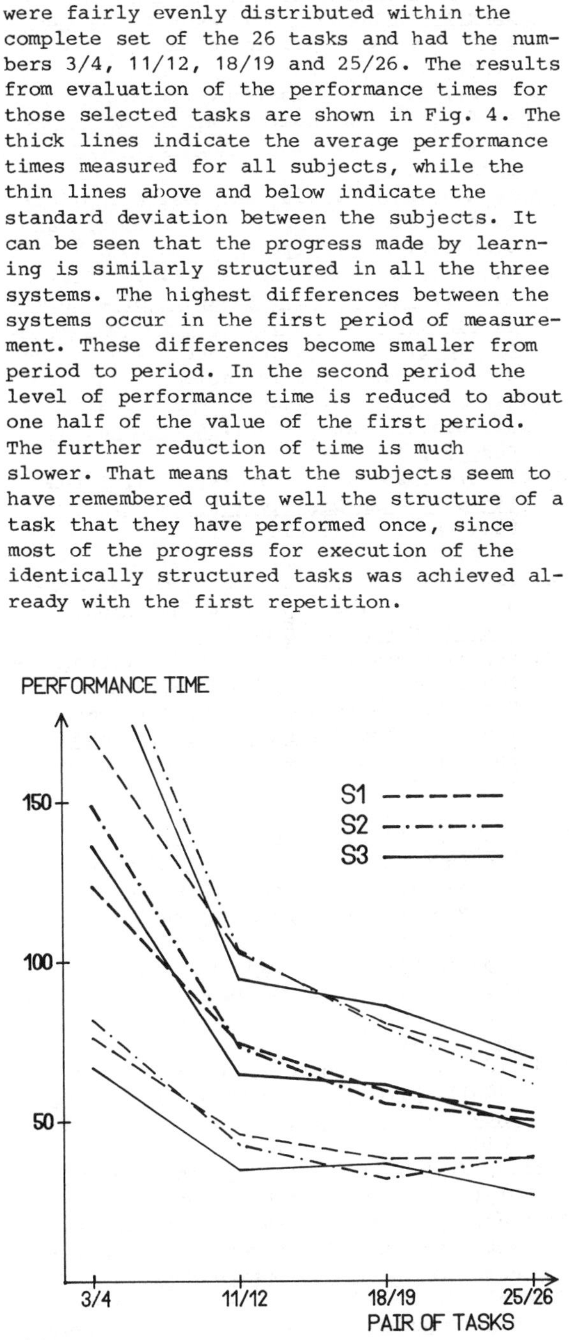

PERFORMANCE TIME

Fig. 4 Performance times for selected sets of tasks (times are scaled with the value of 100 for the overall average of time per 2 tasks)

Comparison between the four user groups shows that the structure of the curves is similar for all groups, although the level of their performance times is different. The subjects of the 'S3'-group perform much faster than the subjects of the other groups, especially those of the 'batch'-group.

RESULTS FROM POST-TEST QUESTIONNAIRE

Asking users for their opinion about features in design of dialogues is often difficult. These users normally have at most the experience with their own system as a background for discussion, and they treat the user interface, especially on the software aspects, as a fact and do not discuss it. Such users have overcome the initial problems of handling and have adapted themselves to which ever kind of dialogue they are involved with. They do not perceive that possibly something in the design could be better, and sometimes they work with a subset of functions without complaint and without looking for more convenient procedures. While working with our systems the subjects have gained experience with some alternatives, and, with that background, the individual judgement could be expected to be somewhat sharper, though in our case certainly restricted to the bandwidth of variation between the three systems.

Some distinct features

The subjects were asked to to give ratings for some distinct features of dialogue design in the three systems. They had to assign a coded number per feature to each of the systems. The code numbers ranged from 1 to 5, code 1 meaning very good and code 5 meaning very bad. The average ratings for some of the features are given in table 4.

Best ratings for the organization of input fields were give to system S1, in which the distance from one input field to another is minimized by the column by column organization. The lower ratings for S3 indicate that the inconsistencies against the line-by-line principle are well noticed and criticized. Comparable low ratings were given for the missing preformatting of S3, while the other two systems are rated somewhat better. It seems to be most useful to have preformatting as contents of the whole field (such as the lines in S1) instead of merely marking the left and right limit. The ratings in regard to the intensity of changing of screen-images indicate that such changes should be used quite restrictively by the designer. Although even in S1 the changes only occurred in accordance with changes in requirements during the dialogue (e.g. looking for account numbers in the dictionary instead of typing the number directly), this system received somewhat lower ratings than the other two.

TABLE 4 Average Ratings for Distinct
        Features in Dialogue Design

Features	average ratings
Organization of input fields	
S1 column by column	2,2
S2 line by line	2,5
S3 line by line with some inconsistencies	3,5
Kind of preformatting used	
S1 Line at the bottom of the field	2,0
S2 '+'-symbol at the limits of the field	2,6
S3 none	3,6
Intensity in changing of screen-images	
S1 high	3,1
S2 medium	2,4
S3 low	2,5

Achievement of goals

Finally the subjects were asked to give ratings for the achievement of the predefined goals in the design of the dialogues (adequate for required functions, capable of explanations, secure in utilization). On the average the ratings for the systems S1 and S2 are nearly the same, for the second and the third goal they are each more than a whole point on the scale higher than those for the compared system S3. The subjects of the 'S3'-group judge their own system to achieve most successfully the first and the third goal, which should bedue to their previous knowledge. Long practice helped them to 'convert' the source documents to dialogue steps immediately and madethem secure in utilization, whereas with the two new systems they could only use some knowledge about the basic structure and the practice in handling of the hardware environment.

CONCLUSION

The results of the experiment prove that the design of dialogues in such applications like online accounting systems leads to a measurable impact upon the behaviour and attitudes of users performing given tasks at a display terminal. The subjects perceived the redesigned systems as more adequate for required functions, more capable of explanations and more secure in utilization. Deeper analysis of the results shows advantages and problems with specific features of the three evaluated dialogues. Users' problems tend to be concentrated upon distinct situations that arise during operation. Of course the scope of contribution to the research in software ergonomics is limited by the special application environment and the extent of variation of dialogue structure. However some systems for

routine operation in the business environment whose functions are others than just accounting have comparable structure in the kind of online operation and can be treated somewhat similarly with regard to the design of dialogues.

REFERENCES

Cakir, A., D.J. Hart, and T.F.M. Stewart (1979). The VDT Manual. IFRA-Institut, Darmstadt.

Dehning, W., H. Essig, and S. Maass (1981). The Adaption of Virtual Man-Computer Interfaces to User Requirements in Dialogs. Springer, Berlin Heidelberg New York.

Dzida, W., S. Herda, and W.D. Itzfeldt (1978). User-perceived quality of interactive systems. IEEE Trans. Software Eng., 4, 270-276.

Eason, K.D. (1980). Dialogue design implications of task allocation between man and computer. Ergonomics, 23, 881-891.

Eason, K.D., L. Damodaran, and T.F.M. Stewart (1974). MICA Survey: A Report of a Survey of Man-Computer Interaction in Commercial Applications. Department of Human Sciences, University of Technology, Loughborough

Embley, D.W., and G. Nagy (1981). Behavioral aspects of text editors. ACM Computing Surveys, 13, 33-70.

Engel, S.E., and R.E. Granda (1975). Guidelines for Man/Display Interfaces. IBM Poughkeepsie Laboratory Technical Report TR 00.2720.

Gaines, B.R. (1981). The technology of interaction - dialogue programming rules. Int. J. Man-Mach. Stud., 14, 133-150.

Griese, J. (1980). Vergleich des Leistungsumfangs von Softwaresystemen zur Online-Finanzbuchhaltung. In P. Stahlknecht (Ed.), Online-Systeme im Finanz- und Rechnungswesen, GI-Anwendergespräch Berlin April 1980. Springer, Berlin Heidelberg New York. pp. 229-239.

Hebditch, D. (1979). Design of dialogues for interactive commercial applications. In B. Shackel (Ed.), Infotech State of the Art Report Man/Computer Communication, Vol. 2, Invited Papers. Infotech International, Maidenhead. pp. 171-192.

Hirsch, R.S. (1981). Procedures of the human factors center at San Jose. IBM Syst. J., 20, 123-171.

Ledgard, H., A. Singer, and J. Whiteside (1981). Directions in Human Factors for Interactive Systems. Springer, Berlin Heidelberg New York.

Martin, J. (1973). Design of Man-Computer Dialogues. Prentice Hall, Englewood Cliffs.

Reisner, P. (1981). Human factors studies of database query languages: a survey and assessment. ACM Computing Surveys, 13, 13-31

Shackel, B. (1980). Dialogues and language - can computer ergonomics help? Ergonomics, 23, 857-880.

Shneiderman, B. (1980). Software Psychology. Winthrop, Cambridge, Massachusetts.

Stewart, T.F.M. (1976). Displays and the software interface. Applied Ergonomics, 7, 137-146.

A GRAPHICAL HARDWARE DESCRIPTION
LANGUAGE FOR LOGIC SIMULATION PROGRAMS

M. S. Knudsen

*Laboratory for Semiconductor Technology, Technical University of Denmark,
DK-2800 Lyngby, Denmark*

Abstract. A graphical input language for logic simulation programs is presented. By means of this input language, the user can construct gate diagrams of the system to be simulated directly on a graphical screen or a
digitizer/plotter. Thus, the user is freed from the manual numbering of
all circuit nodes; a boring task where errors often occur. A case study
shows, that the manual circuit description by means of node numbers requires
around three weeks of manual work for input and checking when dealing with
a 1100-gate circuit. If the set of diagrams is specified by means of the
graphical input language, the manual work required is cut down to two days
and besides, a very good design documentation is obtained.

Keywords. Computer-aided circuit design, computer-aided logic design, computer graphics, digital circuits, large-scale systems, man-machine systems.

INTRODUCTION

Many electrical engineers are designing digital systems of quite a high complexity. For
reasons of design verification, a breadboard
model of the desired system is often constructed. By checking the performance of
the breadboard model, it is possible to obtain a high degree of user interactivity because most design errors can be detected
and corrected quite rapidly. When dealing
with very large and complex systems, however, breadboard modeling is not so attractive since it is so difficult to keep track
of everything. Instead, logic simulation is
preferred as a design verification tool because systems consisting of many thousand
logic functions can be analyzed by a computer in a reasonably short time.

A common disadvantage of the logic simulators most frequently used today is that simulation programs receive information about
the nature of the system to be analyzed in a
way that is very unfriendly to human beings.
When dealing with digital circuits the designer must assign individual numbers to
each node of the circuit as illustrated by
Fig. 1. Following this a tedious and very
awkward specification must be made of input/
output relationships for each gate (Fig. 2).
It is a very difficult and boring task to
control the correctness of all numbers. If
errors are present, the simulation program
will provide the user with erroneous results.

THE GRAPHICAL INPUT LANGUAGE GRANTS

This paper describes an attempt to solve the
problems outlined above with the development

of the input language GRANTS (GRAphical Network Topology Specifications). By means of
this input language, the user can construct
gate diagrams of the system to be simulated
directly on a graphical screen or a digitizer/
plotter. Figure 3 shows the structure of the
entire program system. By means of GRANTS,
it is possible to separate the diagram of a
complex digital system into several sheets
and assign individual names to the electrical
terminals leading from one sheet to the other.
A merging routine then constructs a total
circuit description based upon the individual circuit descriptions from each sheet and
upon a comparison of all the terminal names
that occur within the sheets.

Figure 5 shows the basic diagram symbols of
the input language GRANTS. With regard to
the flip/flops, it is practical to treat the
inversion symbol as a separate entity. Since
the set/reset inputs of a flip/flop can have
a status of either non-existing, non-invert-
ed, or inverted, a total of 18 different
D-flip/flop types exist (Fig. 6). If the
inversion symbol was not a separate entity,
an additional 18 different D-flip/flops would
need to be included in the symbol list - a
very cumbersome approach.

COMPACT DATA STRUCTURE

When the user builds up a diagram on a graphical screen or on a digitizer/plotter during an interactive session, a compact data
structure for the diagram is formed. As Fig.
7 shows, only the essential graphical information (entity type, location, and angle of
rotation) is stored sequentially in an one-
dimensional array G. No limitations exist

regarding the order by which the entities can appear.

EXPANDED DATA STRUCTURE

Based upon the compact data structure described above, an expansion algorithm constructs an expanded data structure; this structure is a basis for the automatic assignment of node numbers. Based upon the information stored in the compact data structure, the coordinates of all input/output terminals of each diagram symbol are calculated. Figure 8 shows, how the coordinate information associated with the terminals of a D-flip/flop are inserted sequentially in the array E. Certain storage positions are reserved for the node numbers which later will be assigned by the tracing routine.

Line separation. In the expanded data structure, the lines are separated into segments determined by the positions of the connection symbols being associated with each line. In order to avoid tedious searches in the compact data structure G, two arrays of connection symbol pointers are set up. By means of the array X of Fig. 9, it is possible to very quickly determine which connection symbols are graphically related to vertical lines with a specific X-coordinate. When horizontal lines are to be separated, the pointer array Y of Fig. 9 is employed. In order to cut down the storage space requirements associated with the expanded data structure, all redundant line points are deleted (Fig. 4).

Entity pointers. After the compact graphic information of the array G (Fig. 7) has been converted to the expanded array E (Fig. 8) an entity pointer array P is set up. By means of this array it is possible to very quickly determine which entities are graphically related to a given point (X,Y).

Tracing routine. When the expanded data structure is formed, a tracing routine is used for the translation of this structure into a circuit description of the actual diagram sheet. This routine identifies all terminals and gate inputs/outputs which are electrically connected; thus, they can be assigned the same node number. Figure 10 shows the structure of the tracing routine and Fig. 11 contains an example of the graphical entities analyzed by one call of the routine. With reference to Fig. 11, connection symbols only exist at the points B, C and G. The first observation point is A and the terminal DATA is assigned the node number 1 just before the tracing routine is called. Table 1 shows how the observation point is moved, how connection symbols are put in the stack and how the different terminals are assigned the node number 1. The reset input of the D-flip/flop is assigned the status 2 because of the presence of the inversion symbol. A status of 0 would indicate, that the reset input was unused.

TERMINAL NAME COMPARISON

Figure 14 shows the structure of the routine that compares terminal names from different circuit diagrams. This routine constructs a total terminal list associated with the complete circuit to be simulated. The total terminal list is a set of pairs of node numbers and terminal names; names of terminals that are electrically connected are assigned the same node number. Figure 12 shows four

Table 1 - The order by which the tracing routine of fig. 10 assigns node numbers to the entities of fig. 11. The node numbers are initialized to zero.

OBSERVATION POINT :	A	B	C	D	E	E	C	F	C	B	G	G	H	H	G	B	I	B	B	B
STACK(1) :			B	C	C	C	C	C	C	B	B	G	G	G	G	B	B	B		
STACK(2) :				B	B	B	B	B	B			B	B	B						
LINE 1 :	0	1	1	1	1	1	1	1	1	1	1	1	1	1	1	1	1	1	1	1
LINE 2 :	0	0	1	1	1	1	1	1	1	1	1	1	1	1	1	1	1	1	1	1
LINE 3 :	0	0	0	1	1	1	1	1	1	1	1	1	1	1	1	1	1	1	1	1
LINE 4 :	0	0	0	0	0	0	0	1	1	1	1	1	1	1	1	1	1	1	1	1
LINE 5 :	0	0	0	0	0	0	0	0	0	0	0	1	1	1	1	1	1	1	1	1
LINE 6 :	0	0	0	0	0	0	0	0	0	0	0	0	0	0	0	1	1	1	1	1
LINE 7 :	0	0	0	0	0	0	0	0	0	0	0	0	0	0	0	0	0	1	1	1
CONNECTION B :	0	0	1	1	1	1	1	1	1	1	1	1	1	1	1	1	1	1	1	1
CONNECTION C :	0	0	0	1	1	1	1	1	1	1	1	1	1	1	1	1	1	1	1	1
CONNECTION G :	0	0	0	0	0	0	0	0	0	0	0	1	1	1	1	1	1	1	1	1
DATA TERMINAL :	1	1	1	1	1	1	1	1	1	1	1	1	1	1	1	1	1	1	1	1
OUTPUT TERMINAL :	0	0	0	0	0	0	0	0	1	1	1	1	1	1	1	1	1	1	1	1
INVERSION SYMBOL :	0	0	0	0	1	1	1	1	1	1	1	1	1	1	1	1	1	1	1	1
RESET INPUT :	0	0	0	0	0	1	1	1	1	1	1	1	1	1	1	1	1	1	1	1
RESET STATUS :	0	0	0	0	0	2	2	2	2	2	2	2	2	2	2	2	2	2	2	2
INVERTER INPUT :	0	0	0	0	0	0	0	0	0	0	0	0	0	0	0	1	1	1	1	1
NOR GATE INPUT :	0	0	0	0	0	0	0	0	0	0	0	0	0	1	1	1	1	1	1	1
NAND GATE INPUT :	0	0	0	0	0	0	0	0	0	0	0	0	0	0	0	0	0	1	1	1

small digital circuits with individual terminal lists which were determined by the
translations from individual diagram sheets
to circuit descriptions. Table 4 contains
the total terminal list constructed by the
terminal comparison routine from the four
individual terminal lists. The Table 2 and
3 represents intermediate situations where
only two and three of the terminal lists
have been merged. Figure 13 shows a circuit
which is equivalent for the four circuits

Table 2 - Total terminal list related to the circuits 1 and 2 of fig. 12.

Node Number	Terminal Name
1	CARRY
2	SUM
3	A
4	B
5	OUT.B
6	OUT.A
7	RESET
8	SET

Table 3 - Total terminal list related to the circuits 1, 2 and 3 of fig. 12.

Node Number	Terminal Name
1	CARRY
2	SUM
3	A
4	B
5	OUT.B
6	OUT.A
7	RESET
8	SET
8	DATA
9	OUT.D
1	OUT.C

Table 4 - Total terminal list related to the four digital circuits of fig. 12.

Node Number	Terminal Name
1	CARRY
2	SUM
3	A
4	B
5	OUT.B
6	OUT.A
7	RESET
3	SET
3	DATA
8	OUT.D
1	OUT.C

and terminal list of Figure 12.

DIAGRAM MERGING

When the total terminal list is constructed,
it is quite straightforward to set up the
total circuit description which is based upon
the description of the individual diagram
sheets. All nodes being connected to terminals are renumbered in accordance with the
total terminal list; afterwards, the internal
nodes of each diagram sheet are assigned different numbers higher than the numbers occur-
ing in the total terminal list.

LOGIC SIMULATION PROGRAM

The total circuit description is a basis for
a nine valued logic simulation program described by Knudsen (1). This program is able
to detect hazard phenomena and critical races.
Experiments show that the analysis time is
not related to the circuit size. Thus, the
program is suitable for the analysis of large-
scale systems. Figure 16 shows the gate diagram of a test circuit and Figure 17 contains
the results of a simulation of the circuit
with input signals at node 226 and 227.

PRACTICAL APPLICATIONS

The input language GRANTS exists in two different versions. A small-scale version runs
on the HP 9825 desktop calculator with the
HP 9872 digitizer/plotter. The software is
written in HPL and the capacity is approximately 1500 gates/modules. A large-scale version is being implemented on the IBM 3033
machine. The programming language is PASCAL
and the capacity is estimated to be very
large - a rough guess is 100 000 gates/modu-
les.

During a case-study carried out with the
small-scale version it has been observed th~t
the traditional circuit specification by
means of numbers (Figure 2) requires around
three weeks of manual work for input and
checking when working with a circuit of 1100
gates. If the set of diagrams is specified
by means of the graphical input language
GRANTS, this manual work requirement is cut
down to two days. Furthermore, a very good
design documentation is obtained together
with a total elimination of user-frustration.

ACKNOWLEDGEMENT

The author is indebted to the Danish Council
for Scientific and Industrial Research for
supporting this work.

REFERENCES

1. Knudsen, M.S., 1982, "A compact nine-
    valued logic simulation algorithm",
    IEEE International Symposium on Circuits
    and Systems, Rome, May 10-12 1982.

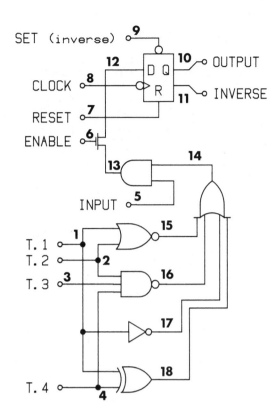

Fig. 1 Example of a digital circuit diagram with node numbers which were assigned manually.

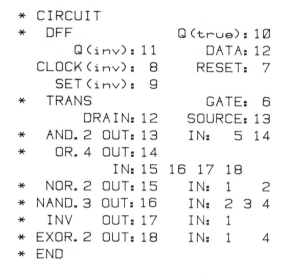

```
*  CIRCUIT
*    DFF                    Q(true):10
         Q(inv):11           DATA:12
     CLOCK(inv): 8          RESET: 7
       SET(inv): 9
*    TRANS                   GATE: 6
         DRAIN:12          SOURCE:13
*    AND.2 OUT:13      IN:   5 14
*    OR.4 OUT:14
             IN:15  16 17 18
*    NOR.2 OUT:15      IN: 1     2
*    NAND.3 OUT:16     IN: 2 3 4
*    INV   OUT:17      IN: 1
*    EXOR.2 OUT:18     IN: 1     4
*    END
```

Fig. 2 Example of an alphanumeric circuit description of the circuit of Fig. 1.

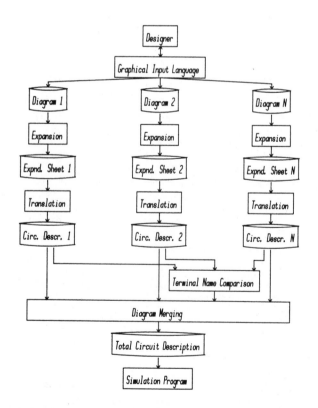

Fig. 3 Program structure of the graphical input language GRANTS.

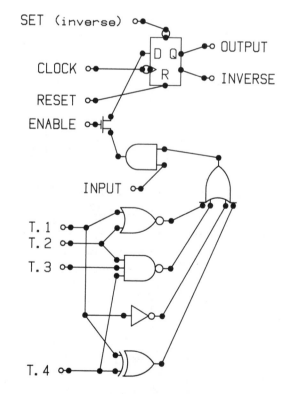

Fig. 4 Illustration of the information stored in the expanded data structure.

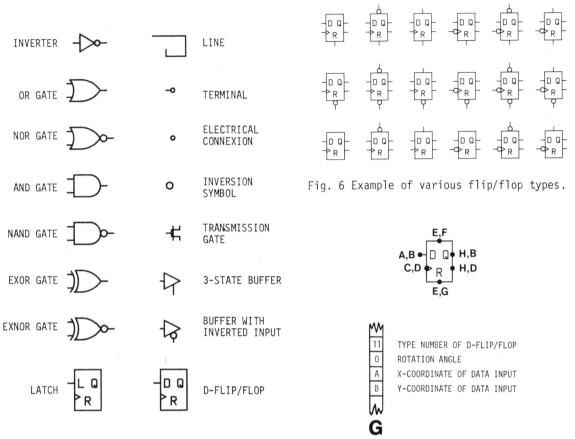

Fig. 6 Example of various flip/flop types.

Fig. 5 The basic symbols of the input language GRANTS.

Fig. 7 Representation of a D-flip/flop.

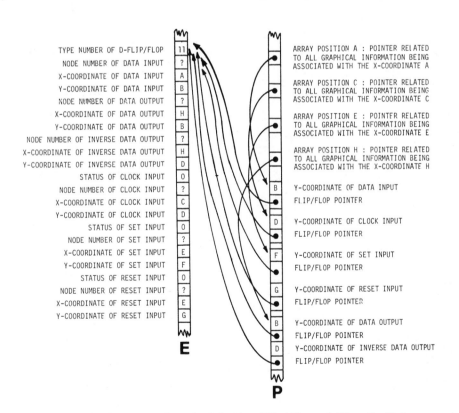

Fig. 8 Expanded data structure associated with the flip/flop of Fig. 7. This expanded structure serves as a basis for the translation from circuit diagram into a circuit description suitable for simulation programs.

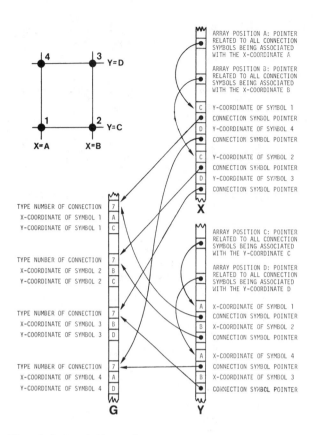

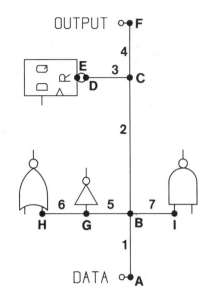

Fig. 11 Example of graphical entities
analyzed by one call of the
tracing routine.

Fig. 9 Intermediate data structure related to
four connection symbols.

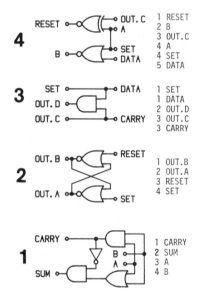

Fig. 12 Four small digital circuits with
their terminal lists.

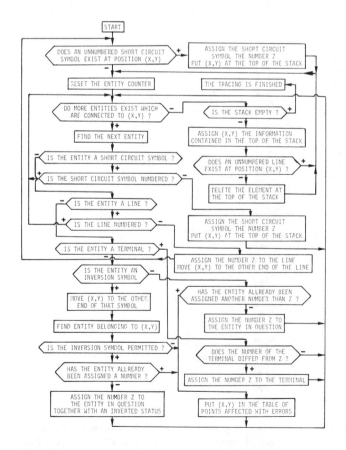

Fig. 10 Structure of the tracing routine.

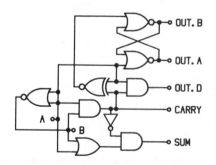

Fig. 13 Circuit equivalent to Fig. 12.

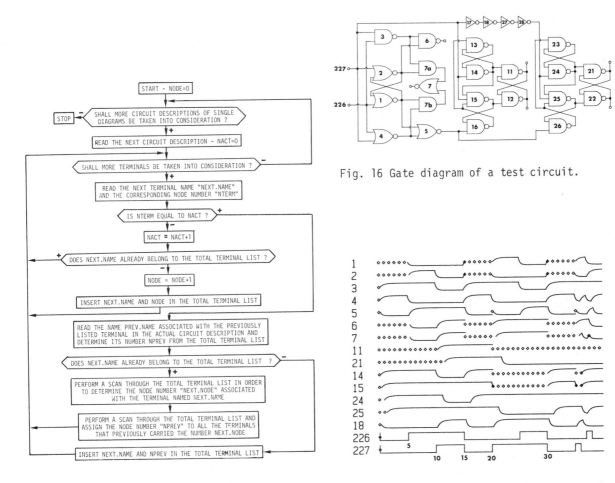

Fig. 16 Gate diagram of a test circuit.

Fig. 14 Structure of the routine which
compares terminal names.

Fig. 17 Output plots of a simulation of the
test circuit of fig. 16.

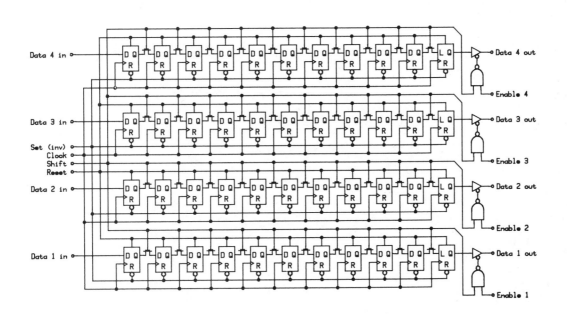

Fig. 15 Example of a diagram constructed by means of the graphical input language GRANTS.

KNOWLEDGE BASED MAN-MACHINE SYSTEM

S. Ohsuga

*Institute of Interdisciplinary Research, Faculty of Engineering, The University of Tokyo,
Tokyo, Japan*

Abstract. A new method to increase user's gain in man machine interaction is discussed. In particular, an attention is directed to describing model in the computer. An intelligent system provided with the knowledge base and the inference mechanism is proposed for the purpose. It is shown that the system can describe both structure and attributes of the real object in the same framework.

Keywords. Model description; knowledge base; inference; predicate logic; database; data structure.

INTRODUCTION

Man machine interaction is a technique to support human user with a problem in mind to achieve his goal. In using computer, he will want to achieve maximum expected return by the least number of interactions. Thus we postulate that the user is balancing costs with the gain.

Two approaches are contributory to achieve the better interactive system. The first is to make systems more friendly to the user to decrease cost. Making the computer's response more natural, designing a comprehensive set of commands, providing the computer with the capabilities to accept erroneous commands and to resolve the elliptic and anaphoric references, automatic error correction, flexible parsing, focus tracking etc.are involved in this approach.

The second approach is to increase the user's gain by making the computer's response more informative to diminish his uncertainty about his problem. The objective of this paper is to discuss the latter approach.

In discussing the issue, we consider the role of computer to aid user in solving problem in man machine environment.

Ordinaly, in solving problem, we make a model of the given problem. A model is an abstraction of real object. It is often described in the mathematical term such as graph, differential equation, stochastic process etc. or, in the other cases, it is represented in a diagram such as engineering drawing, circuit diagram, block diagram, etc.. Many researches have been and are being made on how to analyze and evaluate the model with given structure. Then, once a real problem is represented in one or combination of these abstracted models, the results of these studies can be used to solve problem.

Today, the model and the goal of given problem resides always in the human user's side. He decides what procedure to execute, what data to access and how to use them to achieve his goal based on them. Then, he should express his requests in a highly restricted command language to use computers in this problem solving process, while the computer performs only predefined primitive functions. Consequently the user has to carry out a number of interactions until he obtains information he wanted and reaches goal. A probable break-through to this problem is to give the computer the model and also the capability to decompose user's request, phrased not in terms of computer's functions but in terms of his own task, into the set of commands to the computer's built-in functions. We consider that knowledge base system is very suited for this purpose.

KNOWLEDGE BASED SYSTEM

In many real problems, a data structure is defined ad hoc to represent a specific object structure The user has to decide himself what data structure should be used to the problem and also he has to write down programs to construct and process it. Alternatively some people intend to use the conventional database. Then he can use a set of functions the database system provides him, but because the structure today's data model can support is so simple that he often feels inconvenience in representing the object with complex structure.

Every object in the real world shows its own attributes. An object as the physical existence and its attributes are inseparable. On the other hand, no model in information today has this nature but man has to give the description on attributes every time he defines a new object. The author believes that this is not the intrinsic nature of the model itself but because of our software technique for model building being unriped yet. In the following, we propose a knowledge base system to improve the situation.

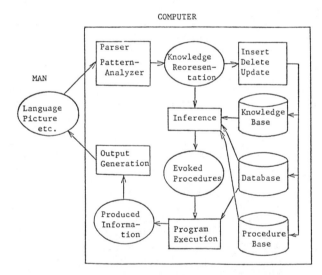

Fig.1 Organization of knowledge based system

Fig. 1 shows the organization of the knowledge based interactive system we are working to realize, named KAUS (Knowledge Acquisition and Utilization System).

There are three different information bases: the knowledge base, the data base and the procedure base. The knowledge base (KB) includes a collection of task specific knowledge as well as general rules, theories, facts etc.. The knowledge is represented in a special form called

the knowledge representation. The database(DB) is the collection of data. The relational database is adopted here because of its basic simplicity and generality. The procedure base (PB) is a collection of built-in procedures. The procedures are evoked at the evaluation stage of the request by the inference mechanism.

In the figure, an ellipse denotes information in the non-procedural form while a block denotes the procedure that converts the information from one form to another. The user uses language and/or picture to express his request. It is transformed into the intermediate language (knowledge representation) by the parser. Then the inference mechanism transforms the request into another form until it can be resolved into a set of built-in procedures, which, in turn, generate answers. The result is displayed by the output generator.

REPRESENTATION OF KNOWLEDGE AND INFERENCE

Predicate Logic and Its Inference

Various knowledge representation languages or schemes have been proposed; the production rule (Newell, 1973; Shortliffe, 1976), the semantic network (Hendrix, 1975; Myloponles, 1975), the conceptual dependency (Schank, 1974), KRL (Bobrow; 1977a, 1977b), the predicate logic (Green, 1968; kowalski, 1974) and so on.

In our system, the so called many sorted logic is adopted as the basic framework, which is expanded in the latter section. In this logic, the predicate represents 'the relationship among objects'. The collection (or the set) U, of all objects which we have interests is called the universe. Then a variable representing objects is defined over certain subset of U, which we call its domain. For example, we formalize the expression, "If the melting point of a metal is lower than a temperature, then it is in the liquid state at the temperature." as

$(\forall x/metal)(\forall y/real)(\forall z/real)$
$[(MELT\text{-}POINT\ x\ y)\cap(LESS\text{-}THAN\ y\ z)$
$\Rightarrow(LIQUID\ x\ z)]$ ($1)

where $(\forall x/metal)\text{-}\text{-}\text{-}(\forall z/real)$, called the prefix, means, "for all x in the domain 'metal', for all y in the domain 'real' and for all z in the domain 'real'". The symbols $\cap$ and $\Rightarrow$ are used to mean 'and' and

'imply' respectively. Besides them, U, ∼ and ⇔ are used meaning 'or', 'not' and 'equivalent' respectively. ∀ and ∃ are the quantifiers meaning 'for all' and 'for some'. A predicate, including no variable is a proposition. For example,
(MELT-POINT #Cu #1084,5) : melting
    point of #Cu is 1084,5(°C)   ($2)
(LESS-THAN #1084,5 #1100) :
    1084,5 is less than 1100   ($3)
are both true propositions. The symbol # denotes that the term is a constant.

Suppose a query;"Is the cupper in the liquid state at 1100°C?" is presented to the system holding the formulas, ($1),   ($2)   and ($3).   It is represented (LIQUID #Cu #1100(° C))? There is no information that matches directly with the query. However, by substituting #Cu and #1100 into x and z of ($1) respectively (it is possible because ($1) holds for every x and z), the consequence-part of ($1) matches with the query. Therefore, if its condition-part is evaluated and proved true, then the query is also true. That is
(∃y/real)[(MELT-POINT #Cu y)
     ∩(LESS-THAN y #1100)]   ($4)
is deduced to replace the original query. The quantifier of y changes from ∀ to ∃ in this process. We can prove that the formula ($4) is true by using the formulas ($2) and ($3) and by substituting 1084,5(°C) into y.   This process is the inference.

In this example we assumed that the very simple relations between individual objects such as ($2) and ($3) are given as the formula. These were necessary to evaluate the corresponding predicates in the formula ($4) through the matching function of the inference mechanism. It is, however, impractical. Because (LESS-THAN a b) is the well defined relation, we can do it more efficiently by providing the system with the specific procedure to evaluate it.

There are a number of similar predicates to the one above other than binary relations; arithmetic operations such as x+y=z denoted (ADD x y z), mathematical functions such as sinx=z denoted (SIN x z) and so on. We prepare an evaluation procedure to each of these predicates. The collection of the procedures forms the program base of Fig.1. The procedure is evoked by the inference mechanism when the predicate appears in the query and becomes evaluable.   A predicate is evaluable if independent variable(s), specified in advance to every

predicate (underlined in above examples), is substituted by the specific value(s). The variable(s) besides the independent variable(s), if any, is the dependent variable(s). Then the procedure is defined to make computation using the independent variable(s) and then to substitute the result(s) into the dependent variable(s).

The formula ($2) is still impractical to represent in the logical form, because there can be a good number of similar relations for possible combinations of objects. But, in this case, we can not define any procedure to evaluate it. They are collected and compacted in the different form, e.g., in the tabular form. This is very the relational database. For example, the collection of the formulas such as (MELT-POINT #Cu #1084.5), is replaced by the relation ALPHA(metal, melting point) shown in Fig. 2. The linking between the knowledge base and the database is performed at the run time by the normal inference operation by providing a special predicate for each relation(file). Let "relation u contains a tuple (x y,...,z)" be a predicate and be formalized to (RELATION u; x y...z). Then, the above relation and its meaning, "melting point of x is y", are linked to form the formula,
(∀x/metal)(∀y/real)[(RELATION #ALPHA
    :x y)⇒(MELT-POINT x y)],   ($5)
meaning that "if the relation ALPHA contains the tuple (x, y), then it means that the melting point of x is y". The RELATION is the procedural predicate to which the procedure is defined to read the file u, where u is a variable of which the value is the identifier of the relation. The predicate RELATION turns to evaluable when the variable is substituted by the specific identifier. Hence, if 'the melting point of x' is asked, the inference algorithm changes the request into the file access procedure using the formula ($5). In case of the example before, the inference mechanism will deduces
(∃y/real)[(RELATION #ALPHA :#Cu y)
     ∩(LESS-THAN y #1100)]   ($6)
from   ($4) and ($5).   Then the formula is evaluated procedurely by evoking procedures corresponding to predicates in the formula. By means of this technique of linking database and knowledge base, a graph is very effectively represented, where the

Ag	961.9
Au	1064.4
Cr	1890.0
Fe	1535.0
Mn	1244.0
.	.
:	:
Ni	1455.0

Fig.2 A relation

set of arcs is listed in a table, say
#FARC, and the formula
($\forall$ x/node)($\forall$y/node)[(RELATION
                    #FARC; x y)$\Rightarrow$(ARC x y)]
defindes that any tuple (x y) in the
table denotes that " there is an arc
from node x to node y".

EXTENTION OF THE FRAMEWORK

Multi Layer Logic (Ohsuga, 1978, 1980,1982)

Though the many sorted logic
possesses a number of remarkable
features, its system is still too
simple to represent every concept
involved in many real applications.
For example, consider an expression
"the average of u is v" denoted
(AVERAGE u v). The syntax seems to
be the same as (MELT-POINT x y).
There is, however, a substantial
difference in the mathematical
structure between them. In the
mathematical term, this is the
difference between $u \subset U$ and $x \in U$.
In order to accomodate both of them
in the same framework, we introduce
the powerset, 2^U, of U and expand
the universe from U to $U \vee 2^U$, where
the symbol $\vee$ means "union" of sets.
Then we apply the many soted logic.
For example, the expression "for
every (set) u, there is an average of
u" can be represented
($\forall$u/2^U)($\exists$v/real)(AVERAGE u v)   ($7)
In fact, the empty set should be
excluded from the powerset. In the
following, we denote the powerset but
the empty set being excluded as *U
introducing the new symbol *. By
definition, some variable u in a
predicate is an element of *U and, at
the same time, can be a domain of the
other variable(s). For example,
"some element of a set is less than
the average of the set" is
represented,
($\forall$u/*U)($\exists$x/u)($\exists$y/real)[(AVERAGE u y)
        $\cap$(LESS-THAN x y)]   ($8)
where u is an element of *U and at
the same time the domain of x.

It is easy to extend the idea upward
to define a high order powerset
denoted, $*^n U$, n>0. It is defined
recursively, $*^n U = *(*^{n-1} U)$. It implies
that the predicate logic is expanded
to include the hierarchically
structured entity as the objects of
description. In the hierarchical
structure, a n-th level entity is
composed of n-1th level entities.
The set of the 0-th level entities
are referred to the base set. Any
substructure of the structure is
itself an entity. If, in Fig.3,
level0, level1, level2 and level3
entities represent the points, the

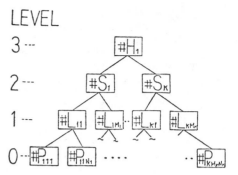

Fig.3 A hierarchical structure

lines, the surfaces and the polyhed-
rons respectively, then the structure
represents a three dimensional geometric model. Arbitrary description
can be made on the structure or any
of its substructures. For example,
the expression "every polygon of the
polyhedron H1 has some line of length
a" is represented
($\forall$x/H1)($\exists$y/x)(LENGTH y a)   ($9)

The more interesting is the case
where the structure is incompletely
specified. For example, the expression, "Is there a polyhedron composed
of the set of vertices X, of which
every surface has a line of length
a?", asking a structure that
satisfies the given condition, is
represented,
($\exists$u/$*^3$X)($\forall$x/u)($\forall$z/x)($\exists$y/x)[(F$_L$ z)
    $\cap$(F$_S$ x)$\cap$(F$_L$ u)$\cap$(LENGTH y a)]? ($10)
*X denotes a set of all subsets of
vertices in X. A line is an element
of it, that satisfies the condition
as the lines (the number of vertices
included in the element is just two).
F$_L$ "represents the condition. $*^2$X
denotes a set of all subset of *X. A
surface, if any, is an element of it,
that satisfies the condition as the
surface, F$_S$ (composed of lines, lines
in the element forms the closed link
and are co-planar, and so on). $*^3$X
denotes a set of all subset of $*^2$X.
A polyhedron, if any, is one of its
element that satisfies the condition
as the polyhedron, F$_H$ (composed of
surfaces which forms a closed body,
and so on). We call the extended
version of the predicate logic multi-
layer logic.

The extention of the formalism causes
the modification of the inference
algorithm. We don't describe it here
but we mention that even though the
algorithm for the multi-layer logic
is a little more complicated in the
handling of the variables than that
of the simple many sorted logic, the
reward far exceeds the investment in
its model building capability.

Representation And Manipulation Of Data Structure

(a) A hierarchical structure

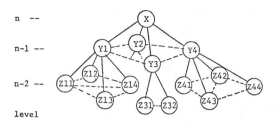

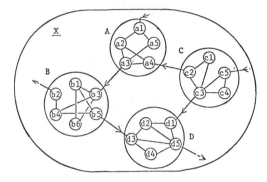

(b) A recursively defined graph

Fig.4 Examples of typical structures

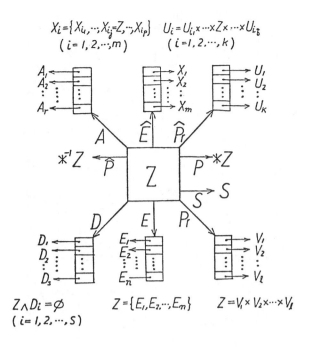

$$X_i = \{X_{i_1}, \cdots, X_{i_j} = Z, \cdots, X_{i_P}\} \quad U_i = U_{i_1} \times \cdots \times Z \times \cdots \times U_{i_g}$$
$$(i = 1, 2, \cdots, m) \qquad (i = 1, 2, \cdots, k)$$

$$Z \wedge D_i = \phi \qquad Z = \{E_1, E_2, \cdots, E_m\} \qquad Z = V_1 \times V_2 \times \cdots \times V_l$$
$$(i = 1, 2, \cdots, S)$$

Fig. 5   Skeleton Structure Cell

The capability to represent and manipulate the hierarchical structure in combination with the graph in the form of relation as described before offers us the powerful means to represent model structure. Note that both schemes for structure representation are completely formalized. Hence, we can give the meaning to and the theoretical basis for manipulation of the data structure. Fig. 4 illustrates a few examples of structures involved in the technique. These structure will cover: the chemical compounds, the geometrical model, the system assembly and so on.

KNOWLEDGE STRUCTURE

The model structure as shown above is, in fact, an instance of a generic structure which is itself a part of a universal knowledge structure (ref. Fig. 5 & 6). In this structure, every entity, generic or specific, is given a node, a fixed length cell in the memory. A node can be linked to the other nodes by means of different but finite types of arcs labelled P(powerset), E(element), D(disjoint), Pr(product set), S(structural component), A(Description, i.e., formula of which the node is a domain of variable in the formula) and the reverse pointers of P, E and Pr. For example, let X be a node and nodes Y, x_i, (i=1, 2, ..., n), Z and W be adjacent to X connected by the arcs

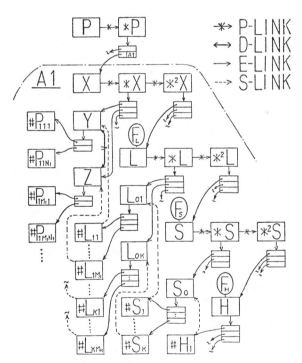

Fig.6 Knowledge structure

P, E, D and S respectively. Then the nodes are in relations $Y = *X$, $x_i \in X$, $Z \wedge X \neq \phi$ (where Z and X are the subsets of the same super set) and $W \in X$ where $W \in X$ denotes that W is the set of constituents composing X in a specific hierarchical structure. A logical expression which contains a concept (entity) X as a domain of variable is linked to the node X by A so that the expression can be reached via the node. A 1-to-n relation like an arc E is, in fact, implemented by a pointer table. All types except d-type are the directed arcs (ref. Fig. 5).

Various kinds of set-theoretical relation are composed as the compounds of basic relations. For example, $X \supset V$ is decomposed to $X \overset{B}{\supset} Y$ & $Y \overset{E}{\supset} V$.

This structure is a part of knowledge representation in the multi-layer logic. Some logical tests appearing in the inference procedure of the logic are replaced by the test for set-theoretical relations either of $X \supset Y$ or $X \wedge Z \neq \phi$. The system, given two domains of variables, can achieve the tests swiftly by following the links in the specified ways. The other advantages of using the structures are as follows:
(1) Given a query, logical formulas relevant to this query can be retrieved, by using the domain of variables, selectively and swiftly.
(2) Suppose a new node Y is defined as being in relation $X \supset Y$ (or $X \ni Y$) to X. Then Y automatically inherits all the descriptions given to X. Thus, for example, if a specific structure is defined as an instance of polyhedron, then it inherits all the attributes as the polyhedron.
(3) A dictionary is provided, in which a word, in the character string, is accompanied with the pointer to the node which the word denotes. The parser, using this mechanism, transforms the natural language into the logic. Thus the system can understand natural language.
(4) A user can give a special identifier to a node. Then only user who designate the identifier can access to the node. If the node is a set and a structure is organized by making use of the set as a base set, then the identifier denotes the structure (Al in Fig. 6). This mechanism is effective when user wants to confine the scope of an expression to a part of universe.

SIMPLE EXAMPLE

The system thus obtained satisfies the basic requirements for being an intelligent interactive system. We have implemented an experimental system named KAUS (Knowledge Acquisition and Utilization System). The first version of the system was based on the many sorted logic and was named KAUS-1. Then it was revised to KAUS-2 which was augmented with the capability to access databases. We are now implementing a new version, named KAUS-3, based on the multi-layer logic. Before going into the discussion of the new version, we show a very simple example by KAUS-2, in order to show the knowledge base system works.

Suppose a set of knowledge elements (1) through (5) shown in Fig.6 exists in the knowledge base, where (1) and (2) represent facts such as "everything lighter than fluid float on the fluid" and "the specific gravity of an object can be obtained by dividing the weight of the object by its volume" respectively.

The system is assumed to have three different files representing the relations holding between "sample (material) number and its (measured) weight", "sample (material) number and its (measured) volume" and "sample (liquid) number and its specific gravity" respectively. Formulas (3) through (5) are descriptions of these files in the logical form. These are a portion of the system's knowledge and the user is not required to be aware of them. He can only ask the system questions like "Which sample material floats on all liquid samples?" (see Q in Fig.7). Since there is no data file that contains the reply to this query directly, the system must try to resolve the query into formulas with equivalent meanings. "Float on" is equivalent to a relation that

Wff's in the knowledge base:
(∀x/physob) (∀y/fluid) (∀z/real) (∀u/real)[(SPC-GRV x z)
∩(SPC-GRV y u)∩(LESS-THAN z u)⇒(FLOAT x y)]   ----(1)
(∀x/physob) (∀y/real) (∀z/real) (∀u/real)[(MASS x y)
∩(VOLUME x z)∩(DIV u: y z)⇒(SPC-GRV x u)]   ----(2)
(∀x/physob) (∀y/real)[(RELATION Fm: x y)⇒(MASS x y)]   ----(3)
(∀x/physob) (∀y/real)[(RELATION Fv: x y)⇒(VOLUME x y)]   ----(4)
(∀x/liquid) (∀y/real)[(RELATION Fl: x y)⇒(SPC-GRV x y)]   ---(5)

Query:
(∃x/physob) (∀y/liquid) (FLOAT x y)?   ----(Q)
   (Which material floats on all liquid sample?)

Deductive procedure:
(∃x/physob) (∀y/liquid) (∃z/real) (∃u/real)[(SPC-GRV x z)
∩(SPC-GRV y u)∩(LESS-THAN z u)] (from (1) and (Q)) ---(Q1)
(∃x/physob) (∀y/liquid) (∃z/real) (∃u/real) (∃v/real) (∃w/real)
[(MASS x v)∩(VOLUME x w)∩(DIV z: v w)∩(SPC-GRV y u)
∩(LESS-THAN z u)]   (from (2) and (Q1)) ---(Q2)
(∃x/physob) (∀y/liquid) (∃z/real) (∃u/real) (∃v/real) (∃w/real)
[(RELATION Fm: x v)∩(RELATION Fv: x w)∩(DIV z: v w)
∩(RELATION Fl:y u)∩(LESS-THAN z u)] (from (3),(4),(5) & (Q2))
                                                    ---(Q3)
                                          ---end---
Note: fluid⊃liquid

Fig.7 A simple example of query processing.

holds between the specific gravities of two objects(Q1). The system tries to find the answer to this new query but fails again because there is no data file on the specific gravity of the sample materials. Therefore, it resolves the query again and replaces the specific gravity of the sample material by its equivalent definition, i.e., the relation between the weight and the volume of the materials (Q2). This time, both weight and volume of materials are given in the files, and Q3 is obtained. The deductive process terminates here and a program including the file manipulations is obtined. This is converted to data base operations.

Fig. 8 shows the results of this example obtained by KAUS-2. This result is optimal in the sense that no ineffective file operation is included.

```
# 1602G

[EX/PO][AY/LIQ](FLT,SX,SY)?
[1]     (#FGET  #WT   PO1  RNUM9   *WRK1)
[2]     (VPRJ  *WRK1  RNUM9  *WRK1)
[3]     (#FGET  #VOL  PO1  RNUM10  *WRK2)
[4]     (VPRJ  *WRK2  RNUM10  *WRK2)
[5]     (AND  *WRK1  *WRK2  *WRK3)
[6]     (EXPD  *WRK3  RNUM4  *WRK3)
[7]     (#DIV  RNUM4  RNUM9  RNUM10  *WRK3)
[8]     (PRJ  *WRK3  RNUM9  *WRK3)
[9]     (PRJ  *WRK3  RNUM10  *WRK3)
[10]    (#FGET  #SW  LIQ2  RNUM5  *WRK1)
[11]    (VPRJ  *WRK1  RNUM5  *WRK1)
[12]    (VPRJ  *WRK3  RNUM4  *WRK3)
[13]    (MULT  *WRK1  PO1  *WRK1)
[14]    (MULT  *WRK3  LIQ2  *WRK3)
[15]    (AND  *WRK1  *WRK3  *WRK2)
[16]    (#GT  RNUM5  RNUM4  *WRK2)
[17]    (PRJ  *WRK2  RNUM4  *WRK2)
[18]    (PRJ  *WRK2  RNUM5  *WRK2)
[19]    (DIV  *WRK2  LIQ2  *WRK2)
[20]    (VPRJ  *WRK2  PO1  *WRK2)
[21]    (OUT  *WRK2  *TTY)
```

Fig. 8 Result of the example of Fig. 7.

In this figure, FGET, PRG (and VPRG), AND, MULT, and DIV denote operations ; to read file, to do projection operation (in the sense used in the relational database, or simply RDB in short), to make intersection, to make cartesian product, to do division (also in the sense used in RDB) respectively[1]. It is not difficult to transform this results farther into the data manipulation language of the particular databases.

Note that, in using computers today, user is required to generate himself a sequence of operations such as shown in Fig. 8. He needs much time

---

[1] Note that the sequence MULT-MULT-AND as appeared from [13] through [15] in Fig. is equivalent to join operation in RDB.

and efforts to get information he wants. As was seen in this example, it is anticipated that knowledge base system will change this situation. It will decrease considerably user's burden in using computers, if it is provided with appropriate knowledge on the domain he concerns. In other words, it will ensure greater user's gain relative to user's effort at interaction.

In fact, however, most of present knowledge base systems including KAUS-2 are not powerful enough to be used for application fields in which very dynamic processes with complicated data are included. Aa a typical example of such applications we refer to engineering design, a very dynamic process of try-and-error involving both analytic and synthetic procedures to the object with complex structure. Most of knowledge base systems seem to be powerless to this class of problems because these systems lack the capability to define and manipulate complex data structure to represent object structure. This is the reason why we need a new theory, multi-layer-logic, which is the logic system involving objects with complex structure. Because, in this system, the data structure is formalized, we can express even synthetic operations in the logical formula. ($10) was such an example.

Fig. 8 shows a few examples of formulas written in the syntax of KAUS-3 representing shapes and geometrical relations of objects in three dimensional space. Those with asteriques are the formulas defined procedurely (corresponding procedures reside in the procedure base). Making use of these primitive formulas as the basic concepts, we can define and construct more complex concepts successively. Then any concept that can be expressed in this syntax, is ensured to be processed by the inference mechanism of KAUS-3.

CONCLUSION

In this paper, we have discussed a method to increase user's gain in man machine interaction. We focused our attention on describing model in the computer. We proposed a knowledge base system for the purpose and showed that both structure and attributes of an object we had concerns could be described within the same framework. Still, by introducing the inference algorithm, we can decompose user's request, allowed to phrase not in terms of computer's function but in terms of user's own task, into the set of queries to the computer's build-in functions by referring to the knowledge. This capability of the system increases considerably user's gain.

REFERENCES

Bobrow, D.G., and T. Winograd (1977). An overview of KRL a knowledge representation language. Cognitive Science, Vol. 1, No. 1.

Bobrow, D. G., T. Winograd, and others (1977). Experience with KRL-Ø one cycle of a knowledge representation language. Pro. 5th Int'l. Joint Conf. on Airtificial Intelligence.

Green, C. (1968). The use of theorem proving techniques in question answering system. Pro. 23rd National Conf. ACM. Brandon Press.

Hayes, P., E. Ball, and R.Reddy (1981). Breaking the manmachine communication barrier. IEEE Computer, pp. 19-3Ø.

Hendrix, G.G. (1975). Expanding the utility of semantic networks through partitioning. Proc. 4th Int'l. J.Conf. on Artificial Intelligence.

Kowalski, R. A. (1974). Predicate logic as programming language. Proc. IFIP 74. North Holland. pp. 569-574.

Myloponles, J., P. Cohen, A. Borgida, and L. Suger (1975). Semantic networks and the generation of context. Proc. 4th Int'l. J. Conf. on Artificial Intelligence.

Newell, A. (1973). Production system: Models of control structures. In H. Gallaire, and J. Minker (Ed.), Visual Information Processing. Academic Press. pp. 463-526.

Ohsuga, S. (1978). Toward intelligent interactive systems. In R.A. Guegj and others (Ed.), Methodology of Interaction. pp. 339-36Ø.

Ohsuga, S. (198Ø). Perspectives on new computer systems of the next generation —— a proposal for knowledge-based systems. J. of Information Processing 3-3. pp. 171-185.

Ohsuga, S. (1982). A new method of model description -- use of knowledge base and inference. (to appear in) Proc. IFIP WG.5.2., Working Conf. on CAD system framework.

Schank, R., and C.J. RiegerII (1974). Inference and the computer standing of natural language. Artificial Intelligence, Vol. 5. pp. 373-412.

Shortliffe, E. H. (1976). Computer Based Medical Consultations :

MYCIN. American Elesvier.

Relationships between points;

* (AT P X Y Z):Point P is at (X Y Z)
* (ON-PP P1 P2):Points P1 and P2 are at the same location
* (DISPLACE-P P1 P2 X Y Z):Point P2 is defined as P1 displaced by X Y and Z
:

Relationships between lines;

(CONTACT-L L1 L2 P1 P2):Lines L1 and L2 are in contact with each other at P1 of L1 and P2 of L2

$(\forall L1,L2/line)(\forall P1/L1)(\forall P2/L2)$ [(AT-P P1 P2)$\Rightarrow$(CONTACT-L L1 L2 P1 P2)]

(TRANSLATE-L L1 L2 X Y Z):Line L2 is defined as L1 translated by X Y and Z

$(\forall L1/line)(\forall L2/*point)(\forall X,Y,Z/real)(\forall P1/L1)(\exists P2/L2)$[(LINE L2)$\cap$(DISPLACE-P P1 P2 X Y Z)$\Rightarrow$(TRANSLATE-L L1 L2 X Y Z)]

(MK-CONTACT-L L1 L2 L3 P1 P2): Line L3 is defined as L2 translated toward L1 until P1 of L1 and P2 of L2 coincide

$(\forall L1,L2/line)(\forall L3/*point)(\forall P1/L1)(\forall P2/L2)(\forall X Y Z/real)$[(DIFF-P P1 P2 X Y Z)$\cap$(TRANSLATE-L L2 L3 X Y Z)$\Rightarrow$(MK-CONTACT-L L1 L2 L3 P1 P2)]

(JOIN-L L1 L2 CL P1 P2):Join lines L1 and L2 to form a complex CL

$(\forall L1,L2/line)(\forall L3,Cl/*point)(\forall P1/L1)(\forall P2/L2)(\forall P3/L3)$[(MK-CONTACT-L L1 L2 L3 P1 P2)$\cap$(UNIFY-P P1 P3)$\cap$(COMPLEX-L L1 L2 CL)$\Rightarrow$(JOIN-L L1 L2 CL P1 P2)]

* (COMPLEX-L L1 L2 CL):CL is a complex formed of L1 and L2
:

Relationships between surfaces;

(P-CONTACT-S S1 S2 P1 P2):Surfaces S1 and S2 are in contact with each other at P1 of S1 and P2 of S2

$(\forall S1,S2/surface)(\forall L1/S1)(\forall L2/S2)(\forall P1/L1)(\forall P2/L2)$[(ON-PP P1 P2)$\Rightarrow$(P-CONTACT-S S1 S2 P1 P2)]

:

* Procedural type atom

Fig.9 A few examples of descriptions on 3-D configurations

SYMBIOTIC, KNOWLEDGE-BASED COMPUTER SUPPORT SYSTEMS

G. Fischer

*Research Group on Man-Machine Communication, Department of Computer Science,
University of Stuttgart, Azenbergstrasse 12, D-7000 Stuttgart, Federal Republic of Germany*

Abstract. The full benefit of computers as tools of thought can come only when we learn to dissect intelligence into a portion which is best suited to the human being and a portion which is best suited to the computer and then find a way to mesh the process. The limiting resource in future man-machine systems will be the human (with respect to time, attention, complexity management) and not the computer. There is a need for symbiotic, knowledge-based computer support systems which will make communication with computers easier, more rewarding and turn the computer into a convivial tool.

Theories, methodologies and tools are needed to make systems of this sort broadly available. In our research project INFORM we are designing, implementing and evaluating a knowledge-based **I**nformation **M**anipulation **S**ystem (**IMS**) as a protypical development.

Keywords. cognitive systems; cognitive ergonomics; computer software; convivial tools; knowledge-based systems; information manipulation systems; man-machine communication; user interface.

INTRODUCTION

Our goal is to design and implement symbiotic computer systems which are convivial tools for the people who use them. It will be shown that a system of this sort has to be a knowledge-based system.

We will describe some of the underlying theoretical aspects which we regard as relevant for this research and illustrate our ideas with examples taken from the work in our research project INFORM (Boecker, Fischer and Gunzenhaeuser, 1980).

COMPUTER-BASED MAN-MACHINE SYSTEMS — HOW THEY ARE AND HOW THEY SHOULD BE

We see the purpose of computer-based man-machine systems to direct the computational power of the digital computer to the use and convenience of man.

There is no doubt that there has been great progress in some fields of computer science. The dramatic price reduction in hardware has opened up totally new possibilities. But other aspects have not kept pace with this progress, especially how easy it is -- not only for the expert but also for the novice and the occasional user -- to take advantage of the available computational power to use the computer for a purpose chosen by him/herself.

Most computer users feel that computer systems are unfriendly, not cooperative and that it takes too much time and too much effort to get something done. They feel that they are dependent on specialists, they notice that software is not soft (ie the behavior of a system can not be changed without a major reprogramming of it) and the casual users finds himself in a situation like in instrument flying: he needs lessons (relearning) after he did not use a system for a long time.

Our goals are to create symbiotic, knowledge-based computer support systems which

    - handle all of their owner's information-related needs; these needs will be quite different for different groups of users

    - make computer systems accessible to many more people and make computer systems do many more things for people

    - help us to gain a better understanding of

the cognitive dimensions and tasks structures

To achieve these goals we need beside technological expertise

   - theories, which take into account knowledge and insights from computer science, psychology, linguistics and sociology and help us to define new design criteria

   - methods which are based on those theories

   - tools which use the computational power of the computer to support the user.

SYMBIOTIC SYSTEMS

Symbiotic systems are based on a successful combination of human skills and computing power in carrying out a task which cannot be done either by the human or by the computer alone. We will illustrate our conception of symbiotic systems by giving examples in different domains.

Software Engineering

We need systems which provide more assistance in all phases of the process (eg problem formulation, design, specification, implementation, testing, verification, documentation and modification) and more automation. This would free the programmer from the clerical parts of his task so he can concentrate more on the difficult aspects of his problems.

There is no doubt that we have made big progress in the area of programming (by having available Assembler, Compiler for high level languages, interactive programming environments (like in LISP and SMALLTALK) so that the labor of programming was reduced by several orders of magnitude during the last twenty years. Yet, despite all this progress, programming a computer to perform a non-trivial task remains a difficult task which is much more complicated and tedious than instructing an intelligent and trained human who would help us as an assistant.

Human communication and cooperation can serve as a model for a symbiotic relationship. What can humans do that most current computer systems cannot do? Human communication partners (eg a programming assistant)

   - do not have the literalness of mind which implies that not all communication has to be explicit; our communication partner can supply additional information which we have forgotten and he can correct simple mistakes; see the DWIM (= "Do What I mean") facility in INTERLISP (Teitelman and

Masinter, 1981) for a first step in this direction

   - can apply their problem solving power to fill in details if we give statements of objectives in broad functional terms; the development in programming from "how" (eg a detailed and exact description in an assembler language how to evaluate an algebraic expression) to "what" (eg to write down this expression in PASCAL) indicates in which direction further progress is needed

   - can interpret meaning and intent, when confronted with the vagueness and informality of natural language;

   - can articulate their own understanding and misunderstanding

   - can provide explanations.

Knowledge-based systems (see below) are one promising approach to equip machines with some of these human communication capabilites.

Help Systems

Help systems (which should be an integral part of every computer system) can be used to illustrate a basic misunderstanding of the real limiting resource in designing computer systems. **The important function of computers is not to multiply information but to analyze it so it can be filtered, compressed and diffused selectively.** Most online assistance systems offer a huge static, tree-structured information structure where it is very time consuming to find a relevant piece of information. What the user really needs are answers to question like:
   - how can I do X?
   - what happens if .....
   - why did Y occur?
   - can I undo the effects of Z?
To be able to give answers of this sort a computer system must have self-knowledge about the functionality and about the dynamic execution state, it must offer a descriptive language so the user can communicate with it and it must provide explanation facilities which are based on a model of the user. A further important characteristic of a symbiotic system is that it should be non-intrusive, ie it should not get in our way if we do not need it.

Further examples of symbiotic computer systems

Computerized axial tomography (CAT scanning; McCracken 1979) is based on a partnership between doctor and computer; the necessary inverse Fourier transformations involve an immense amount of computation and cannot be done without the help of a computer -- and the interpretation of the data requires discrimination between subtle differences in density which is beyond current capabilities in image processing.

Personal computers (eg like the Dynabook; Goldberg 1981) have made it possible to turn everyone into an artist and produce animated pictures, music and other artifacts which in the past could only be produced by the professional because the process was too time consuming and too costly.

In our research project INFORM (in addition to work on software engineering and help systems) we have developed two application systems which serve as prototypes for our thinking about symbiotic systems:

1) a system FINANZ which helps us to fill out finance plans for research proposais; the systems assists us in the planning process, takes over all the clerical details (eg to compute repeatedly the sums of some numbers after one of the numbers has changed), it changes input data to the right format and it warns us if we have made a mistake

2) a system PLANER which helps a graduate student in our department to plan his studies; it generates proposals for time-tables under constraints formulated by the user, it indicates conflicts which will arise in later years by making a specific choice for the next semester and it helps us to resolve these conflicts.

CONVIVIAL SYSTEMS

Illich (1973) has introduced the notion of "convivial tools" which we regard as one of the most important aspects of symbiotic systems. He defines them as follows:

"Tools are intrinsic to social relationships. An individual relates himself in action to his society through the use of tools which he actively masters, or by which he is passively acted upon. To the degree that he masters his tools, he can invest the world with his meaning; to the degree that he is mastered by his tools, the shape of the tool determines his own self-image. Convivial tools are those which give each person who uses them the greatest opportunity to enrich the environment with the fruits of his or her vision.
Tools foster conviviality to the extent to which they can be easily used, by anybody, as often or as seldom as desired, for the accomplishment of a purpose chosen by the user."

Illich's thinking is much broader and tries to show alternatives for future technology-based developments and their integration into society. We have applied his thoughts to information processing technologies and systems and believe that conviviality is a dimension which sets computers apart from other communication technologies. All other communication and information technologies (eg television, videodiscs, interactive videotex) are **passive,** ie the user has little influence to shape them to his own taste and his own tasks. He has some selective power but their is no way that he can extend system capabilities in ways which the designer of those systems did not foresee.

We do not claim that current existing computer systems are convivial. Most systems belong to one of the following two classes (which constitute opposite ends along the dimension of conviviality):

- turn-key systems: they are easy to use, no special training is required but they can not be modified by the user

- general purpose programming languages: they are hard to learn, they are often too far away from the conceptual structure of the problem to be solved and it takes too long to get a task done or a problem solved.

There are promising ways starting from both ends to make systems more convivial. Good turn-key systems contain features which make them modifiable by the user without a necessity to change the internal structures. There are text processing systems which allow the user to define his own keys ("key-board macros"; Fischer, 1980) and there are display systems which allow the user to create and manipulate windows at an abstract and easy to learn level (Bauer, Boecker and Fischer, 1981).

We believe that the development of convivial tools will break down an old distinction: there will be no sharp border line any more between programming and using programs -- a distinction which has been a major obstacle for the usefulness of computers. Convivial tools will take away the impossible task from the "meta-designer" (ie the person which designs design-tools for other people) that he has to anticipate all possible uses and all people's needs.

Another idea related to the concept of convivial systems is a **"toolkit"** which provides a set of components and set of tools (by means of which these components can be viewed and manipulated) that can be used to create many different but related things. This approach has been successfully exploited in other areas (see technical construction systems; Fischer and Boecker, 1981) and has as its goal to protect the user from the full complexities of a general purpose system. Examples in the world of information processing systems are developments from the Learning Research Group at Xerox with SMALLTALK (Goldberg, 1981) and the operating system UNIX which allows the user to create complex procedures not by writing large programs from scratch, but by interconnecting relatively small predefined components.

There is a crucial difference whether a computer user feels that he is computerized or whether he is using a computer as a powerful tool. Decision support systems ("expert systems") in medicine (eg Mycin; see Shortliffe, 1976) can be seen by a physician

    - as a robot doctor which replaces him

    - as a tool (like a X-Ray machine, a more easily consulted reference book or a computer tomograph).

What it really is lies mainly in the mind of the physician, but we (as system designers, as working out the theoretical foundations for them) should do our best that our systems are tools and their users have the possibility and the feeling that they are the **controlling agents.**

With convivial tools we pursue the goal to encourage the user to be actively engaged and generate creative extensions to the artifacts given to him and we hope that they will be tools for everyone and not only the private domain of a few highly educated people.

KNOWLEDGE-BASED SYSTEMS

There is a difference between data bases and knowledge bases. In a **data base** the conceptual organization is simple and the main problems is size and efficiency (an example would be all the records of a company with 10,000 employees). For a **knowledge base** there is no definition what should be included and under which circumstances and in which form the stored knowledge becomes relevant (examples would be the knowledge of a chess master to make a move, the knowledge of a doctor to find a diagnosis and the knowledge of a programmer to explain what his program does).

Knowledge-based systems are a very active research area (see the plans in Japan for a ten years research project to develop "5th Generation Computer Systems" as knowledge-based systems). Our own efforts are centered around the question how knowledge-based systems can be used to improve man-machine communication and contribute towards the creation of symbiotic systems.

The traditional model of man-machine communication is shown in Fig.1.

Fig.1: The traditional model of man-machine communication

It can be shown that there is no way to achieve the characteristics of symbiotic systems which we have described above with this model. Based on an analysis of human communication processes we have developed the model in Fig. 2 which is more suited to fulfill the requirements. It contains a knowledge base which can be accessed by both communication partners; this implies that the necessity to exchange all information explicitly does not exist any more.

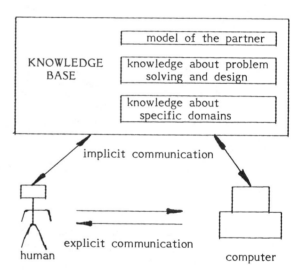

Fig. 2: Extended Model of man-machine communication

We want to illustrate the consequences and the possibilites of these two models with a concrete example of a computer program which could serve as a travel assistant (see Fauser and Rathke, 1981):

Computer: What time do you want to leave?
User: I must be in Berlin before 10 AM.
We can observe that the answer has at a surface level nothing to do with the question but that it is definitely relevant for the question if the computer has the following knowledge: being somewhere is the consequence of going somewhere and the time of leaving determines in a systematic way the time of arrival. It is obvious that a system based on our second model is needed to cope in a meaningful way with this situation.

INFORMATION MANIPULATION SYSTEMS (IMS)

We have used this term to indicate that the original meaning of "programming" as the art of finding and coding of algorithms is too restricted to cover the many possible uses of a computer by a human being.

One of the obstacles computer systems present to the user is the diversity of different languages and conventions which a user has to know to get a certain task done. To write an ordinary program in a conventional system the user has to know a large number of <u>different</u> languages, sublanguages and conventions, eg:
  * the programming language itself (with conventions for specifying the control flow, external and internal data description etc)
  * the operating system (job control language, linkage editor and loader)
  * the debugging system (diagnostic system, symbolic assembler etc)
  * the text processing system (editor and formatter)

The need for an <u>integrated</u> system is obvious to anybody who has tried to struggle through the idiosyncracies of the different systems mentioned above.

An **IMS** offers uniformity in two dimensions to cope with this problem:

- <u>linguistic uniformity</u>: all tools (eg the programming system and superimposed modules as well as more specific creations of the user) are made from the same material and thus part of the same conceptual world. This has the sociological benefit that the system's implementor and users share the same knowledge. Each module in the system can be regarded as a **"glass-box"**, ie it can be inspected by the user and the system can be explored all to the edges. This gives the user an amount of control over his environment which is not reachable in other systems.

- <u>uniformity of interaction</u>: this is based on a good interface, which provides a uniform structure for finding, viewing and invoking the different components of the system. The crucial aspect for this interface is the use of the display screen, which allows the real-time, direct manipulation of iconic information structures which are displayed on the screen. Each change is <u>instantly</u> reflected in the document's image, which reduces the cognitive burden for the user. The screen should be regarded as an extension of the limited capacity of our short term memory (ie it provides a similar support like pencil and paper for the multiplication of two five digit numbers).

The structure of an **IMS** is illustrated in Fig. 3. The most crucial issue in the design is the

<u>integration</u> of the different components with each other.

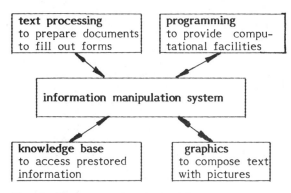

Fig. 3: The structure of an IMS

We regard IMSs as prototypical systems which are an effort towards the goal to construct truly symbiotic systems. IMSs can be used as software production systems, as application systems for end users (eg in office automation) and as a testbed for the construction of modules to improve man-machine communication.

COGNITIVE SCIENCE

To make full use of a symbiotic relationship implies that each partner can take advantage of his strength and gain some support for his weaknesses. Cognitive Science (Norman 1981) is an interdisciplinary research discipline which tries to understand the principles of intelligent systems (independently where and how these systems exist).

In the last years a substantial body of knowledge has been accumulated which can serve as design guidelines for symbiotic systems (this body of knowledge provides a starting point for the new discipline of **cognitive engineering** which tries to construct artifacts along cognitive dimensions). Some of the strong parts as well as the mental processing limits of the human information system have been identified and prototypical systems have been constructed which try to acknowledge these contributions. Examples are (see Newell and Simon, 1972 and Simon, 1981):

- the visual systems is the most efficient chunking method in the human information processing system; this means that two-dimensional representation, iconic instead of symbolic programming (eg screen-oriented editing), forms which support content (eg by using different fonts for different things) are important characteristics in systems design.

- recognition memory is greater than recall memory; this provides an incentive to construct menu-based systems instead of keystroke systems.

- the scarce resource in man-machine systems is not information but human attention; this implies that techniques for intelligent summarizing, prefolding of information etc have to be developed.

- our artifacts have reached a complexity that they can not be understood any more without adequate tools. These tools must be better than pencil and paper and the challenge is to make computer systems better than pencil and paper. Very modest beginnings of this claim have been demonstrated, eg a computer allows to generate dynamic views of a document (a reference manual and a primer are different, but they can be generated from similar knowledge structures), which eliminates the necessity of a static predetermined view.

- our short term memory is limited, ie there is a very finite number of things which we can keep in mind at one point of time. To overcome these limitations we have to construct systems which give us not only reasoning support but also memory support. Levels of abstraction and mnemonic names are able to reduce the cognitive burden of the user.

- our systems must be consistent and uniform for the user. Consistency must be based on an adequate model of how a system will be used.

To develop this preliminary collection further and to demonstrate how to make all these principles operational is an important topic for future research in man-machine communication.

COGNITIVE ERGONOMICS

Research in ergonomics investigates and analyses the effect of technologies for human work and it tries to develop adequate tools to make life easier.

In the past properties of systems were investigated which could be measured with methods from physics (eg the design and layout of a keyboard). This approach is not sufficient to evaluate information processing technologies (Moran, 1981). Modern computer systems try to support the human in decision making, planning, design and other cognitive activities. To evaluate these intelligent tools we must extend ergonomics research to pay attention to the conceptual and cognitve skills of people.

At the current stage of development, research in cognitive ergonomics cannot be a discpline which compares finished products, but it has to take an active part in the design and integration of cognitive dimensions in our information manipulation systems.

PROJECT INFORM

In our research project INFORM (Boecker, Fischer and Gunzenhaeuser, 1980) we are designing, implementing and evaluating an integrated information manipulation system as a prototype for a symbiotic, knowledge-based computer support system which is based on the framework outlined in the previous section. We choose knowledge representation and visualization as two examples of our work and show how they are relevant for the construction of a specific application programs: a system to support program synthesis and analysis (Fischer, Failenschmid, Maier and Straub, 1981).

Knowledge Representation

Our work on knowledge representation is based on OBJTALK, an object-oriented language (Fischer and Laubsch, 1979) which is modelled after SMALLTALK (Goldberg, 1981).

OBJTALK is build on top of LISP and allows the user to define objects which can be grouped into classes. To take advantage of the fact that most complex systems are hierarchical systems and therefore the knowledge to describe them can best be organized as a hierarchical framework, OBTALK has an inheritance mechanism between classes.

This knowledge representation machinery provides a good framework for the implementation of the knowledge base of a computer system to support program synthesis and analysis. There are different sources of knowledge which are important for program synthesis and analysis:

- knowledge about the programming language (which may be used for syntax-driven program construction, for real-time identing of the program text, etc)

- knowledge about programming constructs and techniques (to give assistance in building adequate data structures)

- knowledge about complex artifacts (eg what is the relationship between different parts; can be used to drive the editor and to check the consistency after changes have been made)

- knowledge about the design process (to assist in exploring alternative worlds, to retain the whole development process and not only the finished product)

- knowledge about the semantics of the problem domain (which can provides links between semantic names and the concepts which they denote)

In the example given, the basic knowledge units

are organized around the concept of a function (since we work in a functional language like LISP). Fig. 4 shows a simplified version of a knowledge structure created for a function which is used to compact and cleanup a database in a simple database system.

```
Conventions:
   1) Bold: slot names
   2) underlined: user-provided information
   3) normal font: data supplied by the system
   4) CAPITALS: "guesses" of the system, ie
      derived information which  is incomplete
   5) Bold and underlined: inherited information


(Function: Cleanup
 (Status: defined)
 (Code: . . . omitted . . .)
 (Package: Database-Functions)
 (iscalled by: ())
 (calls:   (as command: Initbuffer)
           (as operation: Intersection)
           (as predicate: Member?))
 (type: command)
 (parameters: 0)
 (local variables:
       (counter1  (type: number LIST)
                  (used as: COUNTER)
                  (possible values: (32 0)))
       (filename1  (type: list)
                ...............
 (free variables: (interupt.enable))
 (history:   (defined: 20/6/1980)
             (modified: 10/10/1980)
             (programmer: HDB)
             (reasons for change: "user
                provided information"--omitted-))
 (side effects: (variable interrupt.enable))
 (Error routines: interrupt.handler)
 (purpose:"updates a knowledge base by com-
        pacting it and getting rid of garbage")
    (algorithm: "copies active parts of a file
                   into a new file"))
```

Fig. 4: Example of a knowledge structure for a LISP function

Most of this context structured knowledge base is simultaneously compiled while the user is working interactively on his problem. In addition it includes information supplied by the user, eg a purpose slot and a natural language description of the algorithm; these slots are not used by the interpreter but they serve an important role in providing memory support for the human programmer.

After a function is defined it is translated immediately into the internal representation of Fig. 4 which is used for all further analysis. Therefore it is important that the internal representation contains all the information. From this example it should be quite obvious that in a knowledge-based system a program is **more** than its listing.

Visualization

One of the greatest steps forward in man-machine communication was the possibility to use the display as a truly two-dimensional medium. But new inventions introduce new problems. We are now confronted with a new set of **display management problems** that did not arise with the teletype terminal: what information should be diplayed? How should it be displayed? How can we direct the users attention so he will notice an important message? Which new techniques are possible?

It is quite obvious from the knowledge structures in Fig. 4 that we do not want to display the information in this form. We want to have a system which supports multiple windows (Bauer, Boecker and Fischer 1981) so we can see selected views of the complex structure (see Fig. 5). To be convivial the **user** should have control to define filters so he can determine the relevancy of the information.

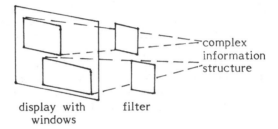

Fig. 5: Example for multiple perspectives using multiple windows and filters

Generating context-sensitive, multiple perspectives each of reduced complexity is a common technique among designers which is used to dissect otherwise intractable entities (compare the use of maps that picture economic, political and hydrological perspectives of a country). The use of windows supports this approach and allows us to regard a program as a complex information structure, of which only parts are of interest to the programmer are displayed at a certain time.

A window system is also a prerequisite to implement a good browser (Goldstein and Bobrow, 1981). A **browser** is a display-based interface which allows a user to examine a complex environment without prior knowledge of its exact structure. It allows the user to navigate in a complex space and to focus on his area of interest. Browsing capabilities can be used to filter information and summarize it into semantic units.

In Fig. 6 we show one of our browsers (for details see Fischer, Failenschmid, Maier and Straub, 1981). With the help of this browser we can traverse a hierarchical network by sequentially selecting items starting in the leftmost top window (selection is shown by underlining). Based on our selection the contents of the next window changes accordingly.

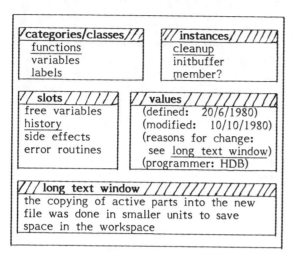

Fig. 6: An example of one of our browsers

User interfaces are a specific, but important issue in our research on man-machine communication within the project INFORM. One empirical finding of our work so far is that a knowledge based system is of little use if the information cannot be delivered to the user in a way which takes the cognitive limitations of the human into account.

CONCLUSIONS

In our work we have found that symbiotic, knowledge-based systems are valuable tools which enhance our work in many dimensions (eg reducing the mental load; pushing off clerical details to the machine so we can concentrate on the difficult conceptual parts of a problem; enlarging our strategies to tackle a complex problem) and our research has provided us with a deeper understanding of the design issues for future symbiotic, knowledge-based systems. The substantial computing power which will be available in future systems should be used to implement new dimensions in man-machine systems and should contribute to the important goal that computers can be regarded as tools to expand human potential.

References

Bauer, J., H.-D. Boecker and G. Fischer (1981): "Entwurf und Implementation eines Systems multipler Fenster", in Heft 6 der Notizen zum Interaktiven Programmieren, Fachgruppe Interaktives Programmieren in der Gesellschaft fuer Informatik, Oldenburg, pp 90-99

Boecker, H.-D., G. Fischer and R. Gunzenhaeuser (1980): "Die Funktion von integrierten Informationsmanipulationssystemen in der Mensch-Maschine Kommunikation", MMK Memo, Institut fuer Informatik, Universitaet Stuttgart

Fauser, A. und C. Rathke (1981): "Studie zum Stand der Forschung ueber natuerlichsprachliche Frage/Antwortsysteme", BMFT-FB-ID 81-006, Institut fuer Informatik, Universitaet Stuttgart

Fischer, G. (1980): "Cognitive Dimensions of Information Manipulation Systems", in R. Wossidlo (ed): "Textverarbeitung und Informatik", Springer Verlag, pp 17-31

Fischer, G. and H.-D. Boecker (1981): "Understanding Design", Proceedings of the 3rd Annual Meeting of the Cognitive Science Society, Berkeley, Calif

Fischer, G. and J. Laubsch (1979): "Object-oriented programming", in Heft 2 der Notizen zum Interaktiven Programmieren, Fachgruppe Interaktives Programmieren der Gesellschaft fuer Informatik, Stuttgart, Februar 1979, pp 121-140

Fischer, G., J. Failenschmid, W. Maier and H. Straub (1981): "Symbiotic systems for Program Development and Analysis", MMK Memo, Institut fuer Informatik, Universitaet Stuttgart

Goldberg, A. (ed, 1981): "SMALLTALK", Special Issue, BYTE, Vol. 6, No. 8, August 1981

Goldstein, I. and D. G. Bobrow (1981): "Browsing in a Programming Environment", in Proceedings of the 14th Hawaii Conference on System Science, January 1981

Illich, I. (1973): "Tools for Conviviality", Harper and Row, New York

McCracken, D. (1979): "Man + Computer: A new Symbiosis", CACM, Vol 22, No 11, pp 587-588

Moran, T. (ed, 1981): "The Psychology of Human-Computer Interaction", ACM Computing Surveys, Vol. 13, No. 1, March 1981

Newell, A. and H. Simon (1972): "Human Problem Solving", Prentice Hall, Englewood Cliffs, New Jersey

Norman, D. (1981): "Perspectives on Cognitive Science", Lawrence Erlbaum Assoc., Hillsdale, New Jersey

Shortliffe, E. H. (1976): "Computer-based Medical Consultations: MYCIN", Elsevier, New York

Simon, H. (1981): "The sciences of the artificial", MIT Press, Cambridge, MA, 2nd edition

Teitelman, W. and L. Masinter (1981): "The Interlisp Programming Environment", Computer, April 1981, pp 25-33

A KNOWLEDGE BASE SYSTEM FOR DECISION SUPPORT USING COGNITIVE MAPS

K. Nakamura, S. Iwai and T. Sawaragi

Department of Precision Mechanics, Faculty of Engineering, Kyoto University, Kyoto, Japan

Abstract. This paper presents an man-computer system for decision support using *cognitive maps*. we first summarize the method of documentary coding and derivation of cognitive maps. Second, we construct a knowledge base by joining cognitive maps of five documents relevant to the traffic problem in Japan and indicate that some kinds of useful information are generated in the knowledge base through the process of joining. Third, we propose three retrieval modes from knowledge base. Those retrieval modes provide the following knowledge according to types of the decisionmakers' requests: 1) Skeleton maps indicate overall causal structure of the problem; 2) Hierarchical graphs give detailed information about parts of the causal structure; 3) Sources of causal relations are presented when necessary, for example when the decisionmaker wants to browse the causal assertions in documents. Using them, we describe an illustrative example of how the knowledge base system assists the decisionmaker in grasping a complex problem.

Keywords. Decisionmaker; information retrieval; knowledge base; computer-aided instruction; cognitive systems; urban system; transportation control.

INTRODUCTION

Recent complex problems involve many aspects and so should be analyzed from those points of view.

Experts' documents describe the various aspects of the problems and are an important source of knowledge used for the analysis. That is, a decisionmaker creates his understanding of the problem by collecting such knowledge from the documents. The understanding is an extremely important part of the decision process since it implicitly defines the set of potential alternatives for solution.

However, the amount of such documents is growing and they are becoming too many for the decisionmaker to investigate by himself. Current document retrieval systems can not provide such knowledge effectively. To permit the decisionmaker to grasp the knowledge described in documents, the knowledge should be extracted and integrated in computers.

Recently the need for intelligent interfaces between such integrated knowledge and the decisionmaker has arisen. The interface consists of a knowledge base into which the knowledge in documents are integrated and an interactive program dealing with the knowledge base.

In this paper, we propose a knowledge representation in computers using *cognitive maps*, which were proposed by R. Axelrod (1976) and are a graphical representation of causal relations described in documents. Then, using it, we construct a knowledge base system for decision support in which all causal structures described in many documents are joined together. The system interactively presents relevant information about causation to decisionmakers according to their requests in decision process.

A REVIEW OF COGNITIVE MAPS

The cognitive maps were proposed by Axelrod and others in international politics and are an attempt to utilize policy experts' beliefs and cognition about foreign policy for policy analysis, forecasting, decisionmaking, and so on. The constituents of cognitive maps are nodes called *concepts* and signed arrows representing causal relations between concepts. The policy alternatives, all of the various causes and effects, the goals and the ultimate utility of the decisionmaker can all be adopted as the concepts. The arrows are also encoded according to experts' causal assertion.

The procedure of derivation of cognitive maps is called *documentary coding*. Documentary coding, as a general rule, transformation of causal assertion in documents into causal relation consisting of cause concept, effect concept, and signed arrow. The concepts must be represented as variables, which have the potential to take on different values, for example " security of Japan " and " consensus of inhabitants ". Entities such as " Japan " and " habitants " are not concepts, since they can take on no other values. The value of concepts is not necessarily continuous or metrical but may be a categorical one such as " large " and " little ". The sign of arrow is determined as plus + when the causal assertion states that the cause concept increases or promotes the effect concept and as minus - when it states the contrary. The coding is rather easy when the sentence consists of the definite subject, object, and verb. For example, the following sentence " Renewed protectionist pressures also threaten the general U.S. policy objective of an open world trading system. " is coded as shown in Fig. 1.

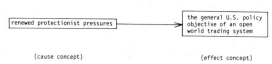

(cause concept)   (effect concept)

Fig. 1.   Coding example of Causal Assertions.

The indirect effect of a single path is calculated as follows. Suppose there are direct effect, i.e., causal assertion, x from concept A to concept B and direct effect y from concept B to concept C, as shown in Fig. 2 (a) and then there is a path from A to C through B. This path carries an indirect effect from A to C. The indirect effect is defined by the multiplication xy. Also, when there are two paths from concept A to concept B as shown in Fig. 2 (b), the total effect from A to B is defined by such addition * of the indirect effects from A to B as follows.

$$x*y = x \quad \text{if } x = y \in \{+, -\} , \qquad (1)$$

$$x*y = u \quad \text{if } x = + \text{ and } y = - . \qquad (2)$$

In this case, two relations + and - exist, i.e., relation u is a set {+, -} and is called *universal*.

Fig. 2.   Mathematics of Cognitive Maps:
(a) calculating indirect effect of a single path xy, (b) determining total effect of different path x and y.

CONSTRUCTION OF KNOWLEDGE BASE BY JOINING COGNITIVE MAPS

Experts' documents concerned with a complex problem contain various expertise from their own, different points of view and the cognitive maps derived from the documents are different in concepts. However it is natural that several concepts are common to some maps as far as the documents are concerned with the same problem. Using the common concepts, we join the cognitive maps to form a knowledge base as shown in Fig. 3 .[1]

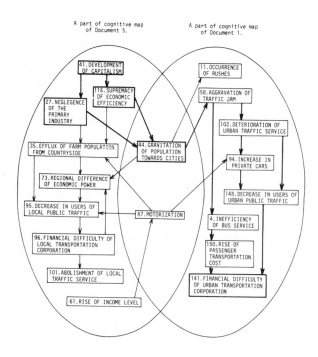

Fig. 3.   Joining of Cognitive Maps to form a Knowledge Base.

The common concepts play a role of relaying nodes in joining. We coded cognitive maps of five documents relevant to " traffic problem of Japan " and constructed a knowledge base concerning the problem. The knowledge base consists of 152 concepts and 265 direct causal relations and is stored in a FACOM-M200 computer.

Generation of Useful Information in Knowledge Base

The knowledge base is considered integration of various cognitions concerning the problem and can provide the decisionmakers with such overall and detailed information about the problem as follows.

Expatiation of causal structure by supplementation of intermediary concepts. Between concepts " A4. motorization " and " A 17. increase in automobile traffic ", only a

_____

[1] Arrows without sign denote positive causal relations and only arrows with sign " - " denote negative causal relations.

direct causal relation is asserted in Document *1.* as shown in Fig. 4 (a). However, by adding the cognitive map of another Document *3,* two intermediary concepts B20 and B21 are inserted into the above causal relation as shown in Fig. 4 (b). By the insertion of

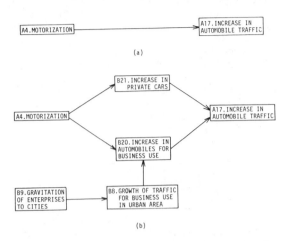

(a)

(b)

Fig. 4.   Expatiation of Causal Structure by
          Supplementation of Intermediary
          Concepts in the process of
          Joining Cognitive Maps.

concepts, we can recognize that " A4. motorization " promotes increase in two kinds of automobiles, private and for business use, and that " increase in automobiles for business use (B20) " is also promoted by " gravitation of enterprises to cities (B9) " through " growth of traffic for business use in urban area (B8) ". As above, supplementation of intermediary concepts expatiates causal structure of cognitive map and make causal relations asserted by an expert easier for decisionmakers to grasp.

Derivation of long causal paths from one cognitive map to another. Figure 3 shows parts of cognitive maps derived from Document *1* and *5.* While Document *1* is only concerned with *urban traffic,* Document 5 describes *traffic related to local problems.* As shown in the map of Document *5,* the author of the document indicates that " 44. gravitation of population towards cities " is caused by " 41. development of capitalism " through. " 116. supremacy of economic efficiency " and " 27. negligence of the primary industry ". In the map of Document *1,* its author asserts that " 44. gravitation of population towards cities " forms a cause of " 141. financial difficulties of urban transportation corporations ". The two indirect causal relations include concept 44 in common. Consequently, concept 44 plays the role of relaying node connecting the two indirect relations in joining the two cognitive maps and a new long causal path is formed from concept 41 to concept 141. The long causal path gives us information that " 41. development of capitalism " is a remote cause of " 141. financial difficulties of urban transportation corporations ". Apparently, the information cannot be got from the map of Document *1* nor *5* but

only from the joined map. Thus, in the joined map, cognitions of various experts (authors) are integrated and an overall grasp of the problem *traffic* is constructed, which each of the experts can never hold.

The above two kinds of useful information in this knowledge base of joined cognitive maps seem to be usually generated by decision-makers in their mind when they read experts' documents and integrate experts' knowledge. The knowledge base can help decisionmakers to make up such integrated grasp of the problem without reading experts' documents.

Search of Corresponding Concept in Joining
Cognitive Maps

Terminologies used in documents vary with author. The same notion may be represented by various terms. Therefore, when we join a cognitive map of a new document to the knowledge base, it seems a hard task in construction of large knowledge base how to find a concept representing the same notion as a concept of the new map from the knowledge base. We propose the following two methods to make it easier to retrieve corresponding concepts in content from the knowledge base.

Search method with key words. A coder selects a key word representing a new concept and enters it to the system. A concept is represented by a string of several words. The system prints out all the concepts including the key word from the knowledge base. Suppose a new concept " drift of population to big cities ". This concept is not word-for-word identical with concept " 44. gravitation of population towards cities " in knowledge base but is nearly the same as it in content. The coder selects POPULATION as a key word representing the new concept and enters it. The system prints out the following three concepts: " 44. gravitation of population towards cities ", " 120. sprawling of population in cities ", and " 46. security of public service in depopulated area ". Consequently, the coder easily finds concept 44 corresponding to the new concept. This system can deal with some inflection and synonyms. For example, in the above case, concept 46 is printed out because it contains term DEPOPULATED instead of key word POPULATION. Also, since CAR and WHEELS are synonyms for AUTOMOBILE, concepts including the former two key words are also printed out by search with key word AUTOMOBILE.

Search method by category of concepts. Some concepts agree with each other in content though no term is common to them. For example, a concept " influx of people into urban areas " has no term common to concept 44 but represents closely similar notion to it. To deal with such a correspondence between concepts, we propose a search method by categories of concepts. We set several categories and classify all the concepts included in knowledge base into the categories. Each category is relevant to an aspect of the problem

(traffic problem in Japan). In our knowledge base, the concepts are classified into 15 categories. A concept may be included in some categories. For example, concept 44 is put in two categories URBAN PROBLEM and QUALITY OF LIFE. When the coder searches a concept corresponding to the above concept " influx of people into cities ", he selects a category closely related to the concept from the 15 categories, for example URBAN PROBLEM and looks over the concepts of the category to find the corresponding concept 44.

After it is found that there is a concept corresponding to a new concept, the new concept is merged into the corresponding concept and all the direct relations incident in and out the new concept are added to the corresponding concept in the knowledge base. When corresponding concepts are not found, new concepts are added to the knowledge base and entered in some categories. The two methods may fail to find concepts corresponding to some new concepts though the corresponding concepts have been already stored in the knowledge base. Consequently, the knowledge base includes the two concepts representing very similar notion. In such cases, the long causal paths which corresponding concepts should relay are left disconnected and so the knowledge supplemented by the connection is missed.

RETRIEVAL METHODS FROM THE PROPOSED KNOWLEDGE BASE AND DECISION SUPPORT EXAMPLE

The knowledge base contains so many kinds of concepts and causal relations between them. For real decision support using the knowledge base, we propose a retrieval system to present relevant information from the knowledge base to decisionmakers according to their requests.

Three Retrieval Modes Corresponding to Types of User's Requests

The retrieval system has the following three retrieval modes according to types of user's requests.

Presentation of skeleton maps. The system displays skeleton maps comprising some significant concepts and causal relations between them, relevant to the user's subject of interest. Axelrod and others (1976) regarded *centrality* of concepts in cognitive maps as the significance of concepts. The centrality is measured by *total degree* of concepts. The total degree of a concept is the number of direct relations incident in and out the concept. In the implemented system, not more than eight most significant concepts according to the centrality and causal relations between them are displayed. The sign of the relations in the skeleton map is that of the shortest paths between the concepts. The user selects one of three codes A (category), B (key word), and C (document) and enters it

to the system. A (category): the user chooses a category relevant to his subject of interest from the before-mentioned 15 categories and a category WHOLE TRAFFIC PROBLEM, and the system selects the significant concepts from the category. B (key word): the user chooses some key words representing his subject of interest and enters them. The system selects the significant concepts containing at least one of the key words. C (document): the user chooses a document from the list of documents whose cognitive maps are stored in the knowledge base and then the significant concepts are selected from the concepts contained in the document. The skeleton maps give the user overall information relevant to his subject of interest and help him to grasp the subject in its totality.

Display of hierarchical graphs of detailed causal structure. When the user wants to know detailed causal structures of causal relations presented in the skeleton maps, the system displays the detailed structure, which is printed out in the form of hierarchical graph. The user selects a causal relation from the skeleton map and enters starting node and ending node of the causal relation. The system retrieves all the direct and indirect relations between the two concepts from the knowledge base, arranges them in the form of a hierarchical graph, and prints out the graph.

Retrieval of sources of causal assertions. The user sometimes wants to know by whom and how the causal relations in the displayed graph are asserted in source documents, for example when the causal relations are difficult for him to understand or accept. In such cases, the system presents the source of the causal relations by document number, page, and line and then the user can browse relevant parts of the sources.

Illustration of Decision Support through Interactive Retrieval with the Knowledge Base

Suppose that a decisionmaker wants to know causal structure of problems concerning public transportation. The decisionmaker selects code A and enters category *1* (PUBLIC TRANSPORTATION) to the system to get a skeleton map on public transportation. Figure 5 shows the skeleton map printed out by the system. In the figure, the positive arrow from concept 55 to concept 10 suggests the decisionmaker that " demand for transportation of passengers " promotes " increase in facilities for public transportations ". When the decisionmaker wants to know why and how the demand increases the facilities for public transportation, he enters the two concepts 55 and 10 to the system to investigate the detailed causal structure between them. Figure 6 shows the hierarchical graph representing the detailed structure. The figure indicates not only a path carrying positive indirect effect $55 \rightarrow 9 \rightarrow 10$ but also two paths carrying negative indirect effect

```
WHICH CODE DO YOU CHOOSE? PLEASE SELECT A,B, OR C.
     A.CATEGORY
     B.KEYWORD
     C.DOCUMENT
:
A
WHICH CATEGORY ARE YOU INTERESTED IN? PLEASE SELECT THE NUMBER.
           1.PUBLIC TRANSPORTATION
           2.AUTOMOBILE TRANSPORTATION
           3.FACILITY FOR TRANSPORTATION
           4.QUALITY OF LIFE
           5.ENVIRONMENT
           6.ECONOMY
           7.LOCAL PROBLEM
           8.SAFETY
           9.URBAN PROBLEM
          10.INDUSTRY
          11.TECHNOLOGY
          12.POLICY
          13.CULTURE
          14.DISTRIBUTION
          15.THE OTHERS
          16.WHOLE TRAFFIC PROBLEM
:
i
```

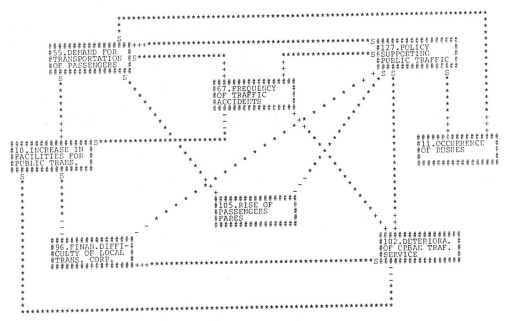

Fig. 5.   Skeleton Map relevant to category PUBLIC TRANSPORTATION; " $As**+++B$ " denotes
$A \longrightarrow B$, " $A---**+++B$ " denotes $A \longrightarrow B$ and $B \Longrightarrow A$.

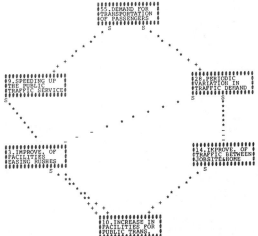

Fig. 6.   Hierarchical Graph Structure from
concept 55 to concept 10 in Fig. 5.

$55 \rightarrow 28 \rightarrow 3 \rightarrow 10$ and $55 \rightarrow 28 \rightarrow 14 \rightarrow 10$; that is, the indirect relation from concept 55 to concept 10 is universal. From the figure, the decisionmaker can recognize that increase in the demand for transportation promotes " speeding up the public traffic service (9) " and then increases the facilities for public transportation, while the increase in the demand causes " periodic variation of traffic demand (28) " and then the variation impedes " improvement of facilities easing rushes (3) " and " improvement of traffic between jobsite and home (14) " and consequently decreases the facilities for public transportation. By entering other significant concepts to the system again, the decisionmaker gets information about them.

As above, the decisionmaker can form his overall and detailed understanding of public transportation through the process of interactive retrieval with the knowledge base. Since the knowledge base is constructed based on causal assertions described in experts' documents, the information indicated in the skeleton maps and hierarchical graphs

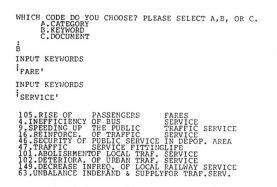

Fig. 7.   Output of Sources of Direct Causal
           Relations displayed in Fig. 6.

is also according to the experts' assertions. So far, decisionmakers get such information by reading documents for themselves. However, those documents are usually so many and it is hard for decisionmakers to read all or even most of them. Our knowledge base system provides decisionmakers with the information without their reading the documents, although the information is only concerned with causation, and helps them to grasp complex problems based on many experts' knowledge described in their documents.

Also, if the decisionmaker wonders how the universal causal relation in Fig. 6 is coded, he requests the sources of the direct causal relations composing the universal relations to know by whom and in what context the causal relations are asserted. Figure 7 shows the output of sources and indicates that the two direct relations composing the positive path 55 → 9 → 10 are asserted in Document 2 while the five direct relations composing the negative paths 55 → 28 → 3 → 10 and 55 → 28 → 14 → 10 are asserted in Document 1. According to the printed out document numbers, pages, and lines, the decisionmaker browses the indicated parts of the documents and knows that the author of Document 2 grasps the effect of concept 55 on concept 10 only in transportation capacity and

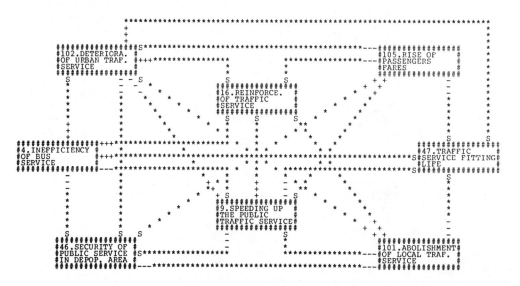

Fig. 8.   Skeleton Map relevant to key words SERVICE and FARE.

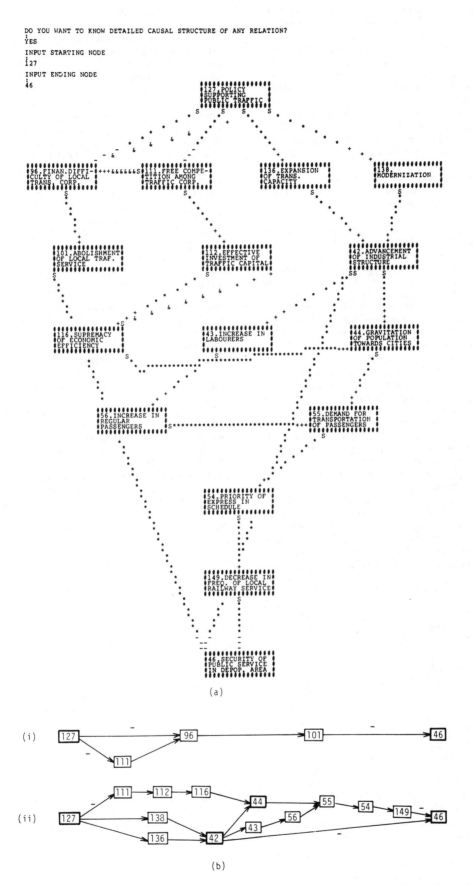

(a)

(b)

Fig. 9.   Causal Effects of Policy supporting Public Traffic (concept 127) on Security of
Public Service in Depopulated Area (concept 46); (a) Hierarchical Graph from concept
127 to concept 46, (b) Two Causal Structures constituting the Hierarchical Graph.

and considers the effect positive while the author of Document 1 considers the effect negative by taking convenience of public transportation into account. The browsing according to the information about sources suggests to the decisionmaker that the universal causal relations stems from the difference between the two authors' cognitions. As above, our system can inform decisionmakers which of many documents include the very information they need and which parts they should read in the documents. This means that the third retrieval mode can be used also for an advanced document retrieval method which can find documents concerned with the very subject the user requests.

Furthermore, suppose that the decisionmaker wants to analyze relationship between two important factors in the traffic problem, service and fare. In such a case, he selects retrieval code B and enters the two key words SERVICE and FARE to the system. The system prints out a skeleton map whose eight concepts include key words SERVICE or FARE, as shown in Fig. 8. In the skeleton map, he focuses on a serious problem of local traffic, i.e., " 46. security of public service in depopulated area " and investigates how the " (national) policy supporting public traffic ", which is a significant concept 127 in Fig. 5, affects the concept 46 in Fig. 8. He enters the two concepts and the system prints out a hierarchical graph from concept 127 to concept 46 shown in Fig. 9 (a). The graph is found to consist of two causal structures (i) and (ii) shown in Fig. 9 (b). The former structure (i) indicates positive effects of the " policy supporting public traffic (127) " on " security of public service in depopulated area (46) " by easing " financial difficulties of local transportation corporation (96) " and controlling " free competition among traffic corporations (111) ". The latter structure (ii) is a causal network mainly carrying negative effects of the policy and includes two pivotal concepts " 42. advancement of industrial structure " and " 44. gravitation of population towards cities ". The structure shows that the policy promotes " expansion of transport.capacity (136) " and " modernization (138) " and then causes " the advancement of industrial structure (42) " and " the gravitation of population towards cities (44) ", which have negative effects on " the security of public service in depopulated area (46) ". Consequently, as far as the problem of security of public service in depopulated area is concerned, only such policies should be taken as to ease the financial difficulties (96) and control the free competition (111) under fundamental policies controlling the advancement of industrial structure and the gravitation of population.

CONCLUSION

The system presented in this paper searches only causal structures relevant to the user's request from the knowledge base and displays them and sources of causal assertions. The system will be improved so as to present the following useful information for decision support by introducing weights to the direct causal relations (e.g., occurrence of the relations in documents): the most significant causal path from one concept to another, cause concepts which have the most significant effects on a specified concept, and pivotal concepts in a causal structure.

As mentioned in previous sections, the problems related to the duplication of similar concepts and the transitivity of causation must be solved for implementation of advanced decision support system with deductive inference mechanism.

Though the documentary coding is coders' manual labor, it will be necessary to develop automated coding systems based on techniques for natural language processing in order to deal with a large amount of documents.

The knowledge base presented here is a particular type of semantic network dealing with only causation between concepts. However, it is necessary to organize other kinds of relations described in documents (e.g., sentiment relation and unit formation[2] F. Heider, 1946; O. Katai and S. Iwai, 1978) into a semantic network, so that in the decision support of real world problems the knowledge base will serve as an intelligent interface between the decisionmakers and the documentary data stored in computer memory.

REFERENCES

Axelrod, R. (1976). Structure of decision: the cognitive maps of political elites, Princeton Univ. Press.

Heider, F. (1946). Attitudes and cognitive organization, J. Psycology, 21, 107-112.

Katai, O., and S. Iwai (1978). On the characterization of balancing processes of social systems and the derivation of the minimal balancing processes, IEEE Trans. SMC-8, 5, 337-348.

---

[2]Sentiment relation includes such relations as " like " and " dislike ", and unit formation implies such a cognitive relation as " owning ".

DMS — A SYSTEM FOR DEFINING AND MANAGING HUMAN-COMPUTER DIALOGUES

R. W. Ehrich

*Department of Computer Science, Virginia Polytechnic Institute and State University,
Blacksburg, Virginia 24061, USA*

Abstract. As the complexity of human-computer interfaces increases, those who use such interfaces as well as those responsible for their design have recognized an urgent need for substantive research in the human factors of software development. Because of the magnitude of the task of producing software for individual human-computer interfaces, appropriate tools are needed for defining and improving such interfaces, both in research and in production environments. DMS (Dialogue Management System) is a complete system for defining, modifying, simulating, executing, and metering human-computer dialogues. It is based upon the hypothesis that dialogue software should be designed separately from the code that implements the computational parts of an application, and different roles are defined for the dialogue author and the programmer to achieve that goal.

Keywords. Computer software, dialogue, DMS, human-computer, tools.

INTRODUCTION

Everyone who has used an interactive computing system is familiar with the frustration of not knowing the right commands or command syntax, not knowing the state of the system, receiving meaningless error messages, not having the right commands, reading poorly formatted displays, and having to use far more effort than necessary to communicate a task to the computer. In the past there has been a failure on the part of many software vendors to distinguish between software functionality and software useability. Functionality refers to the capability of a system to perform given software tasks, while useability refers to the ease and directness with which such a task can be achieved.

There are several reasons why software vendors continue to produce software products with inferior human interfaces.

1. Human engineering is expensive.

2. Software authors are not always aware of the poor human engineering of their products.

3. Many software authors don't know how to do a good job of human engineering.

4. Vendors assume that special training is essential for the use of their products and therefore don't really care about the human interface.

There is not much to be done about software vendors who don't care about the human engineering of their products, but as computer users become more demanding, one would hope that the competition from better products would exert economic pressures. On the other hand, there is much that can be done to reduce the cost of quality human engineering and to provide intelligent environments and methodologies by means of which human interfaces may be improved.

DMS is an attempt to provide an improved environment for managing human-computer interfaces, and it has three principal components:

1. An execution environment

2. Software tools

3. Design methodology

One of the most important issues arising from our research involves the relationship between the dialogue and computational components of a software system. These two components seem to have similar control structures, program environments, and design methodologies. It is proposed that these two components be created by different specialists; the dialogue author

This research was supported by the Office of Naval Research and ONR Contract Number N00014-81-0143, and Work Unit Number NR SRO-101. The effort was supported by the Engineering Psychology Programs, Office of Naval Research, under the technical direction of Dr. John J. O'Hare.

is to be a specialist with experience in the design of human-computer interfaces, and the programmer of the computational components is to be the computer science specialist. In most software systems there is a complex relationship between the dialogue and computational components, and we are studying these relationships in order to find a methodology for separating and designing them.

There are several important reasons for introducing formal tools and methodologies into the task of producing well-engineered human-computer interfaces. One reason is that in the current state of the field there are too many unsupported, sometimes conflicting, design "principles." Carefully designed tools enforce consistency and encourage the dialogue author to use techniques selected for their effectiveness on the basis of behavioral evidence.

A more important reason is that current methodology requires interface design logic to be treated as though it is the tedious detail of the software system it is designed to serve. As a consequence, dialogue is woven into the computer software fabric in such a way that software vendors and designers simply become committed to inferior interfaces because they are too complex and expensive to reprogram. A simple example illustrates the point. Most programmers using a high level language tend to specify input/output formatting at the point where the input/output statements reference those specifications. These details are usually totally irrelevant to the computational task whose logic was interrupted by the occurrence of the input/output statements. Later, when the formatting needs to be altered, it is frequently nearly impossible to locate the code that produced the erroneous format. In fact, the occurrence of the input/output statements themselves are frequently disruptive to the logic of a computational program because they reference files, unit numbers, conditions, and specifications that have nothing to do with the subject of the module within which they occur.

DMS forces the programmer to think in a different way about dialogue. The programmer must recognize the need for dialogue but does not deal with the way in which it is carried out. Interaction with the human is through dialogue defined in dialogue modules, and in the same way, computation is done only in computational modules. Control is either concurrent, or it alternates among the modules.

SOFTWARE STRUCTURE

The preceding example, while explicit and motivating, does not illuminate the very complex general problem of separating

dialogue from computation. In the first place, computation is usually required for purposes of display formatting; usually one selects the modules in which such computations are performed on the basis of the clarity of the software decomposition that results. Those computations that deal with the application are separated from those related to display formatting or other communication modes. An example that is somewhat at the other extreme is a screen editor in which almost every keystroke alters the internal memory state of the display system. In such a software system almost all the code may be attributed to the human interface, and the only parts that may be cleanly decomposed are the modules that belong to the storage system.

From these examples it may be seen that depending upon the nature of the software task, there may be a very close relationship between the design of the human-computer interface and the design of the computational software. Thus, the dialogue author and the computational programmer may need to work jointly on significant portions of the design specifications.

It is possible to distinguish two extreme types of software, called computation dominant and dialogue dominant software. The two cases are distinguished on the basis of the control logic of the software system and on the basis of the nature of the global contextual environment that evolves during the course of the operation of the system. In the case of computation dominance, the control logic is entirely contained within the computational component so that whatever dialogue logic exists is distinct and easily decomposed from the computational component. In this case the dialogue modules do not call one another but are invoked by the computational component. Also, the storage environment for the dialogue modules are local to those modules. Dialogue dominance is just the opposite; in this case the software control logic is entirely contained within the dialogue component. The text editor in the previous example would be an example of a dialogue dominant software system.

Most other software is somewhere inbetween, and its design requires the cooperation of the dialogue author and the programmer of the computational component. What is important is not the classification of the software task but the recognition of the distinct roles of the dialogue author and the programmer in the software production process. Each is a specialist in a particular domain, each works with software tools designed for that domain, and the problems that each must solve are distinct, yet related by the common goal.

There are several ways to achieve the separation of computational modules from dialogue modules. One that was considered was the isolation of dialogue code by placing it in procedures, not a new concept at all. Another is simply to physically isolate the different code components while keeping them both in the same code body. The trouble with both of these alternatives is that there is still a close physical bond between the dialogue and computational code components, and these methods require linking the entire software system whenever the smallest change is made in either component. There are ways around this, too, such as dynamic linking. One technique still under consideration is to provide special dialogue language processors that interpret dialogue code stored in a special database that contains all the human interface definitions. In our case, we are dealing with extremely complex interactive structures involving multiple users, multiple concurrent dialogues, and multiple input/output devices including keyboards, graphics, joysticks, trackballs, touch panels, and voice. Our solution was to make use of a language-independent multiprocessing environment.

One of the most powerful concepts for implementing locally independent but globally interrelated software units is the process. A process provides the environment, services, and resources necessary for running a program, which will be called an image to distinguish the bound executable module from source or object code. A process exists solely for the purpose of executing an image, which is both a logically and physically distinct software entity, much like that which a dialogue author or programmer would produce. The physical realization of the global links between dialogue or computational software modules is the interprocess communication facility, and it is partly the role of DMS to provide well engineered interprocess communication constructs within the context of the host operating system. While multiprocessing is well known to real-time processing specialists, it is not in common use for implementing human-computer interfaces.

The principal consequence of DMS methodology is that a task that might be implemented as a single program under conventional software methodology will normally be implemented as a set of independent communicating programs, each of which executes in a separate process. Such a set of communicating programs will be referred to as a program complex. There are numerous benefits from this type of program decomposition. Module interactions are minimized, and that facilitates the design of the programs which run under the various processes. Concurrencies among modules become a reality, and since

dialogue and computational code may well require different implementation tools, more specialized tools can be applied. The way in which a task is decomposed depends upon the control structures within the algorithm that carries out the task. For example, dialogue tasks are separated from computation tasks, and the interrelationships between them depends upon whether the overall task is dialogue or computation dominant. Thus, modifications to a software unit may be made by modifying any of the programs of the complex individually or by altering the communication among the programs.

In order to account for the different possible relationships among the programs of a complex it is necessary to distinguish special types of programs called executors and coprograms. These are distinguished on the basis of their internal structure, and they may implement either dialogue or computation, depending upon whether or not they are permitted to communicate with a user. Thus there may be dialogue programs, executors, and coprograms, and computation programs, executors, and coprograms. Executors have a particularly important structure; they are collections of modules that do not call one another and have no shared global storage environment. The module interrelationships are defined by the programs that invoke them. Few programs are organized as pure executors, but executors have a structure that is important because of the independence of the component modules.

Programs and executors are used in situations where there is clear computation or dialogue dominance. The dominant task is implemented as a program that makes specific requests of the executor, which achieves logically distinct subtasks and returns information to the requesting program. While specific examples are shown later, the idea is that the executor is subordinate to the requesting program, and it exists to satisfy the program's requests. The protocol that has been implemented looks roughly as shown in Fig. 1. REQUESTs and ACCEPTs, and RECEIVEs and SENDs are complementary. The program makes a request of an executor which is acknowledged when the executor reaches its accept state. Then the program sends data to the executor and, when it desires the executor's response, collects the responses by issuing receives. The executor selects the appropriate service (e.g. dialogue module) specified by the request, issues matching receives and sends to obtain and return information, and performs any necessary input, output, or computation. Either side of the diagram in Fig. 1 might contain computational or dialogue code. One possible organization of the coprogram relationship is illustrated in Fig. 2. This relationship applies when there is no clear computation or dialogue dominance.

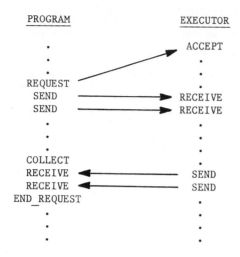

Fig. 1. Program-Executor Relationship

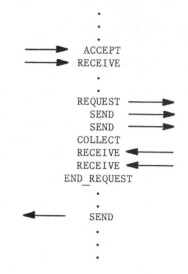

Fig. 3. Third Party Request

DMS DESIGN CONSIDERATIONS

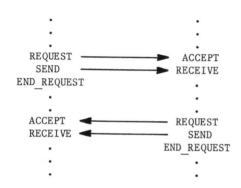

Fig. 2. Coprogram Relationship

Interprocess Communication

DMS has been implemented using a variation of the rendezvous concept (USDOD, 1980) in which two processes must synchronize before an interprocess dialogue can be initiated. While there are some requirements that are better served by asynchronous communication, it was felt that synchronous constructs would be easier to implement and much less confusing to nonspecialists in concurrent programming.

In this case information is simply shuttled back and forth between the coprograms which execute in a quasi-synchronous manner. Of course, any program, coprogram, or executor may issue requests to any number of programs. Each program is identified in the request by the image name. One other structure called a third party request is possible. This situation, shown in Fig. 3, occurs when an executor inserts a request to another executor in the middle of its service to a program. Such a request may be made only after the executor completes its RECEIVE sequence and before beginning its SEND sequence. It is the responsibility of the dialogue author and the programmer to avoid circular 3rd party requests that may lead to a deadlock. In the same spirit it is their responsibility to agree upon request names, to ensure that RECEIVEs and SENDs are properly paired, and to ensure that the correct amount of data is transferred.

Since this is a VAX/VMS implementation, the interprocess communication mechanism is called a mailbox, which is a portion of non-paged physical memory. Mailboxes are treated as ordinary input/output devices, and each process has one mailbox. In each process except the one running ENTRY, mailboxes are serviced by software interrupts called asynchronous system traps (AST's); thus mailboxes are read almost instantaneously.

In this implementation, an image called ENTRY is run under the user's login process. ENTRY is the central executive of DMS, and all DMS processes are subprocesses of the process running ENTRY. Included in ENTRY's responibilities are:

1. Conducting the default dialogue for initiating the execution of a complex.

2. Maintaining the image name, process identification code, mailbox name, and status for each DMS process.

3. Supervising process and mailbox creation.

4. Supervising mailbox deallocation and process termination for both normal and abnormal termination.

5. Servicing queries about processes that permit images to locate one another and establish communications.

Two processes are required to synchronize at the beginning of each interprocess dialogue. If process A requests a dialogue with process B, A sends B a request and hibernates until B gives A its attention. Then A and B can communicate freely, but with two constraints. In order to receive information from B, A must issue a collection request to B at some point before information is desired from B. This permits A to initiate concurrent dialogues with several processes and then order the processes from which it desires responses. The second constraint is that if B makes a third party request to C, it may only do so when A has finished transmitting data to B. Otherwise B would not be able to distinguish data received from A from data received from C. With the exception of the preceding caution, A and B may freely exchange data in either direction as long as they choose.

As an executor receives requests, they are queued internally and then serviced according to their priority. Once a request is serviced, the rendezvous is in effect until both processes decide to end the dialogue. With such an implementation the organization of an executor is quite simple, since it queues only requests and carries out only one interprocess dialogue at a time.

There are two ways in which processes are removed. If the DELETE_PROCESS service is called, ENTRY broadcasts a notice to all processes to cancel records of the deleted process, and then it deletes the process. If any image terminates normally or by accident, ENTRY learns of the event by receiving a message in a special mailbox called a termination mailbox. When such a message is received, ENTRY deletes all processes, sends any error message to the standard error device, and returns to the default dialogue so that the user can execute another complex.

Input and Output

The multiprocess DMS organization is well-suited for supporting multiple concurrent input and output devices required in complex dialogues. The multiprocess organization of a program complex running under DMS is shown in Fig. 4. Each input/output device is controlled by a program executing in a separate process. This organization makes it easy to design separately the control programs for each device, and it is not necessary to know in advance which devices are being used, since

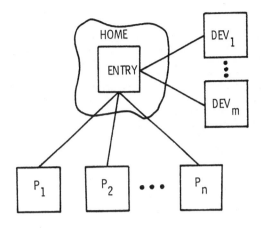

Fig. 4. Organization of a Complex

no process need be created until the corresponding device is referenced. Furthermore, it facilitates the design of intelligent screen handlers and graphics systems, since it is easy to keep track of the device state in a dedicated process.

Concurrency

Up to now the merits of the multiprocess environment for DMS have been argued on the basis of the functional decomposition of programming tasks. However, there is far more involved. There is no difference in the theory if the individual processes were separate hardware processors. All of them may function concurrently with only weak constraints imposed in part to make the programmer's job easier. Most restrictive is the constraint that a process issuing a REQUEST will hibernate until that request is acknowledged. However, assuming that the required executors or co-programs are free, a program may initiate multiple dialogues that execute concurrently with one another and concurrently with its own code.

There exist several programming environments that provide multiple processes and interprocess communication facilities, among them ADA (USDOD, 1980), concurrent PASCAL (Brinch Hansen, 1975), and UNIX (Ritchie and Thompson, 1974). These differ from one another in their constructs and implementation, and since their definition is standard, one does not have the freedom to adapt them to the specific needs of dialogue management. DMS is largely language independent and has an architecture that can easily be modified to meet the needs that arise as one learns how to provide better and better tools for constructing human-computer interfaces.

Debugging

The major problem with concurrent programming is that debugging is difficult unless debugging tools are provided. Even with tools such as DMS one cannot expect everything to be implemented correctly the first time. When an error occurs all processes typically wait for one another, and it is difficult to determine the cause of the problem. The DMS environment has built in debugging tools that make it easier to handle this problem. It is possible to determine the state of all processes by entering a request from the user's login terminal, and if that is insufficient, a complete transcript can be obtained that gives the execution history of each process and the dialogue between each pair of processes.

THE DMS PROCEDURES

DMS consists of only 10 procedures, and every attempt has been made to simplify them without sacrificing functionalily or flexibility.

Process Control

1. CREATE_PROCESS (image_name)
    Creates a subprocess that executes the specified image.

2. DELETE_PROCESS (image_name)
    Deletes the process executing the specified image.

DMS Debugging

3. STATUS_REPORT
    Returns a report that gives the status of interprocess communication for each process.

4. TRACE (on_or_off)
    Produces a transcript of all interprocess communication in DMS.

Communication

5. ACCEPT (request_name)
    Executed by a program to determine the name of a service request issued by another program.

6. COLLECT (image_name)
    Executed by a program to initiate data return from a program to which a request has been made.

7. END_REQUEST
    Terminates an interprocess dialogue initiated by a REQUEST.

8. RECEIVE (data_reference,nbytes)
    Receives data from another process.

9. REQUEST (image_name,request_name,prior)
    Makes the specified request of an image

in another process at an arbitrary priority level.

10. SEND (data_reference,nbytes)
    Sends data to another process.

A FORTRAN EXAMPLE

```
c  This program gets a number and doubles it.
c  According to DMS methodology, this program
c  may not communicate with the user.
c  This program executes in a subprocess of
c  the user's login process.

   call REQUEST ('fexec','get_number',0)
   call COLLECT ('fexec')
   call RECEIVE (n,nbytes)
   call END_REQUEST

   m=2*n

   call REQUEST ('fexec','print_numbers',0)
   call SEND (n,4)
   call SEND (m,4)
   call END_REQUEST

   call REQUEST ('fexec','shutdown',0)
   call END_REQUEST

   end
```

Fig. 5.   Program FDEMO

```
c  This dialogue executor does the i/o for
c  FDEMO.  This program executes in a
c  subprocess of the user's login process.

   CHARACTER*31 request_name

10 call ACCEPT (request_name)

   if (request_name.eq.'get_number') then
     type *, 'Enter an integer'
     read *, n
     call SEND (n,4)
   end if

   if (request_name.eq.'print_numbers') then
     call RECEIVE (i,nbytes)
     call RECEIVE (j,nbytes)
     type *, '2',i,' equals',j
   end if

   if (request_name.eq.'shutdown') call exit

   go to 10

   end
```

Fig. 6.   Program FEXEC

A brief example will illustrate the use of the interprocess communication facilities as well as the structure of a dialogue executor. Looking first at FDEMO, the first request is for an input value to be used in a simple computation. Since FDEMO has nothing to send the dialogue executor, it calls COLLECT and obtains the value from FEXEC. Then, a request is made for FEXEC to print the input and output values. Since these values are not necessarily adjacent in storage, two calls are made to SEND, each having 4 bytes since the default FORTRAN integer data type is INTEGER*4. Finally, the program complex is terminated by the REQUEST for the shutdown service. Since requests at the same priority are executed on a first come, first served basis, the shutdown request which causes all programs in the complex to terminate will not be executed by FEXEC until it has finished printing out the results for the user.

FEXEC begins with an ACCEPT statement, and once a request_name is obtained, the corresponding request is executed. The only unusual service is shutdown, which causes a program exit, which, in turn, causes the entry process to delete all processes in the complex and return to the default dialogue.

THE DMS TOOLKIT

Our goal is to provide the dialogue author with specialized tools for defining human-computer interfaces within the DMS environment. In addition, DMS will have built-in metering services to aid the dialogue author in assessing the performance of a functioning interface. It is a philosophical question whether the dialogue author will be doing programming; certainly constructing a human-computer interface involves designing a logical progression of precisely defined events. However, it is our belief that languages and language systems can be constructed by means of which the interface design task can be greatly facilitated. Each tool in the author's toolkit provides specialized assistance for designing some aspect of the human-computer interface. Those that we currently envision include graphics, command language, menu, form filling, text formatting, and voice design tools. These tools shield the dialogue author from the need to use standard programming languages, and it is hoped that in many cases the dialogue author might also be shielded from the details of interprocess dialogues of the type that have been discussed.

Each tool has the same basic structure that includes a design system and an executor by means of which actual dialogue transactions are carried out. The dialogue author uses the design system in a high level interactive language to design particular dialogue interactions. These interactions are represented in a database in an intermediate language that is interpreted by an executor at the time of an actual dialogue transaction. Using the design facility of a particular tool, the dialogue author specifies dialogue interactions, simulates those that have already been designed, alters those that need improvement, and may even insert metering into the interaction to monitor performance.

The tool of greatest interest to us at the present time is the command language design tool. We have recognized that all the tools, including the language designer, require the same type of language interface to the user which would be provided by the language designer. Command languages are very rich both in structure and in the minute details that often distinguish successful languages from inferior ones. Language design is a major issue in the human factors of human-computer interfaces, and in our view it is unreasonable to require a human factors specialist to master specification systems such as BNF in order to design languages. In command languages the deep structures implied by a BNF specification are of much less importance than the surface structures.

The designer itself is an interactive program that produces a human-readable intermediate representation of the syntax of a language. At the time the end user of the software system is to interact with the system in that language, the system makes a procedure call to the language executor, specifying the name of the language. The language executor retrieves the language syntax from a database and interprets it as the user responds to the system. Objects produced by the parse are returned in a data structure.

The language executor also needs to have a considerable amount of flexibility. For example, it ought to be capable of limited forms of validity checking, such as testing the range of numeric input values. The executor should also be capable of responding in a variety of ways to user input including command completion, spelling correction, exact syntax checking, and prefix decoding.

NOTES

DMS is a large continuing experiment in the psychology and application of language. The author is particularly grateful to H. Rex Hartson for countless discussions of the theory underlying DMS, to John Roach for many discussions about language, to Shuhab Ahmed for his work on the DMS execution environment, and to Bob Fainter for providing the first serious test of DMS in the 6 process air traffic control generic task environment known as GENIE.

REFERENCES

Brinch Hansen, P. (1975).   The programming
    language   concurrent   Pascal.   IEEE
    Trans.   Software   Engineering,   1,
    199-207.

Ritchie, D.M., and K. Thompson. (1974). The
    UNIX timesharing system.   Comm. ACM,
    17, 365-375.
United   States Department   of   Defense.
    (1980). Reference manual for the Ada
    programming language.   DARPA.

DISCUSSION OF SESSION 5

5.1 S MAN-MACHINE INTERACTION IN COMPUTER-AIDED DESIGN SYSTEMS

Rouse: It seems to me that a primary bottleneck in CAD is transferring research results from the primary journal literature to design-oriented data bases. Do you agree?

Hatvany: This is a question of time and money. It is slowly happening.

Rijnsdorp: You have mentioned the problem of integrating the various phases in CAD (requirements, design and analysis, documentation). How will the integration with CAM look like? Is it an order of magnitude more difficult?

Hatvany: Yes, it is!

5.2 T HUMAN-COMPUTER DIALOGUE DESIGN CONSIDERATIONS

Tainsh: Your survey covered over 500 experimental reports. My experience is that much experimental laboratory data often has little operational relevance, i.e., the same results cannot be replicated under operational circumstances or show no direct operational benefits. Would you care to comment on this opinion.

Williges: The report you are referring to is merely a compilation of dialogue design considerations and is not intended to be a dialogue designer's handbook. Additionally, not all of these considerations are based on experimental data in that many represent opinions and personal experiences of the authors of the source documents. I do believe, however, that a truly useable database of dialogue design principles must be based on, and verified by, empirical research. Several considerations must be made in conducting this behavorial research. First, realistic computer-based tasks must be used so that the results will generalize to the operational environment. Second, the results of the empirical research must be translated into principles that designers can use. For example, the design principle for the HELP study that I presented is not the specific mean values from the various statistical comparisons. Rather, it is the concept of providing the novice user with the capability of browsing and making comparisons when using HELP. Third, the results of the empirical studies on one task must not be overgeneralized to all possible human-computer tasks and, finally, the design principles must be cast in a readily retrievable database as discussed in the last sections of our written paper.

5.3 T ONLINE INFORMATION RETRIEVAL THROUGH NATURAL LANGUAGE

Rouse: Does your "generalization from experience" include generalizing across different individuals?

Tasso - Guida: IR-NLI provides a "generalization from experience" capability that applies at two different levels:
- choice of the most suitable approach
- activation of tactics.
In both cases, experience of how problems occurring in previous sessions were solved, is used to face the current situation in the most effective and appropriate way. This clearly allows the system to gather experience from sessions with different users. Generalization from experience can be viewed, in other words, as kind of automatic learning of reasoning strategies.

5.6 I COMPUTERIZED TEXT PROCESSING: NEW DEMANDS AND STRAINS?

Chairman's Discussion Summary:

Question: Are the technologies A to E a scale?

Answer: A to E are representing discrete categories of devices with increasing technology. A represents an electric typewriter while on the other end E represents a VDU workstation with computer controlled text processing.

Question: How many questions are in the AET supplement?

Answer: AET: 216 items; supplement 149 items.

Question: What are the reasons for not accepting new technologies?

Answer: People often first reject a new technology, but really they reject the varied organisation of their work. If an equipment with new technology causes a secretary, e.g., that she has to work in a pool, she will sometimes not accept the new technology. But, the true reason for this attitude will be the changes in work content and personal status due to the new work organisation.

Question: Can you give us recommendation about how to avoid the reject of new technologies by the user?

Answer: The work organisation should be changed towards job enlargement. Hardware and software should take care of human characteristics.

Question: One point was missed in the lecture. The linkage between man and the effects of his actions.

Answer: Fig. 3 illustrates the linkage between man and machine. The degree of freedom a person has to organize his own work is an indicator for work content. If the computer leads the person, there will be only little freedom to the user to organize his work process. Man should have possibilities to control the work process and to correct his input.

ADE-L

5.8 I CONTROL OF INPUT VARIABLES BY HEAD MOVEMENTS OF HANDICAPPED PERSONS

Chairman's Discussion Summary: A question was raised by Mr. Stein of FAT about the speed of learning. He was assured that the device required effectively no learning time.

5.10 I EXPERIMENTAL STUDIES OF MAN-COMPUTER INTERACTION IN FINANCIAL ACCOUNTING SYSTEMS

Chairman's Discussion Summary: In the paper, two dialogue systems (S1, S2) were proposed and compared with an existing dialogue system (S3). The main point in the discussion was the question why these systems S1 and S2 are better than system S3, e.g., which features of S1 and S2 are the reason for better performance and increased acceptance. The author referred to long list of differences between the systems, he mentioned for example the help-system of S1 and the possibility given by S2 to use the line number for the selection of the coded variables of a whole line. The advantage of the approach described in the paper, namely, to put together different features suggested by experience and dialogue-design rules and to validate the resulting dialogue systems experimentally, is to give improved dialogue systems for practical use in a limited amount of time. But a major drawback is that a confounding of the different features occurs. Therefore, it is difficult to interpret the experimental results in terms of the reasons for the improvements measured. Several participants stressed the need for guidelines for dialogue design which are valid in different application areas.

5.11 I A GRAPHICAL HARDWARE DESCRIPTION LANGUAGE FOR LOGIC SIMULATION PROGRAMS

Chairman's Discussion Summary: The paper dealt with the automation of the manual circuit description for logic simulation. In the discussion, the author explained the remaining man-machine interaction, which consists mainly of entering the circuit diagram by means of a light pen. The user may define macro-statements for those circuits, which have to be entered frequently. Furthermore, there are edit-functions available in order to erase components and to reinsert other ones. The software package is portable to other computers having a PASCAL compiler. The future work is directed to an integrated design system for the design of integrated circuits (CMOS).

5.12 I KNOWLEDGE-BASED MAN-MACHINE SYSTEM

Chairman's Discussion Summary:
Tainsh: How do you distinguish between knowledge bases and databases?
Ohsuga: In complex applications, several knowledge bases can be combined into complex databases. In particular, relational data-

bases are of utmost importance. Knowledge representation appears to be most efficiently performed by predicate logic.

5.13 I SYMBIOTIC, KNOWLEDGE-BASED COMPUTER SUPPORT SYSTEMS

Chairman's Discussion Summary:
Question: How do you determine what the level of uniformity should be?
Fischer: I only wanted to demonstrate the concept of uniformity. There are many levels and aspects of uniformity.
Williges: What is the role of behavioral research in the development of new dialogue systems for the non-expert user?
Fischer: Research directed toward determining underlying behavioral principles is extremely important. Additionally, the behavioral research can be quite fruitful in improving further specification of prototype systems.

5.14 I A KNOWLEDGE BASE SYSTEM FOR DECISION SUPPORT USING COGNITIVE MAPS

Chairman's Discussion Summary:
Williges: What is the most difficult aspect of developing cognitive maps?
Iwai: The value judgments of the coders of the information sources are extremely critical. But, their reliability is quite high according to our research results. In fact, we have developed certain metrics to test this reliability.

5.15 I DMS - A SYSTEM FOR DEFINING AND MANAGING HUMAN-COMPUTER DIALOGUES

Tainsh: Is it not true to say that your DMS system is more concerned with the contents and structure of the display formats than the process of man-computer dialogue?

Ehrich: We are concerned with both dynamic and static aspects of interactions, not only with displays, but with all types of input-output devices. Although we ourselves are not primarily concerned with the content of specific dialogues, we are also working on the design of expert tools to aid the dialogue author in preparing high quality dialogue interactions.

Chairman's Discussion Summary:
Trispel: How do you design the dynamics of a dialogue? Are there software tools available for this system?
Ehrich: This can be extremely complicated and currently we do this iteratively. Perhaps we need another tool, which takes all the static components of dialogue design and puts everything together into a dynamic process.
Tainsh: These tools really are directed toward helping the author build dialogues, but are you providing any tools in directing the authors how to build dialogues?
Ehrich: Yes, we are beginning to consider and develop expert systems which will aid the author in how to design dialogues.

MEASURING, MODELING AND AUGMENTING RELIABILITY OF MAN-MACHINE SYSTEMS

T. B. Sheridan

Department of Mechanical Engineering, Massachusetts Institute of Technology, Cambridge,
Massachusetts 02139, USA

Abstract. This paper reviews analytical and empirical trends in dealing with
human reliability. It reflects on the nature of human error and man-machine
reliability, and poses some serious problems of definition which affect numerical results. Techniques for enhancing man-machine system reliability are
presented.

Keywords. Man-machine systems. Reliability. Human error. Models. Displays.

INTRODUCTION

"To err is human; to forgive di-
vine" (Alexander Pope, 1711)

Most of us agree that people will always err.
Thus our nuclear and fossil power, oil and
chemical, air transport, military and other
increasingly complex, high-cost and high-risk
technological systems had better be designed
to be forgiving when their operators make
errors.

> "Let me make a statement about the
> indications. All you can say
> about them is that they are de-
> signed to provide for whatever an-
> ticipated casualties you might
> have. --- If you go beyond what
> the designers think might happen,
> then the indications are insuffi-
> cient, and they may lead you to
> make the wrong inferences. In
> other words, what you are seeing
> on the gage, like what I saw on
> the pressurizer level-- I thought
> it was due to excess inventory --
> I was interpreting the gage based
> on the emergency procedure --
> Hardly any of the measurements
> that we have are direct indications
> of what is going on in the system."
> (Testimony of Three Mile Island
> nuclear power plant operator before
> hearing of US Congress, 1979).

There is consensus among man-machine systems
engineers that we should be designing our
control rooms and consoles so that they are
more "transparent" to the actual working
system, so that the operator can more easily
"see through" the displays to "what is going
on". Often the operator is locked into the
dilemma of selecting and slavishly following
one or another written procedure, each based

on an a priori anticipated casualty. Often
the operator is not sure what procedure, if
any, fits the current not-yet-understood
situation.

> "In the analysis of accidents, the
> human element is the imp of the
> system.----- The variability and
> flexibility of human performance
> together with human inventiveness
> make it practically impossible to
> predict the effects of an opera-
> tor's actions when he makes errors,
> and it is impossible to predict
> his reaction in a sequence of acci-
> dential events, as he very probably
> misinterprets an unfamiliar situa-
> tion" (J. Rasmussen, 1978).

> "-----There is no possibility for
> a definition of units of behavior
> whose reliability can be deter-
> mined.-----Put simply we do not
> know how the reliability of a be-
> havioral sequence can be synthe-
> sized from the reliability of its
> parts" (J. Adams, 1982).

Many reputable analysts of human behavior
have thought hard about human error and human reliability - and have been frustrated
and discouraged thereby. There is persis-
tent skepticism about whether the human operator can be treated as just another element in an otherwise mechanical system. Yet
the discipline of man-machine control and
decision systems has progressed precisely
because engineering models have been adapted,
through experimental calibration, to human
behavior.

> "To err is human factors" (J. Egan,
> 1982).

Some observers believe human factors engineering as it is practiced is little more

than management's way of disguising its inability to administer effectively and to negotiate fairly with union workers, plus everyone's inability to cope with interpersonal problems - the real provocation for human error.

Whether in spite of or because of all the above difficulties, the subject of human error and human reliability has never been of greater concern in the scientific/engineering community.

This paper first reviews current analytical and empirical trends in dealing with human reliability, then reflects on the nature and meaning of human errors and man-machine systems reliability, and finally suggests some promising means for enhancing reliability of man-machine systems.

MODELING RELIABILITY OF MAN-MACHINE SYSTEMS

There are two good reasons for modeling human reliability and that of man-machine systems: (1) to make predictions, and (2) to achieve better understanding through the disciplined effort of modeling.

Three approaches to modeling man-machine systems have been identified (Embry, 1976): (1) the discrete failure combinatorial model; (2) the time-continuum failure model; (3) the Monte Carlo failure simulation. I add another, (4) the normal performance continuum model.

Discrete Failure Combinatorial Model

The two leading proponents of this approach have been Swain, best known for THERP, Technique for Human Error Rate Prediction (Swain and Guttman, 1980) and Meister (1964). THERP, based on standard combinatorial mathematics combined with artful use of tabled human error rates (Table 1) is the method currently being recommended by the U.S. Nuclear Regulatory Commission (American Nuclear Society, 1981).

Some elementary combinatorics are inherent in THERP: Assume estimates q_i can be made for probability of human error on operation i (by subjective judgement or empirical records, see later discussion). Assume the same for probability f_i that such a human error will not be corrected in time to avoid a system error. For two such events A and B which are independent their joint occurrence $q_{AB} = q_A q_B$. Then $Q_j = 1 - (1-f_iq_i)^n$ is the probability of failure in n independent operations i in class j. The total system failure probability is $Q_T = 1-[(1 - Q_1) (1-Q_2)-----(1-Q_i)-----(1-Q_k)]$ for k independent classes.

In dealing with multiple error events one must be particularly concerned about the dependence of one error, say B, on another,

say A. With zero dependence prob $(B|A)$ = prob (B) and the above combinatorial equations are true. With complete dependence, prob $(B|A)$ = 1. In general prob (A,B) = prob $(B|A)$ prob (A).

The steps for conducting a human reliability analysis of a nuclear power plant based on THERP have been stated by Bell and Swain (1981):

1. Visit the plant, survey the control room, interview the operators.

2. Review information available from systems analysis about critical operator interactions with plant systems.

3. Talk-or walk-through various critical procedures step-by-step with a trained operator in the control room or a simulator or a mockup.

4. Do a task analysis for various critical situations, formally listing, diagramming and interrelating task components on paper.

5. Develop initial event trees. Completed event tree is illustrated in Figure 1.

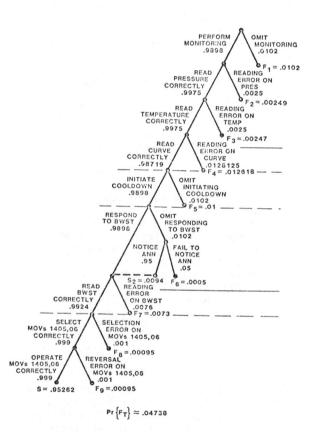

Fig. 1. Event Tree for Human Errors on a given task (Bell and Swain, 1981).

ERROR RATE	ACTIVITIES
10^{-4}	Selection of a key-operated switch rather than a non-key switch (this value does not include the error of decision where the operator misinterprets situation and believes key switch is correct choice).
10^{-3}	Selection of a switch (or pair of switches) dissimilar in shape or location to the desired switch (or pair of switches), assuming no decision error. For example, operator actuates large handled switch rather than small switch.
3×10^{-3}	General human error of commission, e.g., misreading label and therefore selecting wrong switch.
10^{-2}	General human error of omission where there is no display in the control room of the status of the item omitted, e.g., failure to return manually operated test valve to proper configuration after maintenance.
3×10^{-3}	Errors of omission, where the items being omitted are embedded in a procedure rather than at the end as above.
3×10^{-2}	Simple arithmetic errors with self-checking but without repeating the calculation by re-doing it on another piece of paper.
10^{-1}	Personnel on different work shift fail to check condition of hardware unless required by check list or written directive.
5×10^{-1}	Monitor fails to detect undesired position of valves, etc., during general walk-around inspections, assuming no check list is used.
.2 - .3	General error rate given very high stress levels where dangerous activities are occurring rapidly
~1.0	Operator fails to act correctly in the first 60 seconds after the onset of an extremely high stress condition, e.g., a large LOCA.
9×10^{-1}	Operator fails to act correctly after the first 5 minutes after the onset of an extremely high stress condition.
10^{-1}	Operator fails to act correctly after the first 30 minutes in an extreme stress condition.
10^{-2}	Operator fails to act correctly after the first several hours in a high stress condition.

Table 1. A sample of Human Error Probabilities from Wash-1400. Swain and Guttman (1980) regard these as still valid but have made a much more extensive list.

6. Assign from tabled values appropriate nominal human error probabilities (HEPs) for component events.

7. Estimate the relative effects of performance shaping factors such as stress, training, motivation, fatigue, etc. and adjust HEPs (see Embry, 1976).

8. Assess dependence factors and adjust HEPs. Usually these are only considered for adjacent events on the trees; higher order dependencies are neglected.

9. Determine success and failure probabilities for whole sequences of events, neglecting recovery factors f_i.

10. Determine effects of recovery factors. If the failure rates are sufficiently low without the recovery factors those sequences can be ignored anyway, so there is no need to bother with f_i which would decrease failure rates further.

Normally the HEP's for human operator sequences are turned over to reliability systems analysts who incorporate those results into still more complex analyses including equipment failures, weather and seismic conditions, etc.

If warranted, sensitivity analysis may be
done for individual or combined HEPs, i.e. to
get a ratio of the partial derivatives: ∂
(probability of core melt or other calamitous
"top event")/ ∂(probability of some component
"base-event" human error). Hall and his colleagues (1981) made extensive analyses of how
changes in HEP's might affect system unavil-
ability, core melt probability and radiation
release probability. They started from
nominal probabilities for salient events
from the U.S. Nuclear Regulatory Commission
Reactor Safety Study WASH-1400 (1975) for
various accident scenarios. They considered
all human error probabilities to be at least
10^{-5}, some of course much larger. Then, using a special computer program, they determined how making all HEP's 3, 10 or 30 times
smaller or 3, 10 or 30 times larger would
affect the "top events" within those accident scenarios. Figures 2 and 3 show some
typical results. Note that sensitivities to
changes in HEP's are roughly comparable
across various accident scenarios, that these
taper off quickly on the negative side as machine factors dominate. The apparent diminishing effect on the positive side is an artifact of the logarithmic scale. Actually
as the human errors predominate the top event
probability changes tend to become linear
with HEP changes. Table 2 shows results of
a different analysis: how HEP's in different
generic classes affect core melt probability.

Time Continuum Failure Model

There are those risk investigators (Askren
and Regulinski, 1969) who believe a much more
appropriate way to consider system reliability is in terms of time until failure, or if
one wishes to use empirical statistics, mean
time between failure (MTBF). Confidence limits may be established for the MTBF based
typically on a skewed distribution of times
until failure such as log normal. Assuming
the human operator is good at recovery or repairing one can incorporate data on mean
time to repair (MTTR) and generate statistics
for fraction of time a given system is available:

$$\frac{MTBF}{MTBF \ \& \ MTTR} = availability$$

Monte Carlo Failure Simulation

An approach to human reliability somewhat
different from direct calibration of the
above models by empirical data uses a computer-based Monte-Carlo technique (Siegal
et al, 1975; U.S. Navy 1977). Probabilities
of error are assumed for human behavior elements such as finding the right display,
reading and interpreting the display correctly, making the right decision, finding
the right control, or operating it correctly. Each of these is conditioned on task or
situational properties such as number of alternatives, size, distance, familiarity etc.
Then the appropriate combination of behavior-

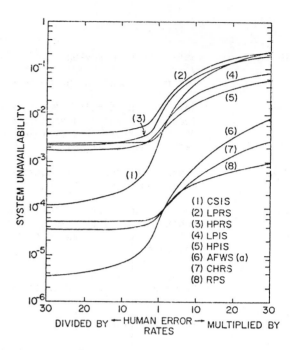

Fig. 2. Hall et al Sensitivity Results.
Changes in Availability due to changes
in all Human Error Rates, for Various
Accident Scenarios. Unity error rate
is for Standard tabled values.

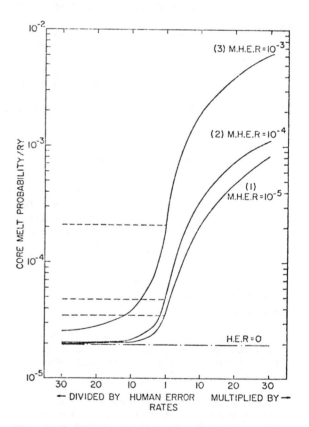

Fig. 3. Hall et al Sensitivity Results.
Changes in Core Melt Probability due
to Changes in all Human Error Rates,
for various Minimum Human Error Rates,
unity for others is for Tabled Values.

GENERIC CLASSES OF HUMAN ERROR	FACTOR INCREASE/DECREASE IN CORE MELT PROBABILITY					
	GENERIC CLASS OF HERs DIVIDED BY			GENERIC CLASS OF HERs MULTIPLIED BY		
	30.0	10.0	3.0	3.0	10.0	30.0
T&M ERROR	1.33	1.31	1.21	1.59	4.18	17.07
TESTING ERROR	1.27	1.25	1.18	1.50	3.62	11.87
MAINT. ERROR	1.23	1.21	1.15	1.39	2.90	10.55
OPERATOR ERROR	1.22	1.20	1.14	1.37	2.67	6.49
SAMPLING ERROR	1.0	1.0	1.0	1.01	1.04	1.13
PRE ACCIDENT	1.35	1.32	1.22	1.61	4.28	17.65
POST ACCIDENT	1.22	1.20	1.14	1.37	2.65	6.31
OMISSION	1.41	1.37	1.26	1.67	4.57	18.85
COMMISSION	1.17	1.16	1.11	1.30	2.35	5.38
CONTROL ROOM	1.26	1.24	1.17	1.43	2.92	7.25
NON CONTROL ROOM	1.29	1.27	1.19	1.54	3.97	16.53

Table 2. Hall et al Results for Sensitivity of
Core Melt Probability to Changes in
Generic Classes of Human Error Rates.

al elements and task situation properties is assembled and "run" a large number of times in a computer, and distribution functions are generated for success (or failure) of the whole sequence of events. Such models can, of course, he made to fit empirical data by adjustment of their plentiful parameters. The latter characteristic also limits their usefulness for prediction purposes.

Normal Performance Continuum Simulation

This is the most common type of behavioral model. Output variables characterize what the human operator does in time, space or other attribute of response without reference to what characteristics of response make it a success or a failure. For example such a simulation outputs how long a person takes to do some act, or what his accuracy is. The criteria of success or failure can be specified extrinsically. The probability of success is thereby determined by the performance resulting from the inputs and initial conditions.

EMPIRICAL RELIABILITY MEASUREMENT

The usefulness of any of the above models depends on its correspondence with empirical events measured under circumstances which correspond to those for which predictions are to be made. From one viewpoint the

models are nothing more than alternative ways of summarizing such empirical observations.

In the nuclear power industry there are three available sources of such observations - from which a "data-base" may be constructed:

1. anecdotal accounts from the memory of operators or other participants;

2. operating logs and formal accident or "event" reports;

3. runs on training simulators in which error provoking conditions are present and errors are automatically measured and recorded.

Of human errors in valve and switch operations noted in official licensee event reports from commercial nuclear power plants in the U.S., Luckas and Hall (1981) estimate human error rates to approximate those used by Swain, i.e. about 10^{-3}.

The credibility of human error data from simulators is most often questioned on the basis that realistic stress, boredom and similar sources of behavior variability are lacking. Other neglected sources of variability are subjects' understanding of what they are supposed to do or what the criteria of success or failure are. Nevertheless, as well proven by aircraft piloting simulators, this approach may be the best way to estimate human reliability in complex operational situations.

REFLECTIONS ON WHAT HUMAN ERROR IS

In the human reliability models and data-base-building exercises described above the assumption is too often made that it is clear what human error is. I contend that a lack of consensus on definition is the greatest source of variability in both modeling and empirical measurement of human error (Sheridan, 1980).

Whether a discrete error has occurred, i.e. whether a person's behavior has gone outside the multi-dimensional bounds of successful performance, is obviously an artifact of where that discrete success-failure boundary is drawn. Varying degrees of success or failure are disallowed. A comparison with other forms of human endeavor is interesting. Ordinary human discourse tolerates infinite variation and shades of inference about human behavior. Psychiatry seeks at least qualitative categories of behavior. Psychometrics requires continuous quantitative scales of behavior. Human reliability analysis reduces behavior to a single binary discrimination.

It is common in nuclear power plant simulator training exercises, for example, that trainees perform certain procedural steps in an arbitrary order because they know "it doesn't matter". However a computer programmed by strict adherence to the procedures will score these steps as erroneous. In normal practice some steps may be delayed, or omitted altogether, or performed to a different criterion because that's the way a particular operator was trained or understands his task - but these will be scored as errors. Depending on the setpoint of an alarm, or a threshold of performance used for scoring and never explicitly revealed, a human-controlled continuous variable may make a single slow excursion just across the line and back, and be counted as one error. Or, it may cross just over and back several times and be counted as several errors. Or, it may hover just short of the criterion and not be counted at all - though to the operator all three behaviors are imperceptably different. An operator may knowingly allow a variable to trigger an alarm,

knowing full well he can easily recover, but nevertheless he is scored in error.

Many and varied taxonomies or classification schemes for human error have been proposed by various investigators   (J. Rasmussen, 1981). Among the alternative ways to classify are:

1. errors of omission; errors of commission;

2. errors in sensing; memory; decision; response;

3. errors in deciding what one intends ("mistakes"); errors in implementing those intentions ("slips") (see Norman, 1981);

4. erratic errors (boredom, "carelessness"); forced errors (task demands exceed physical capabilities).

For a time, when behaviorist psychology prevailed, it was not regarded as "scientific" to consider cognitive events - only stimuli and responses existed. Now computer-science and artificial intelligence have made "cognition" fashionable. Now there is less consensus than ever regarding the candidate list of "behavioral units" (Adams, 1982), and the same is true of human error causation.

An old, but I believe still credible, theory is that people make errors because they don't get enough feedback. That is, people are multi-dimensional feedback control systems, continually moving off-track and correcting themselves in a hierarchy of feedback loops encompassing: manipulations, whole body locomotion, observations, tactical decisions and strategic plans (Figure 4). If the appropriate feedback is lacking, because of sensory limitations or lack of training, or because the task/environment is poorly designed, then the organism will wander off course to the point where an "error" is made.

Since the human operator anticipates or previews his upcoming tasks, that "course" of tasks must be evident to him. A common type of error noted by Rasmussen and other investigators, is where an operator is well practiced in some task sequence ABCD but occasionally intends to do EFBG, i.e. both sequences require the common element B. The operator does EFBCD and then discovers his error (Figure 5). Somehow his conditioning to

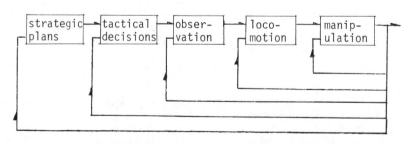

Fig. 4. Hierarchy of Feedback Loops in which
Operator Corrects His Errors.

Fig. 5. Task Sequence Error due to Common Element of More Familiar Sequence.

do C following B was overwhelming. The correct but abnormal B → G course was not sufficiently evident at the critical time.

"Stress" and "mental workload" are two constructs which the non-behaviorist sometimes invokes to account for human error. From everyday experience these terms have meaning, but their use scientifically is plagued with difficulty. Both terms may be defined operationally in terms of time constraints, high risk or problem complexity attributes of the task, not properties of the behavioral response. Thus even a robot would be "stressed" or have "mental workload". Alternatively definition of such words can be defined in terms of physiological response (heart rate, pupil diameter, voice frequency, blood chemistry) or in terms of experimentally imposed "secondary tasks" (the better the performance the more spare capacity, the less the mental workload). Subjective ratings are regarded by some, including the writer, to be the most direct and reliable measure. Subjective behavioral response measures, I believe, are more sensitive anticipators/predictors of decrement and "error" in operator performance than are direct measures of performance.

One emerging theory of error causation appeals to cognitive constructs, but has plenty of precedent in control and signal theory. This is the idea of an "internal model", an input-output simulation of the controlled process or task which, when forced by process input, can be used to "observe" variables that are not convenient to measure directly. The computer internal model can also be repetitively updated with initial conditions and run in fast-time to predict "what-will-happen-if" any particular input is used. The related psychological hypothesis is that from experience operators build up such input-output models in their heads. Presumably errors occur if the wrong mental model is used, or if the parameters are miscalibrated relative to reality. This theory implies that training as well as the design of displays and controls should correspond to an appropriate mental model of the task.

With respect to causation of operator errors it is useful to make a comparison with computer errors. For both, there are both "hardware" and "software" errors, and in both categories there are both "failures" and "limitations". Table 3 shows the two-way correspondence.

Most reliability analysts shy away from considering acts of intentional malevolence as a source of system error. While overt attacks and sabotage are properly the domain of guards and professional security investigators/analysts, there probably exists a large "gray area" of carelessness and neglect by operators as well as by maintenance and administrative personnel provoked by social ill-feelings or apathy.

IN COMPUTER

HARDWARE

Failure. Component burns or breaks. Current stops flowing.

Limitation. Rate of incoming pulses too fast or at wrong magnitude. Bits overflow available registers.

SOFTWARE

Failure. Conventional software bug. Instructions improper. Computer gets "hung up". Actions are other than those intended.

Limitation. Program not intended to accommodate this request or handle this data.

IN HUMAN OPERATOR

BODY

Failure. Injury or disease. Body stops functioning normally.

Limitation. Input is beyond threshold. Reach or load capacity is exceeded.

MIND

Failure. Slip of behavior. Action is other than what this operator intended.

Limitation. Person not trained or motivated to carry out what another person or system intended.

Table 3. Hardware and Software Failures and Limitations for Computer and Human.

REFLECTIONS ON RELIABILITY OF MAN-MACHINE SYSTEMS

It is commonly appreciated that people and machines are rather different and thus, somehow, a combination of both has greater potential for reliability than either alone. It is not commonly understood how best to make this match.

People are erratic. They err in surprising and unexpected ways. Yet they are also resourceful and inventive and can recover from both their own and the equipments' errors in creative ways. Machines are more dependable, which means being dependably stupid when minor change of behavior would prevent a failure in a neighboring component from propagating.

The intelligent machine can be made to adjust to what at the outset must be an unknown - an identified variable whose importance and relation to other variables are sufficiently well understood that control loops or computer software can be designed to deal with this variable.

The intelligent human operator still has usefulness however, for he can respond to what at the design stage may be termed the unknown-unknown - situations which were never anticipated, so that there was never any basis for equations and computers and software.

Reliability analysts of nuclear power plants are now struggling to include human operator errors and human operator initiated recovery factors in their analyses. This is laudable but unfortunately still insufficient. This is because human error/recovery pervades the performance of these large systems in many locations and at many stages - not just operations. There is initial planning and design, plant construction and fabrication of equipment by vendors, installation, calibration, maintenance, administration and management. Operations may even now be the activity most free of serious human errors.

Further, reliability is always in tradeoff with cost. The question is - how much safety is acceptable - short of "100% safe" as recently demanded by a famous American senator who should have known such statements make no sense.

We need to find better sources of data by which to calibrate our human error models. Anecdotal error data are always available, but always suspect. Formal accident reporting schemes such as police reports or licensee event reports in nuclear plants are likely to be both simplistic and "sanitized", i.e. data which are embarassing to any of the parties involved are deleted - including most of what is useful for reliability research. The U.S. National Aeronautics and Space Administration's Ames Research Center developed an excellent reporting scheme in which the anonymity of the person reporting is protected. As noted earlier simulators may well be our best

hope in understanding man-machine reliability in operations, but they offer little hope for providing data on the types and stages of human "error" outside of operations.

Finally, no matter how intelligent the human operator is, in many modern systems he is almost totally dependent upon displays and controls and mediation by computers. In view of computer-aided-design, computer-aided manufacturing, computer-aided-test, and computer-aided-management systems this computer-dependence is gradually coming to be true of participation by people all along the way. Identifying whether an "error" was human or machine is coming to be more difficult, not less.

ENHANCING RELIABILITY OF MAN-MACHINE SYSTEMS

Reliability of man-machine interaction can be enhanced by:

1. discovering through reliability analysis which sources of error are most critical;

2. reviewing the control boards or consoles and correcting obvious human-engineering deficiencies;

3. training operators to be aware of certain error-causative factors;

4. training operators on simulators to cope with emergencies they haven't seen before;

5. acknowledging of social factors which cause neglect and sloppy work;

6. providing special computer aids and integrative displays which show which parts of the system are in what state of "health".

In present nuclear power plants there is a serious problem with the excessive number of alarm lights. In an emergency the control board lights up and blinks like a Christmas tree, which is of little help to the operator in diagnosing what has happened. For example, I observed one new control room where, in a simulated loss-of-coolant accident, 500 lights went on or off within the first minute and 800 within the second minute. As a consequence there is now a trend toward "safety parameter display systems" (SPDS), nominally integrative displays to the operator of approximately ten of the most important plant variables in order to do (6) above. Clearly, meaningful integration of these variables should be more helpful to the operator than separate presentation of the raw data components, but initial SPDS designs have proven less than completely satisfying. Under "abnormal transient" conditions the operator needs still more help in integrating measurements. This is embodied in a new concept called Disturbance Analysis and Surveillance System (DASS).

One form of DASS is based on the conventional a-priori reliability analyses of (1) <u>fault trees</u> (logical AND and OR combinations which determine how component event probabilities affect probabilities of major events like loss of power, loss of major subsystem functions, etc.) and (2) <u>event trees</u> (how one event probabilistically precipitates other events and consequences in time).

Tsach, Tzelgov and I (1982) are developing another form of DASS (Figure 6). It utilizes a dynamic computer-based internal model as discussed above. Effort and flow variables ("power bonds") are measured at various key points in the plant, at least one such pair for every state variable. The model is cut into two submodels at any of these points, with the flow forcing the one submodel and the effort forcing the other submodel (as determined by causality in the real system). The model co-variables are then compared to the actual measured co-variables, and if there is significant discrepancy that indicates a failure somewhere in that submodel. By successive cuts and comparisons at different places in the system a failure may be located to any degree of precision.

A human operator may set thresholds for automatic operation of this system, then confirm or countermand the computer's detection/location decisions based upon raw signal data. Bayesian updating to yield failure/no failure odds-ratios has proven to be useful also.

From simulation experiments we have shown that this technique accommodates multi-loop non-linear systems, noisy measurements, imperfect models and can locate a failure quickly before the variables stray too far from their normal ranges and the model is no longer valid.

CONCLUSIONS

1. Current efforts to measure, analyze and model human errors and reliability by any of several methods may be helpful in understanding relative importances of error sources, but are fraught with difficulties of defining and measuring just what is human error and what is not.

2. Human error causation is an active theoretical field with little consensus. Nevertheless designers should already know to provide adequate feedback to correct improper actions, to help operators to anticipate proper steps especially when they branch away from familiar procedures, and in their designs should try to conform to and reenforce a correct operator's mental model of the task.

3. "To err is human, to forgive devine." People will always err, not only operators, but designers, maintainers and managers. By providing to the operator some computer aids for on-line detection of, diagnosis of and recovery from human errors or mechanical failures, systems can be made to be more forgiving and overall reliability can be enhanced.

REFERENCES

J.A. Adams, Issues on Human Reliability, <u>Human Factors</u>, Vol. 24, No. 1, p. 1-10, Feb. 1982.
American Nuclear Society, IEEE, <u>PRA Procedures Guide: A Guide to the Performance of Probablistic Risk Assessments in Nuclear Power Plants</u>, draft NUREG/CR 2300 Sept. 1981.

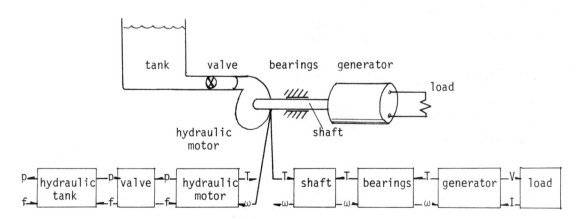

Fig. 6. Failure Detection/Location by Comparison of Measured Power Co-variables to corresponding variables of Computer Model (Tsach et al 1982).

W.B. Askren and T.L. Regulinski, Quantifying Human Performance for Reliability Analysis of Systems, Human Factors, 11, 393-396, 1969.

B.J. Bell and A.D. Swain, A Procedure for Conducting a Human Reliability Analysis for Nuclear Power Plants, Sandia Laboratories Rep. 81-1655 (NUREG/CR-2254), Dec. 1981.

J.Egan, to Err is Human Factors, Technology Review, 1982.

D.E. Embry, Human Reliability in Complex Systems: An Overview, National Centre of Systems Reliability (U.K.) Rep. 10, July 1976.

R.E. Hall, P.K. Samanta, A.L. Swoboda, Sensitivity of Risk Parameters to Human Errors in Reactor Safety Study for a PWR, Brookhaven National Laboratory Rep. 51322 (NUREG/CR-1879) Jan. 1981.

W.J. Luckas, Jr. and R.E. Hall, Initial Quantification of Human Errors Associated with Reactor Safety System Components in Licensed Nuclear Power Plants, Brookhaven National Laboratory Rep. 51323 (NUREG/CR-1880), Jan. 1981.

Meister, D., Methods of Predicting Human Reliability in Man-Machine Systems, Human Factors, 6, 621-646, 1964.

D.A. Norman, Categorization of Action Slips, Psychological Review, 88, 1-15, 1981.

Alexander Pope, Essay on Criticism, Part 2, 1711.

J. Rasmussen, Notes on Human Error Analysis and Prediction, RISO National Laboratory, Denmark, Rep. M-2139, Nov. 1978.

J. Rasmussen et.al, Classification System for Reporting Events Involving Human Malfunctions, RISO National Laboratory, Denmark, Rep. M-2240, Mar. 1981.

T.B. Sheridan, Human Error in Nuclear Power Plants, Technology Review, Feb., 1980.

A.I. Siegel, J.J. Wolf, and M.R.A. Lautman. A Family of Models for Measuring Human Reliability, Proc. Ann. Reliability and Maintainability Symposium, IEEE, 1975.

A.D. Swain and H.E. Guttman, Handbook of Human Reliability Analysis with Emphasis on Nuclear Power Plant Applications. Sandia Laboratory NUREG/CR-1278, Apr. 1980.

U.Tsach, T.B. Sheridan and J. Tzelgov, A New Method for Failure Detection and Location in Complex Dynamic Systems, Proc. 1982 American Control Conf., Arlington, VA, in press.

U.S. Congress Oversight Hearings (Serial 96-8, Vol. 1, p. 138) Testimony of the Three-Mile-Island Operators, 1979.

U.S. Navy, Sea Systems Command, Human Reliability Prediction System User's Manual, Dec. 1977.

U.S. Nuclear Regulatory Commission, Reactor Safety Study: An Assessment of Accident Risks in U.S. Commercial Nuclear Power Plants, Wash-1400 (NUREG-75/014) Oct. 1975.

HUMAN RELIABILITY AND SAFETY EVALUATION
OF MAN-MACHINE SYSTEMS

T. Terano*, Y. Murayama** and N. Akiyama***

*Department of Control and Instrument, Hosei University, Koganei, Tokyo, Japan
**Ship Research Institute, Mitaka, Tokyo, Japan
***University of Electro-Communications, Chofu, Tokyo, Japan

Abstract. The safe operation of plant is strongly affected by human factors. However, there are few data of human reliability and there is no effective way evaluating man-machine systems because of the complexity of human. In this paper, the authors introduce the concept of fuzzy-set into fault-tree-analysis. Since probability can be considered as a special case of possibility[1], there is no theoretical contradiction in treating machine-reliability probabistically and human-reliability subjectively at a time.

Next, they study the fuzziness of human-reliability through experiments especially when the information is redundant and also when the workers are redundant.

Keywords. Man-machine systems; Human factor; Reliability theory; Ergonomics; Fuzzy set; Fault tree.

1. INTRODUCTION

The faults of men play a weighty part in the causes of serious accidents of plants. Men, in this case, include not only operators but also the personel connected with management and maintenance. Plant should be planned as a man-machine system in order to keep the safety, but the planning and the evaluation of man-machine system are very difficult, because of the lack of data on human reliability and the incompleteness of theory.

In many researches[2][3], a hypothesis that men can be dealt with in the same way as machine-elements has been built up because any intricate human works can be resolved into simple jobs by task analysis. However, such an idea is not suited. The reasons are as follows. (1) A single job is not assigned to a specific man. (2) Multi-input is given to man and multi-output goes out from him. His function is flexible. (3) Man does not always act logically and takes measures suited to the occasion experientially. (4) Man has many functions such as detection, judgement, feedback- or sequential-operation and can use them properly. (5) He can detect quantity such as smells and noise which can not be measured by meters.

Moreover, the usual method of reliability engineering in which system is completely resolved into units is not suited for complex machine systems. It will be more efficient to consider macro-event by putting some units in order. This is called fuzzy event.

The purpose of our study is to suggest a new way capable to be applied to the analysis and the evaluation of man-machine system, and also to obtain some data of human reliability in plant operation. We introduce fuzzy set theory which includes probability theory and is convenient to deal with the vagueness of human characteristics.

2. FUZZY FAULT TREE

In ordinary analysis of safety, only whether a certain unit of a system is fault or not is considered as event. However, it is necessary to express the degree of abnormity as to a group of elements or man. Since the degree of abnormity is not definite, it is best to express it in natural language and then replace them with fuzzy set. In this paper, the abnormity is simply expressed by fuzzy set of real number of which membership function is an equilateral triangle and takes the maximum value 1 as illustrated in Fig. 1. The value corresponding to the maximum represents the degree of abnormity and the length of the base shows fuzziness.

The relation of cause and effect between two fuzzy events is expressed in fuzzy implication of type 2. That is

"If A then B" is R.   (1)

The A and B are fuzzy events of cause and effect respectively. R is numerical truth value of type 2, which expresses the degree of certainty of the sequence. Since A, B and R are all vague, we use fuzzy numbers to express them. (Fig. 2)

The expression of formula (1) includes some logical relations among the events, such as AND and OR, beside the strength of relation. Though the reliability of system-element may be represented in combinations of A, R and B, it is considered most proper that R includes the reliability in the ordinary sense. Because the degree of abnormity hardly includes the frequency of occurrence from the viewpoint of the definition. The reliability of machine-elements is the case when the fuzziness of R is zero. In case of a group of machine-elements, R expresses the macroscopic transition between two fuzzy events. If there are some logical relations among fuzzy events, formula (1) will be changed as follows.

AND: "If A_1, then if A_2, ..., then B" is R.   (2)

OR : "If A_1, else if A_2, ..., then B" is R.   (3)

In order to express the human reliability, the following three kinds of idea are applicable, and a suitable one shall be used according to the situation.

(1) If A is the initial state of plant and B is the consequent event resulting from the operation of man, R represents the reliability of operator.
(2) When man is a basic event in fault tree, human characteristics are expressed by the combination of A and R. A can be regarded as seriousness of an error committed by man, and R as its frequency.
(3) When man is in a middle event, R and B represent him. In this case, B is considered the extent of effect of faults, and R the frequency of faults.

Man-machine system should be analysed and fuzzy events are selected. A system model is completed by expressing the fuzzy events and the fuzzy relations in the formulae (1), (2) and (3), and allotting fuzzy numbers to A, B and R. In this fuzzy fault tree, the boundary of events and the sequence of cause and effect are vague. Therefore, this can be conveniently used even when the characteristics of elements or the situation of abnormity are changed from the initial supposition slightly.

2.2 Algorithms

When cause event A, effect event B and transition R are given, the effect B' caused by A' which is a little different from A can be inferred. This is called fuzzy inference. The degree of the top event in a fault tree can be calculated by such an inference from the basic events. The procedure is as follows.

First, A, B, R and A' are given in fuzzy number of which membership functions are denoted by μ_A, μ_B, μ_R and $\mu_{A'}$ respectively.

(i) Truth value τ_A which satisfies A'=(A is τ_A) is found in the following formula. (fuzzy converse truth qualification)

$$\mu_{\tau_A}(v) = \underset{x=\mu_A^{-1}(v)}{\text{Sup}} \mu_{A'}(x) \qquad (4)$$

(ii) Truth value τ_B of B is found from τ_A and R. (fuzzy truth qualification)

$$\mu_{\tau_B}(v) = \mu_R(\mu_{\tau_A}(v)) \qquad (5)$$

(iii) B' which satisfies B'=(B is τ_B) is obtained in the following formula.

$$\mu_{B'}(x) = \mu_{\tau_B}(\mu_B(x)) \qquad (6)$$

Now we know the fuzzy result B'. But $\mu_{B'}$ is usually not a triangle. Therefore, we must represent it with a fuzzy number approximately in order to continue the calculation.

The next problem is how to compare two fuzzy numbers for the system evaluation. For this purpose, we define a function f as for a normal convex fuzzy set A as follows[5].

$$f(A) \overset{\Delta}{=} \int_0^1 (X_L(\alpha) + X_H(\alpha))\alpha \; d\alpha \qquad (7)$$

where $A_\alpha = (X_L(\alpha), X_H(\alpha))$, $0 < \alpha < 1$.

Any fuzzy number can be compared each other. That is, f(A)<f(B) => A<B. One example is shown in Fig. 3.

In fuzzy fault tree, we can define α-cut sets, α-pass sets, possibility importance and structural importance corresponding to those of ordinary fault tree, but they are omitted so far.

HUMAN RELIABILITY IN MAN-MACHINE SYSTEM

How a man-machine system is dealt with logically has been mentioned in the preceding chapter. It is necessary to know human reliability, R in order to apply it to actual systems. Experiments were performed to find it as follows.

Reliability of the individual man as an element of a man-machine system and redundance of elements as a means to improve the man-machine system reliability are dealt with in the experiments.

With regard to basic ways of duplication of communications between man and machine, the following three ways can be thought of, as shown in Fig. 4.

(1) Duplication of the path from machine to man.
(2) Duplication of men.
(3) Duplication of machine factors.

(1) can be evaluated in the method of machinery redundancy if there is no interference between machines or between men. If there is interference, the same kind of problem as that of (2) or (3) will arise.

An operation by two operators according to information from one machine applies to (2). In this case, interference between the two men and the allotment of task naturally arise in process of a series of actions, the detection - judge - operation, conducted in each individual. Therefore, it is expected that the reliability of operators cannot be calculated so simply as in the case of machine redundancy.

To cite an instance of (3), an operator receives multiple informations of different types from one plant. In this case, how the reliability of a series of actions, the detection - judge - operation, is influenced by multiple informations cannot be calculated in the same way as in case of machine system.

Two kinds of experiments corresponding to cases of (2) and (3) were conducted in this research in order to examine the reliability of operators and the effect of redundancy of communication of man and machine on reliability. In one experiment, two operators carry out the same task on the basis of the same information simultaneously, which is called Experiment-I in the following. In the other experiment, how an operator judge the information from two meters showing the same signal to which different noise are added is examined, which is called Experiment-II.

4. HUMAN RELIABILITY EXPERIMENT

4.1 Experiment-I (Odd-Even Judgment Experiment)

Subjects are supposed to judge the sum of three numbers of figure indicated on CRT display odd or even within two seconds and to push a button corresponding to the judgment, in this test. A test consists of 100 consecutive trials and the success rate of 100 trials are regarded as the reliability of a subject in the test. The value is noticed to the subject after each test end.

In order to control the tension of the subjects, tests were performed in three modes, that is, "individual", "cooperation" and "competition". Trials in each mode are conducted as shown in Table 1.

There is a random waiting time of from 0.5 to 1.5 seconds between each trial and the next one, during which the indication of three numbers in the previous trial remains and they are replaced by new three numbers for the next trial without a signal when another trial starts. Therefore, subjects must continue to watch the three numbers in order to know the beginning of the next trial.

Experiment Series A. With regard to individual subject, four "individual" tests, three "cooperation" tests and three "competition" tests, that is, 10 tests in total were carried out for a day, 1000 trials being made a series of experiment. The total number of subject persons was 22.

The result of the experiment as follows.

(1) Subjects' personal reliability varies according to experiment mode. Subjects' personal reliability in case of the "cooperation" mode is lower than that in cases of the "individual" and "competition" modes, as shown in Table 2. It seems the reason is that subjects are strained in cases of the "individual" and "competition" modes but that the tension is eased by depending on a partner in case of the "cooperation" mode.
(2) As subjects become skillful, the phenomenon of lowering personal reliability in case of the "cooperation" mode disappears. As the reliability of the same pair in each mode when they were unskillful and after they became skillful is indicated in Table 3, the difference between all cases of three modes disappeared after they became skillful. It seems that this is because a subject does not depend on a partner as each trial becomes easy for him after he becomes skillful and because his judgment and operation becomes mechanical due to the skill and he ignores the partner.
(3) Even if a man does not become a subject actually, he becomes skillful only by observing tests. Fig. 5 is a part of the experiment record of a certain day. It is known that the reliability of the following subjects is rising due to study effect while they look at the tests of the first subject in the "individual" mode.
(4) Factors of difference between individuals, tension degree and skill degree influence the reliability in Experiment Series A.

Experiment Series B. Experiment Series B was conducted in order to measure the reliability of men who organized a team with a partner of a computer. As shown in Table 4, 10 tests in total including "individual" mode experiment and "cooperation" mode experiment in which a computer is a partner of a subject were carried out with each subject for a day by making the reliability of the computer a partner. A series of this experiment consisted of 1000 trials and the total number of subject persons was 13. The subject were informed of the operation reliability of the computer before each test started.

The results of the experiment are following.

(1) A phenomenon of lowering individual reliability which occurs in case of the "cooperation" mode where a man is a partner does not appear in the "cooperation" experiment in which a computer is a partner.
(2) When the operation reliability of the computer as a partner is of the same grade as a man, his reliability rises. (Table 5)

It is known from this that a computer is ignored as a partner and that dependence on the partner by cooperation is not recognized while individual reliability rises due to competition in spite of the "cooperation" mode when a computer of the same-degree reliability is made a partner.

This may be because there is no communication between man and computer as a partner. In case of a pair of men, a partner's intention is transmitted even if there is no communication by words but since this is not the same in the case of a computer, it would not be recognized by a man as a partner.

In Series B, besides the test load toward subjects was increased by imposing an odd or even judgment task and another task to watch and operate a machine so that numbers which increase and decrease might not exceed limits on the subjects, and the process of their depending on a computer was examined.

The difference between the "individual" and "cooperation" modes was not recognized as a result of examination of the reliability of the operation of the former task only. However, there was a phenomenon that each subject tried to assign either of the tasks mainly on the computer.

Discussion. It was found in the system duplicating human elements that the reliability lowered transitorily due to tension and relaxation which occurred according to interference between operators. The phenomenon was not recognized when the partner was a computer but when the partner was a computor capable of communicating with men, it is expected that the same phenomenon as that between men will occur. This can be expected from the phenomenon that operators try to depend on the computer when load is large like 2-task experiment.

The coefficients of variance of intervals between faults found from the time series of operators' faults through the Experiment-I, almost approach 1 as shown in Table 6 and it is known that the occurrences of the operators' faults on the actions from detection to operation is random process. Therefore, if the mean fault interval of an operator is known by tests, the probability distribution

of an operator's making mistakes during a certain period can be expressed by Poisson distribution. This provides a basis for fuzzy expression of human reliability which is necessary ofr calculation in Chapter 2.

4.2 Experiment-II (Offset Cognition Experiment)

In this experiment, one of the meters shown in Fig. 6 was indicated on a graphic display by a computer. Signal which is put in to the meter is shown in Fig. 7 and it is normal noise $N(0,\sigma^2)$ whose mean value is 0 and variance is σ^2 when no offset signal is raised. When a random time elapses after a trial commenced, an offset signal which increases in proportion to time and whose polarity is randomly selected at the beginning of the offset is added to the input signal. (i.e. the mean value of the noise is increased or decreased gradually depend upon the selected polarity of the offset.) The case is assumed in this experiment that a subject is an operator of a plant whose job is monitoring meters on a console panel to find out abnormity of the plant and to stop the plant if he find out the abnormity. The subject was required not to mistake the noise for a offset and to find out the offset as soon as possible. If he find out it, he should respond by pushing a designated key on the computer. One trial ends when the subject responds or the indicator of the meter overruns its scale by the offset. Then the offset signal is eliminated and another trial begins. One test consists of 20 trials. In a series of experiment, subjects participated in several tests for a day.

The two main parameters of the experiment are σ of noise and the meter indication shown in Fig. 6. With regard to σ of noise, any value of 10, 30 or 50% of the full scale of the meter was used. With regard to the meter indication, one of (a), (b), (c) or (d) was used.

The data recorded during the experiment are response time (t_r in Fig. 7) and the number of wrong operations. The following cases are judged wrong operations.

  i) Though the offset does not begin, the subject responded by the designated key.
 ii) The subject responded by the key which is designated as a key for opposite polarity of the offset
iii) The indicator of the meter overruns its scale because the subject does not respond for a considerable time although the offset started.

By the meter indications of (b), (c) or (d), two input signals can be indicated simultaneously. The purpose of this is to simulate the duplication of information. The only difference between two signals is that each signal has independent noise source, but the each noise has same mean value and variance.

Experiment Series C. In order to examine any differences between (a) and (b) of the meter indication in the response time and the number of wrong operations, and also to examine the relation between the Offset Cognition Experiment and Odd-Even Judgment Experiment described in 4.1, following parameters were selected for this series tests.

1) Meter indication: (a), (b).
2) σ of noise: 10, 50% of fullscale.
3) Mode: "Individual", "cooperation", "competition".

For the result, the following null hypothesis H_0 being build up, analysis of variance was applied on three factors of the repetition of trials S, the meter indication M and σ of noise N.

H_0: The response time is not changed by the factor S, M and N.

From the result of analysis of variance on 16 subjects, H_0 could be rejected about N at the significant level of 1% for 10 subjects, also about M for 3 subjects. About S, H_0 could not be rejected at the significance level of 5% for all subjects. Interaction between N and M is recognized. These results is listed up in Table 7.

Experiment Series D. In Experiment Series C, the effect of the meter indication was recognized on 4 subjects at significance level of 5%. Therefore, Experiment Series D was conducted in order to examin further effect on the meter indication by selecting following parameters.

1) Indication of meter: (b), (c).
2) σ of noise: 10, 30, 50% of fullscale.
3) Polarity of offset between two indicators: Same and opposite polarity.

Table 8 shows the result of analysis of variance for the result of this series. + and ++ indicate the rejection of H_0 at the significance level of 5% and 1% respectively.

Experiment Series E. Experiment Series E was carried out in order to examine whether a man watches a meter as a pattern or deals with the meter as a time sequence when he keeps watch on the meter. The parameter used are the following.

1) Meter indication: (a), (b), (d).
2) σ of noise: 10, 30, 50% of fullscale.

In this series, the meter indication (a) indicated the average value of two independent noises. That is, the σ of noise indicated by (a) is $1/\sqrt{2}$ times of others.

Table 9 shows the result of analysis of variance in a same way as Experiment Series D.

Discussion. It cannot be said that the meter indication (a), (b) and (c) and polarity of the offset between two indicators have some effect on response time from the result of Experiment-II. It is well known that the duplication of machine-element is effective in improving the reliability of the systems but the meter indications (b) and (c) was not effective to duplicate the information recognized by man. The meter indication (d), however, recorded minimum response time in Experiment Series E for almost of subjects. Although the difference of response time between the subjects in experiment Series E, it seems that this means

human tends to detect the offset not as a time sequence but as a pattern.

With regard to the number of wrong operations, since the number of the data obtained in the experiment was small, a statistical analysis was not made. From an overview of data obtained, there is a tendency that wrong operation increase soon after tests begin and in the tests in which σ of noise is large.

5. CONCLUSION

The importance of human factors in the safe operation of plants is well understood today, but unfortunately we have no effective way for analysis and the evaluation of man-machine systems.

We suggested, in this paper, a way of combining the probabilistic treatment of machine-elements and the fuzzy treatment of human operators by introducing fuzzy set theory into the reliability engineering such as fault-tree analysis or event tree analysis.

Concerning this method, the fuzziness of human reliability is studied experimentally. Especially the effects of the redundancy of information and of human are examined. Since the human reliability is effected so many factors, we can conclude that it can be more conveniently represented by fuzzy number than probability.

ACKNOWLEDGEMENT

The authors are indebted to Mr. M. Aizu and Mr. H. Wada, graduate students of T.I.T. and U.E.C. respectively, for their contribution to the experimental studies. Without their continuing efforts the laboratory could not be operated. Special thanks go to Kajima Foundation that supports this project through Research Grant.

REFERENCES

[1] Zadeh, L. A. : Fuzzy Sets as a Basis for a Theory of Possibility, Fuzzy Sets and Systems, 1, 3-28 (1978).
[2] Williams, H. L. : Assigning a Value to Human Reliability, Machine Design, July 4, 102/110 (1968).
[3] Swain, A. D., Guttmann, H. E. : Handbook of Human Reliability Analysis with Emphasis on Nuclear Power Plant Applications, NUREG/CR-1278, NRC (Oct. 1980).
[4] Chaudhuri, B. B., Majumder, D. D. : Fuzzy Sets and Possibility Theory in

Reliability Studies of Man-Machine Systems, from Fuzzy Sets, Wang, P. P., Chang, S. K. (Ed.), Plenum Press (1980).

[5] Tsumura, Y. : Fuzzy Fault Tree Analysis, MS. Thesis, Dept. of System Science, Tokyo Inst. of Tech. (1982).

[6] Rouse, W. B., Hunt, R. M. : A Fuzzy Rule-Based Model of Human Problem Solving in Fault Diagnosis Tasks, IFAC/81 World Congress, 75-4, Kyoto (1981).

TABLE 2 Personal Reliability in Each Mode

Mode	Reliability
Individual	0.91 ± 0.0062 *
Cooperation	0.87 ± 0.0081
Competition	0.92 ± 0.0065

* Confidence interval in level of 95 %

TABLE 1 Trial in Each Mode

Mode	Individual	Cooperation	Competition
The Number of Subject	1	2	2
Indication of Trial Results In case of success in pushing a correct button within two seconds	Indication of "SUCCESS"	Indication of "SUCCESS" on the side who succeeds	
In case of fault	Indication of "......."	Indication of "......." on the side who fails	
Buzzer	In case of fault	In case both fail	In case at least one fail
Notice before the test	Require to do his best	Require to cooperate with each other	Require to compete with each other in the success rate

TABLE 3 Varying in Personal Reliability

Mode	Before becoming Skillful	After becoming Skillful
Subject A		
Individual	0.96 ± 0.027 *	0.99 ± 0.011 *
Cooperation	0.95 ± 0.025	0.995 ± 0.0098
Competition	0.97 ± 0.019	0.99 ± 0.014
Subject B		
Individual	0.97 ± 0.017	0.97 ± 0.019
Cooperation	0.93 ± 0.029	0.98 ± 0.019
Competition	0.99 ± 0.011	0.97 ± 0.024

* Confidence interval in level of 95 %

TABLE 4 Tests for an Operator in Experiment Series B

Mode	Rel. of Computer	Number of Tests
Individual		4
Cooperation with weak comp.	$P=r/2$	2
Same Degree Computer	$P=r$	2
Strong Computer	$P=r+(1-r)/2$	2

r: Reliability of Subject in Individual Mode
p: Set Reliability on the Computer

Note for TABLE 7-8
H: Subject
M: Meter indication
N: σ of Noise
D: Offset polarity between two indications
S: Repetition of trial
+ : Rejection of H_0 (5%)
++: Rejection of H_0 (1%)

TABLE 7 Result of Analysis of Variance (Series C)

Factor Subject ID	M	N	S	MN	NS	MS	Factor Subject ID	M	N	S	MN	NS	MS
01		+					10		+		+		
02							11		++				
							12		++				
							13		++				
05	++	++		†			14						
06		++				+	15		++				
07	++	++		+			16	++	++		++		
08		+					17		++				
09							18	+	++				

TABLE 5 Reliability of Operator Cooperating
with a Computer

Mode	Partner	Rel. of Operator
Individual		0.91 ± 0.0086*
Cooperation	Weak computer	0.91 ± 0.0095
	Same degree computer	0.93 ± 0.012
	Strong comp.	0.89 ± 0.010

* Confidence interval in level of 95 %

TABLE 6 Coefficient of Variation
of Fault Intervals

Mode	Coefficient of Variation
Series A	
Individual	0.97 ± 0.29*
Cooperation	1.01 ± 0.43
Competition	0.94 ± 0.31
Series B	
Individual	1.04 ± 0.39
With weak comp.	0.89 ± 0.31
With same deg.	0.75 ± 0.35
With strong	0.93 ± 0.39
(2 tasks)	
individual	0.93 ± 0.10
With comp.	1.07 ± 0.35

* Confidence interval
in level of 95 %

TABLE 8 Result of Analysis of
Variance (Series D)

Subject ID	Noise σ	M	D	S	MD	DS	MS	N
06	10							
	30							
	50							
07	10							
	30							++
	50							
19	10	+						
	30	+						++
	50				+			
20	10							
	30							
	50				++	+	+	
21	10							
	30							++
	50							
22	10							
	30							++
	50							
23	10							
	30	++	+					++
	50							
24	10							
	30				++			++
	50							
25	10							
	30							
	50							

TABLE 9 Result of Analysis of
Variance (Series E)

Noise Factor	10	30	50
H	+ ++	+ ++	+ ++
M	+ ++		+ ++
S			
HM	+ ++		
HS			
MS	+		

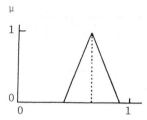

FIg.1. Fuzzy Number.

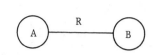

Fig.2. Fuzzy Implication.

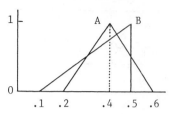

Fig.3. Comparison of F.N.

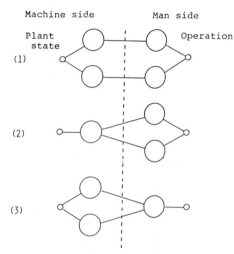

Fig.4. Duplication in
        Man-Machine system.

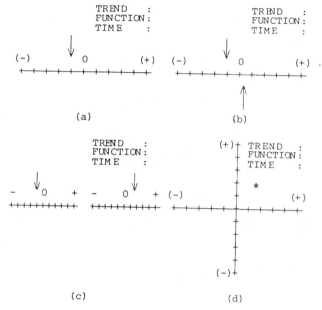

Fig.6. The meter indicaion.

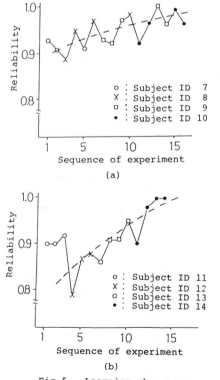

Fig.5. Learning phenomenon
        by observation.

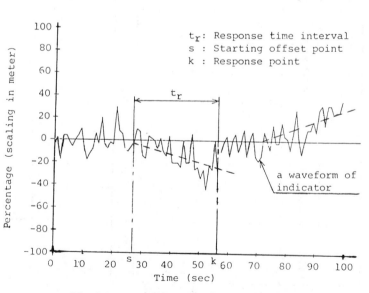

Fig.7. The time sequence of the trial.

DISCUSSION OF SESSION 6

6.1 S MEASURING, MODELING AND AUGMENTING RELIABILITY OF MAN-MACHINE SYSTEMS

Johannsen: You have shown cuts in the model of the system partitioning this into parts, in order to separate failed parts. With different cuts, redundancy can also be achieved. How do you measure and display the many variables generated through the proposed technique for the human operator?

Sheridan: The display is computer-graphic, one or more VDUs. A mimic of the whole system is shown. The operator may select which "cut" he would like to make and sees both the raw data discrepancy for each side of the cut as well as a Bayesian odds ratio of failure/no-failure for each side.

Nzeako: How do we know at which points (parts of the system) to select feedback information to be presented to the human operator, and how is the feedback error weighted against the consequence (economic, safety, and/or personal consequence) of the error (human error) ? In other words, what is the relationship between the magnitude of the feedback error and the consequence of the human error?

Sheridan: This is a good point. Somehow the importance of a discrepancy from the model must be weighted. Some parts of the system may fail or cause damage with smaller "error" than others.

6.2 T HUMAN RELIABILITY AND SAFETY EVALUATION OF MAN-MACHINE SYSTEMS

Question: Do you think that the differences in subjects' reliability between "cooperation" mode and "competition" mode are dependent on nationality or culture? Would you expect other results if the same experiments were carried out outside Japan?

Terano: In my personal opinion, there exists more or less the correlation between human reliability and nationality. Generally speaking, the Japanese aspires to promote himself as much as he can. In such a case, European people intend to change the circumference surrounding them, but the Japanese tries to achieve the goal by increasing his ability. This attitude may be the reason why the subjects competed even in the cooperation mode. We have studied the relation between the human reliability and the personality by checking the results of psychological tests, but no clear relations were obtained. One of the interesting trends is that the older subjects were much more cooperative than the younger one, anyway.

DESIGN AND EVALUATION OF A WIND SHEAR INDICATION SYSTEM FOR TRANSPORT AIRCRAFT

F. V. Schick and U. Teegen

Institute for Flight Guidance, DFVLR, Braunschweig, Federal Republic of Germany

Abstract. Wind shear can be a serious hazard for transport aircraft
operation on take-off and landing, thus threatening flight safety.
A questionnaire campaign concerning airline pilot's experiences and
views of the problem provided basic information for the design of a
wind shear indication system for manual flight. Particular attention
was given to human factors design requirements, especially in the
display layout. In a flight simulator experiment, a human factors-
oriented display version and a flight mechanics-oriented display version
were evaluted, in relation to the conventional cockpit instrumentation.
Twelve airline pilots performed up to 45 landing approaches under wind
shear conditions with the different display versions. Results are based
on flight data, pilot visual scan behaviour and judgement data. The
human factors-oriented version showed advantages over the flight
mechanics display version and conventional instrumentation.

Keywords. Human factors; display systems; aircraft instrumentation;
wind shear; experimental evaluation.

INTRODUCTION

Several severe aircraft-accidents due to wind
shear within the last decade caused an inten-
sification of research efforts on this phenomenon. To increase flight safety, e.g. by improvement of technical aids for flight guidance, national research programs were initiated all over the world, starting in the early
seventies (Shrager, 1976) up to this study,
which was initiated by the Federal Ministry
of Transportation, the theoretical research
being generally supported by the Deutsche
Forschungsgemeinschaft (DFG). Three institutions participated in this study, Technical
University of Braunschweig, Bodenseewerk
Gerätetechnik Überlingen and DFVLR Braunschweig.

FUNDAMENTALS OF WIND SHEAR

The term "wind shear" refers to several meteorological conditions, which are generally
characterized by low-frequency changes of
wind-speed or wind-direction, or both, relative to a defined track, e.g. the flight path
of an airplane. As long as an aircraft is
crusing in high, 'safe' altitudes, wind shear
is not a problem for flight safety, but might
be a handling problem to the pilot. However,
when the aircraft is operating in low altitudes during take-off and landing, wind shear
may become a safety-related issue, if considerable amplitudes of wind shear and aircraft-
or pilot-related delays combine to produce

a hazardous situation.

Meteorological Scenario

Figure 1 illustrates typical weather conditions that are closely associated with the
occurence of wind shear (König, 1980). Within
thunderstorm activity, horizontal and vertical
wind shear is most likely, the combination of
which may lead to hazardous situations for
aircraft, as several fatal accidents indicate
(Whitmore, Cokely, 1976). But wind shear is
not only associated with stormy weather, as
the following conditions point out. Precipita-
tion from high cumulus clouds into hot and
dry air may evaporate and thus produce a cold
downstream, which flows off horizontally near
the ground. Also, cold front activity may

Fig. 1.   Typical meteorological scenario for
wind shear conditions (thunderstorm).

procedure considerable wind shear. Other wind shear conditions include low level jet streams and temperature inversions, or conditions caused by surface roughness and obstacles in the vicinity of airports.

Flight Dynamics Considerations

Effects of wind shear produce essentially airspeed and glide-path deviations. As long as the aircraft is forced above glide-path and/or airspeed increases, there is no particular hazard for aircraft operation. However, airspeed loss and/or operation below glide-path demands an increase of thrust to avoid dangerous situations. It is required to stay close to the flight path under all circumstances, even approaching stall speed, combined with large pitch angles.

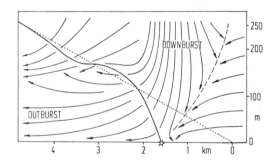

Fig. 2. Effect of severe wind shear on a
         landing approach.

Figure 2 shows a hypothetical example utilizing a thunderstorm condition, that led to a fatal accident in reality. Assuming that the airplane is trimmed according to the initial, regularly constant wind condition, the landing approach of a transport aircraft is simulated under the hypothetical premise of fixed controls, that means under absence of pilot inputs. With no wind changes during approach, the aircraft will maintain a constant airspeed and follow the prescribed flight path (dotted line). A severe wind shear, as outlined in Fig. 2, would, however, affect the aircraft's flight path in the indicated way (solid line), and lead to a 'short landing' crash before runway threshold. The glide-path deviations result from airspeed differences to reference airspeed due to horizontal shear, and from down-draughts. Airspeed deviations in this case reach values of about ±20 kts. This particular example demonstrates the necessity of informational aids, which the pilot can utilize for the compensation of wind shear effects. Only manual flight control will be considered here, because automatic flight control, which provides in general a means to compensate wind shear effects, is indeed often rejected by the pilot, for operational and passenger comfort reasons.

Recognition of Wind Shear

Time delays due to the pilot's unawareness of wind shear or his decision about how to react, will intensify the effects of wind shear. Thus, even relatively harmless wind shear conditions may become hazardous. The immediate recognition of wind shear by the pilot is therefore necessary. As can be shown under automatic control, the pilot is in principle, provided with information for recognition of wind shear. However, this information may be confusing because of the apparent incompatibility between the readouts and the pilot's own experience. The lesson is, that there exists a lack of ergonomic design in aircraft instrumentation, for enabling pilots to cope with wind shear by manual control of the aircraft. Warnings, e.g. by Air Traffic Control or other aircraft crews, as well as recognition of meteorological or local conditions associated with a high wind shear probability, provide the pilot with information about the fact, in general. However, quantifiable cues about its effects on the aircraft, or about pilot reactions required, are not given. An ergonomic design of aids for wind shear indication must include these considerations.

ANALYSIS OF PILOT JUDGEMENTS ON WIND SHEAR

The analysis of human operator judgements and opinions in this context was expected to be useful in two respects. Subjective judgement techniques are not only beneficial for investigations on the acceptance of, or satisfaction with, an existing man-machine interface, but also of considerable value for the design process, as they can aid decisions on display concepts and on the definition of valid experimental procedures for the evaluation of these concepts.

Two successive phases, with different procedures for judgement data collection and analysis, were made here. The first one dealt with wind shear phenomena in general, and the associated operational and ergonomic problems in the cockpit, as they have been experienced by airliner crews. The second phase dealt with pilot acceptance of an onboard wind shear indication system, where judgements were made on the basis of flight simulator experience with this equipment.

The latter will be treated in the context of the simulator tests, whereas the following describes the first phase.

In the beginning of the research program it was recognized that, apart from theoretical considerations on the flight mechanics and physics of wind shear, knowledge of the general attitudes of aircrews towards wind shear, including the identification of variables of wind shear situations, as they are perceived by cockpit crews, was another prerequisite of the study. For this purpose, inquiries of a large sample of cockpit personnel seemed necessary.

Questionnaire Development

A written questionnaire, with predetermined answer categories, providing a standardized inquiry form was given preference over any informal techniques, such as interviews. In the case of large samples, the superiority of a formal questionnaire is obvious, for ease of application, as well as handling of large data quantities, which can be computerized easily.

The preparation of the questionnaire started with a review of theoretical literature on wind shear, accident reports, and publications on warning devices. This resulted in a preliminary list of important aspects of the topic. The list served as a guideline for an extensive group discussion with four airline pilots, which was held in order to identify problem areas in detail, and to clarify terminology.

The analysis of the discussion facilitated the formulation of a set of questionnaire items, which met the intention to cover the entire problem of wind shear from a pilot's point of view.

The final form of the questionnaire consisted of 63 rating scale items, i.e. 63 statements, each one to be evaluated on a six-point rating scale. The main topics in the questionnaire were

- definition and identification of operationally significant wind shear situations,
- relative estimates of their potential hazards,
- the role of wind shear warnings by radio communication,
- other perceptual cues for the recognition of wind shear,
- identification of behavioral strategies,
- general operational requirements for on-board wind shear indication systems.

Figure 3 gives an example of a questionnaire item.

9. In aviation there are many situations more dangerous than wind shear.

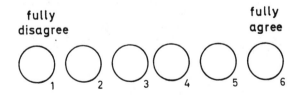

Fig. 3. Questionnaire item.

Additionally, the subjects were asked for a short description of any case of wind shear, which they might have personally experienced.

Questionnaire Campaign and Results

The questionnaire was mailed to all members of the German airline pilots association, the 'Vereinigung Cockpit'. 570 members, all in active airline service, filled in and returned the questionnaire.

The judgements were processed automatically. For each item, the rating distribution was computed, i.e. the frequencies of the ratings on scale points 1 (total disagreement) to 6 (total agreement), and the median of the rating distribution were calculated. Figure 4, below, gives an example of the rating distributions of three items. Furthermore, each distribution was submitted to a significance test, the Binomial Test (Siegel, 1956), in order to check whether there was a general agreement tendency (cumulation of ratings on scale points 4 to 6), or a disagreement tendency (preference of points 1 to 3).

A very clear and differentiated picture of pilot attitudes could be obtained already from an analysis of the total sample. Subgroups - e.g. captains and co-pilots - generally did not show any characteristic differences in their rating behavior.

To pick out from the many significant rating distributions a few items, which were immediately relevant for the further work, there was, for instance, a high rank given to the wind shear problem in general: only 38 % of the total sample agreed with a statement saying that in aviation there are many situations more dangerous than wind shear, whereas 96 % agreed to the statement, that an urgent necessity for research in this area exists.

The potential hazards of wind shear were also rated rather high. 56 % tended to estimate wind shear during take-off to be more dangrous than an engine failure, and 78 % estimated wind shear during the final approach more dangerous than an engine failure.

Headwind shear, which seems to be a particularly significant case, was also rated as being more dangerous on approach than on take-off, by 69 % of the subjects.

Information like this was very helpful for the proper layout of the simulator experiment, particularly for the selection of appropriate wind shear profiles to be simulated. However, still more valuable in this respect were the descriptions of self-experienced wind shear situations. The descriptions, which were given in addition to the ratings, provided an excellent source of detailed information.

The questionnaire items also yielded several significant findings regarding displays. For instance, predominant behavioral strategies were indicated by a significant preference of the airspeed indicator for speed regulation on the approach (96 % agreed), as compared to

the fast-slow indicator (only 28 % agreed).
Moreover, 84 % preferred manual control over
automatic control on approaches with potential
wind shear.

Positive attitudes towards an on-board wind
shear display were expressed directly (96 %
tended to support this idea), as well as indirectly: 71 % agreed to a statement saying
that today's aircraft equipment is not sufficient for always safe passage through strong
wind shear. Although warnings via radio communication were estimated as very useful
(83 %), the use of that type of warning was
judged significant lower, compared to the
potential use of an on-board warning device,
by 82 % of all subjects.

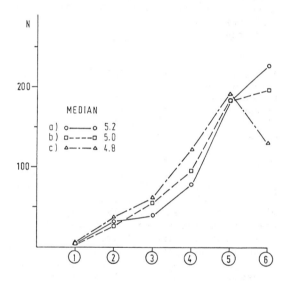

Fig. 4. Rating distributions for three
         questionnaire items.

More detailled suggestions for display layout
were obtained from the following items. For
instance, a display with low information content
(yes/no type) was strongly rejected (90 %). As
regards the presentation of quantitative information, the rating distributions given in
Fig. 4 showed, that (a) a display of raw data
would be accepted most (87 % agreed with the
statement "a wind shear display should indicate the change of wind velocity and direction",
by ticking scale points 4, 5 or 6); (b) a display of processed data would be accepted in
second place (84 % agreed with the statement
"a wind shear display should indicate the effects of wind shear on the aircraft's flight
path"), whereas (c) a display of further processed data would be accepted only in third
place (81 % agreed with the statement "a command display is best to produce immediate
pilot reaction on wind shear"). Although not
statiscally tested, there seemed to be an
inverse relationship between level of signal
pre-processing and pilot preference. That is,
pilots preferred a display which required them
to do more of their own information processing.

MANUAL CONTROL AND DISPLAY CONCEPT

An effective aid for pilot warning and guidance in the wind shear environment basically
contains a control strategy, filter algorithms
and an appropriate display concept. Each of
these has to be adjusted to general ergonomic
and flight mechanics requirements, corresponding to the pilot's desires as far as
possible. Only an optimal solution, in all
these respects, will be acceptable for use in
aircraft cockpit, and will be able to improve
flight safety. The following considerations
are based on on-board measurement of wind
shear, utilizing conventional sensoring and
instrumentation, this allowing, in principle,
an installation of such a wind shear indicating component in today's aircraft generation.

Manual Compensation of Wind Shear

As König and others (1980) point out, airspeed and glide-path deviations due to wind
shear may be most effectfully prohibited by
utilizing a specific energy and energy rate
control concept. Energy rate contains airspeed acceleration and vertical speed error,
indicating an actual lack or excess of thrust.
Energy contains airspeed and height errors,
indicating stationary offsets of these variables, and thus representing a warning signal.
The processing of all these data includes the
relation to flight phase reference data. Additional measurement of wind data is not required, because they are of less importance,
as compared to the energy and energy rate
signal.

An essential problem with the design of a
wind shear indication system arises from the
required filtering of the signals. To avoid
unnecessary warning and reaction of the pilot,
gusts and low amplitude wind shear, which represent no hazard to aircraft, have to be separated from evident wind shear effects. An
unconventional solution proposed by König
(1981) fits the requirements of practical
operation in aircraft most effectively. It
consist of a nonlinear filter algorithm evaluating the energy rate error as a function of
energy error, and thus corresponding to the
hazard.

This control and filter concept has proven
its effectiveness in computer simulations,
with a proportional control pilot model containing a time delay term. Signal processing
is directed to thrust adjustment, compensating
energy error and thus prohibiting dangerous
situations.

Display Versions

The wind shear display design was based on the
utilization of existing cockpit instruments,
because of the lack of free space in the
front instrument panel. The implementation of
the aforementioned control and filter techniques led to a twofold display concept.

- A flight mechanics-oriented display version, which assigns the display of energy and energy rate to seperate existing displays, together with similar information, and thus facilitates a more direct understanding of the physical parameters. Energy rate is indicated by means of a striped second pointer in the Vertical Speed Indicator (VSI, lower right in figure 5). The energy signal is directed to a modified Fast/Slow-Indicator (Top center in figure 5, left vertical scale). Figure 6 shows a more detailed view of the ADI display.

Fig. 5. Flight mechanics-oriented display concept (ADI/VSI).

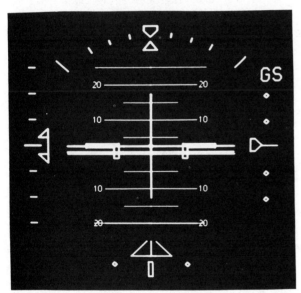

Fig. 6. Attitude Director Indicator (ADI)

- A human factors-oriented display version, which combines both signal parts, energy and energy rate, utilizing only the modified F/S-Indicator within the ADI. That is, the VSI was left in its standard unmodified form. This version is mainly influenced by data on scanning behavior of airline pilots in approaches (Spady, 1978), which indicate less importance of the VSI for flight guidance. Last but not least this provides a direct relation between indicator readout and the required control activity of the pilot.

The Fast-Slow Indicator is modified, because it is used as a command display in both versions. Tolerance limits are given by the fixed spacing between the pointer tips.

The comparison of these display concepts with other wind shear indication systems, some of which are already commercially available, show similarities of design philosophy. The flight mechanical concept is directly related to the proposals of SMITHS INDUSTRIES and SFENA, all of them using the Vertical Speed Indicator, with different layout of pointer and scale, however. An FAA proposal utilizes Fast-Slow indication with modified control characteristics (Gartner and others, 1978).

SIMULATOR EVALUATION OF
WIND SHEAR DISPLAYS

To compare the display concepts presented above in their effects on performance, workload and pilot acceptance, an experimental evaluation in a moving cockpit flight simulator was thought to be best suited for this purpose. This can provide operational conditions which are closer to reality, as well as completely invariant reproductions of flight task scenarios.

The simulator cockpit is equipped with all facilities for manual instrument approaches. Attitude Director Indicator and Horizontal Situation Indicator are on monochromatic CRT displays, to permit simple modification of experimental conditions.

Design of the Experiment

Factors. The different display versions, in combination with different wind shear situations resulted in a two-factor design. The display factor consisted of three levels: the flight-mechanical display version (ADI/VSI), the ergonomic display version (ADI), plus a conventional display version with no wind shear indication provisions (CONV). The latter was added to obtain baseline reference data. Four experimental levels of a wind profile factor were specified: an inversion with moderate gusts, and horizontal wind shear only (INV); a thunderstorm situation with severe gusts, and extremly strong horizontal and vertical wind shear (TS1); another thunderstorm situation, also with severe gusts, but a more moderate horizontal and vertical wind shear (TS2); an additional profile with gusts only, to provide a comparison baseline (G).

Subjects. Twelve airline pilots were selected as test subjects. The sample space, which was made up by the 570 participants of the questionnaire campaign, had been restricted to those pilots with Airbus A300 experience (67 pilots). Because the aerodynamic model, underlying the aircraft simulation, was of the A300 type, that group of pilots was expected to have no difficulties with getting accustomed to the type-specific control- and movement dynamics of the simulator. Out of the 67, six captains and six co-pilots were selected at random, to serve as test subjects.

Test runs. Following a sufficiently large number of training runs, each pilot then performed between 40 to 45 final approaches. The
experimental factors were completely crossed,
so, all 3 x 4 combinations of the above specified display versions and wind profiles were
included. The approach runs started at 1200 ft
altitude and seven km from the runway threshold. The pilots had been instructed to refrain
from go-arounds, which they would have normally made in actual flight. The runs were performed in three blocks, with one display
version per block. Within these blocks, the
four evaluation wind profiles were presented,
together with additional profiles, in a
pseudo-random sequence, to avoid anticipatory
control behavior from "learning the task"
effects.

Measurements and Data Analysis

Pilot control actions and judgements, as well
as simulation data were recorded as measurements of activity, accuracy of task accomplishment, visual load and pilot acceptance.

Performance and accuracy data. During all
approaches, flight data (such as glide slope
and speed deviation), control inputs (e.g.
elevator- and throttle position), and display
data (e.g. energy error, energy rate error)
were recorded continuously, at a rate of 10
cps, together with the corresponding wind
data (horizontal and vertical wind velocity).
A total number of 23 variables was recorded
this way. At the end of each run, statistical
parameters (mean, variance, minimum and maximum values) of each variable were computed.

As a first step, these parameters were checked
whether any statistically significant differences existed, due to general effects of the
display concepts, the wind profiles, or due
to specific interactions of both factors.
(A more thorough analysis of the time-oriented
records will follow, including correlations
between flight/display data and control inputs).

As could be expected, rather marked effects
of the different wind scenarios were found.
The TS 1 profile produced the most unfavourable data, i.e. the highest flight path and
speed deviations, most control activities,
etc. Unfortunately, there was a very high
variance in these measurements. Therefore,
some trends in the analyses of the display
factor, which indicated a superiority of the
ADI version, were not statistically significant. Interaction effects, which might have
indicated differences of display effectiveness
due to particular wind shear scenarios, were
also absent. The reason for the unduly high
variances was, most probably, an invalid a
priori assumption of homogeneity of pilot
control behavior. The data suggested that individually different control strategies or
techniques were applied by the pilots, which
had been shaped through longterm training and
experience. A classification of three distinct
control techniques seems to be possible: a
general preference of elevator control over

thrust control, the reverse case, and an
equivalent combination of elevator and thrust
control. Analyses of the time-line records are
expected to confirm this classification and
to provide a basis for more differentiated
investigations on the relative effectiveness
of the display concepts.

Visual behavior data. Each pilot's visual
scanning behavior was recorded during four
of his approaches, by means of an Eye Mark
Recorder (NAC model IV). This eye-point-of-
regard system provided registration of the
subject's eye fixation point, marked as a
light spot on the instrument panel, which was
recorded on video tape. The records permitted
computation of eye fixation frequencies and
dwell times on all instruments, and rate of
look transitions between instruments (Schick,
Radke, 1981).

As already found in the performance and accuracy data, analyses of variance of the visual
scan parameters also showed a high degree of
variability between the individual subjects.
Since the scanning sequences were recorded
continuously, an analysis of correlations
with flight data and control input data will
be possible.

For the moment, it can be stated that, generally, the visual scan data corresponded closely to those found by Spady (1976) on manual
instrument approaches: the Flight Director
was used overwhelmingly as the primary instrument, occupying between 67 and 72 per
cent of dwell time (see Table 1), and look
transitions occured nearly exclusively between
the Flight Director and other instruments,
whereas transitions between other pairs of
instruments were very rare events.

TABLE 1 Mean Percentages of Visual Dwell
Time on Flight Instruments

Display Concepts

	CONV	ADI	ADI/VSI
Flight Director	70.8	72.3	66.8
Fast-Slow	8.3	10.1	7.1
Glide Slope	4.8	4.2	4.0
Airspeed	5.9	4.3	5.0
Course/Heading	1.4	0.5	0.6
Vertical Speed	1.0	1.0	8.0+)
Altimeter	0.8	0.8	0.5

+) Significant differences between means (p<.01)

As shown by table 1, the time spent for monitoring the Vertical Speed Indicator increased
significantly from one to eight per cent,
under the wind shear display version ADI/VSI,
with the second pointer in the VSI, apparently
at the expense of all other flight instruments,
as compared to conventional instrumentation.
Consequently, the pattern of look transitions
was also changed significantly under the
ADI/VSI concept, with the transitions between
Flight Director and Vertical Speed Indicator
increasing from about three per cent - observed
both under conventional instrumentation and
the ADI wind shear display concept - up to 26
per cent. So, the visual behavior data in

general suggest that the ADI version requires only minor changes of the scan pattern, which has been learned and habitually applied with conventional instrumentation. The ADI/VSI version, however, implies more far-reaching consequences, due to the necessity of additional monitoring of the Vertical Speed Indicator, which is normally of marginal importance.

Judgement data. In contrast to the earlier large-scale questionnaire campaign, only twelve subjects took part in the simulator experiment, but their judgements were made on the basis of actual experience with the particular wind shear equipment. This required a different method for the collection and analysis of judgement data. From the aspect of pilot acceptance, namely the layout of the display versions in detail, and the percieved effects on flight task accomplishment, a list of items was set up, which was presented to the pilots in the form of open-answer questions. A few additional items were included, to check the acceptance of the simulator environment and the selected flight tasks, and thus the validity of the experiment.

The pilots had to give written comments in their own words, on each of the 21 questions. The comments were evaluated by means of content analysis, which was used to classify the comments in three categories (positive, negative, ambivalent), and to identify sample comments, i.e. multiple quotations of equivalent contents.

Regarding the simulation tasks, the comments revealed a high acceptance. For instance, most pilots commented that they had not missed any further important case of wind shear in the experiment. Although the judgements on the simulator environment were generally positive, too, eight pilots criticized the absence of colour on the CRT displays, and four subjects complained about restricted legibility of the engine instruments in the simulator.

With respect to the effectiveness of the two wind shear displays, both versions were judged to be a major improvement of the normal instrumentation. Both versions were judged to facilitate the observance of reference speed, and the decision to make or not to make a go-around. However, different judgements were made concerning glide slope abservance. While the ADI version was judged positively, the ADI/VSI version was judged generally negative in this respect. Five pilots specified as the reason an inconvenient instrument scan, since the VSI is located far from the primary flight instrument, the ADI.

Concerning the display layout more specifically, pilots totally approved the omission of conventional fast-slow indication in favour of a wind shear scale. A number of negative comments on the ADI/VSI concept were observed, which mainly dealt with the unfavourable instrument scan (nine quotations), the learning process required by that (three quotations), a missing aid for fine adjustement of thrust

(three quotations), and a difficulty to relate the VSI readout to the correct control inputs, when both pointers are moving rapidly (two quotations).

Asked for a statement on possible improvements of the display concepts, more comments were made on the ADI version. So, for instance, some pilots recommended to suppress minor deflections of the wind shear pointer, as well as to widen the range of the underlying scale, and thus to avoid very large deflections of the pointer unless wind shear influence is really extreme.

CONCLUSIONS

Experimental measurements and pilot judgements support the conclusion that a quantitative display of information on wind shear contributes to an improvement of flight safety. A comparative evaluation indicated superiority of a human factors-oriented display version (ADI) over a flight mechanics-oriented version (ADI/VSI). Pilot comments showed a relatively lower acceptance of the ADI/VSI version, and visual behavior data indicated an additional load, since monitoring the VSI consumed visual capacity normally used to monitor the primary flight instruments, and led to an unusual scanning pattern. The human factors-oriented ADI wind shear display had better warning qualities, because of its location within the primary instrument, and, only minor deviations from the usual scan pattern suggested that an upgrading of present-day-aircraft instrumentation by such a display would not require specific pilot training. The supposed advantage of the flight mechanics display concept, i.e.,the presentation of two signal parts in two instruments, together with physically similar information, was not acknowledged in the pilot comments. Rather, the prejudice, found in the first questionnaire campaign, against presentation of highly preprocessed signals, as in the human factors-oriented display version, could be eliminated by the experimental evidence.

However, the layout of this concept is not yet optimal either. Future work on the quantitative weighting and combination of the two signals parts (energy error, energy rate error) in one pointer seems necessary, for optimizing the interrelations between wind shear effect, pointer deflection, and aircraft-specific parameters, with respect to pilot's perception, information processing and control behavior. This would provide a command display for wind shear compensation by well-balanced, manual thrust regulation. This work would require no structural changes of the basic concept, but could be done by modifying the display software. Analyses of sequential correlations of the experimentally obtained flight data, control data, and visual scan data would provide the basis for these modifications. The use of modern, multi-colour CRT-displays is also expected to play an important role in this development.

REFERENCES

Gartner, W.B., D.M. Condra, W.H. Foy, W.D. Nice, and C.E. Wischmeyer (1978). Piloted flight simulation study of low-level wind shear, phase 3. FAA-RD-79-9.

König, R. (1981). Verfahren zur Verbesserung der Flugsicherheit unter Scherwindbedingungen. SFB-Colloquium, Sept. 9th - 10th, 1981, Braunschweig, Germany (not published).

König, R., P. Krauspe, and G. Schänzer (1980). Procedures to improve flight safety in wind shear conditions. 12th congress of the International Council of Aeronautical Sciences, ICAS-80-22.3, October 12th - 17th, 1980, Munich, Germany.

Schick, F.V., and H. Radke (1981). A method for semi-automatic analysis of eye movements. In J. Moraal and K.-F. Kraiss (Eds.), Manned Systems Design, Plenum Press, New York. pp. 221-234.

Shrager, J.J. (1976). Wind shear: literature search, analysis and annoted bibliography. FAA-RD-76-114.

Siegel, S. (1956). Nonparametric statistics for the behavioral sciences. McGraw-Hill Kogakusha, Tokyo. 312 pp.

Spady, A.A. (1978). Airline pilot scan patterns during simulated ILS approaches. NASA technical paper 1250.

Whitmore, C.A., and R.C. Cokely (1976). Wind shears on final approach. 29th Annual International Air Safety Seminar and Aviation Safety Technical Exposition, October 25th - 29th, 1976, Anaheim, California.

ANALYSIS OF RIDER AND SINGLE-TRACK-VEHICLE SYSTEM, ITS APPLICATION TO COMPUTER-CONTROLLED BICYCLE

M. Nagai and K. Iwasa

Department of Mechanical Engineering, Tokyo University of Agriculture and Technology, Koganei, Tokyo 184, Japan

Abstract. Single-track-vehicles are dynamically more unstable and more complicated than automobiles, so that a rider must play a more important role in a man-machine system. The purpose of this paper is to analyze the lateral and directional stabilities of this rider-cycle system, and then to make a laboratory facility for a computer-controlled cycle. A dynamical model is presented in a state space form, as a feedback control system with two inputs, handlebar steering and rider's upper body leaning. A rider control model is a preview output feedback model. This closed loop system is analyzed with the automatic control theory. The experimental bicycle is made, which is controlled by a micro-computer. The state of the cycle is measured with an acoustic method. The cycle is designed to run on a moving belt apparatus. According to the lane change experiment, the cycle runs successfully, and the results of dynamical responses are proved by the theoretical analysis.

Keywords. Vehicles; man-machine system; closed loop system; stability; microprocessors; computer control; acoustic measurement; robots; bicycle.

INTRODUCTION

Single-track-vehicles, i.e. motorcycles and bicycles, present unique problems of stability and control, requiring more or less continuous attention by a rider. Because of dynamical unstability and complication of single-track-vehicles, rider handling behaviour plays a more important role than it does for automobiles. To realize the directional and lateral stability, the rider operates cycles with handlebar steering and rider's upper body leaning. Therefore the handling performance of the single-track-vehicle is a significant factor as much as the mechanical property of it.

The handling of motorcycles and bicycles has long been of practical and theoretical interest. Iguchi (1962) has presented a simple mathematical model to explain fundamentally the directional and lateral stability of the closed loop of rider and two-wheeler system. He has successfully described this problem with a stability theorem, but excluding the rider's upper body leaning performance. Kobayashi (1974) has practically simulated the rider-cycle system with a digital computer from a view point of accident avoidance. Weir (1979) has reviewed the recent researches and studied the analysis of rider behaviour and vehicle dynamics, with a recently developed high-order dynamical model, in order to design desired handling properties. Aoki (1979) has experimented the steering performance by field tests, and estimated the effects of the driver's upper body leaning on the handling

properties.

Although there are many studies about rider and cycle closed loop system, it is not enough to estimate comprehensively the properties because of the openness of the control structure problem and the difficulty of experiment. The purpose of this paper is to realize a driving robot which functions are similar to a human rider, and to present an analysis of stability and controllability of this closed loop dynamical system structure. In order to consider the feedback control structure by automatic control theory, the simplified dynamical model has been presented.

MODEL AND ANALYSIS

Model of Cycle

To deal with the dynamical properties of single-track-vehicles in detail, high-order equations of motions have been proposed, including front steering structure, tire characteristics, frame flexibility, etc. (e.g. Kondo 1963, Sharp 1975). These high-order models are useful for studying the dynamical performance of the single-track-vehicle itself. But in order to find the man-machine system structure, it is not essential to use these models, because of the limitation of human operation. Namely human cannot respond to high frequent vibrations, so that a property in high frequency does not play a significant role in a man-machine stability problem.

In this paper, to extract the fundamental properties, that is, directional and lateral stability, the model of single-track-vehicle is simplified on the assumptions as follows.

   (1) Side slip angle of tire is very small, and the motion in the horizontal plane is due to geometrical relations.

   (2) Roll motion is analoguous to an inverted pendulum.

   (3) All dynamical equations can be linearized.

From these assumptions, equations of motions in the horizontal plane are described as,

$$\dot{r} = B_1 \cdot V \cdot u_s \qquad (B_1 = q/L) \qquad (1)$$

$$\dot{y} = -B_2 \cdot V \cdot u_s - V \cdot r \qquad (B_2 = q \cdot L_r/L) \qquad (2)$$

where u_s denotes a handlebar angle input, r yaw angle, y lateral position and V vehicle forward speed. The equation of roll motion is described as,

$$I \cdot \ddot{\theta} + m \cdot h \cdot \ddot{y} + m_h \cdot h_h \cdot \ddot{u}_1$$
$$= m \cdot h \cdot g \cdot \theta + m_h \cdot g \cdot u_1 \qquad (3)$$

where θ denotes a roll angle, u_1 a lean angle (or equally lateral position of upper body center of gravity). (cf. Table 1)

Let the input variables to the cycle be a handlebar angle u_s and a rider's upper body leaning angle u_1, then the open loop equation of motions is described in a state space form (cf. Appendix).

$$\begin{vmatrix} \dot{r} \\ \dot{y} \\ \dot{v}_1 \\ \dot{v}_2 \end{vmatrix} = \begin{vmatrix} 0 & 0 & 0 & 0 \\ -V & 0 & 0 & 0 \\ 0 & 0 & 0 & 1 \\ 0 & -w^2 & w^2 & 0 \end{vmatrix} \cdot \begin{vmatrix} r \\ y \\ v_1 \\ v_2 \end{vmatrix} + \begin{vmatrix} B_1 V & 0 \\ -B_2 V & 0 \\ 0 & 0 \\ 0 & F \end{vmatrix} \cdot \begin{vmatrix} u_s \\ u_1 \end{vmatrix}$$

$$F = (f_\theta - f_u) \cdot g \qquad (4)$$

Output of Cycle

Although it has not been found exactly which variables are directly perceived by a rider, it may be said that at least two variables must be perceived for stabilyzing both roll motion and directional motion. Directional motion can be visually perceived like an automobile driver, and roll angle of cycle can be vestibularly perceived. It may be possible that other variables, such as accelerations or velocities, are also perceived. But as described later, all state variables are proved to be observable in this 4th-order dynamical model from these two variables. Therefore two outputs are defined here to be a roll angle θ and a lateral position of a previewed point y_f, as follows,

$$\begin{vmatrix} y_f \\ \theta \end{vmatrix} = \begin{vmatrix} -\ell & 1 & 0 & 0 \\ 0 & -w^2/g & w^2/g & 0 \end{vmatrix} \cdot \begin{vmatrix} r \\ y \\ v_1 \\ v_2 \end{vmatrix} + \begin{vmatrix} 0 & 0 \\ 0 & -f_u \end{vmatrix} \cdot \begin{vmatrix} u_s \\ u_1 \end{vmatrix} \qquad (5)$$

where y_f denotes a lateral position of the

point at previewed length ℓ ; $y_f = y - \ell \cdot r$; (by the first order prediction). Now the 2-inputs and 2-outputs linear dynamical system equation is obtained in a state space form.

$$\dot{X} = A \cdot X + B \cdot U \qquad (4)$$

$$Y = C \cdot X + D \cdot U \qquad (5)$$

Compensation of Rider

It is not so difficult to ride on a cycle when it does not run very fast. Especially bicycles are easily driven even by children. The fact means that the riding task may not require a complicated skillfulness for human driver. In the meantime, according to the ergonomic research (Iguchi 1970), proportional compensation is the most important and easiest task among the human continuous operation. Then, for the simple feedback control structure, it is assumed that a rider steers the handlebar and leans the upper body, in proportion to roll angle and lateral deviation of previewed point. This relation is described as,

$$\begin{vmatrix} u_s \\ u_1 \end{vmatrix} = -\begin{vmatrix} h_y & h_\theta \\ g_y & g_\theta \end{vmatrix} \cdot \begin{vmatrix} y_f \\ \theta \end{vmatrix} ; \quad U = -G \cdot Y \qquad (6)$$

This feedback control is namely the output feedback control. On the other hand, the skillful rider can drive a cycle optimally. From an optimal control theory, to minimize a certain cost function, it must be required to feed back the all state variables.

$$U = -G^o \cdot X \qquad (7)$$

where X is the observed state variables from output and input variables Y and U, by e.g. Luenberger observer, and G^o is the optimal gain to minimize the cost function.

Observability and Controllability

It is an important and interesting problem where the structure of the system Eqs.(4),(5) is theoretically observable and controllable. If this system is observable, the whole state variables can be obtained from output variables. If it is controllable, the closed loop characteristics can be changed to any desirable one by state feedback control.

For the system Eqs.(4),(5), a necessary and sufficient condition for observability is that the matrix

$$U_o = (C^t \ (CA)^t \ (CA^2)^t \ (CA^3)^t)^t \qquad (8)$$

should have a rank 4. This condition is satisfied when a forward speed is not zero, i.e. V > 0. In the same way, a necessary and sufficient condition for controllability is that the matrix

$$U_c = (B \ AB \ A^2B \ A^3B) \qquad (9)$$

should have a rank 4. This condition is satisfied when a forward speed is not equal to a critical speed

$$V_c = L_r \cdot w \qquad (10)$$

or when a matrix element B(4,2) in Eq.(4) is

not equal to zero. The critical speed depends
on the time constant of roll motion, $T = 1/w$,
and the longitudinal distance between rear
wheel and center of gravity L_r. The value V_C
is practically a very low speed. This means
that at low speeds it is not so easy to control a cycle in order to realize a desirable
closed loop characteristics without upper
body leaning. But the second condition for
controllability is practically satisfied.
Therefore, even if a cycle runs at low speeds,
the system is controllable not only by handlebar steering but also by upper body leaning.

Stability

Because the system is controllable and observable, it is theoretically possible to
obtain the stable closed loop system. Moreover the optimal performance can be realized
if the optimal control technique is applied
to this system. But when a cycle is controlled by the feedback of output variables only,
it is necessary to study the feedback gain
for realizing the stabe system.

From Eqs.(4),(5) and (6), the characteristic
function of the closed loop system is reduced
to,

$$\left| s \cdot I_4 - A + B \cdot G \cdot (I_2 + D \cdot G)^{-1} \cdot C \right| = 0 \quad (11)$$

Then the stable condition is found by e.g.
the Hurwitz method. Several instructive relations, concerning the feedback gains, are
obtained.

Control with single input; When a cycle is
controlled only by handlebar steering without
upper body leaning, that is $g_y = g_\theta = 0$ in
Eq.(6), then the stability condition is reduced to,

$$h_y > 0 \quad (12)$$

$$h_\theta > P_1 \cdot h_y + P_2/V^2 \quad (13)$$

where P_1 and P_2 are the constants depending
on system parameters. The first inequality
means a "reverse handling", that is, a driver
must steer the handlebar to the right side at
first to lean the cycle to the left, if a
driver intends to change the lane to the left.
The second inequality means that there is a
lower limit of gain h_θ, the ratio of steer
angle to roll angle. If the running speed becomes low, this limit value increases about
in inverse proportion to the second power of
speed. Then the driver's task becomes heav-
ier, as the control gain must be increased.

Control with double inputs; When directional control is done only by upper body leaning, and independently lateral roll control
is done only by handlebar steering, that is
$u_s = - h_\theta \cdot \theta$, $u_1 = - g_y \cdot y_f$, then the stability condition is reduced to,

$$0 < g_y < P_3 \quad (14)$$

$$h_\theta > F(g_y)/V^2 \quad (15)$$

where P_3 is constant, $F(g_y)$ the function of

the gain g_y, respectively depending on the
system parameters. The first inequality means
"lean in". It is to say that a driver must
lean the upper body to the left if he intends
to change the lane to the left. At that time,
there is an upper limit of the gain g_y, the
ratio of lean angle to lateral deviation of
previewed point. It can be said that the system becomes unstable if the rider lean the
upper body too much sensitively. The second
inequality means similarly the same relation
as the inequality (13), concerning to the
forward speed.

These two cases are the fundamental models of
the closed loop control structure. The former
is the case of a single input system, and the
latter is the case of a double inputs system.
The closed loop control structure of this
system is shown in general in Fig.1. There
are two practical stability problem, i.e.
lateral and directional stabilities. These
stabilities are not independent of each
other. But it can be said that the lateral
stability is mainly subject to inner feedback, and that the directional stability is
mainly subject to outer feedback in Fig.1.

EXPERIMENT

The experimental bicycle is made to be operated automatically by a micro-computer, according to the stability analysis. Fig.2
shows the whole view of the experimental apparatus. The bicycle can run relatively on
the moving belt, which is driven by a motor.
Although the cycle is mechanically constrained in the longitudinal direction, other motions are all free from any constraints.
There is an on-board mass which is permitted
to move in the lateral direction. This mass
is driven by a servomotor in order that it
can play the same role as a upper body of
rider. There is another motor for steering
the handlebar. The control signals are transmitted from the micro-computer to these
motors.

As mentioned previously, from the previewed
point deviation $y_f(t+\tau)$ and roll angle $\theta(t)$,
all state variables can be observed. But in
this experiment, lateral deviation of present
point of cycle $y(t)$ and yaw angle $r(t)$ are
directly measured in stead of previewed point
deviation. The lateral deviation of previewed
point $y_f(t+\tau)$ can be obtained from $y(t)$ and
$r(t)$, as there is a relation among these variables as shown in Eq.(5).

To get the lateral deviation $y(t)$, yaw angle
$r(t)$ and roll angle $\theta(t)$, it is necessary to
measure the lateral positions of three different points of bicycle main frame. To measure the lateral position, an acoustic measuring method is used with 40 KHz supersonic
wave. Pulsated acoustic wave are transmitted
from the transmitter attached to the cycle,
to the receiver attached to the wall in the
laboratory. Then the transmitted time is
counted by the new phase method which is spe-

cially developed to suppress the measuring error. With this method, the absolute accuracy of lateral position error is less than ± 0.2 mm. The accuracies of the roll angle and yaw angle are less than ± 0.05 deg.

Two input variables, i.e. handlebar angle and on-board mass position, are also measured mechanically by potentiometers. The position signal of the on-board mass is fed back to the control force of the servomotor for the position control. The measured signals are all sampled and put into the micro-computer (16bit).

The fundamental specifications of the experimental vehicle are shown in Table 1. The speed of the moving rubber belt is changeable up to 10 km/h. The lane change experiment is mainly carried out at a constant speed to get the dynamical transient response.

TABLE 1   Spec. of Experi. Bicycle

Weight of cycle	m	136.0	N
Weight of on-board mass	m_h	9.11	N
Height of cycle c.g.	h	0.334	m
Height of on-board mass	h_h	0.465	m
Moment of inertia	I	2.25	Kgm^2
Caster angle	q	19.0	deg
Wheel base	L	0.665	m
Longitudinal distance between rear wheel & c.g.	L_r	0.313	m

RESULTS AND DISCUSSIONS

At first the experimental bicycle is made to run on a straight line on the moving belt. The speed for designing the feedback gains is 5 km/h in this experiment. At this speed the cycle with micro-computer control can run successfully as shown later.

By the way, according to the model analysis, the dynamical characteristics of the closed loop system depends on a running speed, and moreover it has a tendency to become unstable when a cycle slows down. To make sure this tendency, the cycle is made to slow down gradually from 10 km/h. From the experiment, the cycle system is stable enough to run smoothly at first. But at a certain speed, about 1 - 2 km/h, the cycle becomes oscillatory and then unstable.

Next but mainly, the lane change experiment is carried out. The cycle runs constantly at 5 km/h. The transient responses of input and output variables are observed, when a step input signal is added to the course reference signal (cf. Fig.1). Fig.3 - Fig.6 show the experimental results of transient responses of lateral position y, roll angle θ, handlebar steer angle u_s and on-board mass position u_1. The results of theoretical calculation are also shown in these figures.

Fig.3 shows at first the results when the cycle is controlled only by handlebar steering. The on-board mass is fixed to the cycle and is unable to move. Even in this case, the cycle is able to run successfully according to the theoretical consideration in the stability analysis.

Fig.4 and Fig.5 show the results when the cycle is controlled not only by handlebar steering but also by on-board mass moving. The on-board mass is permitted to move in proportion to the lateral deviation of previewed point, and the feedback gain g_y is changed from zero to 0.1 m/m in Fig.4, and to 0.2 m/m in Fig.5. Other values in Fig.4 and in Fig.5 are respectively equal to those in Fig.3. In these cases, the on-board mass moving can be considered to play the same role as the rider's upper body leaning.

From Fig.3 and Fig.4, the time required for the lane change with on-board mass action is shorter than that without on-board mass action. In other words, an upper body leaning action has a function to shorten the time for the lane change. But on the other hand, there is another tendency that the value of the handlebar steering angle, as well as roll angle, becomes large when the on-board mass is permitted to move. This tendency can be found more exactly, when the results in Fig.5 are compared with those in Fig.3. According to the results of the theoretical analysis, i.e. inequality (14), there is an upper limit of the proportional gain between the upper body leaning and the lateral deviation of previewed point. This theoretical result can prove these experimental results. Then it may be concluded that to lean the upper body in proportion to the lateral deviation of previewed point is useful to shorten the time for lane change, but is liable to get the sytem unstable when an upper body is leaned too much.

Fig.6 shows the results when two inputs are independently determined from two outputs; that is, the handlebar angle is steered in proportion to the roll angle, and the on-board mass is moved in proportion to the lateral position of previewed point. In this case, the cycle is able to run also successfully, as analyzed previously.

From these results, it is possible to realize the directional control by handlebar steering and/or by rider's upper body leaning. But it is impossible to get the roll motion stable only by upper body leaning control. It can be said that the handlebar steering compensation is essential to get the lateral roll motion stable, with the proportional rider compensation.

By the way, rider's upper body is assumed to be a lumped mass in this paper, and its behaviour is assumed to be proportional compensation. On this assumption, the experimental results of the computer-controlled bicycle is very well proved by the theoretical results. But precisely speaking, the practical system may be more complex than this fundamental model, especially when a cycle runs at high speeds. But it is fortunately easy to change the micro-computer program, so that the closed loop system can be improved without any problems.

CONCLUSION

This paper has shown the applications of the vehicle dynamics and manual control to a rider and two-wheeler control system, and to a designing of a computer-control cycle. To get the comprehensive understanding of the lateral and directional stability of this system, a mathematical model is presented as a feedback control system with two inputs, i.e. handlebar steering and rider's upper body leaning.

According to the theoretical analysis, an experimental bicycle has been made, which is controlled by a micro-computer in stead of a human rider. The experimental results of lane change show a good running property and quantitatively a good fitness with the theoretical analysis.

These analysis and computer-control test facility have implications for preferred rider control techniques and possibilities of the objective estimation of the dynamical performance of two-wheelers. In the future, a driving robot is expected to be one of the useful means for the researches of good handling vehicles.

REFERENCES

Iguchi, M.(1962). Dynamics of single track vehicles. J. of Researches on machines, 14-7, 34-38.
Kondo,M., Nagaoka,A. and Yoshimura,F.(1963). Theoretical analysis of two-wheeler stability. J. of S.A.E. Japan, 17-1.
Iguchi, M.(1970). Man-Machine Systems. Kyoritzu.
Kobayashi, T.(1974). Man-motorcycle dynamics. J. of S.A.E. Japan, 28-4.
Sharp, R.S.(1975). Dynamics of single track vehicles. Vehicle System Dynamics, Vol.4, No.2-3.
Weir, D.H. and Zellner, J.W.(1979). Lateral-directional motorcycle dynamics and rider control. S.A.E. Paper. 780304.
Aoki, A.(1979). Experimental study on motorcycle steering performance. S.A.E. Paper. 790265.

APPENDICES

Deformation of dynamical equation

Let the coefficients in Eq.(3) be defined as;

$$w^2 = m \cdot h \cdot g / I$$
$$f_u = m_h \cdot h_h / I$$
$$f_\theta = m_h / m \cdot h$$

and also the state variables v_1, v_2 be defined as;

$$v_1 = y + g/w^2 \cdot (\theta + f_u \cdot u_1) \qquad (A-1)$$
$$v_2 = \dot{v}_1 \qquad (A-2)$$

then the Eq.(4) is reduced from Eqs.(1)-(3). And Eq.(5) is reduced from Eq.(A-1).

Fig. 2. The computer-controlled cycle on moving belt.

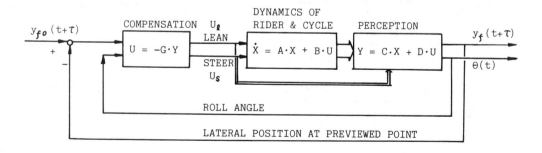

Fig. 1. Scheme of rider and single-track-vehicle feedback control system.

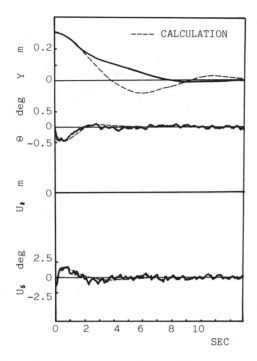

Fig. 3. Transient response with single input
        of handlebar steering (g_y=0 m/m, h_y
        =0.1 rad/m, g_θ=0 m/rad, h_θ= 7 rad/
        rad, ℓ=2 m).

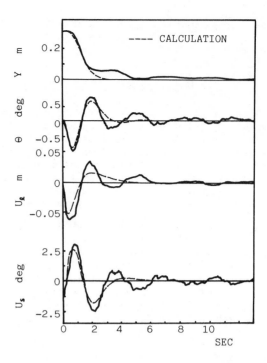

Fig. 5. Transient response with double in-
        puts of handlebar steering and on-
        board mass moving (g_y=0.2 m/m, h_y=
        0.1 rad/m, g_θ=0 m/rad, h_θ=7 rad/rad,
        ℓ=2 m).

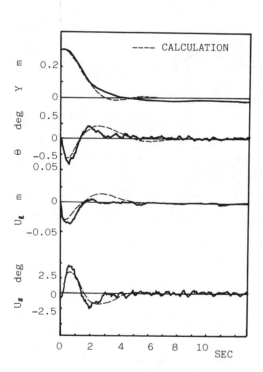

Fig. 4. Transient response with double in-
        puts of handlebar steering and on-
        board mass moving (g_y=0.1 m/m, h_y=
        0.1 rad/m, g_θ=0 m/rad, h_θ=7 rad/rad,
        ℓ=2 m).

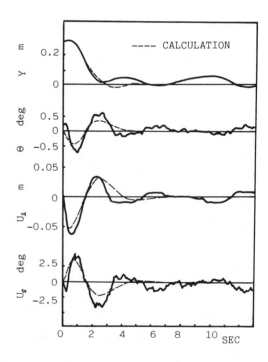

Fig. 6. Transient response with double in-
        puts, independently compensated,(g_y
        =0.2 m/m, h_y=0 rad/m, g_θ=0 m/rad,
        h_θ=7 rad/rad, ℓ=2 m).

PROCESSING OF ON-BOARD ROUTE GUIDANCE INFORMATION IN REAL TRAFFIC SITUATIONS

M. Voss and R. Haller

Fraunhofer Institute for Information and Data Processing, Karlsruhe, Federal Republic of Germany

Abstract. The "ALI" (Autofahrer-Leit- und Informationssystem) guidance and information system for highways has been developed in the Federal Republic of Germany. It is the aim of the research described here to analyze the driver's information processing behaviour with the ALI display in real traffic and to evaluate possible consequences for driving safety. Extensive field experiments and additional experiments in a driving simulator were conducted. To analyze the temporal structure of information perception and processing the measurement of eye movements was an essential tool. The time budget derived from these data shows no overload by time-pressure. The eye movement patterns measured lead to the hypothesis that the information obtained through ALI is processed by queuing. Related to the attention allocation this is not an optimal strategy and should be improved. To minimize false route decisions, the compatibility of ALI symbols with traffic signs should be improved as well.

Keywords. Route guidance systems; eye movements; electrooculography; display systems; driving safety; human factors.

INTRODUCTION

In recent years the technical development provides new possibilities for advanced information systems in cars, e.g. failure diagnosis systems, trip computers, route guidance systems. All over the world different route guidance systems are in the state of planning or realization (e.g. ERGS/ USA, CACS/Japan, AWARE/Great Britain); in the Federal Republic of Germany the guidance and information system "ALI" (Autofahrer-Leit- und Informationssystem) applied to highways has been developed and tested (Brägas , 1979).

The ALI system is interactive: Before starting, the driver has to put in his destination coded as a 7-digit decimal number via a keyboard. During driving he gets the actually optimal route guidance which depends on the traffic flow as prognosticated by a central computer. Before the next decision point (exit, highway crossing) the actual information is transmitted using induction loops in the route and displayed timely on the on-board ALI display.

In the last few years, extensive field testing of the ALI system was performed by Blaupunkt company, Volkswagenwerk (VW) and by Heusch-Boesefeldt, Consulting Engineers. This research work was supported by the Federal Ministry for Research and Technology (BMFT); the research project was accompanied by the Technical Control Board (TÜV e.V., Rheinland). In the context of the field testing, the Fraunhofer Institute for Information and Data Processing (IITB) was ordered to analyze the driver's information processing behaviour with the ALI display in real traffic, and to evaluate the possible consequences for driving safety (Voss, 1982).

EXPERIMENTAL METHODS

Figure 1 shows the route guidance symbols of the ALI display used in the field experiments. The display was mounted right above the dashboard in the experimental car; the distance from subject's eye was about 60 cm, the vertical deviation about 15° down, the horizontal deviation about 25° right. During driving the ALI information was displayed shortly after the first traffic sign, approximately 1000 m before a highway decision point; this was controlled by the experiment leader acting as a front-seat passenger (so, no further infrastructural aids including induction loops, central computer etc. were needed).

The field experiments were conducted on highways near Frankenthal, FRG. General experimental conditions were: No rain, traffic flow intensity about 1000 cars/h, average speed about 100 km/h. The subjects were young and experienced drivers (because

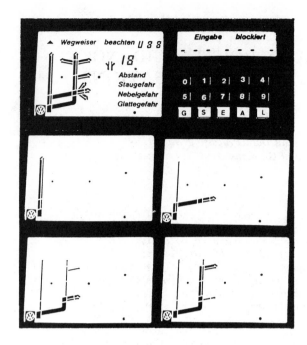

Fig. 1.   Route guidance symbols of ALI
          (above: general view, including
                   keyboard,
           below: 4 symbols used in field
                   experiments).

of the experimental risks); for the interpretation of the results this has to be taken
into consideration. Normally, the route guidance situation was "congruent" (no difference between general destination and local
destination displayed by the ALI symbol)
and "compatible" (no difference between the
topographic structure of traffic sign and
ALI symbol); incongruent and incompatible
situations were considered, too. In most experiments, the subjects were not familiar
with the locality; in an additional experiment the influence of familiarity in an incongruent route guidance situation was also
tested.

In dependence on the different ALI information the following measurements were taken:

- Subjective rating, as an indication of
  mental workload, and for improvement
  suggestions.

- Performance measurements, e.g. route decision error, as a measure of symbol intelligibility.

- On-line measurement of eye movements as
  an essential tool to analyze the temporal
  structure of information perception and
  processing. For this reason the method of
  electrooculography (EOG) was used (Fig.2.);
  the main properties of this method are:
  Sufficient accuracy and resolution, direct
  electrical signal of eye deviation related
  to the head, little impairment of the

subject (only some adhesive electrodes
are needed). The EOG method was already
applied successfully in earlier work (Voss
and Bouis, 1979).

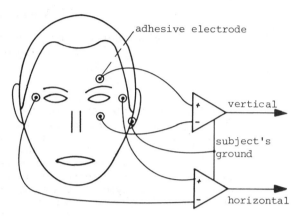

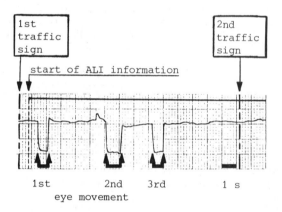

Fig. 2.   Method of electrooculography
          (above: basic principle,
           below: experimental result of a
                   horizontal eye movement).

RESULTS

Figure 3 shows the duration of the eye movements directed to the ALI display and the
intervals between them, measured in the
local area between first and second traffic
signs; these results are independent of the
different ALI symbols. In this area not
more than three eye movements were registered; the frequency of the first eye movement
was 100%, of the second about 70%, of the
third about 30%. Therefore a driver needs
1.5 - 2 eye movements in the mean to perceive the ALI information in this local
area. The duration of these eye movements
is highly constant (0.8 - 0.9 s). Assuming
a speed of 100 km/h, a driver needs therefore about 15% in the mean of the total
time available in this local area; so, no
overload by time-pressure can be seen.

time [s]

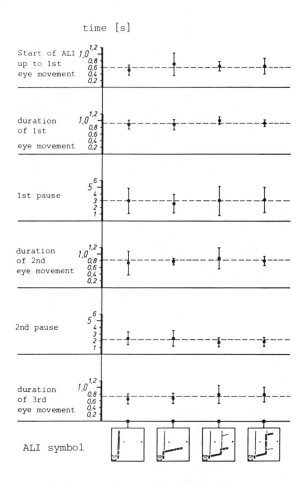

Fig. 3.   Perception of ALI information in
real traffic between first and
second traffic sign (14 subjects
averaged, interindividual deviations as 99%-confidence intervals).

Furthermore it is important to compare
these data with other events of information
perception normally accepted while driving.
Table 1 shows a general view of some reference data measured in field experiments
(17 subjects averaged, rounded).

To analyze driver's information processing
behaviour the number of eye movements with/
without ALI were measured in the following
local areas along the different decision
points, in dependence on the different ALI
symbols. Figure 4 shows the result (8 subjects averaged); in the local area between
first and second traffic sign 1.5 - 2 eye
movements are directed to the ALI display
on an average, as shown in Table 1, too.
Furthermore it can be seen that there are
some additional monitoring eye movements
in the following local areas, in dependence
on the differing complexity of decision
situations.

Further results are: In field experiments
false route decisions were seldom on an
average. They occured with route guidance
information of great complexity only, mostly in situations of incompatibility (i.e.
different topographic structures) of ALI
symbol with traffic sign; the average false
decision rate was 25%, as related to this
case.

In an experiment additionally conducted in
a driving simulator the response time to
unexpected events (which is an important
aspect of a driving task) was measured
with/without ALI; it was independent of the
ALI symbol and slightly increased with ALI
(100 ms in the mean).

Furthermore, the field experiments in incongruent route guidance situations (i.e.
differences between general destination
and local destination actually displayed
by the ALI symbol) have shown that the
drivers needed one additional eye movement
in the local areas before turning off in
the mean, related to situations of congruent route guidance information; this
result was not influenced by subject's
familiarity with the locality.

TABLE 1   Reference Data of Information Perception Events while Driving

Event of information perception	Duration of every eye movement [s]	Number of eye movements
Reading of speedometer	0.5 - 1	1
Reading of an exit sign	0.5 - 1	1 - 2
Reading of ALI display (between first and second traffic sign)	1	1.5 - 2
Tuning the heater fan	1 - 1.5	4 ± 2.5
Tuning the radio	1 - 1.5	6 ± 3

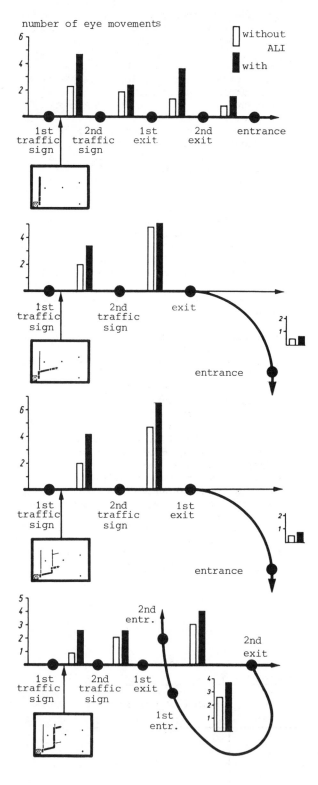

Summing up, the results show that there is no unacceptable risk for driving safety caused by the ALI information, additionally displayed during driving in real traffic: This can be concluded from the eye movement data (mean duration and frequency) which show no overload by time-pressure, and from the relation of these data to reference data of other information processing events normally accepted. This is also confirmed by the response time to unexpected events measured in a simulator experiment which shows no unacceptable distraction of attention caused by route guidance information, and by the results of subjective rating not reported here in detail.

It has to be taken into consideration that the subjects were young and experienced drivers. Under most unfavorable circumstances (e.g. senior drivers, little driving experience, unfavorable sight and weather conditions, high traffic flow intensity) it might be critical to display route guidance information additionally, but this is a general problem of all optical displays in a car.

The first eye movement directed to ALI is strongly triggered by the acoustic signal coupled with the beginning of ALI information. Therefore, the place for displaying route guidance information should be optimized for each decision point individually to minimize the risk for the driver. Furthermore, the signal should be coded for distinguish situations of different complexity as a basic information, e.g. no acoustic signal for driving straight ahead, acoustic signal for turning off, another acoustic signal for incongruent situations. In the latter case additional information (e.g. by speech output) about the background would be helpful for driver's route decision, e.g. a stagnation caused by accident, a 20 min expected delay, recommended deviation amounts up to 20 km.

For processing on-board route guidance information, different strategies are possible: For example, a driver can look at the symbol on-board displayed and, with the help of the first traffic sign, find the destination name, which is the correct for this decision point ("destination name strategy"). So, he has to read the display only once at the beginning, then he can concentrate on the conventional route guidance information outside. Another possibility is to look repeatedly at the topographic structure of the on-board symbol and to transform it into route decisions step by step by means of a queuing strategy. In this case a driver is concentrated more on the on-board display, and his attention allocation is more divided by monitoring eye movements additionally needed. The eye movement patterns measured and averaged in

Fig. 4.   Number of eye movements with/without ALI in the following local areas along different decision points.

following local areas lead to the hypothesis
that the ALI information is processed by
queuing (dependent on the varying symbol
complexity).

Clearly this is not an optimal strategy,
and should be improved by adequate training,
if a symbolic presentation of route guidance
information is preferred. Another possibility is to present the actually right destination name directly, by an optical display,
or by speech output. In this case the
problems of incompatibility with symbolic
structures vanish; on the other hand, problems of incongruent situations (differences
between global and local destination name)
possibly arising must be deactivated by
additional information.

REFERENCES

Brägas , P. (1979). Field testing of a route
     guidance and information system for
     drivers (ALI). Proceedings of the Inter-
     national Symposium on Traffic and Trans-
     portation Technologies, Vol. CI, 371-401.
Voss, M. (1982). Fahrerverhalten bei fahr-
     zeuginterner Zielführung (ALI). Final
     report of the Fraunhofer Institute for
     Information and Data Processing
     (IITB), No. 9616.
Voss, M. and Bouis, D. (1979). Der Mensch
     als Fahrzeugführer: Bewertungskriterien
     der Informationsbelastung; Visuelle und
     auditive Informationsübertragung im Ver-
     gleich. FAT-Schriftenreihe, Nr. 12.

AN ANALYSIS OF THE MAN-AUTOMOBILE SYSTEM WITH A DRIVING SIMULATOR

K. Yoshimoto

Department of Mechanical Engineering, University of Tokyo, Bunkyo-ku, Tokyo 113, Japan

Abstract. To investigate the driving dynamics of automobiles, it is necessary to study the control characteristics of drivers. A self-paced preview
tracking control model was proposed to represent the action of a driver who
was driving an automobile along a curved course, assuming that he could predict the future state of automobile based upon the present state, and the
predicted future deviation and the predicted lateral acceleration were fed
back to the steering force and the vehicle speed command respectively.
In this study, to veryfy the proposed self-pased preview tracking model, the
behaviours of drivers are measured on a driving simulator and are compared
with those of the self-paced tracking control model simulated on a digital
computer. From these comparisons, the proposed self-paced preview tracking
control model should be improved a little and a term of the permissible jerk
is added to the vehicle apeed control loop. Finally, adequateness of the
improved model is confirmed.

Keywords. Man-machine system; manual control; preview control; self-paced
tracking control; driver-automobile system; driver model; driving simulator

INTRODUCTION

If the driving characteristics of automobile
drivers is clarified clearly and their mathematical model is established, it will become
very easy to investigate the driving dynamics
of automobiles or to design the curves of
high-ways, because such various verifications
will be carried out by the digital simulations of man-automobile-road systems making
use of the established mathematical model of
drivers. Many investigators have already
proposed various steering models and have
explained qualitatively the driving stability
employing their models, but their models are
not accurate enough to discuss the tracking
characteristics of automobiles. Hence a
simple useful model which characterizes the
human driver satisfactorily is desired. At
the I.F.A.C. 8th Triennial World Congress in
1981, I proposed a self-paced preview tracking control model to represent the action of
a driver who was driving an automobile along
a curved course, assuming that he could predict the future state of the automobile based
upon the present state, and that the predicted future deviation and the predicted
lateral acceleration were fed back to the
steering force and the vehicle speed command
respectively.

In this study, to verify the proposed self-
paced preview tracking model, a driving simulator is built making use of a mini-computer,
two micro-computers, a television monitor and
some automobile components. The behaviours

of drivers on this driving simulator are compared with those of the proposed driver model
simulated on a digital computer. From these
comparisons, the proposed driver model should
be improved a little and a term concerning
the permissible jerk is added to its vehicle
speed contol loop. Finally, it is confirmed
that the newly improved driver model can represent the behaviours of drivers on this
driving simulator.

I will describe here the outlines of the proposed self-paced preview tracking control
model of drivers, the block diagram of which
is shown in Fig. 1. It is constructed of two
control loops which are the steering control
loop and the vehicle speed control loop.
This model comprises the following assumptions.

At first, concerning the steering control;
(i) Detection of the future error by
    prediction.
The driver can perceive not only the present
position (x,y) and the direction angle ψ of
the vehicle but also the lateral acceleration
a through the visual cues and the motion cues.
He is able to predict the future position
(x^*,y^*) which the vehicle will reach after τ
seconds and to detect the future deviation ε^*
from the desired path (X^*,Y^*) which he can
preview. This τ is called the prediction
time which is presumed to be about $2 \sim 3$ sec.
There is a nonlinear characteristics of
threshold in this detection of the future
deviation. Therefore, for an example, the

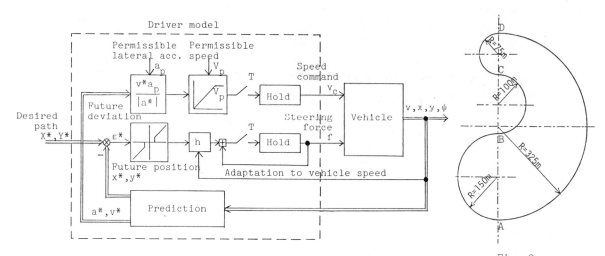

Fig. 1.   Block diagram of the self-paced preview
           tracking control model of drivers.

Fig. 3.
The circuit course
used to the desired
path.

future deviation less than 1 meter is neglected.
(ii) Correction of the steering by the
     discrete-time integral action.
The driver detects the future deviation and
integrates it in a periodic way. If this
sampling period is T sec., this periodic behaviour may be approximated by the conventional human response time of T/2 sec.
However, for such a preview control system
as driving an automobile, this sampled-data
model seems to be more practical. The sampling period is about 0.6 sec. taking the human response time into account. The steering
gain h is adapted to be the optimal value in
accordance with the vehicle speed.
(iii) Steering by the steering force.
The driver's output is not the steering displacement but the steering force, because it
is defficult to drive the vehicle with the
reactionless handle wheel. Therefore, the
influence of the inertia, viscosity, elastic-
ity and the self-aligning torque of the front
wheels must be taken into account.

Secondly, concerning the vehicle speed control;
(iv) Prediction of the future vehicle speed

and the lateral acceleration.
The driver can predict the future speed v^*
and the lateral acceleration a^* of the vehicle after τ sec.
(v) Calculation of the speed command.
The driver tries to drive the vehicle as fast
as possible within the self-determined permissible lateral acceleration a_p and maximum
speed v_p. Then he calculates the value of
$v^* a_p / a^*$ as the speed command v_c. If v_c exceeds v_p, v_c is corrected to v_p.
(vi) Discrete-time correction of the
     vehicle speed.
The driver corrects the speed command in a
periodic way in the same way as the steering
control action.

OUTLINES OF THE DRIVING
SIMULATOR

The driving simulator employed here is constructed of a visual cues generator, an audio
generator, a speedometer, a fixed driving
seat (without motion cue), a steering wheel
with a control feel generator, an accelerator
pedal and a vehicle dynamics simulator as
shown in Fig. 2. and Photo. 1. In this driv-

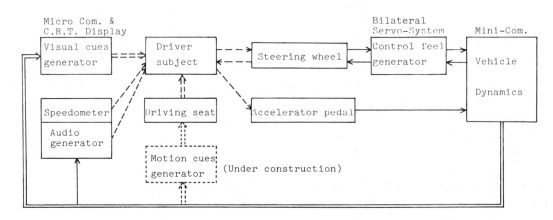

Fig. 2.   Block diagram of the driving simulator.

ing simulator, the driver subject operates the steering wheel and the accelerator pedal watching the foreground image provided on the C.R.T. display. The foreground image is constructed of the two dotted lines which represent the both edges of the desired path perspectively as shown in Photo. 2. It seems as if the path were approaching to this side according to the vehicle speed. The control feel is generated semi-actively by an electric bilateral servo system. The vehicle speed is controlled by the accelerator pedal not only in accelerating but also in decelerating. The motion of the vehicle was fed back to the driver subject mainly by the visual cues mentioned above and auxiliarily by the control feel of the steering wheel and the auditory cue which is the engine noise according to the vehicle speed. The motion platform is now under construction and the driving seat is fixed. However, the skilled driver is able to estimate the lateral acceleration in some degree through the visual cues and the control feel. As the vehicle dynamics, only the horizontal motion is considered. The calculation of the vehicle dynamics is carried out in a period of 0.05 sec. by a mini-computer.

EXPERIMENTS WITH THE SIMULATOR

Making use of this simulator, the data of 22 licensed driver subjects have been gathered. The subjects are male adults. The desired path is the circuit course as shown in Fig. 3. It has two S-shaped curves and is 4 meters in width.

Fig. 4 shows the behaviors of one of the typical subjects. The trace of the lateral acceleration of the vehicle holds a constant value on every corner except the corner whose radius of curvature is 325 meters. All the subjects have showed the similar behaviors. These constant values have been identified with the individual permissible lateral acceleration a_p of the subjects respectively. Similarly, the trace of the speed command holds a constant value on the largest corner while the lateral acceleration does not reach to the limit of the permissible value. This command speed is also identified with the

permissible maximum speed v_p. Fig. 4 shows $a_p = 0.3$ g and $v_p = 30$ m/s.

The upper part of Fig. 5 shows the distribution of the permissible lateral accelerations and the permissible speeds of the driver subjects. Fig. 6 shows the distribution of the maximum lateral deviations of the vehicle from the center-line of the desired path concerning all the subjects. The larger lateral deviations on this simulator than those on the real roads, are considered to be caused on the lack of dangerousness.

COMPARISON WITH THE DRIVER MODEL AND PROPOSITION OF THE IMPROVED MODEL

Applying these identified a_p and v_p to the proposed driver model, some digital simulations have been carried out. Fig. 7 shows the behaviors of the proposed driver model to which the data of the subject U shown in Fig. 6 are applied. Generally, these behaviors agree well with those of the suject U. If examined in detail, at the corner B or C around the contraflaxure of the S-shaped course, the driver model accelerates the vehicle while the driver subject decelerates it. Because, at that corner the direction of the lateral acceleration is reversed and the predicted future lateral acceleration a* becomes very small for a moment. At the moment, the speed command takes the permissible maximum value. Hence, the speed v_h when the vehicle driven by the driver model passes through the contraflaxure is higher than those of the driver subjects as shown in the lower part of Fig. 5.

To solve this problem, the driver model 1 is improved a little and a term concerning the permissible jerk is added to the vehicle speed control loop as shown in Fig. 8. The assumption introduced here is as follows:
(vii) <u>Introduction of the permissible jerk</u>. The driver tries to decelerates the vehicle speed within the self-determined safety speed v_s when the mean jerk during the prediction time $(a*-a)/\tau$ exceeds the permissible lateral jerk J_p. Therefore, he corrects the speed

Photo. 1. General view of the driving simulator.

Photo. 2. Roadway drawn on the C.R.T. display.

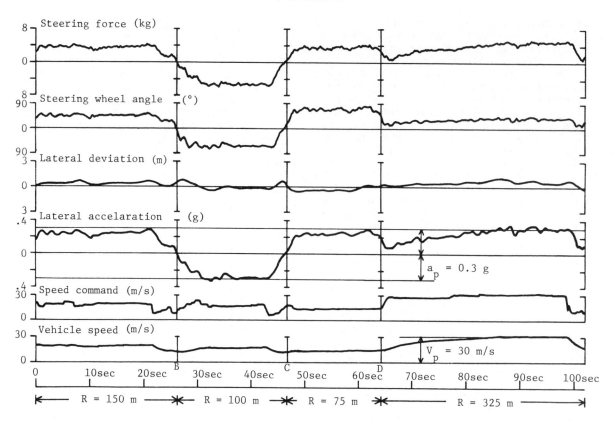

Fig. 4. Behaviors of the driver subject U on the driving simulator.

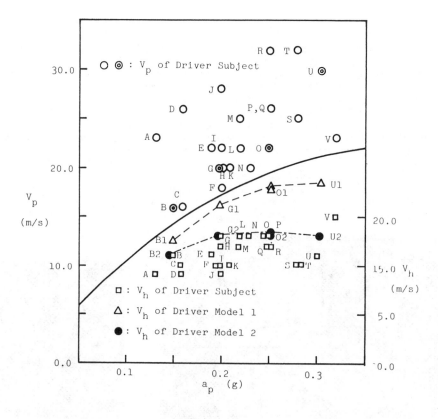

Fig. 5. Distribution of the permissible lateral accelerations a_p, the permissible
maximum speeds v_p and the spees v_h at the corner of contraflexure.

command to the value of $(v_c)_{old} v_s/v$ gradually.

Fig. 9 shows the behaviors of the newly improved model of the driver subject U. Here,

the permissible lateral jerk is choosed to be $J_p = 2$ m/s^3 from the design data of high-ways and the safety speed v_s is identified with the speed v_h 11 m/s at the cotraflaxure. From the comparison of Fig. 4 with Fig. 9, it may be said that close agreement between the behaviors of the subject and those of the newly improved model is obtained. Such agreements are also obtained from the similar comparisons concerning the other subjects.

CONCLUSIONS

In this paper, to verify the proposed self-paced preview tracking control moel od drivers, the behaviors of the driver subjects are measured making use of a driving simulator and are compared with those of the proposed model simulated on a digital computer. From these comparisons, the proposed model should be improved a little and the term concerning the permissible jerk is added to the vehicle speed control loop. Finally, it is confirmed that the newly improved model is able to represent the behaviors of the driver subjects on this driving simulators.

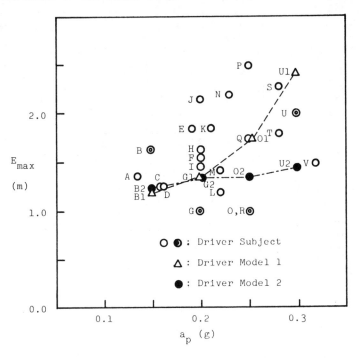

Fig. 6. Distribution of the maximum lateral deviations from the center-line of the desired path.

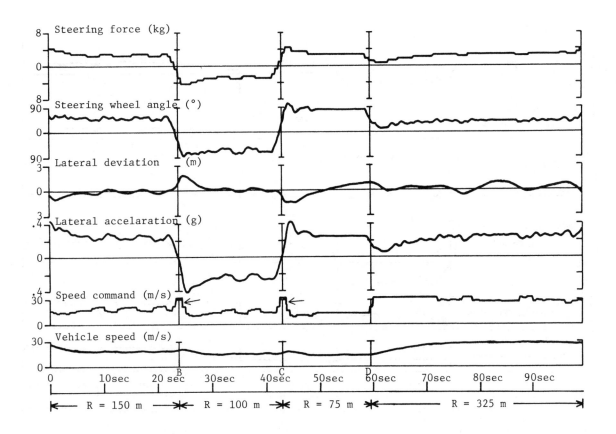

Fig. 7. Behaviors of the propoesd driver model simulated on a digital computer.

ACKNOWLEDGEMENT

Thse experiments with the simulator and the
digital simulations were carried out by my
student, Mr. N. Kitayama making use of the
computer of Tokyo University Computer Center.

The author expresses his gratitude to Mr. M.
Saito for his excellent drawing of figures,
Mr. K.Wakatsuki for his typewriting of papers
and the members of my laboratory for Automatic Control, Department of Mechanical Engineering,University of Tokyo for their assit-
ance Throughout the course of this study.

REFERENCES

Repa, B.S. and Wierwille, W.W. (1976). Driver
    Performance in Controlling a Driving Simu-
    lator with Varying Vehicle Response Char-
    acteristics, S.A.E. papers 760779, 2453-
    2469.
Allen, R.W. and Jex, H.R. (1980). Driving
    Simulation - Requirments, Mechanization
    and Application, S.A.E. papers 800448,
    1769-1980.
Yoshimoto, K. (1981). A Self-paced Preview
    Tracking Control Model of an Automobile
    Driver, Preprints of I.F.A.C. 8th Trien-
    nial World Congress, Vol. XV, 83-88.

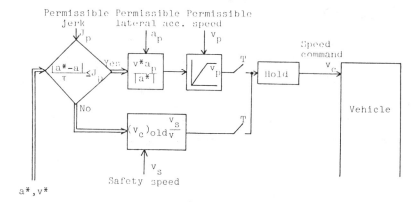

Fig. 8.   Block diagram of the newly improved speed control loop.

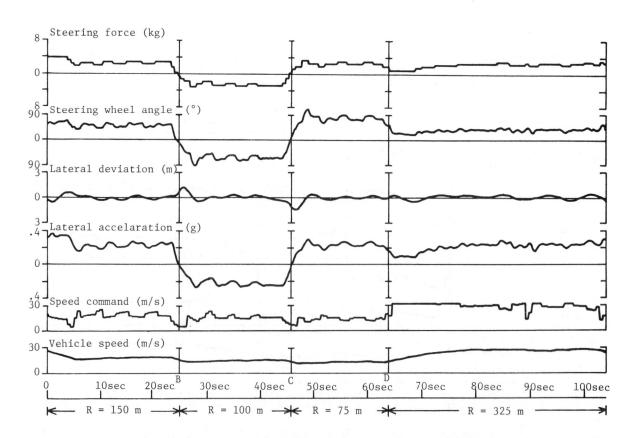

Fig. 9.   Behaviors of the newly improved driver model of the subject U.

CONTRIBUTIONS OF CONGRUENT PITCH MOTION CUE TO HUMAN ACTIVITY IN MANUAL CONTROL

S. Balakrishna*, J. R. Raol** and M. S. Rajamurthy*

*Systems Engineering Division, National Aeronautical Laboratory, Bangalore, India
**Department of Electrical Engineering, McMaster University, Hamilton, Ontario, Canada

Abstract. The contributions of pitch motion cue to the performance of human pilots in compensatory tracking task are evaluated by seperating the visual contributions from the combined motion visual contributions wherein the visual cue is congruent to the motion cue. Analytical human pilot models of the Least Squares (LS) structure are evaluated from the pilot input-output in uniquely planned motion based research simulator experiments. Beneficial contributions of the pitch motion cues in the form of lead and prediction are demonstrated.

Keywords. Motion cues; Pilot modeling; Identification; Compensatory tracking; Least Squares model; Motion Predictor operator.

INTRODUCTION

A human pilot can be expected to perform better in a tracking control task when exposed to multiple sensory cues of the process than a single sensory cue (Seckel and co-workers, 1957). While it is generally realised that visual cue constitute the primary sensory input channel to human pilot, role of inner ear vestibular sensory cue and its importance has not been ignored (Peters, 1969). A number of studies have been carried out to establish the role of motion cues in aircraft like manual control situations (Bergeron and Adams, 1964; Bergeron, Adams and Hurt, 1977; Junker, Repperger and Neff, 1975; Shirley and Young, 1968; Stapleford, Peters and Alex, 1969; Van Gool and Mooij, 1976; Washizu and co-workers, 1977). These include roll motion cue studies, single and two axis tasks and a few pitch motion cue studies. The pilot performance evaluation have been through study of gross tracking indicies (Kuehnel, 1962; Smith, 1966) or through identification of the analytical pilot models. Much of the pilot model structures used in motion cue studies have centered around quasilinear models with the exception of Washizu and co-workers (1977). Recently there have been many efforts to model human

tracking activity using time series model structures (Shinners, 1974; Balakrishna, 1976; Tanaka, Goto and Washizu, 1976; Osafo-charles and co-workers, 1980). Shinners (1974) has utilised Autoregressive/Auto-regressive moving average models to characterise the remnant of quasilinear models, whereas Balakrishna (1976) has considered five different time series pilot model structures. Tanaka, Goto and Washizu (1976) have proposed an Autoregressive model for human pilot which is essentially same as the Least squares model proposed by Balakrishna (1976). Only Washizu and co-workers (1977) have used the time series model structure in analysing contributions of roll motion cues to pilot activity.

This investigation was motivated by a desire to determine the extent of change in performance of pretrained subject when he has to perform only on visual cues just after conducting a combined visual-pitch motion cued run wherein the visual cues are congruent to pitch motion. Under these conditions, the pilot is expected to perform a representative task rather than a maximum performance task.

In analysing the pilot performance,
it is proposed to use the Least Squares
model structure for human pilot
(Balakrishna, 1976), as this structure
is better suited to seperate the
congruent pitch motion cues from the
combined visual-motion cue responses.
Further, the concept of describing the
motion contributions in the form of a
Motion Predictor Operator (Jaegar,
Agarwal and Gottlieb, 1980) are explored.

EXPERIMENTATION

Experiments to seperate the contributions
of pitch motion and visual cues in
human compensatory tracking studies
were performed on a three degrees of
freedom motion based research simulator
developed at National Aeronautical
Laboratory, India.   This facility is
capable of 60 deg/sec pitch velocity
and 400 deg/sec^2 pitch acceleration.
with a fully weighed cockpit.   The
schematic of the compensatory tracking
experiment is shown in Fig.1.   The

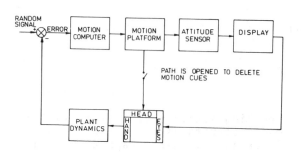

Fig. 1. Schematic of motion cue experiment.

pilot is seated in a cockpit mounted on
the motion platform and is subjected to
angular pitch motion which is the error
of the compensatory loop.   The center
of rotation passes through the pilot
body,   In the enclosed cockpit the
pilot observes an oscilloscope displaying pitch attitude as a line through a
pitch attitude sensor mounted on the
platform but not on the cockpit.   The
pilot uses a central stick which is
loaded with a spring of 0.1 kg/cm
stiffness.

The compensatory tracking experiments
were performed with a band limited
white noise disturbance.   Initially
the pilot performed a compensatory
tracking task while seated in the
cockpit mounted on the simulator,
yielding tracking responses which are
congruent motion-visual (CMV) data.
Immediately after the CMV run, the
cockpit was moved down to the floor
while maintaining the signal connections and the run was repeated.   Two

views of the simulation facility with
the cockpit on the motion platform and
on the ground are shown in Fig.2.

Fig.2. View of the simulator with
cockpit on the motion platform.

Fig. 2. (contd.) View of the simulator
with cockpit on the floor.

Two subjects were pretrained for congruent motion visual runs as well as visual alone runs. Subsequently, in each of the analysis runs the subject performed a CMV run followed immediately (within five minutes) by a visual run with cockpit moved to ground. The analysis data chains thus represent unimodal (visual) and multimodal (CMV) responses of a pilot in visually similar environment.

The compensatory tracking disturbance chosen was a band limited white noise with a corner frequency of 1.2 rad/sec with a roll off rate of 18 dB/oct. The dynamics of the motion system in pitch approximates to a second order system with a natural frequency of 5.0 rad/sec and a damping ratio of 0.6. The pilot input-output data were acquired on-line at a uniform sampling rate of 20/sec over typical length of 30 secs.

The experiments have been performed for two simulated plants K/s and K/(s+2). The dynamics controlled by the pilots thus consists of the platform dynamics and the simulated plant in cascade.

MODELING

In the present analysis, it is assumed that the pilot activity can be represented by a model that relates the observed input-output data through a time series model of the Least Squares structure (Balakrishna, 1976) of the type

$$A(q^{-1})y(k)=B(q^{-1})\,\varepsilon(k)+e(k) \quad (1)$$

where
$$A(q^{-1})=1+a_1q^{-1}+\ldots+a_nq^{-n}$$

$$B(q^{-1})=b_0+b_1q^{-1}+\ldots+b_nq^{-n}$$

and (a_i, b_i) are parameters of an n-dimensional pilot model. It is clear that equation (1) describes a LS pilot model which contains an autoregressed noise component $e(k)$. Given a sequence of $y(k)$, $\varepsilon(k)$ for $k=1,\ldots$ $N+n$ where $N \gg n$, it is easy to estimate the model order n, pilot model a, b such that e is independently and identically distributed noise. Such a model is distinctly different from quasilinear models usually used to describe human pilots. An implication of the LS model is that the pilot responses can be seperated into a sensory and prediction part $H_{SP}(q^{-1})$ and an equalising plus neuromuscular part $H_{EN}(q^{-1})$ directly as shown in Fig.3. Such a structural seperation

was implicitly suggested by Balakrishna (1976). However Washizu and co-workers (1977) have explicitly indicated and demonstrated this seperation. The frequency domain description of LS model breaks up into

$$H_{SP}(j\omega)=B(j\omega)$$

$$H_{EN}(j\omega)=\frac{1}{A(j\omega)}$$

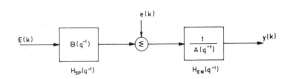

Fig. 3. Structure of LS model.

ANALYSIS AND DISCUSSION

The parameters of the discrete LS pilot model were estimated utilising the pilot input-output data chains after determining the model order through both subjective and objective procedures (Jategaonkar, Raol and Balakrishna, 1982). This resulted in a fourth order model which was subsequently used throughout the analysis.

The results of pilot identification for visual alone and for CMV tracking tasks are presented in two forms. Figures 4 to 7 show the frequency domain descriptions for the subjects A and B. These were directly obtained from parametric models. Some of the discrete models were also transformed into continuous time counterparts by complex curve fitting of the frequency domain descriptions of the discrete model, using modified Levy's method (Santhanam and Koerner, 1963) and the results are tabulated in Table 1.

Following inferences can be drawn from an inspection of Table 1 and Figures 4 to 7, in respect of removal of congruent pitch motion from a combined motion-visual situation.

1. The human pilot models exhibit second order behaviour with damping ratio less than unity and natural frequency varying from 9 to 16 rad/sec. However the removal of motion cue for subject B shows first order behaviour.

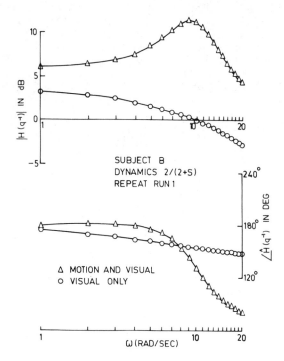

Fig. 4. LS pilot models Sub. B.

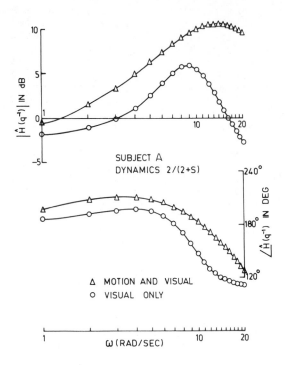

Fig. 6. LS pilot models Sub. A.

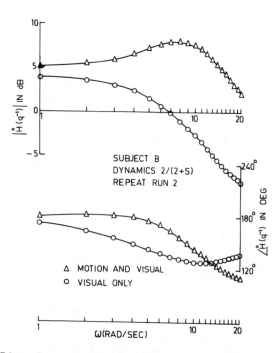

Fig. 5. LS pilot models Sub. B.

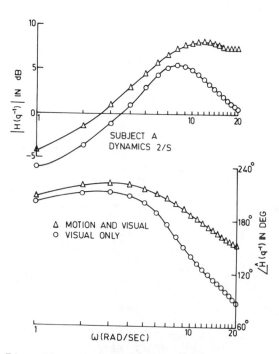

Fig. 7. LS pilot models Sub. A.

TABLE 1 Human Operator's Models

Plant Dynamics	Subject	Motion+Visual Cue	Complex curve fit error %	Visual Cue only	Complex curve fit error %
$\dfrac{2}{s+2}$	A	$\dfrac{0.8(1+0.33s)}{1+\left(\dfrac{2\times0.64}{11.5}\right)s+\left(\dfrac{s}{11.5}\right)^2}$	0.14	$\dfrac{1.3(1-0.023s)}{1+\left(\dfrac{2\times0.52}{15}\right)s+\left(\dfrac{s}{15}\right)^2}$	0.79
	A	$\dfrac{1.1(1+0.28s)}{1+\left(\dfrac{2\times0.58}{13}\right)s+\left(\dfrac{s}{13}\right)^2}$	0.23	$\dfrac{0.72(1+0.24s)}{1+\left(\dfrac{2\times0.45}{9.13}\right)s+\left(\dfrac{s}{9.13}\right)^2}$	0.20
	B	$\dfrac{2.2(1+0.07s)}{1+\left(\dfrac{2\times0.38}{10.7}\right)s+\left(\dfrac{s}{10.7}\right)^2}$	0.42	$\dfrac{1.4(1+0.054s)}{\left(1+\dfrac{s}{7.96}\right)}$	0.03
$\dfrac{2}{s}$	A	$\dfrac{0.47(1+0.56s)}{1+\left(\dfrac{2\times0.47}{9.37}\right)s+\left(\dfrac{s}{9.37}\right)^2}$	0.07	$\dfrac{0.51(1+0.38s)}{1+\left(\dfrac{2\times0.38}{8.5}\right)s+\left(\dfrac{s}{8.5}\right)^2}$	0.24
	A	$\dfrac{0.48(1+0.78s)}{1+\left(\dfrac{2\times0.94}{12.8}\right)s+\left(\dfrac{s}{12.8}\right)^2}$	0.02	$\dfrac{0.77(1+0.2s)}{1+\left(\dfrac{2\times0.4}{9.3}\right)s+\left(\dfrac{s}{9.3}\right)^2}$	1·6
	B	$\dfrac{2.3(1+0.11s)}{1+\left(\dfrac{2\times0.46}{16.4}\right)s+\left(\dfrac{s}{16.4}\right)^2}$	0.5	$\dfrac{1.64(1+0.08s)}{\left(1+\dfrac{s}{13.8}\right)}$	0.02

2. The gain and the lead generated by both the subjects decrease over the entire or a part of their frequency responses. This agrees with the observations made by Shirley and Young (1968), and Washizu and co-workers (1977). Maximum degradation is as much as 10 dB and 75 degrees respectively.

3. Subject B displays more sensitivity to the removal of congruent motion cue. This was confirmed by verbal comments from B. In contrast, subject A is less sensitive to removal of motion cue. Thus when congruent motion cue is removed some subjects may not be able to adapt to fixed base conditions unless they do a maximum performance task. This observation is relevant in the context of changing cues on the overall man-machine system performance.

4. Bandwidth of the models reduce.

5. Lead time constant decreases.

It may be concluded that when congruent pitch-motion is suppressed there is a degradation in the over-all Man-Machine system performance.

The seperation of the model envisaged in the previous section was attempted next. The estimated sensory plus prediction part H_{SP} and the equalising plus neuromuscular part H_{EN} were evaluated and the frequency domain versions of these are presented in figures 8 and 9. The H_{SP} with CMV shows high gain and lead features, whereas H_{EN} generally remains same for CMV and visual alone cases. Similar observations were made by Washizu and co-workers (1977) though second order unstable plant dynamics was considered there. For this case the Index of Seperability defined as the ratio $\sigma^2_{CMV}/\sigma^2_V$ is estimated to be about 4 dB.

Another method of analysing the contributions of CMV over visual alone tasks is to study the improved phase lead. Concept of a 'Predictor Operator' has been proposed in the literature (Jaegar, Agarwal and Gottlieb, 1980). The extra phase lead $\angle H_{SP\ CMV} - \angle H_{SP\ V}$ can be defined as being due to operation of a Motion Predictor Operator' in the case of combined motion visual

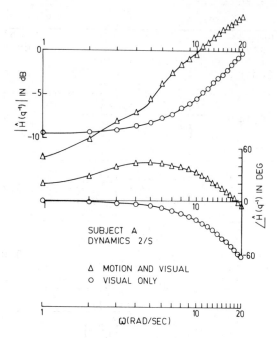

Fig. 8. SP part of LS model Sub. A.

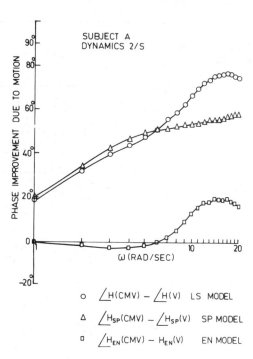

Fig. 10. Phase improvement curves Sub. A.

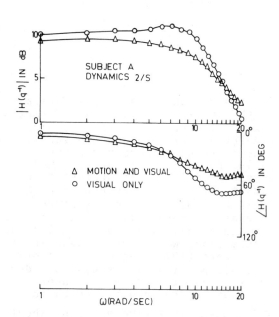

Fig. 9. EN part of LS model Sub. A.

CONCLUSIONS

It has been demonstrated that removal of congruent pitch motion cue to a pilot degrades his performance in a compensatory tracking manual control task.   A unique method of deleting motion cues from a subject after exposing him to combined motion-visual cues has been used in the ground based motion simulator.   In the pilot model analysis a Least Squares model structure has been used which allows the pilot activity to be seperated into sensory plus prediction part and an equalising plus neuromuscular part.   It has been shown that motion cues, through vestibular apparatus, provide considerable prediction or lead capability. Finally, the contributions of motion cues have been explained through the concept of a 'Motion Predictor Operator'.

SYMBOLS

ε	Pilot input
y	Pilot output
e	noise
q^{-1}	Backshift operator
A,B	n th order polynomial of discrete time series model
n	Order of the model
N	Number of observation samples
k	integer index of sampling
s	Laplace operator

cues to pilot.   In Fig.10 the phase improvement that has been generated by the congruent motion cues are shown.   The magnitude of phase improvement is as high as 80 degrees.

ω Frequency in radians per sec.
σ^2 Variance of estimated remnant
CMV Congruent Motion Visual
V   Visual
SP   Sensory and Prediction
EN   Equalising and Neuromulcular
H   Pilot description

REFERENCES

Balakrishna, S. (1976). Time domain and time series models for human activity in compensatory tracking experiments. National Aeronautical Laboratory, India. TN-50.

Bergeron, H.P., and J.J.Adams (1964). Measured transfer functions of pilots during two axis tasks with motion. NASA TN-D 2177.

Bergeron, H.P., J.J. Adams, and G.J. Hurt.Jr. (1977). The effects of motion cue and motion scaling on one and two axis compensatory control task. NASA TN-D 6110.

Jaegar, J.R., G.C. Agarwal, and G.L. Gottlieb (1980). Predictor operator in pursuit and compensatory tracking. Human Factors, 22 (4).

Jategaonkar, R.V., J.R. Raol, and S. Balakrishna (1982). Determination of model order for dynamical system. IEEE Trans. Syst., Man & Cybern., SMC-12, 56-61.

Junker, A.M., D.W. Repperger, and J.A. Neff (1975). A multiloop approach to modelling motion sensor responses. Proc.11th annual conf.on manual control, NASA TM X-62,464, 645-655.

Kuehnel, M.A. (1962). Human pilot's dynamic response characteristics measured in Flight and on a non-moving simulator. NASA TN-D 1229.

Newell, F.D. (1967). Human characteristics in flight and ground simulator for roll tracking task. NASA TN-D 5007.

Osafo-Charles, F., G.C. Agarwal, and W.D.O'Neill (1980). Application of time series modelling to human operator dynamics. IEEE Trans.Syst., Man and Cybern., SMC-10, 849-860.

Peters, R.A. (1969). Dynamics of vestibular systems and their relation to motion perception, spacial disorientation and illusions. NASA CR-1309.

Santhanam, C.K., and J. Koerner (1963). Transfer function synthesis as a ratio of two complex polynomials. IEEE Trans.Autom.Control., AC-8, 56-68.

Seckel, E., I.A.M. Hall, D.T. McRuer, and D.H. Weir (1957)..Human pilot dynamic response in Flight and simulators. WADC TR-52.

Shinners, S.M. (1974). Modelling of human operator performance utilising time series analysis. IEEE Trans. Syst., Man & Cybern., SMC-4,

Shirley, R.S., and L.R. Young (1968). Motion cues in man-vehicle control - Effects of Roll motion cues on human operator's behaviour in compensatory systems with disturbance inputs. IEEE Trans. Man Machine Systems., MMS-9,4, 121-128.

Smith, H.J.(1966). Human describing functions measured in flight and on simulators. NASA SP-128.

Stapleford, R.L., R.A. Peters, and F.R. Alex (1969). Experiments and a model for pilot dynamics with visual and motion inputs. NASA CR-1325.

Tanaka, K., N. Goto, and K. Washizu (1976). A comparison of techniques for Identifying Human operator dynamics utilising time series analysis. Proc. 12th ann. conf. on Manual Control, NASA TM X-73,170, 673-693.

Van Gool, M.F.G., and H.J. Mooij (1976). A comparison of inflight and ground based pitch attitude tracking experiments. Proc. 12th ann. conf. on Manual Control, NASA TM X-73,170, 443-454.

Washizu, K., K. Tanaka, S. Endo, and T. Itoke (1977). Motion cue effects on human pilot dynamics in Manual Control. Proc. 13th ann.conf. on Manual Control, NASA CR-158107, 403-413.

TRACKING PERFORMANCE WITH VARYING ERROR-CRITERIA

T. Bösser

*Psychologisches Institut der Westfälischen Wilhelms-Universität,
Schlaunstr. 2, D-44, Federal Republic of Germany*

Abstract. A compensatory tracking task including an additional
error condition is introduced. A number of experiments show that
nonlinear manual control behaviour can be induced by the weight
given to the error-conditions. Although it represents only a small
part of the output-variance, it still may be of importance for the
stability of a man-machine system. The adaptation of the human
operator can be regarded as the selection of an optimal strategy
on the basis of the weight given toconsequences of actions.

Keywords. Man-machine systems, motivation, nonlinear control,risk,
target tracking.

Human manual control behaviour is mostly
modelled with formal analogies drawn from
control theory and described in the form of a
linear filter. This approach has the advantage of making available a comprehensive
methodology, and it has given satisfacory
results for describing typical tracking
behavior in laboratory experiments. It also
offers a basis for investigating many man-
machine problems in areas, where perceptual-
otor characteristics of the man in the
control-loop limit the system performance.
Nonlinear models have not gained much acceptance, mainly on the grounds that they do not
improve the descriptive properties of the
model much (JOHANNSEN 1972).

Typical behaviour in some manual control
tasks, like driving, can not be very well
modelled by linear describing functions. It
was pointed out before (BÖSSER 1980), that
the typical manual control behaviour may be a
function of the very restrictet task. A model
based mainly on investigations of the behaviour in tracking tasks in the laboratory,
which may induce linear behaviour, will only
be valid for tasks of a correspondigly simple
structure.

What are the aspects of a tracking task
that constitute this simplicity or
complexity? The two main aspects are the
number of concurrent tasks and the importance
given to the error. Almost all conceivable
tasks in real life consist of several tasks
beeing executed concurrently: Driving a car
requires, among other things, keeping the car
in the center of the lane, avoiding excessive
lateral acceleration and vibration in order
to drive safely, use minimal time and
expenses, and not to awaken the dcg on the
rear seat. Some of these tasks overlap, and
some may have undoubted priority, but all
actions are considered under all of the

relevant aspects al all times. Decisions are
made such as to be optimal in regard to all
of those aspects.

Secondly deviations from the set point
do not have equal importance in real
situations, or importance corresponding to
the minimization of a quadratic (RMS-) cost
function. Although the RMS-error criterion
seems a reasonable representation for many
real tasks, where large deviations from the
target value are given unproportionally large
weight, there are many other situations where
there is indifference to the state, as long
as a certain limit is not exceeded, or tasks
may be inherently asymmetric.

The question of concurrent tasks has
been investigated in connection with the
'central capacity', or with 'attention'
(KAHNEMANN 1973), typically in experiments
using a time-sharing paradigm and tasks with
non-overlapping requirements and error-
criteria. NAVON & GOPHER (1979) suggest a
view of human performance which assumes,
similar to economic concepts, the allocation
of 'resources' - independent dimensions of
man's capabilities - to one or more tasks in
order to opimize total performance.

In our view, not in disagreement with
NAVON & GOPHER's, motivation and decision
processes are seen as outer control loops
generating the target values for executing
the manual control task. Motivation represents the evaluation of the outcomes of
alternative behaviour strategies (they are
similar to the 'utilities' of decision
theory) and decision processes represent the
information processing required by the task.
This is a much more restricted definition of
motivation than the usual loose term. Human
decision-making may have some peculiarities
(TVERSKY & KAHNEMANN 1981) in that human is

not a rational decision maker in the sense of normative decision theory, such that his decisions are not entirely predictable on the basis of the utilities. In manual control-tasks motivation and decision generate the set point for the manual control task. These processes govern the adaptation of the manual control performance by the consequences the environment delivers for the alternative actions. They may be the factors underlying the adaptation of manual control behaviour.

In order to investigate some aspects of these assumptions we define a compensatory tracking task with an additional task-dimension. Several experiments are conducted to study the rules underlying the adaptation of the subjects behaviour to these task conditions.

Our hypothesis includes the assumption that nonlinear tracking behaviour may be induced by the experimental conditions. In order to identify the proportion of output-variance due to nonlinear behaviour we regard the output variance of the human operator as composed of
- linear
- nonlinear and
- remnant (not explained). The usual 'describing-function'-view would consider 'remnant' and 'nonlinear' together. Nonlinear characteristics will be identified with the nonparametric identification procedure suggested by WIENER (1958).

Although the experimental situation is not a simulation, it does have similarities to some real-life tasks like speed- and distance control in an automobile, landing an aeroplane, or operationg a crane.

METHOD

A standard compensatory tracking task with proportional control and zero-order control system was augmented by a second error-condition: The oscilloscope display shows a borderline (FIG.1) and the subject is instructed to prevent the error-marker from crossing this line.

The display is generated by computer at a rate of 50 Hz, the disturbance is generated by filtering a 3-level pseudorandom-signal (period of 40.8 secs) with a second-order filter, first break-frquency at 0.1 Hz, the second one varying from 0.2 to 1.4 Hz. (Fig.2)

SUBJECTS were eight psychology students known to be good trackers. They received about 8 hours of training on various conditions before taking part in the experiment.

The first three of four experiments were conducted with all subjects, for a more convenient representation they are presented together.

EXPERIMENT I
Varied the distance D of the borderline

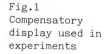
Fig.1
Compensatory display used in experiments

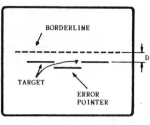

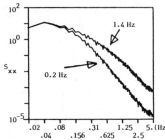
Fig.2
Spectra of track, lowest and highest freq. used. Break-freq. at 0.2 and 1.4 Hz

(0.5, 0.8, 1.1, 1.4 cm) and the spectrum of the disturbance (second break-frequency 0.2, 0.6, 1.0, 1.4 Hz), such as to span the whole range from a very easy to a very hard task. Each task combination was performed by each subject for 10 minutes, with 10 minutes training before, on 5 days of one week. The subjects were instructed in addition to the compensatory tracking task 'to cross the line as little as possible'. Payment was promised based on performance.

EXPERIMENT II
Consisted of five sessios without borderline, otherwise the same as Expt. I.

EXPERIMENT III
Was the same as experiment I, except that the subjects were instructed to give 'highest priority to the line-criterion, but also to do as well as possible to reduce the error'. Payment was promised adjusted accordingly.

RESULTS

All subjects were able to follow the instructions and to perform the task satisfactorily. Fig.3 shows that the relative RMS-error (RMS observed / variance of the track) increases regularly as a function of frequency and decreasing distance of the line, when higher priority is given to the line-criterion the increase in error is much larger (Fig.4).

The time over line increases, when both types of error have the same weight, with higher frequency of the disturbance, and with line closer to the target. (Fig.5) However, when time of crossing the line is given priority, total time over the line can be reduced to a very low value under all conditions (Fig.6). This has the consequence of increasing other types of error, the constant position error (Fig. 7 and 8) becomes much larger. Thus the human is able to adjust his tracking performace according to his criteria for weighting the error.

The distributions of the error (Fig.9)

demonstrate how the subjects adapt to task-demand and accept a different distribution of the error. By FFT-procedures the transfer-function was calculated and the remnant, the part of the variance not due to linear filtering of the input, was calculated. The distributions (Fig.10) show that the proportion of the output not accounted for by the linear transfer function is distributed asymmetrically when this is induced by the priority given to the line-error.

In Fig.11 the error vers. speed of the error-marker are represented as a joint probability-density-distribution. From this representation the strategy of the subjects for adapting to the changing weight given to the error is apparant: In order to prevent the error-marker from reaching the line at the higher frequency of the disturbance, the marker is moved away from the center, thereby accepting a larger RMS-error. In the joint-pdfs a dynamic asymmetry may be suspected, caused by the movements of the subject away from the borderline beeing faster than those towards the borderline.

The nonlinear transfer-characteristics can be determined by the WIENER-method in an analogous way to the linear case by measurement of the higher order kernels. We restrict ourselves to the second order kernel, the higher order kernels requiring much more data and impractical calculation-time. LEE & SCHETZEN(1965) showed that the WIENERkernels can be measured by calculation of the higher-order crosscorrelations, when the input to the systems is white noise. When the spectrum of the input is restricted, a measurement of the higher-order transfer-functions is possible (MARMARELIS & MARMARELIS 1978). Here the second order crosscorrelations between disturbance and the remnant ,as mentioned above for two conditions with the same input-spectrum are shown (Fig.12), they can be compared due to the same input. The peculiar form of the kernels reflects the highly correlated input signal, the large difference in the value of the second order correlations indicates a much larger contribution of the second order kernel when the line is closer and when high priority is given to observing the line.

The second order kernel represents the dynamic asymmetries (HUNG, STARK & EYKHOFF 1979) in the humans control behaviour, which is not an unusual feature in biological control systems (CLYNES 1961). These experiments demonstrate that manual control behaviour adapts to the error-criteria, which the subject is instructed to follow, or which may be inherent in his evaluation of the ccnsequences of his actions. An asymmetric error-criterion induces dynamic nonlinear behaviour.

Further analysis was directed towards identifying the rules according to which the adaptation takes place. In experiments I and III subjects were asked to rate the difficul-ty of each experimenting condition on a scale from 0(very easy) to 10(very difficult). The ratings of experiment I (Fig.13) plotted relative to RMS-error increase for some conditions, but here greater difficulty of the task can still be compensated by increasing capacity (attention) allocated to the task.

The rating of task difficulty as a function of both types of error differs among subjects, Figs.14 and 15 are the data from the same subject. The rating of difficulty agrees largely with the error in both error - dimensions.

EXPERIMENT IV
Another experiment was designed to test the ability for adaptation to different weighting of the error-conditions. One experienced subject participated.

METHOD. There was only one condition, distance of line D=0.8 cm and an input spectrum similar to the one previously used with break-frequency 1.4 Hz.

The experimental condition was the weighting given to the two types of error: The subject was instructed to give different weight, ranging from 100%/0% to 0%/100% (all possible combinations) in steps of 10% to the two types of error. They were presented in random order on the same day, one training session and one experimental session for each condition. Four cycles of this sequence were executed on four consecutive days, with intervals between trials. On the first and second day the subject was asked to do as well as possible, the first day was regarded as training. On the third day the subject was asked to do the task in an 'easy and relaxed' way, on the fourth day he was asked to monitor, in addition to the tracking-task, a video-recording of a traffic-situation and also a noisy, disturbing environment was created by various means.

RESULTS

The results of this experiment are summarized in Fig.16. The performance is represented for all conditions as a function of time over line and of RMS-error. These performance-operating curves demonstrate the limits of the subjects performance under the three instructions. He is able to allocate his resources such as to take into account the different error-criteria in a widely varying range. This holds also, when his capacity is impaired by an additional task and by disturbances in the environment. The POC represents, most likely, the best performance the subject is able to give under the specific conditions.

Which point on the POC, and according to which rule, does the subject select? Among the advantages and disadvantages to be taken into account is the subjective judgment of difficulty the task is given. Whether this represents the proportion of the ressources

394   T. Bösser

allocated, or just the self-evaluation of
performance can not be inferred from this
investigation. The subject in experiment IV
preferred the extreme conditions where one of
the two error-conditions was given overall
priority, so in this case there is no obvious
preference for the linear or nonlinear
strategy. Rather the situation where both
criteria have to be observed equally seems
less preferable.

DISCUSSION

It has been demonstrated that nonlinear
manual control behaviour can be induced by
the experimental conditions, and therefore
also prevented from occurring by those conditions. The argument that only a small
portion of the variance observed can be
attributed to nonlinear control is not valid
for two reasons: The proportion of the
variance is calculated as a part of the total
variance, but may only be of importance in
the 'crossover-region' of the human operators
transfer-function, which is the critical area
for stability and performance in man-machine
systems. In order to determine the exact
nature of the nonlinear properties, experiments better suited to their identification
have to be done.

The dynamical asymmetry observed is a
function of the weight given to certain
consequences of the systems behaviour. Under
real conditions asymmetrical weighting-func-
tions occur frequently, in particular in
conjunction with safety and accidents. 'Risk'
can be understood to be the cost associated
with large and rare deviations - this represents an extreme weighting function. The
stability of the system depends upon staying
away as safely as possible from these extreme
conditions, i.e. the distribution of the
error must be sure to be at or near the value
of zero at certain extreme deviations. This
however, may only be achievable at the very
costly consequence of larger deviations in
another performance-dimension. The availability of an asymmetrical control strategy would
be very efficient, especially if only the
extreme deviations are concerned. Such a
strategy would account for a minimal proportion of the variance in the observable
behaviour, but of highest importance for the
stability (safety) of the system.

There is no reason to assume a
preference for or avoidance of the nonlinear
strategy we observed.

The assertion that man selects from the
possible behaviour strategies an 'optimal'
one is tautological, as long as the optimality criteria are not specified. If however we
prefer to accept this as a convenient axiom,
the empirical problem remains to identify
- the number of independent dimensions
  present to evaluate actions and
- the scaling of the axis of this space,
the scaling of the axis of this space
represents motivation.

In principle the number of consequences
observed at all times is very large, and
there are some very subjective ones involved,
e.g. comfort, sympathy etc. which also serve
as basis for making decisions, but largely
defy measurement.

In most cases, however, the determinants of the situation reduce the number of
relevant dimensions in the decision-space to
a very small number, which also corresponds
with observations from investigations of
human decision making (RESTLE 1961),
indicating that man tends to base his
decisions on as few dimensions as possible.

LITERATURE

Bösser, T. (1980). Effects of Motivation on
    Car-Following. 16. Annual Conference on
    Manual Control, May 5.-7. Massachusetts
    Institute of Technology, Cambridge,
    Mass.
Clynes, M. (1961). Unidirectional rate
    sensivity: A biocybernetics law of
    reflex and humoral systems as
    physiologic channels of control and
    communication. Ann. N. Y. Acad. Sci.,
    92, 946-969.
Hung, G., Stark, L. & Eykhoff, F. (1977). On
    the interpretation of Kernels I.
    Computer Simulation of Responses to
    Impulse Pairs. Annals of Biomedical
    Engineering 5, 130-143.
Johannsen, G. (1972). Development and
    Optimization of a Nonlinear
    Multiparameter Human Operator Model.
    IEEE Trans. Systems, Man and
    Cybernetics, SMC-2, 494-503.
Kahnemann, D. (1973). Attention and Effort.
    Prentice-Hall, Inc., Englewood Cliffs,
    New Jersey.
Lee, Y. W. & Schetzen, M. (1965). Measurement
    of the Wiener kernels of a nonlinear
    system by cross-correlation.
    Int.J.Control,2, 237-254.
Marmarelis, P. Z. & Marmarelis, V. Z. (1978).
    Analysis of Physiological Systems. The
    White- Noise Approach. Plenum Press, New
    York and London.
Navon, D. & Gopher, D. (1979). On the Economy
    of the Human-Processing System.
    Psychological Review. 86, 214-255.
Restle, F. (1961). Psychology of Judgement
    and Choice. Wiley,New York.
Tversky, A. & Kahnemann, D. (1981). The
    Framing of Decisions and the psychology
    of Choice. Science, 211, 453-457.
Wiener, N. (1958) Nonlinear Problems in
    Random Theory. Wiley, New York.

This work was supported by Deutsche
Forschungsgemeinschaft. Elke Melchior, Peter
Schäfer and Martin Schütte contributed to
this research.

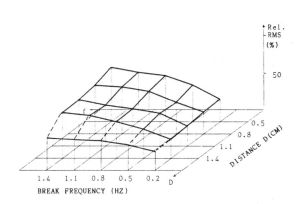

Fig.3   RMS-error in expts. I and II

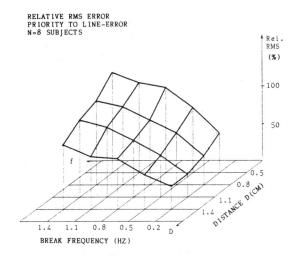

Fig.4   RMS-error in expt III

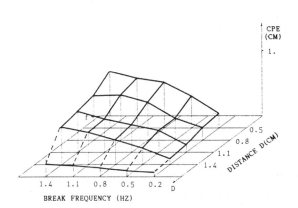

Fig.5   Constant position error in
        expts. I and II

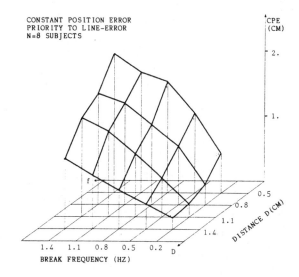

Fig. 6   Constant position error in expt.III

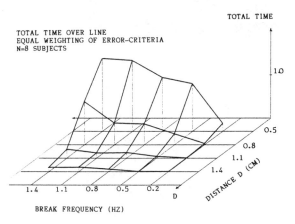

Fig.7   Total time over line in
        expts. I and II

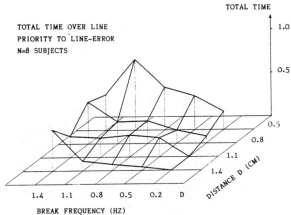

Fig.8   Total time over line in expt. III

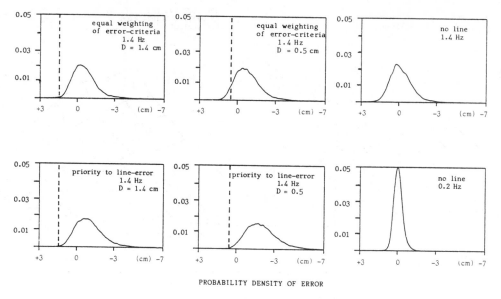

Fig.9   Probability-densities of error in expts. I, II and III

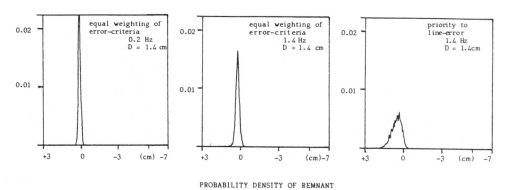

Fig.10   Probability-densities of remnant (subject J.R.) in expts. I, II and III

Fig.12   Second order cross-correlation

$$R_{\tau_1,\tau_2} = 1/n \sum_{i=1}^{n} x_{i-\tau_1} \cdot x_{i-\tau_2} \cdot z_i$$

(z = remnant)

subject J.R., D=1.4cm, 1.4Hz
equal weighting of error

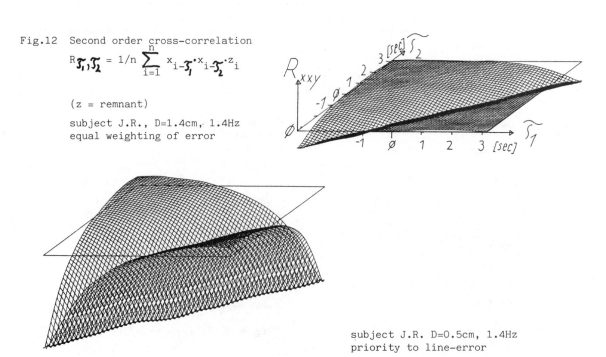

subject J.R. D=0.5cm, 1.4Hz
priority to line-error

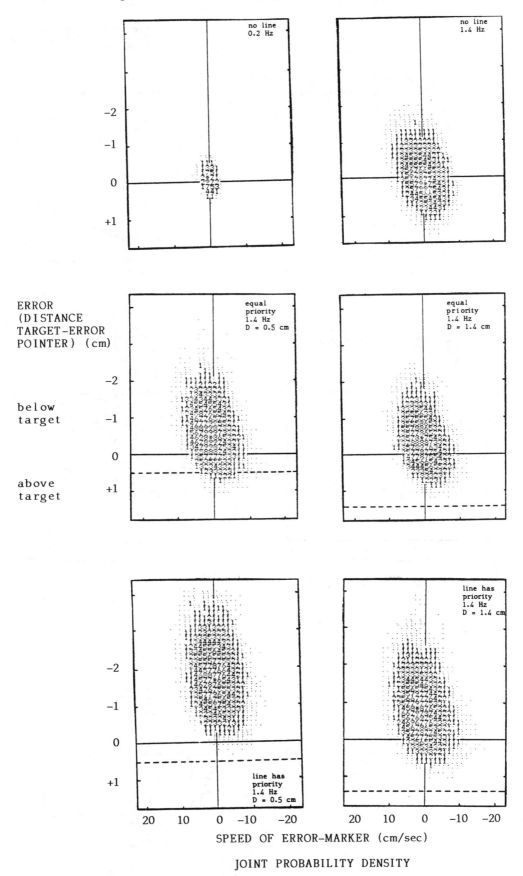

JOINT PROBABILITY DENSITY

Fig.11   Joint probability-densities expts. I, II and III

RATING OF TASK DIFFICULTY VS. TIME OVER LINE
EQUAL WEIGHTING OF ERROR CRITERIA
N=8 SUBJECTS

Fig.13 Rating of task-difficulty
        vers. RMS-error

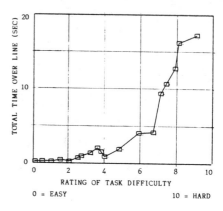

Fig.14 Rating of task-difficulty in
        expt.I (subject J.G.)

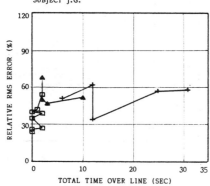

Fig.15 Rating of task-difficulty in
        expt.III (subject J.G.)

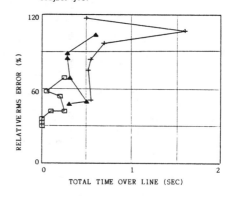

Fig.16 Performance operating
        curve (subject J.G.)
        expt. IV

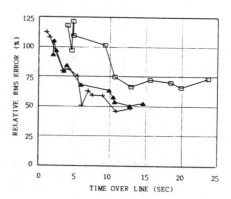

ACCIDENT HAZARDS OF REAR WHEEL STEERED VEHICLES

L. Strandberg, G. Tengstrand and H. Lanshammar

Accident Research Section, National Board of Occupational Safety and Health, S-171 84 Solna, Sweden

Abstract. Many severe accidents with fork lift trucks and loaders may be due to their rear wheel steering (RWS). Tyre nonlinearities inhibit recovering from a rear wheel skid with RWS. Linear analysis reveals some inherent RWS hazards: Asymptotic lateral vehicle stability at every speed requires that the rear wheels have a larger cornering stiffness coefficient than the front ones. Thus a front loading vehicle may become unstable upon unloading; Without lateral king-pin inclination, steering system stability during straight driving demands a positive caster distance, which however may increase the RWS angle if the steering-wheel is released during cornering; The phase lag between steering input and lateral acceleration output is usually larger in RWS compared to front wheel steering (FWS) vehicles. Experiments with an instrumented fork lift truck point at significantly larger response times with RWS compared to FWS and to ordinary automobiles.

Keywords. Automobiles, controllability, dynamic response, fork lift trucks, man-machine systems, stability, vehicles.

1 BACKGROUND

The official statistics of Sweden indicate that vehicles are involved in about half of all fatal occupational accidents. Most of these fatalities occur while the vehicle is in motion. Thus it is important to occupational safety that the driving characteristics are improved at work vehicles.

Unfortunately, certain vehicle design principles, that have been introduced for low speed manoeuvrability, bring about poor stability and poor controllability at high speed. Such incentives and effects from articulation and rear axle steering in truck-trailer combinations were pointed at by Strandberg, Nordström, and Nordmark (1975). Seemingly, accident hazards have developed similarly through rear wheel steering (RWS), which is common in fork lift trucks and loaders.

Quite a few reports on accidents with RWS vehicles have been found, where the events are blamed on driver errors only. In doing so the reports are symptoms of a wide-spread ignorance of the deceptive driving characteristics of RWS vehicles. A recent Swedish report (Strandberg, 1981) on the man-machine dynamics of RWS was considered a novelty by mass media and RWS vehicle manufacturers, thus confirming the urgent need of further knowledge on the Control and Man-Machine System (CS and MMS) properties

of RWS vehicles.

The CS and MMS qualities, elucidated in this paper , have been found to be particularly crucial for the safety of RWS vehicles. These qualities have been selected with evidence from accident descriptions in literature (e.g. Bonefeld and Macheleidt, 1976), in labour inspectorate investigation reports, and in the new computer based Swedish information system on occupational injuries (described by e.g. Andersson and Lagerlöf, 1978, and Strandberg and Lanshammar, 1981).

A typical RWS accident may occur as follows (references to the sections below within parentheses). "After unloading, the vehicle is driven at top speed. Upon a driver steering action or a road unevenness, the ground forces at the steered wheels tend to increase the steering angle (section 5). The lateral acceleration of the vehicle increases in amplitude. The sideslip angle of the rear wheels exceeds its optimal value, when the driver tries to compensate for their skid in the common way by steering in the direction of the skid (section 4). Finally, the vehicle goes out of control and collides or overturns. Afterwards it was discovered that the tyre inflation pressure of the rear wheels (section 3) was much below the recommended value, or that only the front, driven, wheels were studded."

2   VEHICLE MODEL

Only lateral and yaw motions will be considered, since they are decisive of both skidding and overturning tendencies. Therefore, the forces, velocities and positions of the four wheels are calculated as averages for the two axles, and the influence on the side force coefficients from dynamic load transfer as well as from longitudinal tyre forces is neglected.

The vehicle and its load, if any, is assumed to consist of a rigid mass (m) with its centre (O) constituting the origin of a cartesian coordinate system Oxyz (cf. Fig. 1). The corresponding unit vectors are vehicle fixed. If one refers to the current direction of driving they are directed forwards ($\underline{e}_x$), to the left ($\underline{e}_y$), and upwards ($\underline{e}_z$). The front and rear axles are labelled with indices 12 and 34 respectively. Although steering is realized at one axle only, steering angles (δ) are introduced at both axles (index 12 and 34) in the following expressions. Consequently, δ_{12} should be set to zero with RWS and $\delta_{34} = 0$ with front wheel steering (FWS).

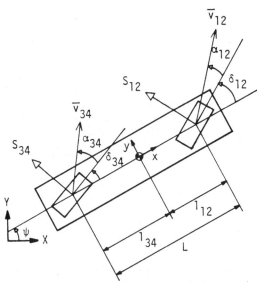

Fig. 1.   Definition of axle positions and positive directions in the vehicle model. Note that the physical side force (S) must be negative if the sideslip angle (δ) is positive.

The origin O has a translational velocity, $\underline{v} = v_x \underline{e}_x + v_y \underline{e}_y$, and a rotational velocity, $\dot{\psi}\underline{e}_z$, where ψ is the so called yaw angle, taken counterclockwise, from a ground fixed X-axis to the vehicle fixed x-axis. The sideslip angle (α) is the angle, taken counterclockwise, from the wheel symmetry plane to the velocity vector of the wheel centre. At each axle

$$\alpha = \arctan\left(\frac{v_y + x\dot{\psi}}{|v_x|}\right) - \delta\ \mathrm{sgn}\ v_x \qquad (1)$$

where the x-coordinate is l_{12} for the front axle and $-l_{34}$ for the rear one. Each side force (S) is acting on the tyre in the ground contact area colinearly to the spin axis of the wheel. If α is assumed to be small ($\alpha \approx \tan \alpha$), two tyre constants may be introduced from SAE (1975): the cornering stiffness (C) and the cornering stiffness coefficient (K). In the linear case one obtains for each axle

$$S = -\ C\alpha = -\ C\left(\frac{v_y + x\dot{\psi}}{v_x} - \delta\right)\mathrm{sgn}\ v_x \qquad (2)$$

$$C \triangleq KN = K\ \frac{mg(L-1)}{L} \qquad (3)$$

where N is the normal force, g the acceleration of gravity (9.81 m/s^2), L the wheel-base, and 1 the distance between O and the axle.

Differentiation of the translation and rotation velocity vectors of the origin yields the lateral force and yaw moment equations

$$m(\dot{v}_y + v_x\dot{\psi}) = S_{12}\cos\delta_{12} + S_{34}\cos\delta_{34} \qquad (4)$$

$$J_z\ddot{\psi} = l_{12}S_{12}\cos\delta_{12} - l_{34}S_{34}\cos\delta_{34} \qquad (5)$$

where J_z denotes the vehicle mass moment of inertia with respect to the z-axis. Below, the steering angles will be considered small. Hence, $\sin\delta \approx \delta$ and $\cos\delta \approx 1$. Linearization is completed by considering the x-component of the mass centre velocity to be constant (v) and forwards directed, i.e. $v_x = v > 0$.

The state equation of the vehicle can now be obtained in its normal form

$$\dot{\bar{\xi}} = A\ \bar{\xi} + B\ \bar{\eta} \qquad (6a)$$

The state vector $\bar{\xi}$, the control force vector $\bar{\eta}$, the system matrix A, and the distribution matrix B are defined below, where the components of the state equation have been solved from the expressions (2), (4), and (5).

$$\xi \triangleq \begin{pmatrix} v_y \\ \\ \dot{\psi} \end{pmatrix} \qquad (6b)$$

$$\bar{\eta} \triangleq \begin{pmatrix} \delta_{12} \\ \\ \delta_{34} \end{pmatrix} \tag{6c}$$

$$A \triangleq \begin{pmatrix} \dfrac{-C_{12}-C_{34}}{mv} & \dfrac{-C_{12}l_{12}+C_{34}l_{34}}{mv} -v \\ \\ \dfrac{-C_{12}l_{12}+C_{34}l_{34}}{J_z v} & \dfrac{-C_{12}l_{12}^2-C_{34}l_{34}^2}{J_z v} \end{pmatrix} \tag{6d}$$

$$B \triangleq \begin{pmatrix} \dfrac{C_{12}}{m} & \dfrac{C_{34}}{m} \\ \\ \dfrac{C_{12}\,l_{12}}{J_z} & \dfrac{-C_{34}l_{34}}{J_z} \end{pmatrix} \tag{6e}$$

3 LATERAL STABILITY CRITERIA

Control systems theory (see e.g. Elgerd, 1967) states that a linear system described by Eqs. (6) is asymptotically stable if and only if none of the eigenvalues of the system matrix A has a nonnegative real part. The eigenvalues are obtained as the roots $s = \lambda_1, \lambda_2$ of the characteristic equation

$$\det(sI - A) = 0 \tag{7}$$

where det denotes the determinant, s the complex frequency and I the identity matrix. The roots of Eq. (7) are

$$\lambda = -\frac{\beta}{2} \pm \sqrt{\frac{\beta^2}{4} - \gamma} \tag{8}$$

where

$$\beta \triangleq \frac{C_{12}+C_{34}}{mv} + \frac{C_{12}l_{12}^2+C_{34}l_{34}^2}{J_z v} \tag{9}$$

$$\gamma \triangleq \frac{C_{12}C_{34}L^2}{mJ_z v^2} + \frac{-C_{12}l_{12}+C_{34}l_{34}}{J_z} \tag{10}$$

Since $\beta \geq 0$, the stability condition is satisfied if $\gamma > 0$. By insertion of (3) into (10) it results that the unequality $\gamma > 0$ is always valid, if the cornering stiffness coefficient (K, equal to the slope of the trajectory at the origin of Fig. 2) is larger at the rear axle than at the front one, i.e. if

$$K_{34} > K_{12} \tag{11}$$

If (11) is not valid, $\gamma > 0$ and stability is

achieved only below the critical speed (v_c)

$$v_c = \sqrt{\frac{K_{12}K_{34}Lg}{K_{12} - K_{34}}} \tag{12}$$

Since tyres usually increase their K value when their load decreases (see Fig. 2 and e.g. Freudenstein, 1961), a front loading vehicle may become unstable upon unloading. This paradoxical effect is even more likely to occur with insufficient rear tyre pressure or if only the front wheels (often the driven and braked in fork lifts) are studded. The decreasing influence on K from longitudinal tyre forces (see e.g. Nordström, Nilsson, and Nilsson, 1974) imposes a hazard on rear wheel driven vehicles, but also implies a possibility to regain stability in conventional fork lift trucks (by increasing throttle or by sudden braking). Finally, the deteriorating influence from a shorter wheel-base, L in (12), should be pointed at.

4 SKID CONTROL

Consider a vehicle turning to the left ($\ddot{\psi} > 0$). If the rear wheels start skidding (i.e. $|\alpha_{34}|$ approaches α_{opt} in Fig. 2), it is well-known that the driver of a FWS vehicle may control the vehicle and make $\ddot{\psi} < 0$ by steering more to the right (decreasing δ_{12}) in the direction of the skid. Then the absolute values of the demanded side force and the sideslip angle decrease at both axles (see Fig. 1), and the driver may succeed in keeping $\alpha < \alpha_{opt}$ (see Fig. 2) through quick and precise steering motions. In a RWS vehicle, on the other hand, recovering from such a rear wheel skid requires larger side forces at the rear axle.

Since S_{12} and S_{34} have the same sign during cornering, this difference in side force demand between FWS and RWS may as well be interpreted from Eq. (5), where

$$\frac{\partial\ddot{\psi}}{\partial S_{12}} > 0 \text{ and } \frac{\partial\ddot{\psi}}{\partial S_{34}} < 0 \tag{13}$$

During forwards driving ($v_x > 0$), it is also clear from Eq. (1) that

$$\frac{\partial\alpha}{\partial\delta} < 0 \tag{14}$$

which means that a negative α_{34} value will become even more negative in a RWS vehicle when the driver tries to recover from a rear wheel skid by changing δ_{34} in the positive direction. Hence, α_{34} may exceed α_{opt} and controllability is lost. With FWS, recovering is achieved by a δ_{12} decrement, thus incrementing α_{12} towards a smaller absolute value.

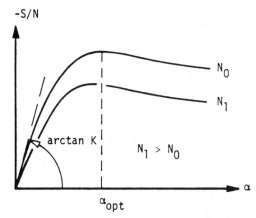

Fig. 2.   Common functional relationship of
          pneumatic tyres between the side
          force coefficient (S/N) and the
          sideslip angle (α) at different
          normal forces (N). The constants K
          and α_{opt} introduced in Eq. (3) and
          section 4 respectively, are indica-
          ted for one normal force (N_0).

The last mentioned relationship between
differences in α_{34} and δ_{34} may even result
in immediate and simultaneous loss of
control and stability when a quick change in
steering angle is imposed on a previously
stable RWS vehicle.

Hence, the driver cannot recover from a rear
wheel skid with RWS, which has been
confirmed during driving experiments on
slippery surfaces.

5   STEERING SYSTEM TORQUE

A safe MMS should return to a stable state
when the human operator becomes inactive.
Thus, the side force moments in a steering
system should tend to return the steering
angle to zero. Therefore, at least two
conditions must be satisfied:
a) The steering system in itself must be
   asymptotically stable;
b) During steady-state cornering, the
   steering-wheel torque (M) required from
   the driver to maintain the steered wheel
   side forces should have the same
   direction (sign) as the current steering-
   wheel angle (θ, considered positive
   during left-hand cornering).

Consider a steering system, where the lateral king-pin inclination and the compliance
are negligible. The steering ratio is deno-
ted by $i = \theta / \delta$ or by $n = |i|$. Note that
$i < 0$ for RWS. The caster distance (ϵ) is
considered positive, if the side force re-
sultant act on the tyre behind the intersection with the ground of the steering axis.
Although the "pneumatic" caster of the tyres
themselves often is positive for small sideslip angles (see e.g. Freudenstein, 1961),
FWS suspensions have usually a "geometric"
caster above zero, as well, to increase the
total self-aligning torque from the side
forces. The required steering-wheel torque is

$$M = J_s\ddot{\theta} + d\dot{\theta} + t\epsilon S /n \qquad (15)$$

where J_s is the equivalent moment of
inertia at the steering-wheel (including the
steering axis moment of inertia of the
steered wheels reduced by n^2), d is the
viscous damping constant, and $t = +1$ for FWS
while $t = -1$ for RWS vehicles. Consider
straight forward driving ($v_y = \dot{\psi} = 0$)
and substitute the wheel side forces (S) in
Eq. (15) from Eq. (2). Then Laplace
transformation yields the transfer function

$$\frac{\theta(s)}{M(s)} = \frac{\dfrac{1}{J_s}}{s^2 + \dfrac{d}{J_s} s + \dfrac{t\epsilon C}{J_s in}} \qquad (16)$$

Since $sgn(i) = t$, the two eigenvalues (λ_s)
have the same expression for both FWS and
RWS:

$$\lambda_s = -\frac{d}{2J_s} \pm \sqrt{\frac{d^2}{4J_s^2} - \frac{\epsilon C}{n^2 J_s}} \qquad (17)$$

Hence, asymptotic steering system stability
during straight driving requires a positive
caster distance ($\epsilon > 0$). This conclusion is
supported by experimental observations by
Perret (1964) but contradicts Mitschke
(1979), claiming that a negative caster is
necessary for the stability of RWS vehicles.
It is interesting to note how two different
approaches to the same problem end up in
contradictory solutions.

Mitschke's analysis (1979) arrives at a
caster criterion for a laterally stable
vehicle model (corresponding to Eqs. (6)
above) by assuming fixed control, i.e.
constant steering-wheel angle, and by
reducing the steering system to a constant
resilience about the king-pins (between the
vehicle body and the wheels, that are
considered free from inertia). If such a
suspension is installed with a negative
caster at the rear axle, it acts similarly
to a conventional roll steer design (see
e.g. Mitschke, 1972), making the condition
$\gamma > 0$ more likely to be satisfied.

The expression (17) on the other hand is
arrived at by considering the steering
system stability only, and disregarding the
lateral motions of the vehicle.

Continuing from the stability and CS
requirement (a) to the MMS condition (b)
above, Eq. (15) and Fig. 1 yield during
steady-state cornering

$$M = t \ \epsilon S/n \qquad (18)$$

$$sgn(S) = sgn (\theta) \qquad (19)$$

Hence, the condition (b) requires that $\epsilon_R < 0$ in a RWS vehicle (index R). Consequently, both the CS and the MMS requirements cannot be fulfilled only through a certain caster distance in RWS vehicles. However, Zomotor (1963) and Perret (1964) found acceptable steering properties when a negative caster was combined with a lateral king-pin inclination, which requires the vehicle to ascend when the steering angle increases.

The caster design conflict in RWS vehicles may be enlightened by the fact that the side forces at the steered wheels change direction during the transition from straight driving to steady-state cornering.

6 LINEAR MODEL STEERING RESPONSE

Even if the vehicle and the steering system are stable during normal driving, the MMS emergency avoidance performance may vary significantly with the vehicle parameters. Cf. Maeda and others (1977). In a number of other studies, as well (e.g. ISO, 1976, and Repa and others, 1977), the steering performance was found to be improved with decreasing time lag between the steering angle input and the output of yaw velocity ($\dot\psi$) or lateral acceleration ($a_y = \dot v_y + v_x\dot\psi$, cf. Eq. (4)). Therefore, the corresponding transfer functions will be evaluated in this section to further illustrate the specific and sometimes hazardous MMS properties of RWS vehicles. Also from a pure CS viewpoint, RWS has been found more demanding than FWS (Ito, Sarumaru, and Ikeya, 1975).

The wheel steering angles are considered the components of the input vector, η in Eq. (6c). The output vector, ζ, is defined below.

$$\bar\zeta \triangleq \begin{pmatrix} a_y \\ \dot\psi \end{pmatrix} = E\bar\xi + D\bar\eta \qquad (20a)$$

where

$$E \triangleq \begin{pmatrix} A_{11} & A_{12} + v \\ 0 & 1 \end{pmatrix} \qquad (20b)$$

$$D \triangleq \begin{pmatrix} B_{11} & B_{12} \\ 0 & 0 \end{pmatrix} \qquad (20c)$$

and where A_{11}, A_{12}, B_{11}, and B_{12} are components of A and B in Eqs. (6d) and (6e). Laplace transforming Eq. (6a) and solving for $\bar\xi$ yields

$$\bar\xi = (sI - A)^{-1} B\bar\eta \qquad (21)$$

The inputs and outputs are related by the transfer function matrix F.

$$\bar\zeta = F\bar\eta \qquad (22a)$$

where F is solved from Eqs. (20a) and (21)

$$F = E(sI - A)^{-1} B + D \qquad (22b)$$

Dividing F into one mutual transfer function, M, and one transfer matrix, G, yields

$$F = MG \triangleq \frac{1}{s^2 + \beta s + \gamma} \begin{pmatrix} G_{11} & G_{12} \\ G_{21} & G_{22} \end{pmatrix} \qquad (23a)$$

where β and γ are defined in Eqs. (9) and (10). If cascaded with M, the components of G represent the FWS (G_{11}, G_{21}) and RWS ($-G_{12}$, $-G_{22}$) transfer functions (for the a_y, $\dot\psi$ outputs respectively), since the steering angle input (δ) here is considered only at one axle for each vehicle. The negative signs for RWS are introduced as a part of the closed-loop gain constant to avoid a 180 degree phase shift at zero RWS frequency, due to the sign convention in Fig.1. Solving for the G components yields

$$G_{11} = \frac{C_{12}}{m}\left(s^2 + \frac{C_{34}l_{34}L}{J_z v}s + \frac{C_{34}L}{J_z}\right) \qquad (23b)$$

$$G_{12} = \frac{C_{34}}{m}\left(s^2 + \frac{C_{12}l_{12}L}{J_z v}s - \frac{C_{12}L}{J_z}\right) \qquad (23c)$$

$$G_{21} = \frac{C_{12}l_{12}}{J_z}s + \frac{C_{12}C_{34}L}{mJ_z v} \qquad (23d)$$

$$G_{22} = -\frac{C_{34}l_{34}}{J_z}s - \frac{C_{12}C_{34}L}{mJ_z v} \qquad (23e)$$

Quite a few work vehicles with steering at one of their two axles (particularly those with a turnable driver's cab) can be driven at the same speed in both directions. This would be possible without influencing the steering response, if the "front" and "rear" axle parameters (index 12 and 34 respectively) could be exchanged without alterations in the transfer function matrix F. An examination of the F components reveals, however, that qualitative FWS/RWS-differences appear between the transfer functions G_{11} and $-G_{12}$ in Eqs. (23a) - (23c).

```
TOTAL MASS                                    7000.00 KG
TOTAL YAW MOMENT OF INERTIA                   7000.00 KG M^2
CM TO STEERED AXLE DISTANCE                      .90 M
CM TO DRIVEN AXLE DISTANCE                      1.02 M
LONGITUDINAL VELOCITY                          8.00 M/S
STEERED AXLE CORNERING STIFFNESS COEFFICIENT   2.80 RAD^-1
DRIVEN AXLE CORNERING STIFFNESS COEFFICIENT    3.80 RAD^-1
```

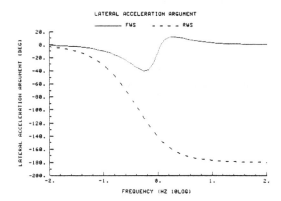

Fig. 3.   Phase plot of MG_{11} (FWS) and
$-MG_{12}$ (RWS) from Eqs. (23).

In the FWS a_y response, G_{11}
compensates for the 180 degree phase lag of
M, by increasing the phase level of the Bode
plot asymptotes with 180 degrees. The RWS
phase asymptote level, on the other hand,
remains unaffected when the frequency has
passed the break frequencies of $-G_{12}$,
since one of the G_{12} real roots is
positive. This MMS inferiority of RWS
compared to FWS is exemplified in Fig. 3,
where the vehicle data correspond to the
fork lift truck in Fig. 4. Note that if also
the steering system influence would have
been considered, much larger phase lags may
have appeared. See e.g Waldmann (1971).

7 DRIVING EXPERIMENTS

The linear model studies above are now being
expanded with computer simulations of
nonlinear models. For validation purposes,
driving experiments are being performed in
March 1982 with an instrumented fork lift
truck. See Fig. 4. Some preliminary data
from these experiments will be presented in
this section.

The vehicle has a turnable driver's cab and
can easily be driven at about 8 m/s in both
FWS and RWS directions. It is equipped with
an outrigger to prevent overturning, an
electric steering machine controlled by a
mobile computer, which also governs the data
recording from the motion and angle sensors:
accelerometers - the laterally oriented one
mounted on a gyro stabilized platform; a
rate-gyro for ψ ; a pulse generator for
velocity; potentiometers for velocity vector
argument and for wheel steer angle.

Fig. 4.   Instrumented fork lift truck.

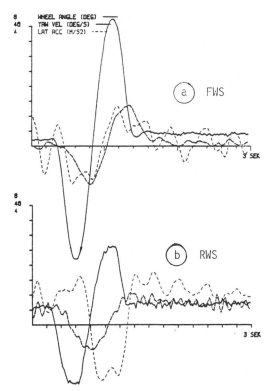

Fig. 5.   Examples of steering response by the
vehicle in Fig. 4. a) FWS, b) RWS.

TABLE 1   Response time examples with
unloaded vehicle in Fig. 4 driven
close to top speed (8 m/s) in
ISO-manoeuvre.

Steering frequency (Hz)	Lat. acc. 1st peak FWS/RWS (m/s2)	Response time (ms) from 1st halfwave; Corresponding cross correlation function (normalized value).			
		Yaw velocity		Lateral acceleration	
		FWS	RWS	FWS	RWS
0.2	3.2 / 2.9	180; 0.999	450; 0.993	220; 0.992	671; 0.977
0.3	2.4 / 3.3	180; 0.998	460; 0.989	205; 0.986	720; 0.980
0.5	2.8 / 2.7	190; 0.996	325; 0.960	145; 0.951	532; 0.961
1.0	1.2 / 1.5	175; 0.989	190; 0.973	85; 0.903	400; 0.990

The steering machine turns the steering-wheel one sine period at the preselected frequency and amplitude. The response time between steering angle at the wheels and yaw velocity (τ_ψ) or lateral acceleration (τ_a) is evaluated from cross correlation functions according to ISO (1981). Table 1 shows such data from the introductory experiments, pointing at RWS response times, that are much above the "safety limit" for automobiles. See e.g. ISO (1976) and Jaksch (1979), presenting shorter response times in general, although their measurements include the steering system lag. Figure 5 illustrates the substantial deterioration of the steering response when changing from FWS to RWS in the same vehicle.

8 CONCLUSION

Vehicle dynamics analysis and experimental data point at fundamental safety deficiencies in the control system and man-machine system properties of RWS vehicles. If possible, FWS should be preferred for normal driving.

ACKNOWLEDGEMENTS

This study has been partly sponsored by the Swedish Work Environment Fund. The driving experiments were performed with support from the Swedish National Road and Traffic Research Institute. The vehicle used in the experiments, was kindly provided by Stocka-Möllan AB, Eslöv, Sweden.

REFERENCES

Andersson, R., and E. Lagerlöf (1978). The Swedish information system on occupational injuries. Memorandum, National Board of Occupational Safety and Health - Arbetarskyddsstyrelsen, S-171 84 Solna, Sweden.

Bonefeld X. and M. Macheleidt (1976). Arbeitssicherheit beim Einsatz von Gabelstaplern. Bundesanstalt für Arbeitsschutz und Unfallsforschung, Dortmund, Forschungsbericht Nr. 147, 1976.

Elgerd, O.I. (1967). Control systems theory.McGraw-Hill, New York.

Freudenstein, G. (1961). Luftreifen bei Schräg- und Kurvenlauf. Experimentelle und theoretische Untersuchungen an Lkw-Reifen. Deutsche Kraftfahrtforschung und Strassenverkehrstechnik, Heft 152.

ISO (1976). Evaluation of open loop test transient handling quality criteria by means of subjective rating. International Organization for Standardization, ISO/TC22/SC9 (Sweden), N103.

ISO (1981). Road vehicles-Transient response test procedure (sinusoidal input). International Organization for Standardization, ISO/TC22/SC9 (Sweden), N 219, April, 1981.

Ito, Y., T. Sarumaru, and N. Ikeya (1975). Automatic steering of rear-wheel-steered fork lift. Bulletin of the JSME, Vol. 18, No. 124, pp. 1109-1116.

Jaksch, F.O. (1979). Vehicle characteristics describing the steering control quality of cars. VIIth ESV Conference, Paris, June, 1979.

Maeda, T., N. Irie, K. Hidaka, and H. Nishimura (1977). Performance of driver-vehicle system in emergency avoidance. SAE Technical Paper Series 770130.

Mitschke, M. (1972). Dynamik der Kraftfahrzeuge. Springer, Berlin.

Mitschke, M. (1979). Comparison of the stability of front wheel and rear wheel steered motor vehicles. In H.-P. Willumeit (Ed.), The dynamics of vehicles on roads and on railway tracks, Swets & Zeitlinger. pp 316 - 322.

Nordström, O., A. Nilsson, and B. Nilsson (1974). Measurements of tyre-road characteristics Low-µ Variable-µ High-µ. Report 6-01, June 1974, PV 3B, AB Volvo, S-405 08 Göteborg, Sweden.

Perret, W. (1964). Lenkkräfte und Lenkwege am Lenkrad von Kraftfahrzeugen und ihr Einfluss auf die Lenksicherheit. Technischen Hochschule Stuttgart, Abhandlung.

Repa, B.S., A.A. Alexandridis, L.J. Howell, and W.W. Wierwille (1977). The influence of vehicle control dynamics on driver-vehicle performance. In A.H. Wickens (Ed.), The dynamics of vehicles on roads and on railway tracks, Swets & Zeitlinger, Amsterdam. pp. 320 - 333.

SAE, 1975. Vehicle dynamics terminology. SAE J670d, Handbook supplement.

Strandberg, L., O. Nordström, and S. Nordmark (1975). Safety problems in commercial vehicle handling. In Symposium on commercial vehicle braking and handling. Ann Arbor, Michigan, May 5-7, 1975. UM-HSRI-PF- 75-6, pp. 463-528, Highway Safety Research Institute & The University of Michigan.

Strandberg, L. (1981). Danger, rear wheel steering. (In Swedish with English summary.) Investigation report 1981:14, National Board of Occupational Safety and Health - Arbetarskyddsstyrelsen, S-171 84 Solna, Sweden.

Strandberg, L., and H. Lanshammar (1981). The dynamics of slipping accidents. Journal of Occupational Accidents, Vol. 3. pp. 153-162.

Waldmann, D. (1971). Untersuchungen zum Lenkverhalten von Kraftfahrzeugen. Deutsche Kraftfahrtforschung und Strassenverkehrstechnik, Heft 218, 1971.

Zomotor, A.(1963). Rechenverfahren zur Ermittlung der Rückstellmomente am Achsschenkel von Kraftfahrzeugen. Automobiltechnische Zeitschrift, Jahrg. 65, Heft, 2, Feb. 1963, pp. 42-49, and Heft 4, Apr. 1963, pp. 101-106.

MODELIZATION OF THE PROCESS OF TREATING VISUAL INFORMATION AND DECISION-MAKING BY A PILOT PERTURBED BY VERY LOW FREQUENCY VIBRATIONS

J. C. Angue

Laboratoire d'Automatique Industrielle et Humaine, Université de Valenciennes,
Le Mont-Houy, 59326 Valenciennes, France

Abstract. The development of automation and computerization in many work, driving, or supervision stations tends to favorise or try to maximalize the operator's visual functions, in his role of the supervisor who can intervene in the case of abnormal behavior of the overall man-machine system. The supervisor, or the pilot, then goes from an alert state to a state of intense activity, where the speed and efficiency of his reactions are essential, while he himself is perturbed by a considerable increase in his work load, due to the quantity of information he must process, or, again, to direct "agressions" he may be subjected to, such as accelerations or vibrations of the vehicle he is in.

An important experimental set-up, comprising, in particular, a modular motion generator with six degrees of freedom (roll, pitch, yaw, X, Y, Z) and a mobile visual information support (crystal and LED displays, and various screens) enables us to reconstitute a variable mechanical and visual environment representative of driving or piloting conditions ; various capters: accelerometers, oculometers (LETI, NAC Eye Marker), and joysticks, allow us to obtain the input and output variables, as well as those pertaining to the perturbations of the man-machine system under study. A synthesis of the results obtained leads to a model of the human operator's behavior when acquiring visual information while being perturbed by very low frequency vibrations, which subsists beyond the filtering devices.

This model integrates the phenomena of visio-vestibulary interactions, as well as certain theories on information concerning the pilot's work load.

Thus, a criterion of manual response time reveals that a rolling motion is the last perturbing, while the deterioration in performance, due to a pitching rotation is greater than that induced by yawing motion.

Furthermore, in the range of less-than-one-hertz frequencies studied the frequency of 0,6 Hertz seems to be a "threshold" from which the perturbing effects are more and more intense.

The means by which the vehicle's vibrations are transmitted to the driver and the law of relative displacement between the operator's eyes and the visual information's support he is observing also turn out to be very important. These results are connected with the phenomena of oculo-vestibulary interactions revealed by measuring the operator's eye movements, and by complex experiments using several axes of perturbation simultaneously.

The modelization of the phenomena observed allows one to consider, in an ergonomic perspective, the proposal of a "feedback" tending to modify, in real time, certain parameters, so as to optimize the performances of the overall man-machine system.

INTRODUCTION

The increasing automation of industrial processes and of driving, piloting and supervising stations tends more and more, to give human beings a role of supervisors, emphasizing the intellectual aspect of the tasks of analysis and decision they must accomplish (Copin 77). The "routine" tasks'being carried out automatically, human intervention can be called for because of an abnormal state of the system, generally arising from a perturbation : an accident in a nuclear reactor, a machine's running off course, etc... and which, moreover, can have a direct effect on the operator (perturbing vibrations, for example). It seems, therefore, important to analyse the behavior and reactions of a human operator when suddenly perturbed, and to attempt a quantification of the influence of these perturbations on his ability to react in a given situation.

In this context, this study lies in the vast field of transportation and air, sea, or overland driving or piloting, where the driver's actions can sometimes be compared to a "reflex" type action, notably, in the case of low altitude, high speed piloting when the margin of error allowed is evidently very small (Aerospace Medical 65) ; or to a "decision" task on the pilot's part, taking into account a grasp and processing of sensory information (Simon 80) (figure 1).

The perturbations considered are of vibratory origin in a range of very low frequencies (less than one Hertz) and high amplitudes (more or less ten centimeters or 20°) which cannot be filtered out and subsist in spite of shock absorbers.

In this environment, the vestibulary system, according to the law of relative displacement between the pilot's head and the visual information display he is observing, may either contribute to stabilizating the direction of gaze, or on the contrary, perturb it (Barnes 78, Angué 80, Benson 60). Thus, various experimental conditions are envisaged where the pilot observes a signal outside his vehicle

or else picks up information from an instrument panel, while the vehicle-pilot and vehicle-display transfer function is modified (figure 1). A criterion of manual response time then enables us to quantify the influence of vibratory perturbations on visual and manual performance, and can, moreover, contribute to a better understanding of the means of information transfer and processing in man (Beyart 79, Boutinet).

After presentation of the possibilities of the experimental set up and the data treatment system, a synthesis of the results prepared as an algorithmic model of the pilot's behavior, allows us to contemplate, in an ergonomic perspective, the proposition of a "loop" tending to modify certain parameters in real time, so as to optimize the performances of the overall man-machine system.

EXPERIMENTAL SET-UP

The whole experimental set-up is composed on one hand of a modular motion generator (Mangin 80) designed to create vibratory perturbations by rotation of the pilot's body around its center of gravity (roll, pitch, yaw, see figure 2) or by translations in a horizontal or vertical plane . A micro-processor safety system (Mangin 80) guarantees that the point of functioning is kept within a certain area of use the characteristics of which are presented in figure 3.

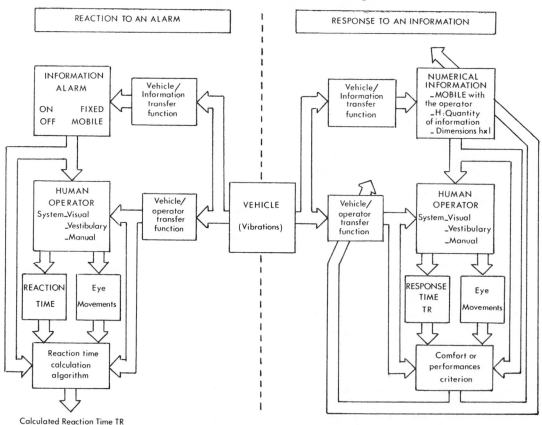

Fig 1a. Task of reaction to a visual alarm.

Fig. 1b. Task of response to a numerical visual information.

On the other hand, a system for generating visual stimuli on various technological supports (electro-luminescent diodes, cathode tubes, plasma screens...) with an adjustable law of relative displacement in space with respect to the operator, allows notably, for the simulation of supervision and piloting tasks for different configurations.

This ensemble is completed by an apparatus for measuring and processing the input, output and perturbation parameters of the man-machine system under study, represented in figure 4, which shows principally :

- The cabin motions (6 potentiometers) ;

- The accelerations at head level (angular and linear accelerometers) ;

- The manual responses ;

- The eye movements.

For this last parameter, various sensors are available, depending on the dynamics of the signals to be measured and on the objective sought. Let us mention particulary the "NAC EYE MARKER" which uses the principle of reflection from the cornea to detect the ocular displacements, and an objective to cover the visual field observed. An automatic device for processing the images obtained (Anguë 82) helps to surmount the main disadvantage of this sensor, which lies in the slowness of manual analysis.

The overall management of the experiments is taken over by an industrial computer, equiped on one hand with peripherals that let it dialogue with the environment, and on the other hand with a system of interruptions allowing a real time management of the tasks of visual

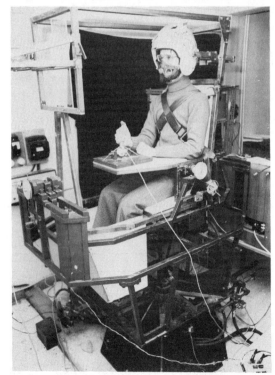

Fig. 2 Triple rotation (Roll, pitch, yaw) platform.

instruction generation, perturbation generation, and coordination of the two.

This automation guarantees the reproductibility of the experiments, which is necessary when considering a statistical study on a population of subjects.

EXPERIMENTAL MODEL OF THE STUDY OF THE PROCESS OF VISUAL DETECTION UNDER LOW FREQUENCY VIBRATIONS

The research objective consists in quantifying the effects of vibrational perturbations on the operators'visual and manual performances.

Without claiming to reduce a pilot's performances merely to his reaction speed to a visual stimulus, nevertheless, this parameter is often decisive with regard to the effectiveness of the ensuing actions undertaken (Malvache 81) ; thus, during the experiment, the subject placed on the platform has the task of watching an alarm light or picking up visual information and reacting to the appearance of the information by pressing a button. The operator's manual performance is measured by the reaction time, in milliseconds, between the visual stimulation and the manual reaction.

The operator is closely strapped to the seat, so as to guarantee the reproductibility of the vibratory perturbations, on one hand, and on the other, to define the relative movements of the visual information display with respect to the subject's head. These movements, in fact, condition the parameters of the visio-vestibulary interactions during watching task.

Moreover, in order to limit as much as possible the undesired variations of supplementary parameters such as room lighting or visual "landmarks" during the phases in motion, the experimental cabin is plunged in darkness.

Each "experimental module", of which two examples are presented in figure 5, is divided into "sequences", during which, on one hand, the frequency of the seat's motion is held constant, and, on the other hand, the operator is subjected to visual stimulations. At the beginning of the experiment, three sequences at rest are juxtaposed so as to gather a larger (60 response times) sample serving as a reference for the operator in the absence of perturbations. Each sequence in motion is separated from the next one by a sequence at rest, in the aim of evaluating separately the influence of the various frequencies tested.

A SAMPLE OF THE POPULATION TESTED AND PRESENTATION OF THE CHOICE OF THE STATISTICAL TREATMENT USED

A minimum of ten common subjects were put through all the experiments, in order to allow comparison of the results as a function of the characteristics of the perturbation produced by the motion generator and of the characteristics of the visual information.

PLATFORM	DEGREES OF FREEDOM	MAXIMUM AMPLITUDE	FREQUENCY RANGE	MAXIMUM ACCELERATION
3 ROTATIONS	ROLL PITCH YAW	±20° ±20° ±20°	0-2 HERTZ	±20 rad/s²
2 TRANSLATIONS	Y(FRONT-BACK) X(RIGHT-LEFT)	±100mm ±100mm	0-2 HERTZ	±1g
1 TRANSLATION	Z(UP-DOWN)	±100mm	0-2 HERTZ	±1g

Fig.3 Area of use of the motion generator for the six degrees of freedom with an acceleration limit of $\pm$ 1 g (or $\pm$ 20 rad/δ^2).

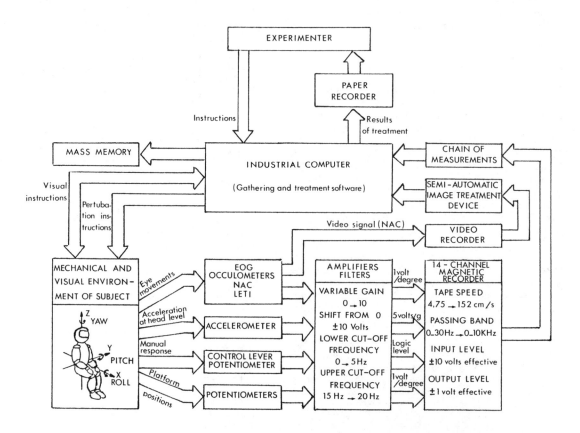

Fig.4 Data gathering and treatment device for the input, output and perturbation variables of the man-machine system.

EXPERIMENTAL MODULE 1																							
SEQUENCE NUMBER	1	2	3	4	5	6	7	8	9	10	11	12	13	14	15	16	17	18	19	20	21	22	23
AMPLITUDE (IN DEGREES)	0	0	0	10	0	5	0	5	0	10	0	10	0	5	0	5	0	10	0	10	0	5	0
FREQUENCY (IN HERTZ)	0	0	0	0.4	0	1.2	0	0.6	0	1	0	0.8	0	0.8	0	0.4	0	0.6	0	0.2	0	1	0
NUMBER OF VISUAL STIMULI	20	20	20	20	10	20	10	20	10	20	10	20	10	20	10	20	10	20	10	20	10	20	10

TYPE OF MOTION	VISUAL INFORMATION
ROLL PITCH YAW	NUMERICAL (2.5.6.9) INTEGRATED WITH THE DISPLACEMENT

Fig.5a. Succession of sequences at rest and in motion, for a total time of about 22 mn.

EXPERIMENTAL MODULE 2													
SEQUENCE NUMBER	1	2	3	4	5	6	7	8	9	10	11	12	13
STATIC LEANING OF BODY	-	-	-	0°	-	10°FORE	-	10°BACK	-	17°5FORE	-	17°5BACK	-
FREQUENCY OF MOVEMENT	0	0	0	1	0	1	0	1	0	1	0	1	0

TYPE OF MOTION					VISUAL INFORMATION
STATIC		DYNAMIC			
AXIS	AMPLITUDE	AXIS	AMPLITUDE	FREQUENCY	
PITCH	0° 10° FORE 17°5 FORE 10° BACK 17°5 BACK	ROLL YAW	±10°	1 HERTZ	NUMERICAL (2.5.6.9) INTEGRATED WITH THE DISPLACEMENT

Fig.5b. Succession of sequences at rest and in motion with a static leaning of the body, for a total time of about 16 mn.

Each pilot's aptitudes are brought out by the variations in his reaction times with respect to his reference as identified during the first three sequences at rest of each experimental module.

The method used for characterizing, the sample of response times in one sequence (Roger 78) is based on the observation of the experimental histograms of reaction times, TR, which leads us to hypothesize that TR follows a Gamma law of three parameters To,α,τ, with:

- Mean time   : $Tme = To + \alpha \ \tau$

- Modal time : $Tmod = To + (\alpha-1) \ \tau$ (1)

- Standard deviation : $\sigma = \sqrt{\alpha} \ . \ \tau$

To,α and τ are then estimated by minimizing the squared difference between the experimental distribution and the Gamma law's theoretical distribution function (Doche 72, Sarhan).

This first finding, added to the formulas in (1), entails that parameters To, Tmod and Tme show similar variations. Under these circumstances, parameter Tmod is chosen, insofar as this value corresponds to the greatest probability of occurence.

Taking into account overall the first three sequences at rest enables us to determine, for each subject, his personal characteristics at the start of the experiment, when he is not perturbed by the seat's movements.

$$Tmod \ (sequences \ 1,2,3) = Tref \qquad (2)$$

It is then possible to evaluate the variations in the human operator's behavior as a function of the parameters of motion and time with respect to this reference :

$$Tmod \ (i) = Tmod \ (i) - Tref \qquad (3)$$

The influence of the perturbation can then be isolated from that of "fatigue" by calculating for each sequence in motion, the deviation of the modal value, not from a fixed reference, but from a reference, Tref (i), that varies itself with time.

One solution is to choose as the variable reference for a sequence in motion, the average of the two modal values of the sequences at rest that frame the sequence in motion, that is :

$$Tref \ (i) = \frac{Tmod \ (i-1) \ + \ Tmod \ (i+1)}{2}$$

for i = 4, 6, 8, 10, 12.

Thas is the formulation adopted for the presentation of results in this article.

EXPERIMENTAL RESULTS

More than 200 experiments were done for this research, first of all on a human operator's

reaction speed when confronted with a visual alarm signal, and then extending this exploration under conditions approaching work in a piloting cabin where the pilot must process more complex visual information. The model presented further on is a synthesis of the results, of which a few examples are presented here :

For a simple task of reaction to an alarm, the curves obtained show the non-significance of the perturbing effects of translational movements in a horizontal plane, and a larger deterioration of performance for pitching than for rolling. Moreover, in these last two instances, the curve of increase in reaction time shows a change in slope near 0,6 Hertz. All these results are connected with the type of simulation of the inner-ear receptors engendered by the vibrations, and they correspond to the frequent situation where the alarm signal to be watched is part of the experimental cabin ;

For a task of reading and grasping visual information, the statistical result (Fig.6) show that, as in the case of a watching task, rolling is the last perturbing motion. Furthermore, the deterioration of performance due to a pitching rotation is larger (an increase of 60 % in TR for $\pm$ 10° amplitude and 1 Hertz frequency) than that engendered by a yawing motion with the same characteristics (40 %).

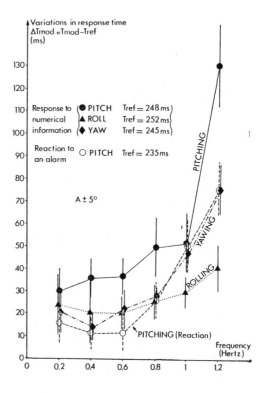

Fig.6 Variations of the manual response time to a numerical information (among four = 2,5 , 6 , 9) as a function of the frequency.

The interpretation of these results is proposed (Angué 80) by studying, on one hand, the
type of stimulation of the inner-ear receptors
for each of the perturbing motions, and, on
the other hand, by analysing the characteristics of the eye movements that result from
them, as well as their consequences on the
process of visual perception which also depends on the luminous and dimensional characteristics of the information to be observed.

Accounting for the entire set of experimental data and integrating them in a model of
control requires an identification of the various laws established and their formulation,
contributing to the establishment of a model
of a human operator perturbed by vibrations.

STRUCTURE OF THE MODEL

Each subject is characterized, at rest (fig.7)
by his three reference parameters, To, α and
τ, and the increase, ΔT, in his reaction or
response time is calculated as a function of
the parameters of the vibrations from the motion generator (axis, amplitude and frequency)
of the spatial, luminous and dimensional characteristics of the visual information presented   of the means of transfer of the vibrations to the operator   and of the length of
the experimentation. The knowledge of these
results supplies the parameters To, α and τ
of the perturbed operator, which allow the
generation of a response time, TR, corrected
for the time between visual stimulations, TES
(figure 7).

The following paragraphs show the results of
the identification of the experimental data
with a view to numerical simulation of the
model.

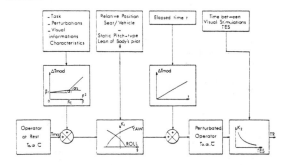

Fig.7 Algorithmic model of the perturbed
       operator

IDENTIFICATION OF THE MODEL'S
      PARAMETERS

The analysis of the influence of the instantaneous parameters of the vestibulary perturbation at the moment the visual stimuli appear
has shown that it is not significantly different, in the range of vibrations considered,
from that of the vestibulary considered
overall (Roger 78). This fact justifies the
choice of the method consisting in characte-
rizingthe vibrational perturbation by its frequency (at constant amplitude) and in identi-
fying its influence by the quantity $\Delta T \, \text{mod}(F)$.

Furthermore, the generally parabolic air of
all the ΔT mod (F) curve obtained, whatever
the experimental module used, incites us to
get out a plane, ΔT mod (F^2) allowing a linear alignment by parts, in the form :

$$Tmod\ (F) = Sup\ F\{\alpha 1 F^2 + \beta 1, \alpha 2 F^2 + \beta 2\}$$

This formulation, moreover, has the advantage
of stating explicity the existence and contin-
gent values of a "break-off" frequency, Fc,
from which the deterioration in performance
increases more rapidly.

The parameters $\alpha 1$, $\beta 1$, $\alpha 2$ and $\beta 2$ of the two
linear regression lines obtained minimizing
a quadratical function of cost are shown in
figure 8, for the various configurations of
visual tasks and vibrational perturbations
tested. The results of these various identi-
fications can be used for effecting the numerical simulation of the pilot's behavior.

NUMERICAL SIMULATION AND COMPARISON
   OF RESULTS MAN MODEL

Knowing the parameters of the pilot's mechanical and visual environment, we can calculate
the parameters αi and βi of the preceding model, and therefore, produce automatically a
reaction or response time for each of the visual simulations of a sequence.

Several curves can be used for comparing the
real and simulated results :

a) The statistical treatment implemented after
   performing an experimental module supplies
   "average" data (mean and standard deviation
   on the population tested) of ΔT mod (F)for
   the discrete values of the frequencies
   tested (figure 9).

b) The parameters αi and βi of the model,
   identified from these data, let us trace
   the curve of the "model of the population"
   which gives an interpolation of the results
   in the range of frequencies tested.

c) Under these conditions, the choice of a
   subject taken in the population has the
   object of illustrating the individual behavior of an element of the population ("real
   subject", figure 9), and implementing the
   program of simulation gives results relating to the same subject ("simulated subject", figure 9).

Among the many possible methods for quantifying the quality of the modelization, the
comparison of the results, man vs machine,
can be done by calculating, for example, the
coefficient of linear correlation for the
"real subject/simulated subject" data in order
to evaluate the model's capacity for representing an individual behavior ; an identical
calculation for the "real subject/population
model" data allows the comparison between a
given individual's behavior and the average
behavior of the population, where as in the
case "simulated subject/population model",

the calculation is used only for purposes of verifying the simulation.

For the set of configurations tested, it was possible to verify the validity of the model by a coefficient of linear correlation always greater than 0.9, as can be seen in the example of figure 9.

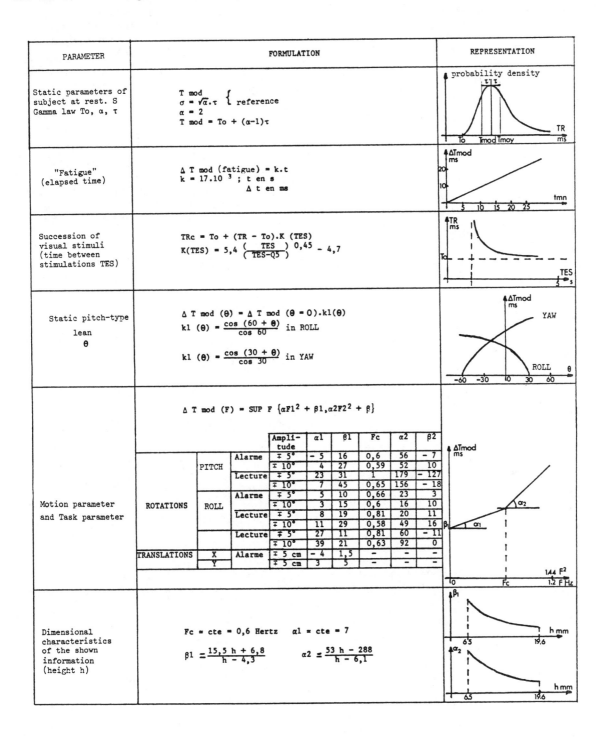

Fig.8   Parameters of the human Operator Model.

Two important consequences stem from these results : on one hand, they validate the structure of the model and the method of fitting adopted, which can then be used when applying, this study to real cases ; and on the other hand, they allow an interpolation, in the range of frequencies studied, between the experimental results obtained for discrete values.

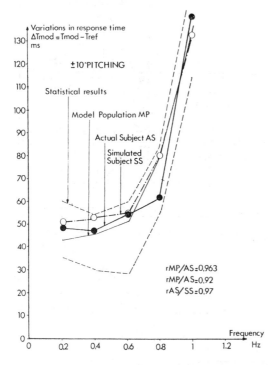

Fig.9 Comparison of person with model for pitching vibrations.

CONCLUSIONS

The synthesis of the results in the form of an algorithmic model of the perturbed pilot's behavior permits the quantitative estimation of the pilot's reaction or response speed for various experimental configurations, with a quality evaluated by a coefficient of linear correlation between the real values and the simulated values higher than 0.9 in all the cases tested. The model thus allows us, for example, to define an individual's behavior with respect to the average behavior of the population, and therefore, to supply possible objective criteria for selection.

Another very important application of the identification of the experimental laws is to supply indispensible elements for restoring a "normal" universe, leading to a minimization of the perturbations arising from vibrations applied simultaneously to the pilot and his environment. These results concern, notably, the law of relative displacement between the operator and his visual environment, as well as the the nature of the visual information presented to him.

REFERENCES

Angué, J.C.(1980). Contribution à la modélisation des effets des vibrations de basse fréquence sur les performances visuelles et manuelles du personnel embarqué, par P. Mangin, J.C. Angué. Journées d'Etudes "Vibrations et Chocs", 10,11,12 Juin 1980. Ecole Centrale de Lyon.

Angué, J.C, Gerber J.(1982). Traitement des signaux vidéo d'un oculomètre pour la mesure semi-automatique de la direction du regard. Journées IRIES. Application des Micro-processeurs et des Micro-Ordinateurs en médecine, Toulouse, 8-12 Mars 1982.

Barnes, G.R, Benson, A.J., Prior A.R.J.(1978) Visual-vestibular interaction in the control of eye movement. Aviation, Space and Environmental Medicine.

Benson, A.G. Compensatory eye movements produced by angular oscillation. Proc. Int. Union Physiol. Sci., 25th Int. Cong. Munich, n°167.9.60.

Beyart, G., Malvache, N.(1979). Communication Homme-Machine : Traitement des informations et applications à l'aide à la décision du pilote. Rapport final. Contrat D.R.E.T. Bis 78/1246.

Boutinet. Recherches sur les stratégies perceptives.

Copin, B.(1977). L'automatique dans les moyens de production d'énergie électrique Congrès S.E.E. Grenoble.

Doche, C.(1972). Etude et application clinique d'une méthode d'acquisition et de traitement des temps de latence dans les circuits sensitivo et sensorio-moteurs et planatoires chez l'homme. Thèse de Docteur Ingénieur, Grenoble.

Mangin, P.(1980). Dispositif de contrôle et de commande d'un système Homme-Machine : Réalisation d'une plate-forme et essais. Thèse de 3ème Cycle. Université de Valenciennes.

Roger, D.(1978). Modelisation du transfert oeil-main chez l'homme soumis à des perturbations vestibulaires d'origine vibratoire à très basse fréquence. Thèse de Docteur Ingénieur Université de Valenciennes.

Sarhan, Greenberg. Contribution to order statistics.

Wanner, J.C.(1978). Introduction à l'étude d'un modèle mathématique de pilote. Biomécanique du pilotage et l'interface Homme-Machine. Toulouse.

DISCUSSION OF SESSION 7

7.1 I DESIGN AND EVALUATION OF A WIND SHEAR INDICATION SYSTEM FOR TRANSPORT AIRCRAFT

Chairman's Discussion Summary:
Mr. Pitrella meditated upon the application
of a CRT display with flight path predictive
information (e.g., "tunnel display"). This,
however, requires a totally different display
concept.
R. Seifert commented on the "not expected" result that in the subjective acceptance of
display versions the "command display" was
rated lower than "display of the mere wind
data" and that of "effects of wind shear".
Here, the fact applies that you should not
design "taking the decision away from the pilot". The pilot wants to be informed on facts
and effects, but the decision on subsequent
actions has to be kept to himself.
R. Seifert commented the results of the simulation concerning visual dwell time (Table 1).
The necessity to gather the "energy" and "energy rate" information from different displays (ADI and VSI) results in higher dwell
time percent, however, in a greater number of
eye movements as well and consequently in
higher mental workload.

7.2 I ANALYSIS OF RIDER AND SINGLE-TRACK-VEHICLE-SYSTEM, ITS APPLICATION TO COMPUTER-CONTROLLED BICYCLE

Seifert: Does the model include terms for including gyro effects of the wheels in higher
speeds and with greater masses (e.g., motorcycle)?

Nagai: I have made at first the cycle model
with the term of gyro effects of wheels. According to the parametric study, I have found
that the effect of it is small on the condition of driving in the city area. Then I neglected that term. From the view point of
stabilizing the unstable roll motion, it can
be easier in the heavy mass situation than
in this case.

Brauser: Do you consider handling quality by
optimization of the bicycle construction
(e.g., wheel diameter, length of cycle, steering wheel lead term etc.)?

Nagai: I intend to use this experimental facility to decide on the optimal quality of

cycle. The experimental facility developed
here is useful to be applied to the optimization of hardware construction.

Rajamurthy: Did you make measurements of
steering and lean angles in an actual riding
situation and compare it with your feedback
control strategy? Is the latter the same as
the strategy adapted by human subjects?

Nagai: I have not yet measured the control
variables in the real riding situation. But
we can get the preferable driving strategies
required for such a control system, from this
stability analysis.

7.3 I PROCESSING OF ON-BOARD ROUTE GUIDANCE INFORMATION IN REAL TRAFFIC SITUATIONS

Question: 1. To what extent does the additional information-processing-load affects
the performance of more complex situations,
i.e., leaving the highway just after passing
another car?
2. Shouldn't the information be given early,
not 1000 m before the exit?
3. The ALI-display-system seems to be of very
little redundancy. What about of giving the
driver the same information as the traffic-
sign contains, i.e., name of exit?

Voss: 1. Clearly, there is a limit to process
information in complex situations, but one
reason for route guidance information internally given is to reduce the uncertainty
(and, as a consequence, the informational
workload as a whole) about route decision.
Another problem is the interference of route
guidance information with the actual situation, but this is much more critical on city-
roads than on highways. Our measurements have
shown no unacceptable risk (e.g., see the
time budget).
2. The ALI symbols are very abstract and need
the corresponding traffic sign as additional
information. An additional signal (e.g.,
acoustic) may be helpful as a prewarning, if
it is not too much before the decision point
(because of forgetting it), but we haven't
examined this. (ALI display and displaying
procedure were given side-conditions for us.)
3. The duration of reading a traffic sign
completely is much longer than 1 s, as we
have measured for reading the ALI display.
If we have a copy of a traffic sign given as

route guidance information internally displayed, then we would have much more distraction of attention allocation.

Brown: Is the system self-correcting for
driver's errors? That is, if the driver mis-
ses a turning, will he be given correcting
information at the next junction?

Haller: Yes, the system can correct false decisions for instance by giving the driver a
turn signal to come back to the decision
point where he made a wrong decision.

Chairman's Discussion Summary:
1. Peters: Did you take into account drivers
wearing glasses? Because they move the head
not only the eyes.
Answer: The test subjects used did not wear
glasses. So, the results do not reflect this
case.
2. Summary of further discussion:
In "incongruent situations" there was no difference in the eye movement results between
drivers being familiar with the locality and
those being unfamiliar.
To display the information "head-up" was not
investigated.
It was stated, that possibly more redundancy
would be required. Background information
would be helpful. Reasons and consequences
of and for a certain situation information
could be displayed. However, the concept is
not yet formulated.

7.4 I AN ANALYSIS OF THE MAN-AUTOMOBILE
        SYSTEM WITH A DRIVING SIMULATOR

Brauser: With respect to your last slide:
The introduced model 2 (jerk feedback) seems
to be "better" (with respect to deviation)
than most of the subjects, only one subject
appears to be better. Are the subjects different in driving experience?

Yoshimoto: Yes they are. The driver model 2
is better than most of the subjects except
the subjects G, L, O, and R. I cannot say
certainly, because the driving experience of
each subject was not investigated in detail.
However, if most of the subjects become experienced in real driving and become familiar
with the driving simulator, their performance
may become closer to that of the driver model
2.

Strandberg: a) What is the objective of the
driver model? Optimizing vehicle design or
what?
b) If so, did you consider using your simulator with different vehicle models and a great
number of human subjects instead of a model
of the driver to evaluate the driver-vehicle
performance?

Yoshimoto: a) The object is the estimation
of vehicles or high-ways.
b) I think we can estimate driver-vehicle
systems by the following steps: (1) We choose
some standard vehicles for the measurement of
the human driver's characteristics. (2) We
gather a great deal of data concerning all
kind of drivers. (3) We can get the distributions of the human drivers' characteristics
and can establish a standard driver model.
(4) We can use this standard driver model to
estimate the vehicles or the high-ways.

7.5 I CONTRIBUTIONS OF CONGRUENT PITCH
        MOTION CUE TO HUMAN ACTIVITY IN
        MANUAL CONTROL

Brauser: Which method did you apply to identify the pilot model parameters (frequency
response) given in the 3d slide?

Rajamurthy: The s-domain models given in
Table 1 were obtained as follows.
1. First, the parameters of the LS model
along with the model order is estimated.
(Refer.-Balakrishna (1976) or Hsia, T.C.,
System Identification-Least Squares Method,
Lexington, Toronto, 1977, 163 pp.)
2. Then the time domain model is converted
to frequency domain

$$H_{LS}(j\omega) = \frac{\sum_{k=0}^{n} b_k \exp(-j\omega\Delta t)}{\sum_{k=0}^{n} a_k \exp(-j\omega\Delta t)} \quad \Delta t = \text{Sampling interval}$$

From this we can get the frequency response.
3. Now to the frequency response: Curve fitting is done by modified Levy's method, by
introducing iteration (in the basic Levy's
method) that gradually eliminates the bias.
(Refer.: Santhanam & Koerner, 1963).

Pitrella: Which sequence were the cue conditions presented to the subjects? (It is
known that operators construct models based
on their first experience conditions which
may not be appropriate with changed conditions).

Rajamurthy: The experiments were conducted
over a long period. The experimental sequence
was always that the subjects performed motion
plus visual (CMV) run first and then the run
with visual cue only. Both the subjects did
have previous experience in tracking tasks
in fixed-base. Here, the subject is either
provided with pitch motion cue or not, and
the visual cue is congruent to pitch. The
sequence hardly influences the pilot performance.

PROBLEMS OF APPLICATION OF HUMAN OPERATOR MODELS IN INDUSTRY

Chairman's Summary of Discussion (with some
modifications by Rijnsdorp)
Panel members: Charwath, Gallagher, Miller,
Prutz, Seifert (chairman),
Veldkamp

The discussion was mainly directed towards
highlighting possibilities, difficulties
and constraints in using human performance
models in industrial design. Possible applications were shown on the blackboard (see
Figure 1).

MODELS	
Dynamic Simulation	Computer Aided
Processes (chemical, power)	Design
	Production (NC)
Machine Operation (vehicles, air-craft)	Process control
	Guidance and control
Multi-Machine Operation (city traffic, ATC)	Problem solving/ decision making
Medium: VDU's	
Levels of Man/Computer Communication: - Dialogue (all above) - Text Processing - Product Support	

FIGURE 1

Gallagher reported a positive example of
applying a decision-making model (by Rasmussen) to power plant control room design.
However, most participants saw constraints
and problems in the applicability of human
performance models in industrial design.
Amongst others, a model should enable der-
viation of qualitative and/or quantitative
human performance data for functions and
their limitations, leading to requirements
for system functioning or equipment specification. For instance, in the area of aircraft handling quality the models do not
yield results for specifying flight control
systems. Consequently they are not applied.

In the discussion, a number of statements
were made which can be seen as a challenge
for further research concerning human per-

formance models:
- many systems require provisions for recon-
  figuration in case of single or double
  failure.
- training requirements should be considered
  in model configuration to enhance habit
  formation in coping with the system and
  its operation.
- in certain applications operating personnel
  have greatly differing skill levels and
  performance capabilities.
- the cross-over from skill-based to rule-
  and knowledge-based operation (depending
  on the state of operation, type of
  function, and type of task) has considerable consequences for model configuration.
- problem solving requirements are widely
  different in cases where the process can
  easily be halted (e.g. rolling mills) and
  where this is difficult or impossible
  (e.g. chemical industry, ATC).

Finally, the relevance was discussed of cognitive models. The present problem solving
and decision making model repertoire is
still a long way off from a comprehensive
model of man's cognitive performance. If
we were able to configure such a model, we
had the key to the human brain, and would
achieve the highest level of automation
thinkable in a wide field of application.

MAN'S ROLE IN MAN–MACHINE SYSTEMS

Chairman's Summary (J.E. Rijnsdorp)

The chairman started with a number of topics
for discussion, after which he invited the
panel members to make a brief statement.
Bainbridge stressed the underutilization of
human resources and capabilities in man/
machine systems. She made a plea for richer
task situations.

White summarized his experiences with human
subjects in laboratory experiments. Training
them for supervisory control tasks took a
very long time. The final hypothesis to be
tested was: presenting more detailed information (on recorders instead of indicators)
leads to fewer control actions. Actually,
the opposite happened. The explanation was
found in post-experimental interviews: the
subjects had never imagined the amount and
frequency of fluctuation before it was
shown to them on recorders. Evidently for
adequate description of system behavior one
has to know what the person in the system
is thinking.

Martin made a plea for avoiding technological
determinism in system design. Person-oriented
criteria should be introduced at the earliest
possible stage. Alternative structures should
be developed for an adequate choice, and user
participation should be incorporated as well.

Terano discussed effects of automation on
human society. The goal should be humane,
not unmanned systems. The trend of developments goes from production to management of
production: thinking automation. In safety
automation one should move from unbelief of
the human (the operator must obey) to unbelief of the machine (the operator can interfere). Office automation will lead to improvement of human life.

Mårtensson stressed the importance of user
participation, by all people concerned in
the company. Presently the computer specialist is considered as the total design
expert. In order to specify their problems
and demands in an adequate way, the users
must have more knowledge. Also, participants
and designers speak different "languages",
which makes communication very difficult.

Rosenbrock started the general discussion
with his experience in dealing with social
scientists: they do not indicate what alternative is better for men, but say only that
the particular question cannot be answered.

Martin remarked that communication between
engineers and social scientists takes much
patience.

Seaman has participated in the building of a
three-dimensional model of a new system in
order to enhance communication. He considered
the operator as the system expert.

Nzeako said that social scientists are too
sentimental. You cannot meet all demands.
More attention should be paid to training,
and to the changing environment. In answer,
Mårtensson suggested to start user participation from the very beginning, in order to
clearly show demands to the engineer.

Nzeako then stressed the cultural differences between users. Petersen offered a design
solution: the customizable man/machine-inter-
face, which the user can adapt to him/herself.

Henning mentioned, Bainbridge confirmed, but
White and Lenior contradicted the lack of
interest among engineering students for
social sciences.

Henning also indicated the lack of money
available for implementing social aspects
in projects. He told that in his group software is made modular in order to enable
user participation even at the last moment.

Strandberg posed a question to the audience:
what is the contribution to operator error
by different sources (design, organization,
the operator him/herself, etc.)? The proper
answer was: all of these sources simultaneously. Modelling human behavior is not
simply a matter of adding up to 100%!

The chairman acknowledges Pisoo's assistance
in preparing this summary.

HIGHLIGHTS AND SUMMARY OF THE CONFERENCE

John E. Rijnsdorp

State-of-the-Art and Future Trends in the Field of Man-Machine Systems

1. Introduction

The first international Conference on man-machine systems, held at Baden-Baden (September 1982) was centered around three key-words: Analysis, Design and Evaluation. These key-words represent different ways of looking at man-machine systems (MMS) which, in a way, complement each other. Hence, it also makes sense to pay attention to their interrelations. This will be summarized in the following sections (Fig. 1), with special reference to presented papers and discussions.

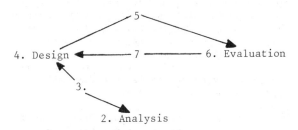

FIGURE 1: Sections in this contribution

2. Analysis of MMS

Compared to the 'machine'[*] relatively little is known about 'man'[**] in the MMS. Hence, analysis is directed to modelling human behavior, in order to arrive at a complete system model.

Initially, attention was focussed on manual control of fast systems, such as airplanes and cars, where human behavior is mainly 'skill-based' (Rasmussen, 1980). More recently, other situations became the object of research, which has led to investigation of 'rule-based' and 'knowledge-based' behavior. In particular, human decision-making is receiving much attention (Rouse, 1982). This trend towards higher levels of human functioning is also being stimulated by computerization of actual MMS and by developments in artificial intelligence

However, human behavior cannot be neatly divided into three compartments: skill-, rule- and knowledge-based. In fact, human activities frequently switch form one level to another (heterarchical behavior; Rouse, 1982).

When performing actual tasks, people appear to be active at all levels, more or less simultaneously (Bainbridge, 1982).

A special topic is human reliability. Only little progress has been made so far to model human error, although its relation to safety and reliability is of great practical importance.

Human error is conceptually quite different from 'machine error', and cannot be reduced to simply assigning error rates to human activities, particularly if rule- and knowledge-base behavior is important. Moreover, the system boundary is ill-defined: many human errors in system operation are consequences of human errors in system design or in work organization.

There are positive and negative relations with training and task variety. Man/machine systems should be designed in such a way that there is freedom for human error (Sheridan, 1982). It is also desirable to analyze the effects of competition and cooperation in men-machine systems (Terano, 1982; please note the plural!).

In the problem-solving field, human compensation for off-normal events is a new topic for research (Rouse, 1982).

3. Interactions between Design and Analysis

In some cases, models for human behavior are being used as system components (Baron and

---

[*] The term 'machine' has a very broad meaning; it can refer to a vehicle, a production process, an administrative system, etc.

[**] The term 'man' is to be understood in its widest sense, encompassing male and female. In some cases, it even has a plural meaning.

Pew, 1982). This adds to the various meanings of the word 'model'.

In the other direction, design practice creates the need for rules of thumb in order to roughly estimate human characteristics, e.g. work load. Especially when there are tight deadlines, there is simply no time for detailed analysis, hence, if nothing rough-and-ready is at hand, human factors cannot be seriously taken into account.

4. Design

Obviously, man-machine systems design depends on the type of 'machine', i.c. the area of technical application. However, it is not so easy to indicate the characteristics which are mainly responsible for these differences.

During a round-table discussion in Baden-Baden some aspects became clear: there is a great difference between cases where realistic system simulation is possible during the design phase (as, e.g., in electric power generation), and where this is impossible due to cost and/or time limitations (as, e.g., often in the chemical industry). Another difference exists between cases where the system can be shut down if there is a failure (as e.g., in rolling mills), and cases where the operators have to keep the system going (as, e.g., in blast furnaces and airplanes). It is very desirable to investigate these differences in more detail.

In the field of man-machine interaction, novel possibilities are becoming feasible. For instance, a display can be automatically 'zoomed' depending on human sitting posture: leaning forward means wishing to see more details; leaning backward can be interpreted as wishing to obtain an overall view (Nemes, 1982).

Much progress has been made in natural language input to the computer (Guida and Tasso, 1982), an area where artificial intelligence techniques are producing useful results for man-machine system design.

In a certain way, man-machine interfaces present their 'personalities' to the user. All too often, however, these personalities appear to be unpleasant. In particular, CAD (computer-aided design) users are very un-happy (Hatvany, 1982).

This situation can be improved when manufac-turers offer 'customizable man-machine inter-faces' (Peterson, 1982): flexible 'user-friendly' interfaces to be adapted by the user him/herself.

5. Evaluation of design

The trend towards high degree of automation is often advocated by stressing the limita-tions of human beings and the superiority of automatic devices. However, ultimately man is reintroduced into the system as a super-visor, evidently to cope with shortcomings in automation, and, possibly, system design. This is one of the 'ironies of automation' (Bain-bridge, 1982). Another one is related to the way human beings cope (or rather fail to cope) with unexpected and rare events:

"BY TAKING AWAY THE EASY PARTS OF HIS TASKS, AUTOMATION CAN MAKE THE DIFFICULT PART OF THE HUMAN OPERATOR'S TASK MORE DIFFICULT."

In highly automated systems, where the human operator is restricted to be a passive observer most of the time, a central issue is 'how to stay trained'. Training and education are also very important in other respects:
- workers are not on an equal level with system designers (Mårtensson, 1982), which reduces their participation in design and can jeopardize their acceptance.
- social scientists and engineers have all to be educated to understand each other (conclusion RT-2, 1982).

6. Evaluation

Man-machine systems cannot be considered in a completely neutral way. Points of view vary widely, depending on how one approaches the human being in the system.

One extreme is to consider him/her simply as a system component, with its inherent charac-teristics and limitations. The other extreme is to deny the possibility of man-machine system studies altogether. Man should freely decide how to interact with the machine and interference by others is an undesirable attack on his/her autonomy.

Intermediate positions are also represented: the 'man' in the system is not merely a system component, but also a person. Conse-quently, the designer should consider job content and job satisfaction. One step further is to accept him/her as a participant in the design process, or even as a co-designer (in a certain sense, the user is the expert; Seaman, 1982).

In fact, in these intermediate points of view the challenge is to harmonize two different points of view: man as a system component and man as a human being in his/her own right.

7. Impact of evaluation on design

The last remarks of the previous section, if taken seriously, leads to modifications in the design process: in order to carry suffi-cient weight, person-oriented criteria should be introduced as early as possible (Martin, 1982). A practical problem is the balancing of these criteria with respect to technical/economic ones. Sometimes this is put in terms of a conflict between 'soft' and 'hard' criteria, with the underlaying assumption that the latter will override the former. However, experience has taught that apparently hard facts sometimes rest on shaky foundations, and apparently soft effects (such as user acceptance or non-acceptance) can kill system performance. 'Design is based on belief, if only in numbers.'

8. Concluding remarks

The situation in the man-machine field resembles the development of culture as interpreted by Van Peursen (1973). He distinguishes three phases:
- the mythical phase.
- the ontological phase.
- the functional (relational) phase.

The mythical phase is characterized by wholeness, without clear distinctions between 'subject' and 'object' (Fig. 2).

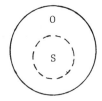

FIGURE 2: Mythical phase

In contrast, in the ontological phase subject and object are separate from each other, with clear-cut distinctions (Fig. 3).

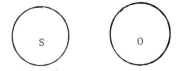

FIGURE 3: Ontological phase

Finally, in the functional phase the relation between subject and object is of primary importance (Fig. 4).

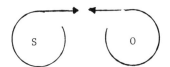

FIGURE 4: Functional phase

Applied to man-machine systems design, the challenge is to move from the 'ontological phase', where the technical designer is subject and the man-machine system is only object, to the functional phase, where all participants are in open relation to each other (Fig. 5).

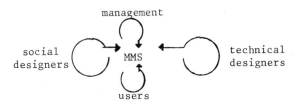

FIGURE 5: The future?

Acknowledgement

This paper is only a partial and personal summary of conference contents and events. It goes without saying that simultaneous participation in all parallel sessions was impossible. The author wishes to apologize for not referring to any other discussions and contributions of great interest for the development of the field.

List of references

Bainbridge, L.; Ironies of Automation[*)]

Guida, G., Tasso, C.; On-line Information Retrieval Through Natural Language[*)]

Hatvany, J., Guedj, R.A.; Man-Machine Interaction in Computer-Aided Design[*)]

Mårtensson, L.; remark during RT-2[*)]

Martin, T.; Human Requirements Engineering for Computer-Controlled Manufacturing Systems[*)]

Nemes, L.; Man-Machine Synergy in Highly Automated Manufacturing Systems[*)]

Peterson; remark during RT-2[*)]

Van Peursen, C.A.; Strategy of Culture (in Dutch) (Elsevier, 1973); being reprinted under the title 'Cultuur in Stroomversnelling'

Pew, R.W., Baron, S.; Perspectives on Human Performance Modelling[*)]

Rasmussen, J.; Some Trends in Man-Machine Interface Design for Industrial Process Plants; In: Aune, A.B. and Vlietstra, J. (Eds.); Automation for Safety in Shipping and Off-shore Petroleum Operations; IFIP/IFAC Symposium; North Holland, 1980

Rouse, W.B.; Models of Human Problem Solving: Detection, Diagnosis and Compensation[*)]

RT-2, Man's Role in MMS[*)]

Seaman; remark during RT-2[*)]

Sheridan, T.B.; Measuring, Modeling and Augmenting Reliability of Man-Machine Systems[*)]

Terano, T., Murayama, Y., Akiyama, N.; Human Reliability and Safety Evaluation of Man-Machine Systems[*)]

[*)] IFAC/IFIP/IFORS/IEA Conference on Analysis, Design, and Evaluation of Man-Machine Systems, Baden-Baden, FRG, 1982

AUTHOR INDEX

424

Seifert, R. 417
Sheridan, T. B. 337
Stein, W. 221
Strandberg, L. 399
Sundermeyer, P. 197
Sykes, R. 71
Syrbe, M. 77

Tasso, C. 247
Teegen, U. 357
Tengstrand, G. 399
Terano, T. 347
Trimmer, P. 71
Trispel, S. 205

Voss, M. 371

Wahlstrom, B. 55
Warnecke, H. -J. 93
Watanabe, T. 137
Weeks, G. D. 255
Wewerinke, P. H. 191, 221
Whitfield, D. 77
Williges, B. H. 239
Williges, R. C. 239

Yoshimoto, K. 377

Zapp, A. 99
Zemanek, H. 121
Zoltan, E. 255